INTRODUCTION

TO

NATURAL PHILOSOPHY

DESIGNED AS

A TEXT BOOK,

FOR

THE USE OF THE STUDENTS

N

YALE COLLEGE.

COMPILED FROM VARIOUS AUTHORITIE

BY DENISON OLMSTED, LL.D.,
PROFESSOR OF NATURAL PHILOSOPHY AND ASTRONOMY.

STEREOTYPE EDITION.

NEW YORK:
COLLINS & BROTHER,
NO. 82 WARREN STREET.
1858.

Stereotyped by
RICHARD C. VALENTINE,

ADVERTISEMENT TO THE FIFTH EDITION

THE publishers of this work have hitherto forborne to stereotype it, in order that the author might more conveniently make such alterations and corrections, in the successive editions, as his own studies, and the communications of his scientific friends who use the work as a text-book, might suggest. Of these means of improvement, he has diligently and carefully availed himself; and having now brought the work to the highest degree of improvement and accuracy that seems at present attainable by him, he has consented that it should be stereotyped, without contemplating any further changes for some time to come.

The more economical mode of publishing thus secured; the adoption of a style of printing somewhat more compact, while it is equally legible, thus rendering it convenient to reduce the two volumes to one; and, above all, the extensive and increasing demand, have enabled the publishers materially to reduce the price of the work, in conformity with what is believed to be the general wish of the numerous literary institutions that use it as a text-book.

Although some passages found in former editions have been omitted in the present, yet others deemed of greater value have been added, so that the amount of matter in the present form has not been diminished, but rather increased.

From many of his brethren of other Colleges, the author has received corrections and suggestions, for which he is indebted in no small degree for the more improved and accurate state of the work as now published. To all of them he would express his grateful acknowledgments. To Professor LOOMIS and Professor SNELL, he is under peculiar obligations.

To blend practical utility with scientific accuracy, was originally, and still continues to be, the aim of this work—a plan which secures to the scholar, along with the discipline of the understanding, habits of philosophical reasoning and observation on the phenomena of nature and art, and information which may be made available in the practical business of life.

YALE COLLEGE, May, 1844.

PREFACE.

The compiler of this work has had two objects constantly in view first, to make the student thoroughly and familiarly acquainted with the *leading principles* of Natural Philosophy; and secondly, to furnish him with as much *useful information* as possible, within so limited a compass. The foundation of all accurate attainments in Natural Philosophy and Astronomy being laid in the science of Mechanics, a large proportion of the work is devoted to this subject. The Mathematical Elements of Mechanics are moreover first considered, separately, that the student may have nothing to divert his mind from the contemplation of these universal and fundamental truths. Still further to render the knowledge of these truths familiar, as well as to supply a most useful intellectual exercise to the student, a great variety of Problems are annexed to each chapter, the utility of which must be obvious to every experienced instructor. Indeed, Problems hold so important a place in the estimation of the writer, that he has introduced them into various parts of the work, wherever the subject appeared to be susceptible of deriving aid from them. Problems put the student on his own resources; they compel him to think for himself; they lead him to a just understanding of the principles demonstrated; and they teach him how to reduce his knowledge to practice. These truths are so obvious, that it is difficult to account for the singular fact, that treatises on Natural Philosophy have, in general, contained few or no problems, although they occupy so large a space in most of the branches of the pure mathematics.

The first part of the following treatise on mechanics, comprising the "Mathematical Elements," is taken chiefly from *Bridge's Mechanics.* This work was peculiarly adapted to our purpose, partly because it is written in a style well suited to the average capacities and attainments of college classes, and partly because it is enriched with a finer collection of problems than any similar work with which we are acquainted. We have aimed to select such parts as promised the most practical utility; and in order to adapt the treatise to the purposes of recitation, the propositions are more distinctly enunciated than in the original work, and various alterations, and occasional additions, are introduced, and explanations added by note or otherwise, with the view of suiting it better to the course of instruction adopted in Yale College.

In Part II, the *Practical Applications of the Principles of Mechanics to the Arts and to the Phenomena of Nature,* are pursued as far as our limits would permit, and further perhaps than some instructors will deem necessary; for we are aware that some maintain the expediency of occupying the attention of the student almost exclusively with *general principles,* and leaving him to make the application for himself. According to our experience, however, the student who is furnished with the knowledge of abstract principles merely, seldom acquires the necessary readiness in reducing them to practice. It has appeared to us no less necessary to initiate the learner in the habit of philosophizing, than in the doctrines of philosophy. In this manner, he will indeed acquire the knowledge of fewer principles; but he will know much better how to use his acquisitions.

We cannot, however, agree with those instructors who have yielded to the spirit of the age, (which is still hunting after a "royal road" to knowledge,) so far as to forsake demonstration altogether, and substitute for the mathematical elements of Natural Philosophy, text-books in which the principles of the science rest on no better basis than mere popular illustration, although works of this kind may, indeed, furnish us with very useful materials for exhibiting the *applications* of these principles. In philosophy, as in morals, the most important principles are usually characterized by great simplicity: in this respect the "golden rule" and the law of gravitation resemble each other.

The subsequent parts of the work are compiled from various authors. In the belief that the truths most important to be inserted in a text-book on Natural Philosophy, are, in general, such as have long been known, no effort has been made to conform the style, either of the propositions or the demonstrations, to a modern dress; but authors, old and new, English and French, have been consulted and used indiscriminately Aiming solely at preparing such an elementary work, as would be most useful to the academic student, we have not aspired to the praise of originality, nor felt at liberty to consult the pride of authorship. Among the truths, however, found in the wide field of Natural Philosophy, a vast difference exists in regard to their value; and no small acquaintance and familiarity with the subject is requisite in the writer of a text-book, in order that he may be able to cull the choicest truths, and present them to the young learner in their native beauty and simplicity. For this purpose, no powers of original investigation, or gifts of genius, can compensate for the want of the experience of the instructor.

Since the publication of the first edition of this work, the compiler has been favored with the opinions of a number of his brethren of other Colleges, who have used it as a text-book. A few would prefer to have a more strictly mathematical complexion preserved throughout, in the place of those parts which are written in a more popular style, since the method of expressing philosophical truths by mathematical formulæ, is more concise and comprehensive, and better suited to mental discipline, and the cultivation of a mathematical taste. A greater number, however, of those who have been so good as to communicate their views to the writer, have approved of the method adopted,—namely, to establish the *general principles* of the science by rigorous mathematical demonstration, or by precise experiments, but to rehearse the *applications* of those principles to the arts, and to the phenomena of nature, in a style divested, as far as possible, of technical phraseology. They believe that such a method renders the study of Natural Philosophy peculiarly attractive to the young learner, and conducive to the formation of habits of philosophical observation, while they rely more on other parts of the scientific course, particularly on the pure mathematics, to fulfil the purposes of mental discipline, and to inspire a mathematical taste. We have desired to accomplish, as far as possible, these several objects of a philosophical education,—to improve the faculties of the mind itself, to imbue it with a love of rigorous demonstration, and to commence the formation of habits of philosophical observation, which shall be carried forward, beyond the pale of academic study, to be confirmed and strengthened throughout the period of after life. The variety of subjects comprehended under Natural Philosophy, some admitting of strict geometrical and analytical reasoning, and others conducted wholly by experimental research, is well adapted to the attainment of these important objects; and it is the prerogative of this science, at once to enlarge the mind by the most profound inquiries, and to conduct it through the most delightful and varied fields of experiment and observation.

In a work necessarily so limited as this, (when compared with the vast extent of the subject,) many topics must be treated with extreme conciseness, and many others, essential to a complete philosophical education, must be omitted altogether. While we aim to furnish the student with a knowledge of the *great laws of nature*, and to exemplify and illustrate them by numerous applications, we can claim nothing more than an "Introduction to Natural Philosophy," suited to beginners.

It is recommended to the student to make a free use of the *Analysis*, especially in reviewing. Let him submit each of the particulars indicated in this outline, to deliberate and repeated reflection, and he will not only fully possess himself of the contents of the work, but will lay up in the mind a system of heads, or "common-places," under which he can conveniently and usefully arrange all his future acquisitions on similar subjects.

The course of instruction in Natural Philosophy pursued in Yale College, to which this treatise is adapted, proceeds as follows. The *mathematical* part of mechanics is first recited, in the same manner as a branch of the pure mathematics. With the *practical* part commences a series of familiar Lectures,* designed to amplify the text, and to illustrate it by numerous experiments. These are continued during the perusal of the remainder of the work. To the same class is afterward delivered a select course of Lectures, which are chiefly devoted to the discussion of the great principles of Philosophy and Astronomy, and especially to such subjects as require a fuller attention than they can receive in the elementary course.

---

* Should any teacher who may use this work, think it better to connect the Lectures and Practical Applications immediately with the theoretical part, it will be easy to do so, by giving occasional lessons in Part II. while the student is reading Part I.

ANALYSIS.

☞ *These Outlines are intended to aid in reviewing, (answering the same purpose as a series of questions,) and it is earnestly recommended to the Learner to make free use of them*

CENTER OF GRAVITY.

COLLISION OF BODIES.

THE LEVER.

FLUIDS AT REST.

SPECIFIC GRAVITY.

LIQUIDS IN MOTION.

ELECTRICAL APPARATUS.

LEYDEN JAR.

ELECTRICAL LIGHT.

ELECTRIC BATTERY.

MECHANICAL EFFECTS OF ELECTRICITY.

CHEMICAL EFFECTS OF ELECTRICITY.

MOTIONS OF THE ELECTRIC FLUID.

EFFECTS OF ELECTRICITY ON ANIMALS.

DIRECTIVE PROPERTIES OF THE MAGNET.

Page.

METHODS OF MAKING ARTIFICIAL MAGNETS.

LOCAL ATTRACTION OF VESSELS.

OPTICS.

REFLEXION OF LIGHT.

IMAGES FORMED BY MIRRORS.

REFRACTION OF LIGHT.

DECOMPOSITION OF LIGHT.

NATURE OF LIGHT.

COLORS IN NATURAL OBJECTS

INFLEXION OR DIFFRACTION OF LIGHT.

DOUBLE REFRACTION AND POLARIZATION.

POLARIZATION OF LIGHT.

VISION.

MICROSCOPES.

TELESCOPES

NATURAL PHILOSOPHY.

PART .—MATHEMATICAL ELEMENTS OF MECHANICS.

[ON THE BASIS OF BRIDGE'S MECHAN 'S.]

PRELIMINARY DEFINITIONS.

ART. 1. NATURAL PHILOSOPHY *is the science which treats of the Laws of the Material world.*

This is the primitive signification of the term Natural Philosophy; but the vast extension given to inquiries into the laws of nature, rendered a division of them necessary. Hence, those laws of nature which relate to *masses* of matter were retained by Natural Philosophy, (which has been farther divided into Mechanical Philosophy and Astronomy,) while those which relate to *particles* of matter, and to the changes of constitution produced by their action on each other, were assigned to Chemistry.

The term *law*, as here used, signifies *the mode in which the powers of nature act.* Laws are *general truths*, comprehending a great number of subordinate truths.

Natural or Mechanical Philosophy is divided into Mechanics, Hydrostatics, Pneumatics, Electricity, Magnetism, and Optics.

MECHANICS *is that branch of Natural Philosophy, which treats of the Equilibrium and Motion of bodies.* As the changes which occur between masses of matter, involve the idea of motion, hence, the causes which produce motion, or which prevent it, and the manner in which it takes place, (its *laws*,) constitute the great object of inquiry in mechanical philosophy.

Body is any collection of matter existing in a separate form.

Force is any cause which moves or tends to move a body, or which changes or tends to change its motion. Every force produces *actual* motion if it is not counteracted by contrary forces; but if it remains counteracted, the motion which it *tends* to produce is called *Virtual.*

That part of Mechanics which relates to the action of forces producing *equilibrium* or *rest*, in bodies, is called *Statics;* that which relates to the action of forces producing *motion*, is called *Dynamics.**

---

* In the following treatise, it is found convenient to disregard this distinction

The science of Mechanics comprehends those laws of equilibrium and motion only, which are common to all bodies in the universe, and to bodies in every form, whether solid, fluid, or aëriform; but the laws of equilibrium and motion undergo certain additional modifications in consequence of the peculiar properties of *fluids*. Hence that branch of Mechanics which treats of the peculiar mechanical properties of fluids in the form of water, is called *Hydrostatics;* and that which treats of the peculiar mechanical properties of fluids in the form of air, is called *Pneumatics*.

2. The two *essential* properties of matter, both of which are inseparable from it, are *extension* and *impenetrability*. Extension, in the three dimensions of length, breadth, and thickness, belongs to matter under all circumstances; and impenetrability, or *the property of excluding all other matter from the space which it occupies*, appertains alike to the largest body and the smallest particle. The word *particle* is much used in writings on physical subjects. In Natural Philosophy we mean by particles, the *smallest parts* into which a body may be supposed to be divided by mechanical means, without any reference to the different elements of which such particles may be composed. Inquiries respecting these belong to Chemistry.

The quantity of matter which a body contains, is called its *Mass;* the space it occupies, its *Volume;* its relative quantity of matter under a given volume, its *Density*. All bodies have empty spaces denominated *Pores*. In solids, we may often see the pores with the naked eye, and almost always by the microscope; in fluids, their existence can be proved by experiment. The ratio of the space occupied by the pores of a body to that occupied by the solid matter, is not known; but there are reasons for believing that even in the densest bodies, the amount of solid matter is small compared with the empty spaces.* Hence it is inferred that the particles of matter touch each other only in a few points.†

Although extension and impenetrability are said to be the *essential* properties of matter, because they are inseparable from its very existence, yet there are also several other properties which are known by experience to belong to all matter, as *gravity, inertia,* and *divisibility;* and others still which belong not to matter universally, but only to certain classes of bodies, as *elasticity*, or the power a body has of recovering itself when compressed; *malleability*, or the power of being extended into leaves or plates; and *ductility*, or the power of being extended in length, as when drawn into wire.

In *Geometry*, we conceive figures to possess extension only without solidity; or to occupy space without excluding other figures

* See Newton's Optics, Lib. II, iii, Pr. 8.

† Playfair.

from it; but in *Mechanics*, we take objects as they occur in nature, viz. not only extended, but *impenetrable.**

CHAPTER I.

OF MOTION AND THE LAWS OF MOTION.

3. Attraction is the tendency which one portion of matter has toward another, and exists both between *particles* and between *masses* of matter. *Aggregation* is the union of particles of the *same kind* in one body; as of the particles of lead in a musket ball. *Affinity* is the union of particles of *different kinds* in one body; as of particles of copper and of zinc in brass. *Cohesion* is the union of *compound* particles in one body; thus a particle of copper and a particle of zinc are united to form a particle of brass by affinity, and the particles of brass are united by cohesion. In mechanical philosophy, however, the term cohesion is usually employed to denote the union of particles of all sorts, whether simple or compound, leaving to Chemistry all inquiries respecting the composition of bodies. *Gravity* is that property by which all terrestrial bodies tend toward the center of the earth. It is in this sense that gravity is understood as a force in Mechanics. But in order to give the learner correct views of this important subject, we subjoin a few other particulars respecting it.

4. *Gravity is a property of matter, universally; and the force of gravity in any body is proportioned to its quantity of matter.*

Since every particle of matter is endowed with this property, it follows that the force of gravity is proportioned to the mass or quantity of matter.† We do not say what gravity *is*, but what it *does*,—namely, that it is something which gives to every particle of matter a tendency toward every other particle. This influence is conveyed from one body to another without any perceptible interval of time.‡ Gravity extends to all known bodies in the universe, from the smallest to the greatest; but the consideration of the subject, in this extent, belongs to Astronomy. We at present contemplate gravity only as it affects *terrestrial* bodies. By it all bodies are drawn toward the center of the earth, not because there is any peculiar property or power in the

---

* Whewell's Mechanics, p. 2.

† A decisive proof that the force of gravity is always proportioned to the quantity of matter, is furnished by the pendulum, its vibrations (which depend on gravity and measure its force, as will be seen hereafter) being always performed in the same time, of what material soever it is made.—*Francœur*, Mech. p. 63.

‡ If the action of gravitation is not instantaneous, it moves more than fifty millions of times faster than light.—*Fourier*, Eulogy on Laplace.

center, but because, the earth being a sphere, the *aggregate* effect of the attractions exerted by all its parts upon any body exterior to it, is such as to direct the body toward the center; as will be more fully explained hereafter.

5. This property discovers itself, not only in the motion of falling bodies, but in the *pressure* exerted by one portion of matter upon another which sustains it; and bodies descending freely under its influence, whatever be their figure, dimensions, or texture, are all *equally accelerated* in right lines perpendicular to the plane of the horizon. The apparent *inequality* of the action of gravity upon different species of matter near the surface of the earth, arises entirely from the resistance which they meet with in their passage through the air. When this resistance is removed, (as in the exhausted receiver of an air-pump,) no such inequality is perceived; bodies of all kinds there descend with equal velocities; and a guinea, a feather, and the smallest particle of matter, if let fall together, are observed to reach the bottom of the receiver exactly at the same instant.

6. The *weight* of a body is the force it exerts in consequence of its gravity, and is measured by its mechanical effects, such as bending a spring, or turning a balance. The force thus exerted by a given mass of matter, (as a cubic foot of water,) being taken as the standard, called 1000, and accurately counterpoised in a balance by some substance easily susceptible of division, (as a mass of lead for example,) multiples or aliquot parts of this standard weight, afford the means of estimating the weights of all other bodies. We weigh a body by ascertaining the force required *to hold it back*, or to keep it from descending. Hence, weights are nothing more than *measures of the force of gravity* in different bodies; but since the force of gravity is proportioned to the quantity of matter, (Art. 4,) weights are also measures of the comparative quantities of matter in different bodies.

7. *Gravity at different distances from the Earth, varies inversely as the square of the distance from its center.*

The total amount of attraction exerted by the earth upon bodies exterior to it, is the same as though that force were all concentrated in the center. (Art. 4.) But a force or influence which proceeds in right lines from a point in every direction, is diminished as the square of the distance is increased. For, let S be the center of the earth; and since the force of gravity acts in right lines directed towards that center, whatever be the nature of gravity, its influence at the distance SA, will be equally diffused over the surface ABCD; and at the distance SE, it will be equally diffused over the surface EFGH. Therefore its intensity or force will be as much less at the point

Fig. 1.

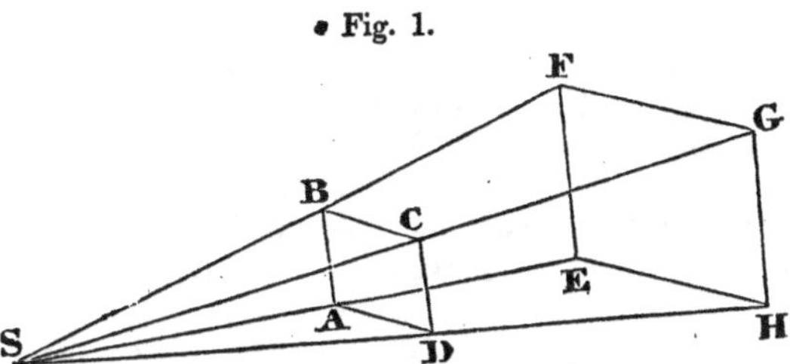

E than at A, as EFGH is greater than ABCD; that is, the force of gravity at A, is to the force of gravity at E, as EFGH is to ABCD. But,

$$EFGH : ABCD :: EF^2 : AB^2 :: SE^2 : SA^2.$$

$\therefore$ Force of Gr. at A : Force of Gr. at E :: $SE^2 : SA^2$; or the forces of gravity at A and E are inversely as the squares of the distances from the center.

8. The weight of a body, therefore, will vary at different heights above the earth's surface. Thus, at twice the distance from the center, or at the height of about 4000 miles above the earth, the force of gravity is only one fourth as great as at the surface, and a given body would weigh only one fourth as much as at the earth. The moon being 60 times as far from the earth's center, as the distance from that center to the surface, the attraction of the earth upon the moon is 3600 ($=60^2$) times less than upon bodies near the earth. But the heights at which experiments are commonly made upon the weights of bodies, bear so small a ratio to the radius of the earth, that this variation is commonly imperceptible. At the height of *half a mile*, the diminution does not amount to more than about $\frac{1}{4000}$th part of the weight at the surface. For, let r=the radius of the earth=4000 miles, nearly; and let x be the height of the body, W its weight at the earth's surface, and W′ its weight at the height x. Then,

$$W : W' :: (r+x)^2 : r^2 :: r^2+2rx+x^2 : r^2.$$

$$W : W-W' :: r^2+2rx+x^2 : 2rx+x^2 \therefore W-W'=\frac{W(2rx+x^2)}{r^2+2rx+x^2} \text{(A)}$$

But when x is a small fraction of r,* x^2 may be neglected, and then,
$$W : W' :: r^2+2rx : r^2 :: r+2x : r,$$

$$W : W-W' :: r+2x : 2x \therefore W-W'=\frac{W\times 2x}{r+2x} \text{(B)}$$

Let x be *half a mile;* then $\frac{W\times 1}{4000+1}=\frac{1}{4001}$th part of the whole weight; or, a body would weigh so much less at the height of half a mile than at the surface of the earth. But if the height

* If, for example, $x=\frac{1}{8000}$ of r, or $\frac{1}{2}$ a mile, then $x^2=\left(\frac{1}{8000}\right)^2=\frac{1}{64,000,000}$, a quantity so small that it may be neglected.

were 100 miles above the earth, then $\frac{100}{4000}=\frac{1}{40}$; and the square of this $=\frac{1}{1600}$ of the radius of the earth, a quantity too large to be neglected; and the difference of weights at the surface and at the height of 100 miles, will be found by formula (A.)

What loss of weight would a body sustain, by being elevated 500 *miles above the earth?* Ans. $\frac{17}{81}$. Were x^2 neglected, then the loss by formula (B) would be $\frac{1}{5}$, which differs from $\frac{17}{81}$ by only $\frac{4}{405}$.

9. *A body situated within a hollow sphere, would remain at rest in any part of the void.**

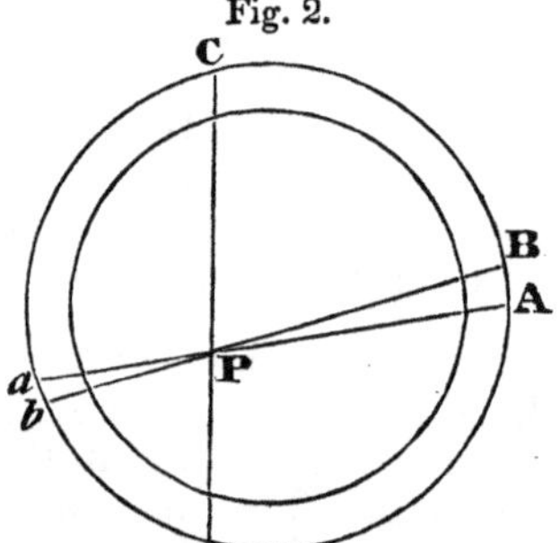

Fig. 2.

Let CAD represent the surface of a hollow sphere, and P any point in the void. Through P let the plane CD pass, dividing the sphere into any two segments CAD and C*a*D.

Let PBA and P*ba* represent two cones meeting in a very small angle at P, and having their bases in the surface of the sphere; which bases, being indefinitely small, may be considered as plane figures, and being circles, they are to one another as the squares of their diameters; that is, as AB^2 to ab^2. But when BA and *ba* are indefinitely small, the lines PB and PA may be considered equal, as also P*a* and P*b*. Therefore, the two triangles PAB and P*ab* are similar, and $AP^2 : aP^2 :: AB^2 : ab^2$. But, putting Q for quantity of matter, D for distance, and G for gravity,

$$AB^2 : ab^2 :: \text{base of } PAB : \text{base of } Pab :: Q : q.\dagger$$

$$\therefore AP^2 : aP^2 :: Q : q :: D^2 : d^2\ddagger \ \therefore Q \propto D^2 \ (1).$$

But since the force of gravity varies directly as the quantity of matter, and inversely as the square of the distance,

$$G \propto \frac{Q}{D^2} \ \therefore (1) \ G \propto \frac{D^2}{D^2}, \text{ or G is a constant quantity.}$$

Hence, the point P (or a body at P) is equally attracted towards AB and *ab*, and the same will be the case with all the corresponding portions of the two opposite segments. The same reasoning evidently applies to all the concentric surfaces or laminæ of which the shell of the sphere may be supposed to be made up; therefore a body situated anywhere within the shell, being attracted by equal and opposite forces, would remain at rest.§

10. *The force of gravity below the earth's surface is, at different distances from the center, directly as those distances.*

Were a body placed at the center of the earth, being attracted

* The solid part of the sphere is supposed to be throughout of uniform density.
† The surface of the sphere is here considered as a thin lamina of matter
‡ Since here D and *d* are the same as AP and *a*P.
§ See Newton's Principia, Book I, Pr. 70.

equally in all directions, it would evidently remain at rest; and were it situated at any point between the center and the surface, the force of gravity toward the center would be diminished by the loss of the attraction of the matter exterior to it; for the matter exterior to it would, by Article 9, have no effect upon it. Thus, if a body were to fall through a hole bored from the earth's surface to its center, the gravity of it would constantly diminish, until, at the center, it would become nothing.

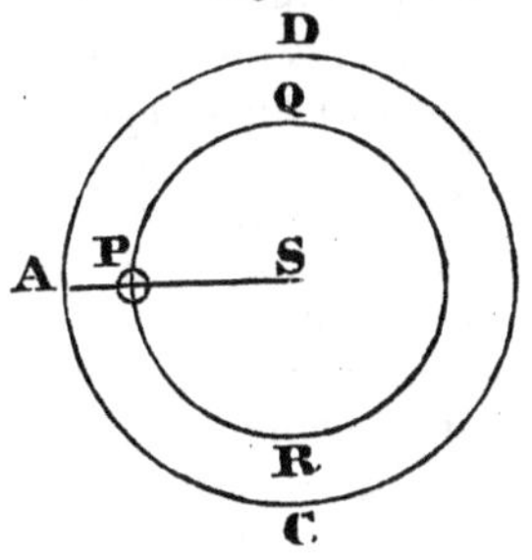

Fig. 3.

Let P be a body situated within the earth at any distance D from the center. Then it appears by the last article, that the gravity of P toward the center would not be affected at all by the shell exterior to PQR, and that P would gravitate only by the force exerted by the sphere PQR. But this force G is directly as the quantity of matter in PQR, and inversely as the square of the distance from the center. That is, $G \propto \frac{Q}{D^2}$. But $Q \propto D^3$* $\therefore G \propto \frac{D^3}{D^2} \propto D$. Therefore, the gravity of P varies as its distances from the center of the sphere.†

11. The Inertia of matter is its *resistance to a change of state whether of rest or motion.* The inertia of a body at rest is the resistance to be overcome to bring it to a given velocity; or, in common language, "to start it;" and the inertia of a body in motion, is the resistance it makes to being stopped, after the moving force is withdrawn. Thus the inertia of a steamboat, while getting under weigh, requires a great expenditure of force to bring the boat to its final velocity; but its inertia carries it still forward after the engine is stopped. Since every particle is endued with this property, *the inertia of a body is proportioned to its quantity of matter, and of course* (Art. 6) *to its weight.* Inertia, however, is a more sure criterion of the quantity of matter in a body than weight is; for inertia is always, and under all circumstances, the same: whereas weight, being merely the measure of gravity, is diminished as gravity is diminished; so that it is less on the tops of high mountains, than at the general level of the earth.

12. In observing the phenomena‡ connected with the actual motion of a body, we consider the *space* over which it moves, the

* The quantity of matter in spheres being as the cubes of their radii.

† Principia, Book I, Pr. 73.

‡ The word *phenomena* is much used in Natural Philosophy. It is thus defined: The phrase Natural Phenomena, in its widest acceptation, denotes any effects in the material part of the creation, addressed to one or more of the senses. (Parkinson's Mechanics, p. 1.)

time of its motion, and the *velocity*. A body is said to move with *uniform* velocity when it describes *equal spaces in equal times.* When the spaces described by it in equal portions of time continually increase, it is said to move with an *accelerated* velocity; and with a *retarded* velocity, when those spaces continually decrease. If its motion is so regulated, that it receives equal increments of velocity in equal times, then it is said to be *uniformly accelerated;* and *uniformly retarded,* if the body suffers equal decrements of velocity in equal times.

The space described by a body moving with uniform velocity, increases in the compound ratio of the time and velocity.

For, a body moving 10 seconds, at the rate of 40 feet per second, will move over 10×40, or 400 feet; and let x equal the number of seconds for which a body moves uniformly, and y the number of feet described in each second; then it is evident that xy will denote the number of feet (i. e. the space) described by it in x seconds. In general, if S be the space described by a body, T the time of its motion, expressed in seconds, V the uniform velocity with which it moves, expressed by the number of feet described in a second, then, $S=T\times V$.*

This is the fundamental equation of uniform motion, from which the other equations may be derived by the common rules of Algebra. For, since $S=T\times V \therefore T=\frac{S}{V}$, and $V=\frac{S}{T}$: and if $s=$ the space, $t=$the time, $v=$the velocity of any other body, expressed in the same manner; then the relation between S, T, V, and s, t, v, may be expressed by the following proportion;

$$S : s :: T\times V : t\times v.† \therefore S \propto T\times V;\ \therefore T \propto \frac{S}{V}, \text{ and } V \propto \frac{S}{T}.$$

If S be given, then $T \propto \frac{1}{V}$, and $V \propto \frac{1}{T}$.

The laws of Uniform Motion, therefore, are comprehended in the following THEOREMS, which are to be treasured up in the memory.

I. *The* SPACE *equals the product of the time into the velocity;*‡ or (when different spaces are compared) the space *varies as* the product of the time into the velocity.

II. *The* TIME *equals the space divided by the velocity;* or (when different times are compared) the time *varies as* the space divided by the velocity.

III. *The* VELOCITY *equals the space divided by the time;* or (when

* The young learner is apt to be puzzled with such abstract expressions as "*Space* equal to *time* multiplied into *velocity;*" but it may be observed that by velocity is meant nothing more than the *space* passed over in *one second;* which may evidently be so multiplied as to equal another space.

† Euc. V. 7.

‡ This is a concise mode of saying, The *number* expressing the space, equals the product of the *number* expressing the time into the *number* expressing the velocity.

different velocities are compared) the velocity *varies as* the space divided by the time.

IV. *When the space is given,* the time varies inversely as the velocity.

13. Questions on Uniform Motions.

1. A ball was rolled on the ice with a velocity of 30 feet per second, and moved uniformly 45 seconds: What *space* did it describe? *Ans.* 1350 *feet.*

2. A steamboat moved steadily across a lake 53 miles wide, at the rate of 16 miles per hour: What *time* was occupied in crossing? *Ans.* $3\frac{5}{16}$ *hours.*

3. On the supposition that the earth describes an orbit of 600 millions of miles in $365\frac{1}{4}$ days, with what *velocity* does it move per second? *Ans.* 19 *miles, nearly.*

4. Three planets describe orbits which are to each other as 15, 19, and 12, in times which are as 7, 3, and 5: What are their *comparative velocities?* *Ans.* 225, 665 *and* 252.‡

14. *The* Momentum *of a body is its quantity of motion, and is as the product of its quantity of matter and velocity.* The quantity of motion, or momentum, of each particle evidently depends on its velocity; and therefore the momentum of the whole must depend on the same particles multiplied into the common velocity. By velocity is understood the space moved over in a second. According to this definition, a body at rest cannot be said to have any momentum, but it is then said to have an amount of *inertia* corresponding to its quantity of matter, or *mass.* Inertia opposes the same resistance as momentum of similar amount.

Let M be the momentum of a body; Q its quantity of matter or *weight* expressed in pounds; V its velocity expressed in feet; and let m be the momentum, q the weight, v the velocity, of any other body expressed in the same manner; then the relation between M, Q, V, and m, q, v, will be expressed by the following proportion;

$\mathrm{M} : m :: \mathrm{Q} \times \mathrm{V} : q \times v$; $\therefore \mathrm{M} \propto \mathrm{Q} \times \mathrm{V}$; $\therefore \mathrm{Q} \propto \frac{\mathrm{M}}{\mathrm{V}}$ and $\mathrm{V} \propto \frac{\mathrm{M}}{\mathrm{Q}}$. If Q be given, $\mathrm{M} \propto \mathrm{V}$; if V be given, $\mathrm{M} \propto \mathrm{Q}$; and if $\mathrm{Q} \propto \frac{1}{\mathrm{V}}$, or $\mathrm{V} \propto \frac{1}{\mathrm{Q}}$, then M will be *given.* If $\mathrm{V} \propto \frac{1}{\mathrm{Q}}$, or $\mathrm{V} : v :: q : \mathrm{Q}$, then $\mathrm{QV} = qv$ and $\mathrm{M} = m$.

The following Theorems therefore comprehend the doctrine of Momentum.

‡ Day's Algebra, 360, cor. 1.

I. *The* MOMENTUM *equals the product of the quantity of matter into the velocity;* or (when different momenta are compared) the Momentum *varies as* the product of the quantity of matter into the velocity.

II. *The* QUANTITY OF MATTER *equals the Momentum divided by the velocity;* or (when different masses are compared) the Quantity of Matter *varies as* the Momentum divided by the velocity.

III. *The* VELOCITY *equals the Momentum divided by the quantity of matter;* or (when different velocities are compared) the Velocity *varies as* the Momentum divided by the quantity of matter.

IV. *If the quantity of matter is given, the momentum is as the velocity.*

V. *If the velocity is given, the momentum varies as the quantity of matter.*

VI. *If two bodies move with velocities which are inversely as their quantities of matter, they have equal momenta.*

Thus, if a ship of 100 tons, sailing at the rate of 7 knots, meet another ship of 50 tons, sailing 14 knots per hour, they will encounter each other with equal momenta. This constitutes a fundamental principle in the mechanical action of bodies.

15. QUESTIONS ON MOMENTUM.

1. A ship weighing 336,000 lbs. is dashed against the rocks in a storm, with a velocity of 16 miles per hour: With what momentum did she strike? *Ans.* 7,884,800 *lbs.*

2. On the supposition that Goliath of Gath presented an obstacle of 350 lbs., and that the stone hurled by David's sling weighed two ounces: With what velocity must it have been thrown to have prostrated the giant?

Ans. It must have exceeded 2800 *feet per second.**

3. Wishing to know the velocity of a musket ball weighing 1 oz., I suspended after the manner of a pendulum, a log of wood weighing 53 lbs. The ball on entering the log gave it a motion of 2 feet per second: What was the velocity of the ball?

Ans. 1698 *feet per second.*†

4. If a comet moving at the rate of 1,000,000 miles per hour, were to meet the earth moving 19 miles per second: What ratio will the mass of the comet bear to that of the earth, supposing that they mutually destroy each other's motions?

Ans. 1 : 14.6 ; *or the comet must have nearly* $\frac{1}{15}$ *as much matter as the earth.*

5. Two railway cars have their quantities of matter as 7 to 3, and their momenta as 8 to 5: What are their respective velocities? *Ans. As* 24 *to* 35, *or nearly as* 5 *to* 7.

---

* The maximum velocity of a cannon ball is usually reckoned 2000 feet per second

† It is to be remarked, that the ball moves with the log, and therefore its mass is to be added to that of the log.

16. FORCE *is any cause which moves or tends to move a body, or which changes or tends to change its motion.* (Art. 2.) Forces can, for the most part, be reduced to the three following classes, *attraction, repulsion,* and *animal strength.* Thus, the power of the waterfall can be traced to gravitation; that of steam to the repulsive energies of heat; and that of the horse and the ox to animal strength. Forces are divided into two kinds, according to the manner in which they act. If a force acts *instantaneously,* and then ceases, it is called an impulsive force. A ball, suddenly put in motion by the hand or any instrument, along a horizontal plane, is an instance of the effect produced by an impulsive force. When a force acts *incessantly,* it is called an *accelerating* force, and is either constant or variable; *constant,* when the increments or decrements of velocity caused by it, in equal successive parts of time, are *equal;* and *variable,* when the increments or decrements of velocity thus produced, are *unequal.* The force of gravity near the Earth's surface, is an example of a *constant* force; for it causes equal increments or decrements of velocity in equal portions of time, not by *impulses,* but by *incessant* action. Gravity at different distances from the earth, is a *variable* force, whose variation is estimated in the same manner as that of weight in article 8.

17. *Different constant forces generate velocities, which are as the product of the forces and times.*

Let T denote the time, and F the constant force; and conceive the time to be divided into exceedingly small equal portions; then, since equal impulses, and of course equal velocities, are added to the moving body at each of these instants, the whole velocity acquired, must be proportioned to that of each instant (which is the measure of F) multiplied by the number of instants; or $V \propto F \times T$.

One steam car was propelled by a constant force of 25 lbs. for 10 minutes, above what was sufficient to overcome all resistances, and another was driven by a similar force of 18 lbs. for 7 minutes: What were their comparative velocities?

Ans. As 250 *to* 126; *or the first car had nearly twice the velocity of the second.*

18. There are three great principles of motion, called the LAWS OF MOTION, derived from universal experience and observation, and of such extensive application as to comprehend all the phenomena of mechanics.

19. FIRST LAW.—*A body continues in the state in which it is, whether of rest or motion, until compelled by some external force to change its state.* That a body at rest will continue at rest, is a consequence immediately arising from the inertia of matter.

That a body in motion will continue to proceed uniformly along the right line in which it began to move, until it is acted upon by some external force, is inferred from the fact, that any deviation from uniform rectilinear motion, in a moving body, is observed to be owing to some external force; and that such deviation is diminished in proportion as such external force is withdrawn; hence, were it entirely withdrawn, we infer that the motion of the body would then become uniform, rectilinear, and perpetual. We may see approximations to such a state in a ball rolled successively on the earth, on a floor, and on smooth ice. The most general impediments to motion are friction, resistance of the air, and gravity. But if a small wheel is put in motion round a horizontal axis, the effect of gravity is taken off, (since one side of the wheel gains as much in falling, as the other loses in rising,) and no impediments remain but the resistance of the air and friction, the former of which may be removed by placing the apparatus in the vacuum of an air-pump, and the latter may be greatly diminished by methods to be described hereafter. In proportion as these several impediments are removed, the wheel approximates to a motion which is uniform and continued. A pendulum has been constructed to move with so little resistance as, when barely set in motion with the finger, to continue to vibrate 24 hours.

20. Second Law.—*Motion, or change of motion, is proportioned to the force impressed, and is in the direction of that force.* It has already been observed, that every change in the state of rest or motion in a body must be effected by the agency of some force; this Second Law asserts, that this change will in all cases be proportional to that force, and will be produced in the direction in which the force acts.

That motion or change of motion in a body will be proportional to the force which produces it, is also inferred from observation and experiment, as well as from the known connection between cause and effect. Thus, a ball which moves with a double or triple velocity is found to generate in another, by impulse, a double or triple velocity. Two bodies meeting with equal quantities of motion mutually stop each other. Two forces, which, by acting singly during equal times, produce equal velocities in some third body, are found by acting together during the same length of time, to produce a double velocity. If a new force is impressed upon a body in motion, in the direction in which it moves, its motion is increased proportionally to the new force impressed: if this force acts in a direction contrary to that in which the body moves. it is found to lose a proportional part of its motion: if the direction of this force is oblique to that of the moving body, it gives it a new direction compounded of both. A force which we know to act equally, produces equal increments of velocity in equal times.* Hence it follows, that *the smallest force is capable of moving the largest body.*

With respect to the *direction* in which a body moves, it is evident that when it is under the direction of any given force, whether it be an impulsive one, or one that acts incessantly, the body can have no tendency whatever to deviate to the one side or to the other,† but must proceed along the right line in which the force acts.

21. Third Law.—*When bodies act on each other, action and reaction are equal and in opposite directions.* The meaning of this law is, that when a body imparts motion in any direction, it loses an equal quantity of its own in the opposite direction—that when a body receives a blow, it gives to the striking body an equal blow—that when A presses on B, B returns to it an equal pressure—and that when it attracts or repels B, it receives from B the same influence in the opposite direction.

Fig. 4.

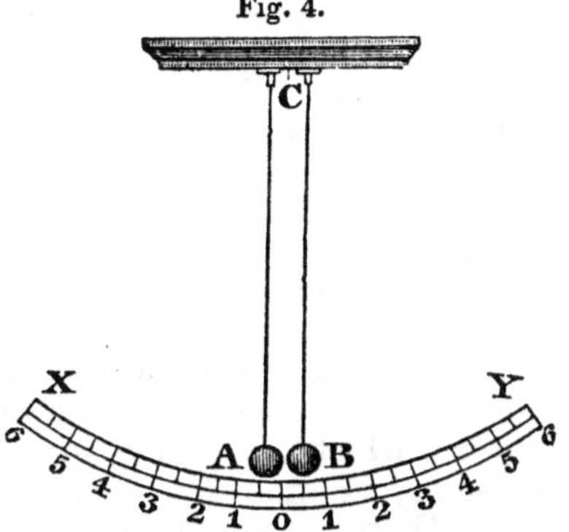

This law is brought to the test of experiment by means of the apparatus represented in Fig. 4, where A and B are two balls of lead for example, suspended at C, by a flexible line, by which A may be drawn out towards X, and let fall upon B. The velocities gained or lost are indicated by the graduated arc XY; and it is found that when A falls on B, whatever motion A communicates to B, is communicated to A in the opposite direction; that is, the same amount is taken from A. Thus, if the two bodies are equal, and A falls on B at rest, they will, after the blow, move on together, with half the velocity of A,—B having acquired, and A having lost an equal amount of motion. If A is greater than B, still it is found that the *momentum* gained by B (ascertained by multiplying its mass by the velocity) is precisely equal to the momentum lost by A; and if A meets B with a momentum greater than that of B, the latter will deprive A of a momentum equal to its own, and return along with A, both bodies having a momentum equal to the difference of their momenta previous to collision. It is a general Law of the Material World, that no body loses motion in any direction, without communicating an equal quantity to other bodies in the same di-

* Gregory's Mechanics, I, 9.

† According to the principle of the sufficient reason, there being no cause why the body should deviate to one side of this line rather than the other · hence it will remain in it. (See Playfair's Outlines, I, 4.)

rection; and conversely, that no body acquires motion in any direction, without diminishing the motion of other bodies by an equal quantity in that same direction.*

Now, the moving force by which A communicates momentum to B, is called the *action* of A; and the tendency of B to diminish the momentum of A, is called the *reaction* of B. Since, therefore, according to this meaning of the words action and reaction, the effect produced by the action of A is equal to the effect produced by the reaction of B, action and reaction are said to be equal during the impact of A upon B. That these effects are produced in "opposite directions," is evident from the very nature of the case.

This law applies not only to the *impact* of bodies, but to every case in which one body acts upon another. It holds good, not only when bodies come into actual contact, but when they act upon one another at any distance whatever. A body, A for instance, is sustained by another body, B, and both bodies remain at rest; if the pressure exerted by the two bodies were not equal, it is evident that some motion would ensue; which is contrary to the supposition. If motion does ensue, then the case becomes, in a great measure, analogous to that of impact; and the effects produced, estimated in a similar manner, are found to observe the same law. The mutual *attractions* of bodies are also subject to this law. Thus, if two equal magnets, connected with two equal and similar pieces of cork, be made to float upon the surface of water, as soon as they come within the sphere of attraction, they are observed to move towards each other in a right line, with equal velocities, and consequently with equal momenta; and as the resistance which each body meets with from the fluid is evidently the same, we infer that their actions upon each other are *equal.*†

22. These fundamental principles of "Mechanics" rest on three different kinds of evidence:—

1. They are conformable to all *experience and observation.*
2. They are confirmed by various accurate *experiments.*
3. The *conclusions* deduced from them have always proved true in fact without exception.‡

Observation and Experiment, then, constitute the basis of the science of Mechanics. Observation is the close inspection, and attentive examination, of those phenomena *which arise in the course of nature.*§ Experiment is an *artificial trial* made for the

---

* Playfair's Outlines, I, 8.

† If two *unequal* magnets, placed upon pieces of cork similar to each other, and proportional to the respective magnets, were made to float in the same manner, they would approach each other with velocities inversely proportional to the quantities of matter moved, and consequently with *equal momenta;* but this experiment is liable to very great inaccuracy, from the different *resistances* which the bodies would meet with.

‡ See Gregory's Mechanics, I, 9. Atwood on Rectilinear Motion, p. 360.

§ Leslie's Natural Philosophy, I, 2.

purpose of learning the powers of nature, or the properties of substances. The most comprehensive results obtained by both these methods, so far as respects mechanics, are expressed in the foregoing Laws of Motion. Upon these, therefore, the science of Mechanics is built. Algebra and Geometry, called in to the aid of these fundamental principles, lead to the discovery of new relations, and bring to light a great number of subordinate truths, of the highest degree of practical utility. *Granting the truth of the Laws of Motion*, as these subordinate truths are attained by principles purely scientific, (namely, those of Algebra and Geometry, and especially the latter,) they are attended with the evidence of demonstration; but since the conclusions can be no more certain than the premises, we can claim for the truths in Mechanics that degree of evidence only, which results from observation and experiment, applied in their greatest perfection.

23. Questions on the Principles of Motion.

1. A bird of passage was observed to fly with a uniform velocity of 19 feet per second: Over what Space would she pass in 24 hours? *Ans.* 310,909 *miles.*

2. A lame man set out to travel round the world. He could walk but two miles an hour for seven hours out of the twenty four. Provided he could go forward, without impediment, on the circumference of a great circle of the globe, (25,000 miles,) what Time would he require to complete the journey?
Ans. 4 *years and* 325$\frac{5}{7}$ *days.*

3. A wind blows uniformly from the equator to the pole, (say 6000 miles,) in 12 days: What is its Velocity per hour?
Ans. 20$\frac{5}{6}$ *miles.*

4. How much weight would a rock that weighs ten tons (22,400 lbs.) at the level of the sea, lose if elevated to the top of a mountain five miles high? (Art. 8.) *Ans.* 55.8952 *lbs.**

5. If the Earth were a hollow sphere, and if, through a hole bored through the center, a man were let down by a rope, would the force required to support him be increased or diminished as he descended through the solid crust, and where would it become equal to nothing? (See Art. 9.)

6. How much would a 44 pound shot weigh at the center of the earth; and how much at a point half way from the center to the surface? (See Art 10.)

7. If a hole were bored through the center of the earth, and a stone were dropped into it, in what manner would the stone move

* The weight would be ascertained, in this case, by the effect on a spring, (Art. 10,) and not by scales, since a counterpoise would sustain a loss of weight in the same degree with the body in question.

in its way to the center and after it reached the center? (See Arts. 9 and 10.)

8. Suppose the battering-ram of Vespasian weighed 5760 lbs., and was found sufficient, when propelled with a certain velocity, to demolish the walls of Jerusalem; and suppose that a 32 pound cannon-ball, fired with a velocity of 2000 feet per second, is found capable of doing the same execution: What was the velocity of the battering-ram? *Ans.* 11.11 *per sec.*

9. Suppose a grain of light, moving at the rate of 192,000 miles per second, to impinge directly against a mass of ice moving at the rate of 1.45 feet per second: What weight of ice would the light stop? *Ans.* 99877.832 *lbs.,* or nearly* $44\frac{1}{2}$ *tons.*

10. If a ball of the same density with the earth, $\frac{1}{10}$th of a mile in diameter, were placed at the distance of $\frac{1}{10}$th of a mile above the earth: What space would the earth move through to meet it, the diameter of the earth being taken at 8000 miles?

Ans. $\frac{1}{80000000000}$ *inch, nearly.*

11. The quantity of matter in the sun being 354,000 times that of the earth, and their distances from each other being 96,000,000 miles: If the two bodies were abandoned to their mutual attraction, where would they meet?

Ans. 271.186 *miles from the center of the sun; or the sun would advance so far to meet the earth.*†

12. Two men are pulling a boat ashore by a rope, one at each end, A being in the boat and B on the shore: How will the time of bringing the boat ashore compare with the time in which A would pull it ashore alone, were the other end of the rope fixed to an immovable post? (See Art. 21.)‡

CHAPTER II.

OF THE LAWS OF FALLING BODIES.

24. Since, when a body moves with a uniform velocity for a given time, the space described is in proportion to that time and velocity conjointly, (Art. 12,) therefore if one side of a right-

* 1 lb. av. = 7000 grains.

† In this problem it is supposed that all the matter of each body is collected in its center, (Art. 4,) and that their own semi-diameters present no obstacle to the approach of their centers.

‡ This problem is intended merely to illustrate the doctrine of action and reaction. No numerical answer is required.

angled parallelogram represents the *time* of a body's motion, and the other the uniform *velocity* with which it moves, the parallelogram itself (whose area is equal to the product of the two sides) will represent the *space* described in that time. Thus, let the line AE be divided into any number of equal parts in the points B, C, D, &c., and from those points draw the equal straight lines AF, BG, CH, &c., at right-angles to AE, and complete the parallelogram AFLE; then if AB, BC, CD, &c., represent equal successive portions of time, and AF, BG, CH, &c., represent the uniform velocity with which a body moves, then will the parallelograms AG, BH, CK, &c., represent the spaces described in those equal portions of time, and the parallelogram AFLE the whole space described in the time represented by AE.*

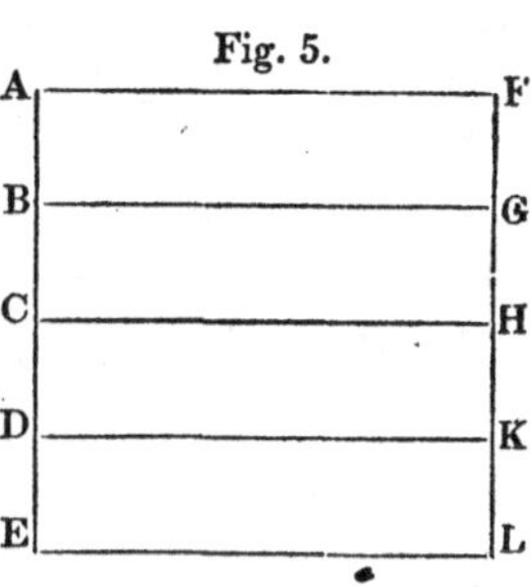

Fig. 5.

25. Suppose now that a body moves uniformly as before, during the equal successive portions of time represented by AB, BC, CD, &c., but at the end of each portion of time receives an *increase* of velocity; for instance, during the time AB let it move with a velocity represented by AF, during the time BC with a velocity represented by BH, &c.; complete the parallelograms AG, BK, CM, and DO, then the space described in the time AB will be represented by the parallelogram AG, in the time BC by the parallelogram BK, &c., and the whole space described in the time AE by the irregular figure AFOE.

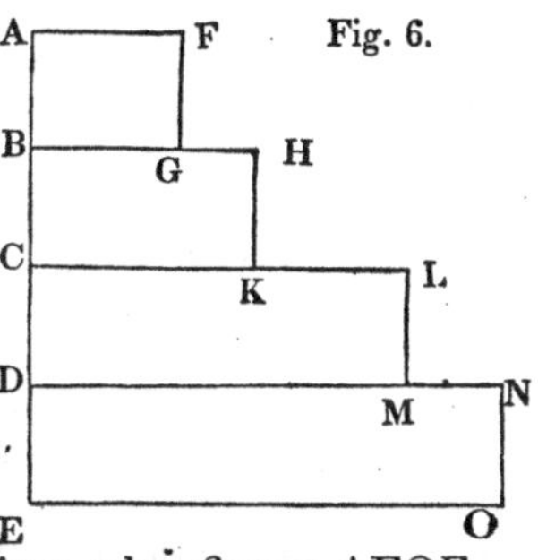

Fig. 6.

26. Let us next suppose that a body receives equal increments of velocity at the end of each successive portion of time, so that

* Since the track described by a moving body is a *line*, how (it may be asked) can the space be properly represented by a *superficies?*

To avoid misconception on this subject, it will be useful for the young learner to recur to a few elementary principles. Quantity (it will be recollected) is any thing which can be *increased* or *diminished*, or which is capable of being *measured*. (Algebra, Art. 1.) Thus *time* is a quantity, whose measure can be expressed in hours, minutes, and seconds. *Velocity* is a quantity, being measured by the number of feet passed over in a second. But these two quantities (time and velocity) have no permanent representatives of their own, like numbers, which are represented by the digits, or like magnitudes, which are denoted by lines, surfaces, and solids. Hence, such quantities as times, velocities, and forces, are denoted by representatives *borrowed* from those of magnitude. In the case before us, the space described by a moving body is represented by a parallelogram, not because the space actually described has any *resemblance* to a parallelogram, but because a parallelogram has the *same relation* to the sides of which it is the product, as space has to the two quantities, time

during the second interval of time it moves with twice the velocity, during the third interval with three times the velocity, &c., then will CK=2BG, DM=3BG, &c. But AC=2AB, AD=3AB, &c. Therefore,

AC : CK : : AB : BG : : AD : DM, &c. Hence, the figures ABG, ACK, ADM, &c., are similar triangles; and if AG, GK, KM, &c., be joined, AO will be a straight line, and the figure AFOE, which represents the space described in the time AE, will differ from the triangle AOE only by the sum of the triangles AFG, GHK, KLM, MNO, which are all equal to each other.

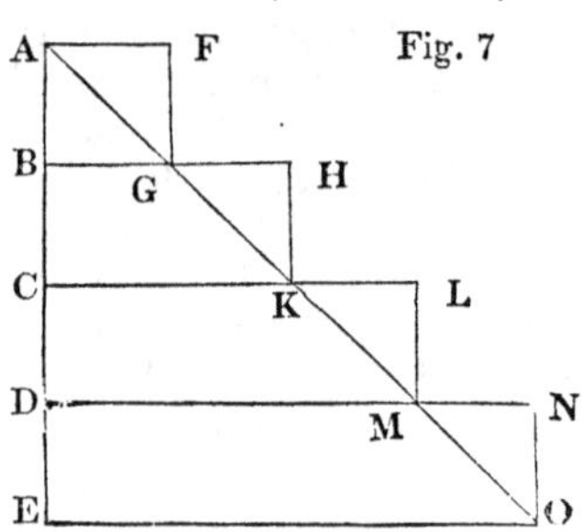

Fig. 7

27. Now let the intervals of time and the corresponding increments of velocity be only half what they were in the former instance. Bisect AB, BC, CD, &c., (Fig. 8,) in *b*, *c*, *d*, &c., and complete the parallelograms as before, then the figure which represents the space described in the time AE will differ from the triangle AOE by the sum of the small triangles A*fg*, *g*FG, G*hk*, &c., which is only *half* the sum of the triangles AFG, GHK, KLM, &c., in the preceding figure. By continually halving these intervals of time and the corresponding increments of velocity, the figure AFOE will approach to the form of the triangle AOE; and when they are diminished *ad infinitum*, A*f* (which represents the velocity with which the body begins to move) will be equal to 0; as will also the sum of the triangles A*fg*, *g*FG, &c.; the space therefore described by a body beginning to move from rest by the continued action of a force which generates equal increments of velocity in equal times, will be accurately represented by the right-angled triangle AOE, one of whose sides AE represents the time of the body's motion, and the other, OE, the last acquired velocity.

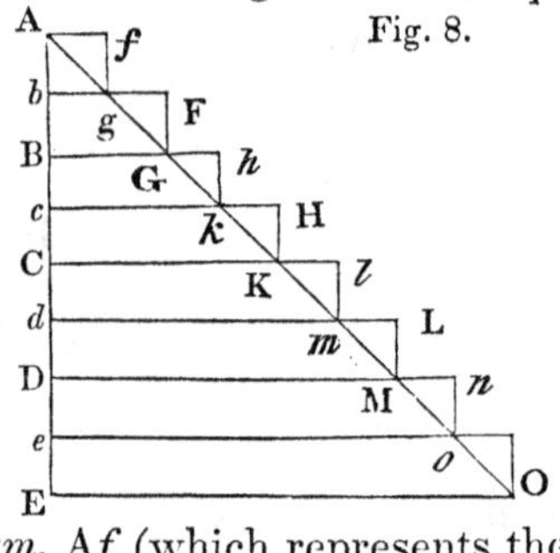

Fig. 8.

28. We have seen, (Art. 24,) that in uniform motion, the time, velocity, and space, have the same relation to each other, as the

and velocity, of which it is the product. An *identity* being thus established between the relations that subsist among magnitudes, and those that subsist among such quantities as have no representatives of their own, the representatives of magnitudes may be substituted to denote the relations of the other quantities; and thus a great number of new relations are frequently discovered to exist among those quantities, because they are known to exist among the magnitudes, as the lines, surfaces, &c., which are taken to represent them. It is thus that Geometry becomes a powerful auxiliary to Mechanics.

sides and area of a right-angled parallelogram. Now if a moving body should retain all the velocity it has already acquired, and take on equal increments at equal successive instants, we see by the last article, that the whole space described would be represented by the sum of the parallelograms described successively. We see, moreover, that the smaller the time is taken, the nearer the whole space approaches to a right-angled triangle. But when a body is descending by the force of gravity, its velocity increases *continually;* the instant is reduced to nothing; and the little triangles which denote, in the other case, the difference between the figure described and that of a right-angled triangle, vanish, and leave the triangle as the proper representative of the space described. The laws of variable motion, however, are more perfectly exhibited by means of the calculus, than they can be geometrically.*

29. *The spaces described by bodies falling from rest under the influence of gravity, are to each other as the squares of the times in which they are described, or as the squares of the last acquired velocities, or as the times and last acquired velocities conjointly.*‡

For let S be the space described, V the velocity acquired by a body falling from rest for the time T; s the space described, v the velocity acquired at any other period t, of its fall; then, from what has already been demonstrated, if the ratio of T : t be represented by the lines AB, Ab, and the ratio of V : v by the lines BC, bc, drawn at right-angles to them, the ratio of S : s will be represented by the triangles ABC, Abc. Now,

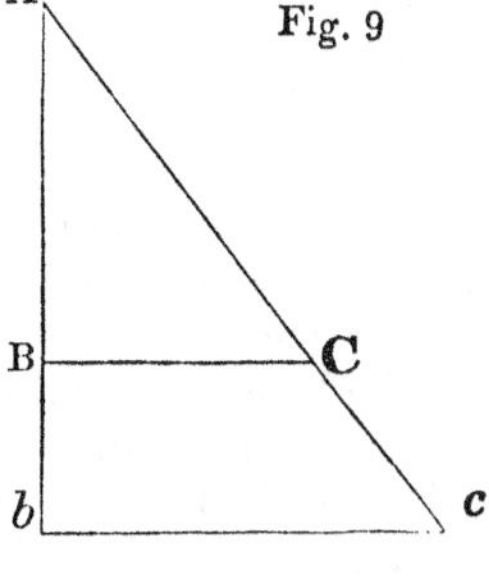

ABC : Abc : : $AB^2 : Ab^2$; or, as $BC^2 : bc^2$; or, as $AB \times BC : Ab \times bc$.† Hence, S : s : : $T^2 : t^2$, or as $V^2 : v^2$, or as $T \times V : t \times v$.

As equal increments of velocity are generated in equal times, it is farther evident that the *velocity acquired varies as the time*: the same conclusion may also be deduced from the similar triangles ABC, Abc; for BC : bc : : AB : Ab, i. e. V : v : : T : t.

Since the spaces described are as the squares of the times; if a body falls from rest for times which are represented by the numbers 1, 2, 3, 4, 5, &c., the spaces described in those times will be as the square numbers, 1, 4, 9, 16, 25, &c.; and the spaces de-

* See *Young*, Elements of Mechanics, art. 108. *Renwick*, p. 46.

‡ The demonstration applies to *any* uniformly accelerating force, as well as to gravity; and hence, although "Falling Bodies" are here under particular consideration, yet the proposition may be predicated of all bodies urged by uniformly accelerating or constant forces.

† For the right-angled triangles ABC, Abc, are to each other both as the squares of their homologous sides, (by Euc. 6, 19,) and in the ratio of the parallelograms of which they are respectively *halves.*

scribed in equal successive portions of time will be as the odd numbers 1, 3, 5, 7, 9, &c., as exhibited in the following table.

Times.	*Spaces described.*	*Spaces described in equal successive portions of time.*
1	1	In 1st portion of time 1
2	4	2d . . . . 4−1 =3
3	9	3d . . . . 9−4 =5
4	16	4th . . . . 16−9 =7
5	25	5th . . . . 25−16=9
&c.	&c.	&c. . . . . &c =&c.

30. *If a body be projected perpendicularly upward, with the velocity which it has acquired in falling from any height, it will rise to the point from which it fell, before it begins to descend again.*

As, in the descent of a body, the force of gravity generates equal increments in equal times, so, in its ascent, equal portions of velocity will be destroyed in equal times. The spaces described in equal successive parts of time, by a body thus ascending, reckoning from the beginning of its motion, will be the same as those stated in the foregoing table, but in an inverted order: thus, if the time be divided into four equal parts, then the spaces described in the descent of the body during these equal times are as the numbers 1, 3, 5, 7, but in its ascent they will be as 7, 5, 3, 1; that is, the space described in the first portion of time, in its ascent, will be the same as that described in the last portion of time, in its descent, and so on, till the body arrives at its highest point.

31. *The space which a body describes from rest in any time, by the action of gravity, is* HALF *that which it would describe in the same time with the last acquired velocity continued uniformly.*

Fig. 10.

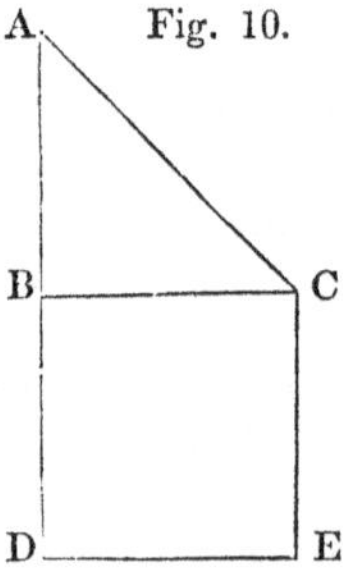

Let the triangle ABC represent the space described by gravity in the time AB, and BC the last acquired velocity; produce AB to D, making BD equal to AB, and complete the parallelogram BCDE; then, if a body moves for the time BD with the uniform velocity represented by BC, the space described in that time will be represented by the parallelogram BCDE, (Art. 24;) but the triangle ABC is half the parallelogram BCDE; hence the space described with the continually increasing velocity during the time AB, is half that which would be described in the same time BD, with the velocity BC continued uniformly.

Since the space described by a body falling from rest, is half that which it would describe in the same time with its greatest velocity continued uniformly, and since a body projected perpendicularly upward rises to the same height as that from which it must fall to acquire the velocity of projection, *the whole space described by a body projected perpendicularly upward, is* HALF *that which it would describe in the same time with its first velocity continued uniformly.*

32. *The space described in any time by a body* PROJECTED DOWNWARD *with a given velocity, is equal to the space which would be described with that velocity continued uniformly for that time, together with the space through which a body would fall from rest by the action of gravity in the same time.*

Let AD represent the given velocity of projection, and AB the given time, and complete the right-angled parallelogram ABED; produce BE to C, and let EC represent the velocity generated by gravity in the time AB or DE, and join DC. Then, according to what has been said in Art. 24, a body moving under the influence of projection alone, with a uniform velocity represented by AD, would describe the parallelogram ABED in the time AB; and, by Art. 27, a body falling from a state of rest during the same time, so as to acquire the velocity represented by EC, would describe the triangle DEC; hence the figure ABCD truly represents the joint effects of both forces, or the whole space described.

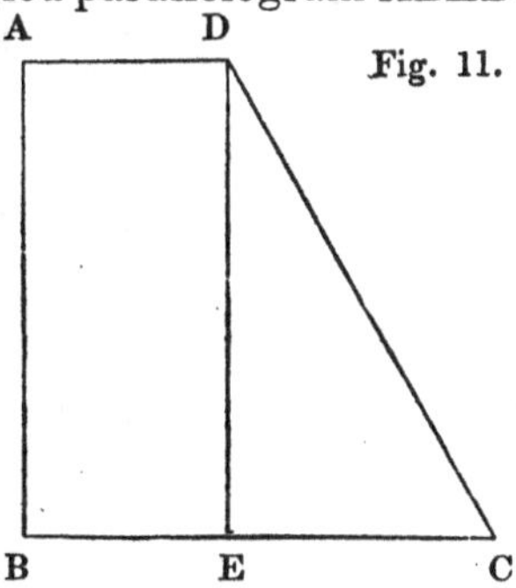

Fig. 11.

33. *The space described by a body* ASCENDING *for a given time, is equal to the difference between the space which would be described by the body moving uniformly for that time with the velocity of projection, and the space through which a body would fall from rest by the action of gravity in the same time.*

Let BC (Fig. 12) represent the given velocity of projection, and AB the time in which it must fall from rest to acquire that velocity; draw BC at right-angles to AB, and join AC, then the triangle ABC will represent the space through which the body must ascend to lose all its velocity. (Art. 30.) In AB take any point *b*, and complete the parallelogram BCD*b*; then will *bc* represent the velocity of the body at the end of the time B*b* of its ascent, and *c*D will represent the velocity destroyed by gravity in the same time. But the velocity *destroyed* by gravity in any time is equal to the velocity *generated* by gravity in the same time, (Art. 30;) hence the triangle CD*c*

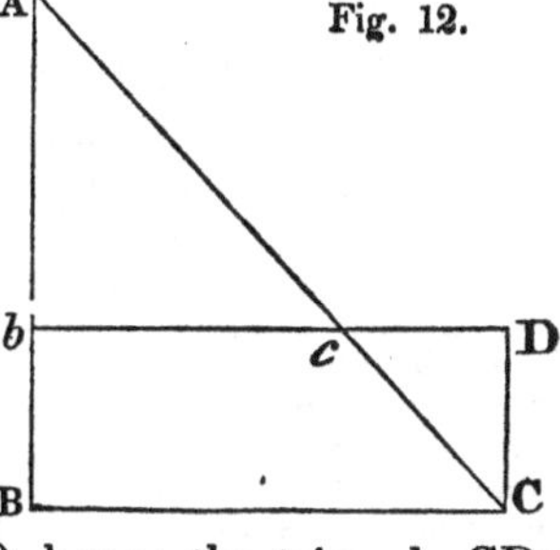

Fig. 12.

will represent the space through which a body would fall from rest in the time CD or B*b*. Now the figure BC*cb* represents the space through which the body would ascend in the time B*b*, and the parallelogram BCD*b* represents the space through which a body would move in the time B*b* with the velocity BC continued uniformly; but the figure BC*cb* is equal to the *difference* between the parallelograms BCD*b* and the triangle CD*c*.

34. The foregoing investigations show the *ratios* between the velocities, times, and spaces of falling bodies; but in estimating the actual motion of bodies descending or ascending by the force of gravity, it is necessary to have recourse to some fixed stand ard of measurement of space and velocity. Now it has been ascertained, by the most accurate experiments, that a body falling freely* from rest describes a space equal to $16\frac{1}{12}$ feet in the first second of its fall; and (Art. 31) a body so falling would acquire a velocity which, if continued uniformly, would carry it over $32\frac{1}{6}$ feet (that is, *twice the space,*) in the same time. If, therefore, $m=16\frac{1}{12}$ feet, m will express the space fallen through from rest in one second, and $2m$ will express the velocity *per second* acquired in that time. Let S be the space described by the body in any other time T, and V the velocity acquired; then since the spaces are as the squares of the times, we have,

† (1) $S : m :: T^2 : 1^2 \therefore$ . . . . $S=mT^2$.

(2) $S : m :: V^2 : (2m)^2=4m^2 \therefore$ . . . . . . $S=\frac{V^2}{4m}$.

(3) $V^2=4mS \therefore$ . . . . $V=2\sqrt{mS}$.

(4) $1 : 2m :: T : V$. (Art. 29) $\therefore$ . . . . . $V=2mT$.

(5) $V=2mT \therefore$ . . . . $T=\frac{V}{2m}$.

(6) $S=mT^2 \therefore$ . . . . $T=\sqrt{\frac{S}{m}}$.

1. A body has been falling for 6 seconds: What *space* has it fallen through in that time, and what is the *velocity* which it has acquired?

$S=mT^2=16\frac{1}{12}\times 36=579$ feet.

$V=2mT=32\frac{1}{6}\times 6=193$ feet per second.

2. *How far* must a body fall to acquire a velocity of 50 feet in a second, and *how long* will it be in falling?

S=38.86 feet.

T=1.55 seconds.

---

* All bodies descending or ascending near the surface of the earth, meet with more or less resistance from the air; so that, strictly speaking, a body can never be said to descend *freely* but in the exhausted receiver of an air-pump. It is in a vacuum that a body describes 16 1-12 feet in a second; the conclusions, therefore, deduced in this section, will approximate to the truth only in those cases where the resistance of the air bears little or no proportion to the weight of the body.

† These are important formulæ, and are to be carefully stored in the memory.

3. A body fell from the top of a tower which was 150 feet high: *How long* was it in falling, and what velocity had it acquired when it got to the bottom?

T=3.054 seconds. V=98.237 feet per second.

4. A body was projected perpendicularly upward with a velocity of 100 feet in a second: *How far* would it ascend before it began to return?

By Art. 30, the height to which the body would ascend is equal to that through which a body must fall from rest to acquire the velocity of projection; here, therefore,

$$V=100, \text{ and } S=\frac{V^2}{4m}=\frac{10000}{4\times 16\frac{1}{12}}=\frac{30000}{193}=155.44 \text{ feet.}$$

5. A body was observed to fall for 3 seconds, and afterward to move uniformly for 2 seconds along the horizon with the velocity which it had acquired by its fall: What was the *whole space* described in its perpendicular and horizontal motion?

The space described in its fall $=mT^2=16\frac{1}{12}\times 9=144\frac{3}{4}$ feet. The velocity acquired $=2mT=32\frac{1}{6}\times 3=96\frac{1}{2}$ feet per second; and as it moved along the horizon for 2 seconds with this velocity, it must in that time have described 193 feet; hence the whole space described from the beginning of its fall to the end of its horizontal motion is $144\frac{3}{4}+193$, or $337\frac{3}{4}$ feet.

6. A cannon ball fired perpendicularly upward, was gone 10 seconds, when it returned to the same place: How *high* did it rise, and what was the *velocity* of projection?

Ans. Height $402\frac{1}{12}$ feet. Velocity of projection $160\frac{5}{6}$ feet per second.

35. Since the spaces described in equal successive parts of time (by Art. 29,) are as the odd numbers 1, 3, 5, 7, 9, &c., and since the space described by a body falling from rest is in the first second m feet, the space described in successive seconds will be m, $3m$, $5m$, $7m$, $9m$, &c. feet.

1. A body had been falling for 5 seconds: Compare the spaces described in the *third* and *fifth* seconds of its fall.

Ans. The space described in the *third* second$=80\frac{5}{12}$ feet; the space described in the *fifth* second$=144\frac{3}{4}$ feet.

2. A body has fallen through 579 feet: What was the space described by it in the *last* second?

Ans. It will be found that the body has been falling 6 seconds; therefore, the space described in the last second is $176\frac{11}{12}$ feet.

36. The method adopted in the last example for finding the

space described by a body in the *last* second of its fall, is only applicable when the time consists of a *determinate* number of seconds; but it is not difficult to investigate a *general* expression for the space described in the last n seconds, whatever be the value of T. For the space described in T seconds$=m\mathrm{T}^2$; and the space described in T$-n$ seconds$=m\times(\mathrm{T}-n)^2=m\mathrm{T}^2-2mn\mathrm{T}+mn^2$; hence the space described in the last n seconds=

$$m\mathrm{T}^2-(m\mathrm{T}^2-2mn\mathrm{T}+mn^2)=2mn\mathrm{T}-mn^2=m(2n\mathrm{T}-n^2);$$

if $n=1$, then the space described in the *last* second$=m(2\mathrm{T}-1)$; which expression will lead to the same results as the method practiced in example 1, and is likewise applicable in cases where the time does not consist of any even number of seconds. For example, let the time of falling be $6\frac{1}{2}$ seconds; then the space described in the last second, namely, from $5\frac{1}{2}$ to $6\frac{1}{2}$ seconds, will be $16\frac{1}{12}\times(13-1)=193$ feet.

If it were required to find the space described in the second *immediately previous* to the last n seconds, we have,

Space described in the last n seconds$=m(2\mathrm{T}n-n^2)$ (A).

Ditto in the last $(n+1)$ seconds$=m\overline{2\mathrm{T}(n+1)-(n+1)^2}$ (B).

Subtract (A) from (B), then the space described in the second immediately previous to the last n seconds$=m(2\mathrm{T}-2n-1)$.

1. What was the space described in the *last* 2 *seconds* by a body which had fallen from the top of a tower 300 feet high?

Ans. The whole time is found to be 4.32 seconds; therefore, the space fallen through in the last 2 seconds is 213.58 feet.

2. A body has been falling for $9\frac{1}{2}$ seconds: What was the space described in the *last* second but 4 of its fall?

Ans. $160\frac{5}{6}$ feet.

37. *To find the space described in a given time by a body projected upward or downward with a given velocity.* Let V be the given velocity with which a body is projected *downward*, and T the time of its motion; then the space described in the time T with the uniform velocity V will be equal to T$\times$V, and the space through which a body would fall by gravity in the same time is $m\mathrm{T}^2$; hence, from what was shown in Art. 32, the space described in the time T by a body projected downward with the velocity V is equal to $\mathrm{T}\times\mathrm{V}+m\mathrm{T}^2$; and applying the same process of reasoning to Art. 33, the space through which a body would ascend in the time T, if projected upward with a given velocity V, will be equal to $\mathrm{T}\times\mathrm{V}-m\mathrm{T}^2$.

1. A body is projected downward with a velocity of 30 feet in a second: *How far* will it fall in 4 seconds?

Ans. $377\frac{1}{3}$ feet.

2. A body is projected upward with a velocity of 120 feet in a second: *How far* will it rise in 3 seconds? Ans. $215\frac{1}{4}$ feet.

38. Miscellaneous Examples.

1. With what *velocity* must a body be projected downward from the height (a), that it may describe it in T seconds?

Let x=the velocity required; then the space described by a body projected downward with velocity (x) in the time (T) is $Tx+mT^2$; hence $Tx+mT^2=a$ $\therefore x=\frac{a-mT^2}{T}$. For instance, let $a=150$ and T=2, then the velocity with which a body must be projected downward from the top of a tower whose height is 150 feet, so that it may arrive at the bottom in *two* seconds=$42\frac{5}{6}$ feet per second.

2. With what *velocity* must a body be projected from the top of a tower 300 feet high, to reach the ground in 4 seconds?

Ans. $10\frac{2}{3}$ feet per sec.

3. The space described by a heavy body in the 4th second of its fall was to the space described in the last second except 4, as 1 to 3: What was the *whole space* described by the body?

The space described in the 4th second=$7m$; the space described in the last second but 4 $=m(2T-2n-1)=m(2T-9)$, where T= the *whole time* of falling; hence from the question

$$7m : m(2T-9) :: 1 : 3, \therefore 2T-9=21, \text{ or } T=15;$$

the whole space described, therefore, $(=mT^2)=16\frac{1}{12}\times 225=3618\frac{3}{4}$ feet.

4. Suppose at the same instant that a body begins to fall from rest from the point D, another body is projected upward from B with a velocity which would carry it to A: It is required to find the point where they would meet.

Fig. 13. (points A, D, C, B on a vertical line)

Let C be the point where the bodies would meet; and let AB=a, BD=b, DC=x; then will AD=$a-b$, AC=$a-b+x$.

Now the time of descending through DC=$\left(\frac{x}{m}\right)^{\frac{1}{2}}$; and the time of ascending through BC (=time down AB−time down AC)=$\left(\frac{a}{m}\right)^{\frac{1}{2}}-\left(\frac{a-b+x}{m}\right)^{\frac{1}{2}}$; but the time down DC must be equal to the time up BC; hence we have

$$\left(\frac{x}{m}\right)^{\frac{1}{2}}=\left(\frac{a}{m}\right)^{\frac{1}{2}}-\left(\frac{a-b+x}{m}\right)^{\frac{1}{2}}, \text{ or } x^{\frac{1}{2}}=a^{\frac{1}{2}}-(a-b+x)^{\frac{1}{2}};$$

$$\therefore (a-b+x)^{\frac{1}{2}}=a^{\frac{1}{2}}-x^{\frac{1}{2}}, \text{ and } a-b+x=a+x-2(ax)^{\frac{1}{2}};$$

$$\therefore 2(ax)^{\frac{1}{2}}=b, \text{ or } 4ax=b^2, \text{ and } x=\frac{b^2}{4a}.$$

5. Suppose a body to have fallen from A to B, (Fig. 14,) when another body begins to fall from rest at D: *How far* will the latter body fall before it is overtaken by the former?

Let C be the point where one body overtakes the other, and let AB$=a$, BD$=b$, DC$=x$; then AC$=a+b+x$. Now time down DC$=\left(\frac{x}{m}\right)^{\frac{1}{2}}$, and time down BC=time down AC$-$time down AB$=\left(\frac{a+b+x}{m}\right)^{\frac{1}{2}}-\left(\frac{a}{m}\right)^{\frac{1}{2}}$; but at the moment when the lower body is overtaken,

Fig. 14.

A, B, D, C

Time down DC=time down BC, or $\left(\frac{x}{m}\right)^{\frac{1}{2}}=\left(\frac{a+b+x}{m}\right)^{\frac{1}{2}}-\left(\frac{a}{m}\right)^{\frac{1}{2}}$;

$$\therefore x^{\frac{1}{2}}+a^{\frac{1}{2}}=(a+b+x)^{\frac{1}{2}}, \text{ and } x+a+2(ax)^{\frac{1}{2}}=a+b+x,$$

$$\text{or } 2\,(ax)^{\frac{1}{2}}=b, \therefore x=\frac{b^2}{4a}.$$

39. Questions on Falling Bodies.

1. From a black cloud a flash of lightning was observed, and 15 seconds afterward it began to rain: On the supposition that the rain began to fall on the instant of the flash, what was the height of the cloud?* *Ans.* 3618.75 *feet.*

2. A meteoric stone fell upon a projecting stick of timber, with a momentum which, from the motion given to the stick, was estimated at 18435 pounds. It occupied in falling, 10 seconds: From what height did it fall, and what was the weight of the stone? *Ans. Height,* $1608\frac{1}{3}$ *feet; Weight,* 57.31 *lbs.*

3. A man fell into a pit 500 feet deep: How long was he in falling, and what velocity did he acquire?
Ans. T=5.57 *seconds;* V=179.17 *feet per sec.*

4. Wishing to ascertain the difference in the depths of two wells, I dropped a pebble into one of them, and heard it strike the water in 6 seconds; and then into the other, and heard it strike in 7 seconds: What was the difference of their depths?
Ans. $209\frac{1}{12}$ *feet.*

5. An archer wishing to know the height of a tower, found that an arrow sent to the top of it, occupied 8 seconds in going and returning: What was the height of the tower?
Ans. $257\frac{1}{3}$ *feet.*

---

* No allowance is here made for the resistance of the air, which, in fact, greatly retards the descent of drops of rain.

6. In what time would a man fall from a balloon three miles high, and what velocity would he acquire?

Ans. T=31.38 *seconds, or about half a minute.*
V=1009.39 *feet per second, or about half the maximum velocity of a cannon ball.**

7. A body having fallen for $3\frac{1}{2}$ seconds, was afterward observed to move along the horizon (with the velocity which it had acquired in its descent) for $2\frac{1}{2}$ seconds: What was the whole space described by the body from the beginning of its fall?

Ans. $478\frac{1}{2}$ *feet, very nearly.*

8. Through what space would the aëronaut (in question 6) fall during the last second? *Ans.* 993.3 *feet.*

9. A body has fallen from the top of a tower 340 feet high: What was the space described by it in the last three seconds?

Ans. 298.957 *feet.*

10. Suppose a body be projected downward with a velocity of 18 feet in a second: How far will it fall in 15 seconds?

Ans. $3888\frac{3}{4}$ *feet.*

11. A body is projected upward with a velocity of 65 feet in a second: How far will it rise in 2 seconds? *Ans.* $65\frac{2}{3}$ *feet.*

12. With what velocity must a stone be projected into a well 450 feet deep, that it may arrive at the bottom in 4 seconds?

Ans. V=$48\frac{1}{6}$ *feet in a second.*

13. Upon a steeple 160 feet high, is a spire of 50 feet; at the same instant that a stone was let fall from the top of the steeple, another was projected perpendicularly upward from the bottom of it, with a velocity sufficient to carry it to the top of the spire: At what point will these stones meet?

Ans. 30.476 *feet from the top of the steeple.*

14. Upon the top of a tower 200 feet high, is placed a flag-staff of 26 feet; a bullet is let fall from the top of this flag-staff; and at the instant of its passing the bottom of it, a stone is let fall from a window 44 feet from the top of the tower: At what distance from the bottom of the tower will the bullet overtake the stone? *Ans.* 137.385 *feet.*

---

* In this problem we have an example of the immense velocity which bodies falling toward the earth from a great height finally acquire, being, in the case supposed, more than eleven miles per minute. To form an idea of the whole progress of a very distant body falling toward the earth, we must conceive of it as at first moving with extreme slowness, and as accelerated by very small increments of velocity. For although, in terrestrial mechanics, gravity is considered as a *constant* force, producing *uniform* acceleration, yet it must be remembered that it is, in fact, a *variable* force, diminishing as the square of the distance increases; and hence, that the acceleration which it produces in a given time, is not only much less at remote distances from the earth than at its surface, but that the *rate* of acceleration itself is constantly increasing as we approach the earth. The principles, therefore, which serve for estimating the time, velocity, and space, of bodies falling near the earth, as in the foregoing examples, do not answer for bodies falling from great distances. The laws governing these are investigated by means of the Calculus. Since the ac-

CHAPTER III.

OF THE COMPOSITION AND RESOLUTION OF MOTION.

40. In the two preceding Chapters, we have considered the motion produced in bodies by the action of only a *single* force. We now proceed to show the manner in which a body would move when acted upon at the same time by *several* forces. Let us first consider the case of a body acted upon by two forces.

41. *Two impulses, which, when communicated separately to a body would make it describe the adjacent sides of a* PARALLELOGRAM *in a given time, will, when they are communicated at the same instant, cause it to describe the diagonal in that time; and the motion in the diagonal will be uniform.*

Suppose a body placed at A to be acted upon by two forces, one of which would cause it to move uniformly over the line AB,

Fig. 15.

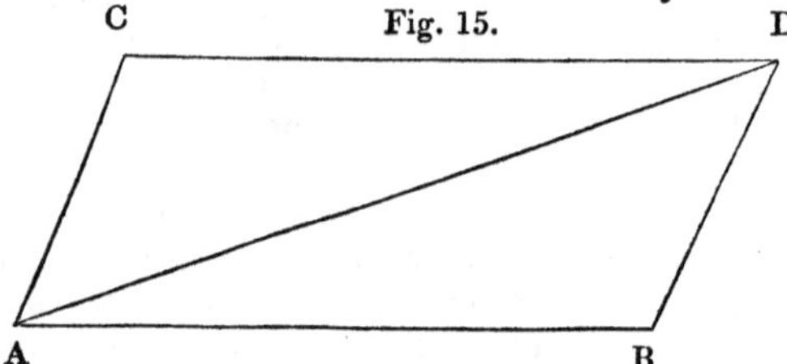

and the other over the line AC in the same given time, then complete the parallelogram ACDB; and if both forces act at the same instant upon the body, it will, by their *joint* action, move uniformly over the diagonal AD in the same time that it would have described either of the sides AB or AC by the forces acting separately. For it is evident that the force which acts in the direction AB, can have no tendency whatever to prevent the access of the body toward the line CD, which is parallel to AB, (2d Law of Motion, Art. 20.)* When both forces act together,

celerating force diminishes so rapidly as the distance from the earth increases, there is a limit to the velocity which a body can acquire by gravity. If it falls from an infinite distance, it can acquire only a velocity of about *seven miles* per second; and half of this is gained within 1354 miles of the earth. Were a body projected from the earth with a velocity of 7 miles per second, it would never return. (See Vince's Fluxions, Sec. VIII, Prop. XL, 5. Young's Elements of Mechanics, Art. 116.)

* "When any force is exerted upon a body already in motion, the motion which the force would produce upon a body at rest, is compounded with the previous motion in such a way, that *both produce their full effects parallel to their own directions.*" (Whewell, 228.) That any force impressed upon a body, already moving under the influence of different forces, has its full effect, either in producing or in destroying motion, is evident from the fact that a given force is found to have *the same effect* upon bodies in different parts of the earth, although, in consequence of the diurnal motion of the earth, bodies apparently at rest are moving with various velocities in different latitudes. (Ib. 231.)

therefore, it will, by the action of the force in the direction AC, arrive at the line CD (but in a *different point* of that line) in the same time as if the force in the direction AB had not acted; for the same reason, the force in the direction AC will have no tendency to prevent the access of the body toward BD, which is parallel to AC; it will arrive therefore at the line BD in the same time as if the force in the direction AC had not acted. Hence the body will arrive at the lines CD and BD at the same instant of time, and consequently will be found at their common intersection D; and as the body, after it leaves the point A, is acted upon by no external force, it must, by the first law of motion, have described the diagonal AD with a uniform motion.

42. As this motion of a body in the diagonal of a parallelogram by the joint action of two forces which (acting separately) would have caused it to describe the two sides, is a fundamental theorem with respect to the composition of motion, let us consider it in another point of view. Let the lines AC, AB be di-

Fig. 16.

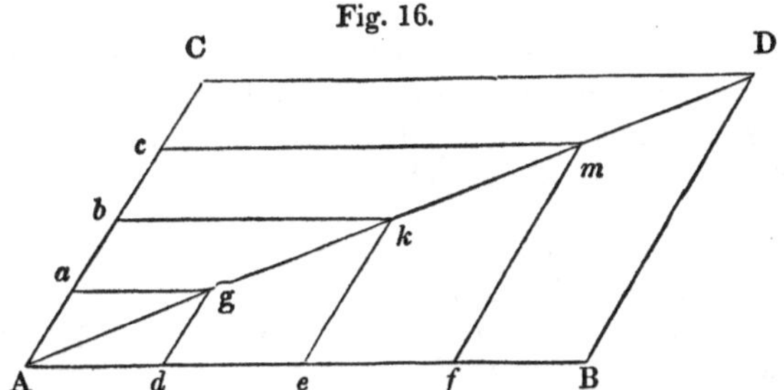

vided into the same number of small equal parts, A*a*, *ab*, *bc*, &c.; A*d*, *de*, *ef*, &c., which will be to each other as the whole lines AC, AB, i. e. A*a* : A*d* : : AC : AB; *ab* : *de* : : AC : AB; &c., and consequently (Alg. Art. 388,) A*a*+*ab* (A*b*) : A*d*+*de* (A*e*) : : AC : AB; &c. If therefore the parallelograms A*dga*, A*ekb*, &c., be completed, then (by Euc. 6, 26,) the points *g*, *k*, *m*, &c., will all fall in the diagonal AD. Now since AC, AB are described uniformly in the same time, the proportional parts A*a*, A*d*; A*b*, A*e*, &c., will be described uniformly in the same time. From what has already been demonstrated therefore, at the end of those different parts of time the body will be brought to the point *g*, *k*, *m*, &c., by the united action of the forces which would have separately made it move over A*a*, A*d*; A*b*, A*e*, &c. Let the number of parts into which AC, AB are divided be indefinite, then the number of points *g*, *k*, *m*, &c., will be indefinite, and the lines, A*g*, *gk*, *km*, &c., will be indefinitely small; the body therefore will begin to move in the line A*g*, and, being found at the end of each successive instant of time in the line AD, it must have moved over that line with the uniform velocity with which it set off.*

43. *If a body be acted upon by two forces, one of which would cause it to move uniformly over one side of a* TRIANGLE, *and the other over another side of the triangle, in the same time, then by the joint action of those forces it will be made to describe the third side, in the same time that it would have described either of the sides by the forces acting separately.*

Thus if a body be acted upon by two forces, one of which would cause it to move uniformly over the side AC, (Fig. 15,) and the other over the side CD, of the triangle ACD, (or over a line parallel and equal to CD,) then, by the joint action of those forces, it would be made to describe the third side AD in the same time that it would have described either of the sides AC, CD, by the forces acting separately. For if the parallelogram ACDB be completed, then, since AB is equal and parallel to CD, a force acting in the direction AB would make a body describe AB in the same time as that in which it would describe CD; but by what has already been proved, if two forces act upon a body, by one of which it would be made to describe AC, and by the other AB, in the same time, it would by the joint action of these forces be made to describe the diagonal AD, which is the third side of the triangle ACD.

Fig. 17.

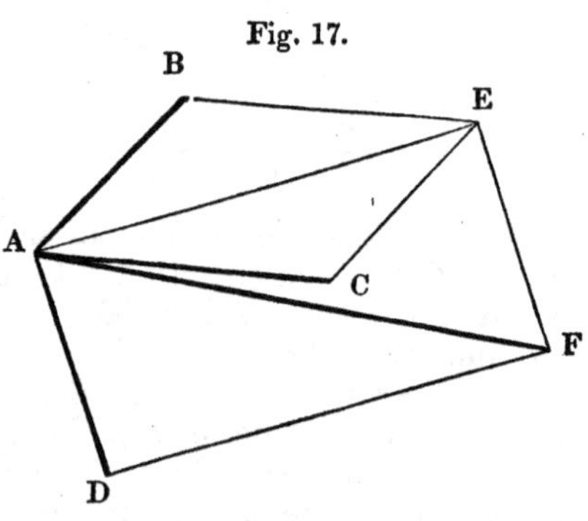

44. Let us next suppose a body placed at A, (Fig. 17,) and acted upon by three forces, by one of which it would be made to describe AB, by another AC, and by the third AD, uniformly in the same time; complete the parallelogram ABEC, and join AE; complete also the parallelogram AEFD, and join AF; then AF will be the line over which the body will move uniformly by the joint action of those forces, in the same time in

* The body will also describe the diagonal AD, when acted upon by uniformly *accelerating* forces. For, let A*e*, A*d*, (Fig. 16,) be two spaces described in the times T, *t*, by one uniformly accelerating force, and A*b*, A*a*, be two spaces described by another similar force, then,

$$Ae : Ad :: T^2 : t^2$$
$$Ab : Aa :: T^2 : t^2$$
$$\therefore Ae : Ad :: Ab : Aa;$$

consequently, the points *g*, *k*, which denote the positions of the body at the end of each successive instant, are all in the same straight line. (Euc. VI, 26.)

which it would have described AB, AC, or AD, by either of the forces acting separately. For by Art. 41, a body acted upon by two forces in directions AB, AC, would be made to describe the diagonal AE; a body placed at A therefore, and acted upon by *three* forces in directions AB, AC, AD, is under the same circumstances as if it was acted upon by two forces, one of which would make it describe AE, and the other AD in the same time; but the line over which a body would move uniformly by the action of two forces in the directions AE, AD is the diagonal AF; AF therefore is the line over which it would move uniformly by the joint action of the three forces in the directions AB, AC, AD.

Since BE is equal and parallel to AC, (Fig. 17,) and EF equal and parallel to AD, it follows, (for the same reason as in Art. 43,) that if a body be acted upon by three forces, each of which acting separately would make it describe in *succession* the three sides, AB, BE, EF of the figure ABEF, (or lines parallel and equal to them,) taken in the order of the letters, A, B, E, F, it would by the *joint action* of those forces be made to describe the fourth side AF in the same time that it would have described those sides respectively when the forces act separately. And since the same mode of reasoning applies to a polygon of any number of sides, we have in general the following theorem.

45. *If a body be impelled by any number of forces, which acting separately, would, in a given time, make it describe all the sides of a* POLYGON *except the last side; when all these forces act at the same instant, it will be made to describe the remaining side in the same given time.*

Fig. 18.

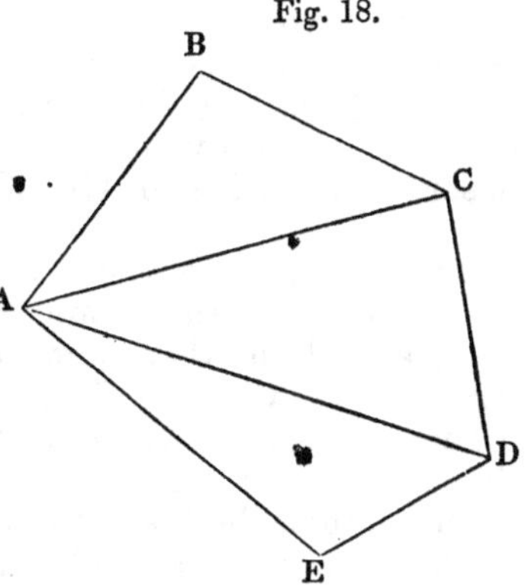

Thus, if a body be impelled by any number of forces which, acting *separately*, would, in a given time, make it describe each of the sides AB, BC, CD, DE of the polygon ABCDE; when all those forces act *at the same instant*, it will be made to describe the remaining side AE in the same given time.

46. *If all the sides of a polygon except the last represent the quantity and direction of several forces, acting at the same instant upon a body, the remaining side will represent the quantity and direction of a force* EQUIVALENT *to them all*

A force is said to be equivalent to any number of forces, when it will, singly, produce the same effect that the others produce jointly in any given time. The single force is frequently called the *resultant,* and the forces that produce it are called the *components.*

By the second law of motion, the space described is proportional to the force impressed; in all these cases, therefore, the spaces respectively described by the body will represent the quantity and direction of the forces by which it is impelled. Thus (see Fig. 15,) if the quantity and direction of two forces be represented by the two sides AB, AC of the parallelogram ACDB, the diagonal AD will represent the quantity and direction of a force equivalent to them both; or if the two sides AC, CD of the triangle ACD represent the quantity and direction of two forces acting at the same time upon a body, the third side AD will represent a force equivalent to them both. With respect also to the forces by which a body is made to describe the sides AB, BC, CD, DE, of the polygon ABCDE, (Fig. 18); if AB, BC, CD, DE, represent the quantity and direction of several forces acting at the same instant upon a body, the remaining side AE will represent the quantity and direction of a force equivalent to them all.*

47. Since the lines which represent the proportion of the forces in these different figures are described in the same time, and since the velocity of a body is proportional to the space described in a given time, these lines will also represent the proportion of the *velocities* with which they are respectively described. Thus (Fig. 15,) the velocity with which the diagonal AD is described is to the velocity with which either of the sides AC or AB is described as AD is to AC or AB; and in the case of the polygon in Art. 45, the velocity with which the side AE is described is to the velocity with which either of the sides AB, BC, CD or DE is described as AE is to AB, BC, CD or DE.

48. Hitherto the forces have been supposed to be such as by their separate action would produce *uniform* velocities; in which case, a body by their joint action will be made to describe a straight line with a uniform velocity. But if two forces act upon a body, by one of which it would be made to describe a straight line with a uniform velocity, and by the other with a *variable* velocity, then the body, by the united action of those forces, will neither describe a straight line, nor will it move with a uniform velocity; but will describe with a variable velocity some curve

* It will be remarked by the learner, that several of these forces acting in opposite directions, partly destroy one another, so that AE represents merely the *resultant,* or what remains after all these mutual actions.

line, the form of which must be determined from the particular nature of the two forces which act separately upon the body. Let us take the case of a body projected obliquely at the earth's surface, on the supposition that it meets with no resistance in its passage through the air. Conceive a body to be projected from the point P, (Fig. 19,) in the direction PN, with such velocity as would carry it uniformly over the line PN in the same time that it would descend by the force of gravity through the space PV. Complete the parallelogram PNVQ; then for the same reason as in Art. 41, the body at the end of that given time would be found in the point Q; having described, not the diagonal PQ, but some *curve line* POQ.* In PN take any point M, and let T, t represent the times of describing PN, PM respectively; make PL equal to the space through which a body would fall by gravity in the time t, and complete the parallelogram PMOL; then O will be the place of the body at the end of the time t; and in the same manner the other points of the curve POQ might be determined.

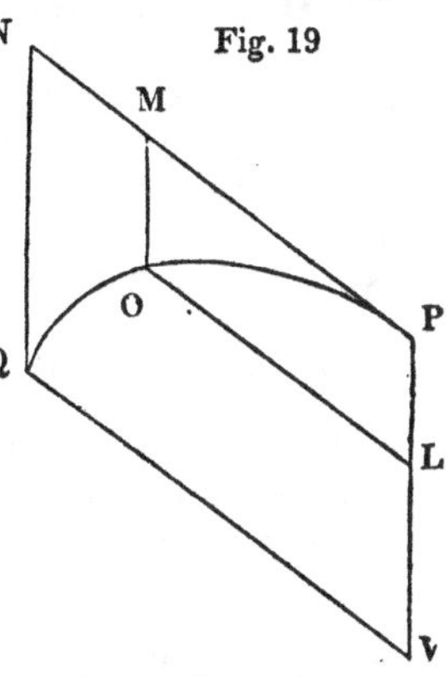

Now since PN is described with a uniform velocity,

$$PN : PM :: T : t \therefore PN^2 : PM^2 :: T^2 : t^2.$$

$$NQ : MO :: T^2 : t^2 \therefore NQ : MO :: PN^2 : PM^2.$$

Hence the curve is such, that $MO \propto PM^2$, which is a well known property of the parabola.† The curve POQ, therefore, is a parabola whose diameter is PV, ordinate QV, and whose parameter to the point P is $\frac{QV^2}{PV}$‡$=\frac{PN^2}{PV}$.

49. Questions on the Composition of Motion.

1. A body is acted upon at the same time by two forces which are to one another as a : b, and their directions are inclined to each other in the given angle A: What is the magnitude of the resultant?

---

* For the body in descending in the direction of PV, recedes from the line PN very slowly at first, but faster and faster as it proceeds; and since the rate of acceleration is augmented *continually*, the body must be constantly drawn further and further from the direction PN. But a constant change of direction, implies that the path is a curve line.

† Bridge's Conic Sec. Art. 27.

‡ By Cor. to Prop. 8 of *parabola*, parameter$\times$PV$=$QV^2, $\therefore$ parameter$=\frac{QV^2}{PV}$. (Bridge's Conic Sections.)

Let AC : AB (Fig. 15,) represent the ratio of $a:b$, and let BAC be equal to the given angle A. Complete the parallelogram ABDC, then AD will represent the resultant. Since CD=AB, AC : CD :: $a:b$; and since CD is parallel to AB, the angles BAC+ACD=180°, ∴ ACD=180°−∠A; hence the problem is reduced to finding trigonometrically the third side AD of the triangle ACD, in which are given the two sides AC, CD, and the included angle ACD.

Let AC : AB :: 2 : 3, and ∠A=60°, ∴ ACD=120°; then* 5 : 1 :: tan. 30° : tan. (½CAD−½CDA) =⅕ tan. 30°; ∴† log. tan. (½CAD−½CDA)=log. tan. 30°−log. 5=9.0624694=log. tan. 6° 35′; hence CAD=36° 35′, and CDA=23° 25′, and

sin. 23° 25′ (CDA) : sin. ACD (120°) :: AC (2) : AD=4.36,

i. e. if two forces which are to each other as 2 : 3 act upon a body at an angle of 60°, the resultant will be proportional to 4.36.

2. From an island in the Straits of Sunda, we sailed S. E. *b* S. (33° 45′) at the rate of 6 miles an hour; and being carried by a current, which was running toward the S. W. (making an angle with the meridian of 64° 12′¼) at the end of four hours, we came to anchor on the coast of Java, and found the said island bearing due north: Required *the length of the line* actually described by the ship, and *the velocity of the current?*

Ans. S=26.4 miles.
V=3.7024 miles per hour.

3. A sloop is bound from the main land of Africa to an island bearing W. *b* N. (78° 45′) distant 76 miles, a current setting N. N. W. (22° 30′) 3 miles an hour: What is the *course* to arrive at the island in the shortest time, supposing the sloop to sail at the rate of 6 knots per hour; and what *time* will she take?

Ans. Course 76° 41′ 4″ S.—Time 10h. 40m. 7 sec.

50. We may likewise find the magnitude of the force compounded of any number of forces, whose quantities and directions are represented by the sides of the given polygon ABCDE, (Fig. 18.) For since AB, BC and ABC are given, AC and BCA may be found; but ACD=BCD−BCA, ∴ AC, CD, and ACD are known, from which AD and ADE may be determined; and in the triangle ADE we have AD, DE, and ADE, ∴ AE is known.

4. Three men are pulling at a boat with equal forces and in the same plane. A pulls at right angles to B, and B at an angle of 45 degrees with C: In what *direction* will the boat move, and

* Day's Trigonom Art. 44. † Ib. Art. 41.

what is the ratio of the *resultant* to the sum of the individual forces?

Fig. 20.

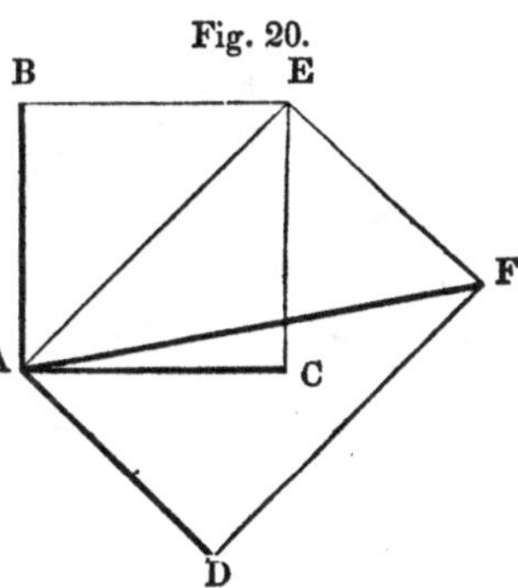

Let the point of application be at A, (Fig. 20,) and let AB, AC, and AD, represent respectively the magnitude and direction of A, B, and C. Then it may be shown that AF is the resultant, which makes an angle with AB of 80° 16′, and bears to the sum of A, B, and C, the ratio of $\sqrt{3}$ to 3.

RESOLUTION OF MOTION.

51. *A given force may be resolved into an unlimited number of others, acting in all possible directions.*

Let AB represent the quantity and direction of some given force; draw any lines AD, AC, and join DB, CB; complete also the parallelograms ADBE, ACBF. Since AB is the diagonal of two parallelograms whose adjacent sides are respectively AD, AE, and AC, AF, it may (by Art. 41) be considered as the resultant of two forces whose quantities and directions are represented either by AD, AE, or AC, AF, i. e. by AD, DB, or AC, CB. The forces represented by AD, DB, or AC, CB, may also be resolved into other pairs of forces, and so on without end.

Fig. 21.

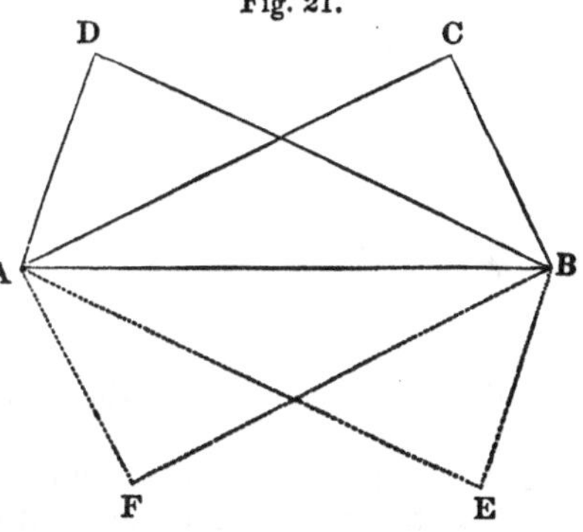

52. Sometimes, however, by the conditions of the problem, the resolved forces are required to make a *given* angle, or to be in a certain ratio, with each other. The method of solving cases of this kind, may be illustrated by a few examples.

First, let a given force AB, (Fig. 22,) be resolved into pairs of forces which shall always act at *right angles* to each other. Upon AB describe a semicircle, ABC, and from the extremities of the base draw straight lines to meet in any point of the circumference. The sides of the triangle thus formed, will severally contain right angles;* and AC and a line drawn from A

* Euc. III, 31.

parallel and equal to CB, will represent two forces equivalent to the given force AB.

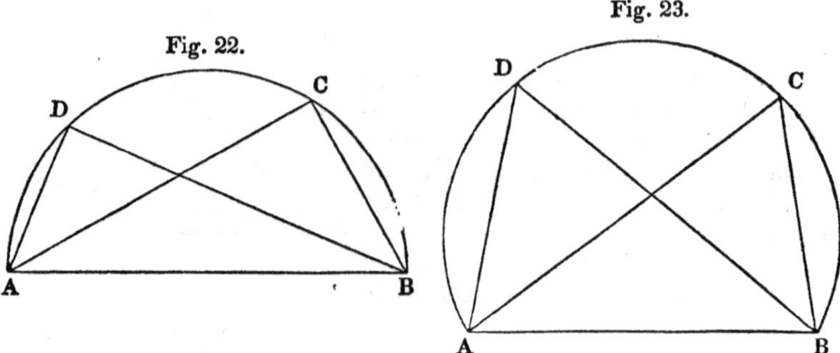

Secondly, let the resolved forces be required to make with each other *any* given angle. Upon AB, (Fig. 23,) describe the segment of a circle, ABC, containing an angle, the supplement of the given angle,* and draw straight lines from the extremities of the base to any point in the circumference.

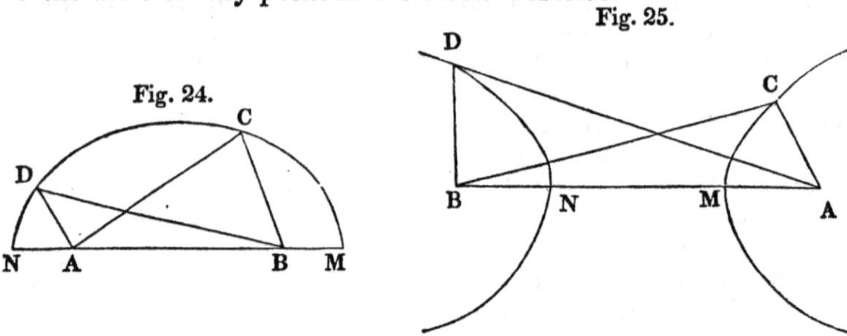

Thirdly, let the *sum* of the resolved forces be required to be equal to a given quantity. Let MN (Fig. 24) be equal to the sum of the forces required and AB be the given force; and upon MN, as the transverse, and with the points A and B (equally distant from M, N,) as foci, describe the ellipse MCN. From A and B draw straight lines to any point in the ellipse, and the sides of the several triangles will form the pairs of forces required.†

Fourthly, in like manner, pairs of forces whose *difference* shall be always equal to the same constant quantity, may be found by making A and B the foci of an hyperbola, as in Fig. 25, and drawing straight lines from these points to the curve.‡

1. *A given force* (*a*) *is required to be resolved into different pairs of forces which shall act at an angle of* 135° *to each other* · *What is the radius of the circle whose segment shall contain pairs of the resolved forces?*

---

* Euc. III, 33. † Bridge's Conic Sect. Art. 8. ‡ Conic Sect. Art. 5.

Let AB=a, and upon AB describe the segment of a circle which shall contain an angle ADB of 45°: then the radius AC=$\frac{a}{\sqrt{2}}$.

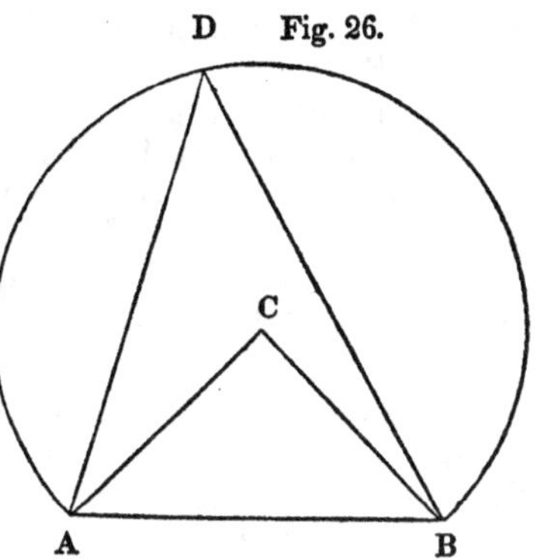

Fig. 26.

2. *To determine the radius of the circle when* AB (Fig. 26) *is required to be resolved into pairs of forces acting at any given angle whose supplement is* A.

Let AB=a, ADB=A; find the center C, and join CA, CB; then BC or radius=$\frac{a \times \text{cos.A}}{\text{sin. 2A}}$.

Let a=10, A=40° } then radius=7.7786.

53. The most obvious consideration with respect to the composition of motion is, that if two equal forces act upon a body in contrary directions, they will destroy each other's effects, and the body thus acted upon will remain at rest; or if any two forces act upon a body in the same straight line, then the effect (or, in other words, the motion) produced will be proportional to the sum or difference of those forces, according as they act in the same or in opposite directions. But if these forces act obliquely to each other, then the resulting force will be some intermediate quantity between that sum and difference, the magnitude of which will increase according as the angle of inclination between the directions of these forces is diminished. For it is evident that the smaller the angle of inclination between two forces is, the more nearly will they conspire together, and consequently the whole effect produced will be greater; on the contrary, as the angle of inclination increases, the two forces will more strongly oppose each other, their whole effect therefore will keep diminishing.

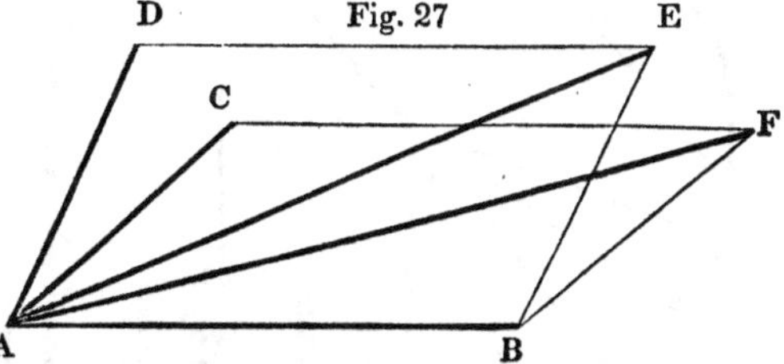

Fig. 27

This latter conclusion may also be drawn from the *geometrical* representation of the forces. Let two forces be represented by AB, AD, or by AB, AC, of which AC=AD; let the angle DAB be greater than the angle CAB, and complete the parallelograms

DABE, CABF; then since DAB is greater than CAB, its supplement ABE must be less than ABF, the supplement of CAB; hence in the triangles ABE, ABF, we have AB, BE equal to AB, BF, and the angle ABF greater than ABE, ∴ AF is greater than AE.* Let CAB=0, then AF=AC+CF=sum of the forces; let CAB=180°, then AF=CF−AC=difference of the forces; in all other cases, AF is of some intermediate magnitude between AC+CF and CF−AC, and keeps increasing as the angle CAB is diminished.

54. In the composition of forces which act obliquely on each other, some force is actually lost; for the sum of the forces before they are compounded together is represented by the two sides AD, DE of a triangle, and after composition by the third side AE. The contrary happens with respect to the resolution of forces; for the two resolved forces being represented by the two sides of a triangle of which the given force is the third, the absolute quantity of the resolved forces must be greater than that of the given force.

Five sailors raise a weight by means of five separate ropes, in the same plane, connected with the main rope that is fastened to the weight in the manner represented in figure 29. B pulls at an angle with A of 20°; C with B, at 19°; D with C, at 21° 30′; and E with D, at 25°. A, B, and C, pull with equal forces, and D and E with forces one half greater: Required the magnitude and direction of the resultant, and the loss of force occasioned by the forces acting partly against each other.

Let the sides of the polygon (Fig. 28) represent the several forces in magnitude and direction, then AF will be the resultant

Fig. 28. Fig. 29.

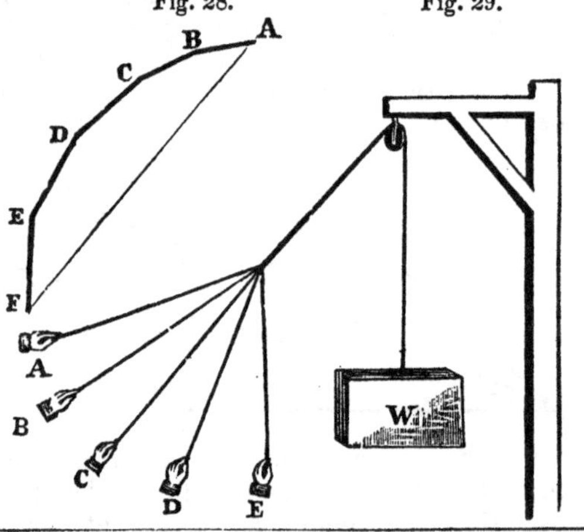

* Euc. I, 24.

The angle at B=160°; at C, 161°; at D, 158° 30′; at E, 155°. Hence,

1. The resultant makes an angle with AB=46° 33′ 10″.
2. Its value is 5.1957, (that of all the components being 6,) and it falls between C and D.
3. The loss of force is .1341, or about $\frac{13}{100}$ of the whole.

55. *A body acted upon at the same time by three forces, represented in quantity and direction by the three sides of a triangle taken in order, (or by lines parallel to these,) will remain at rest.*

Fig. 30.

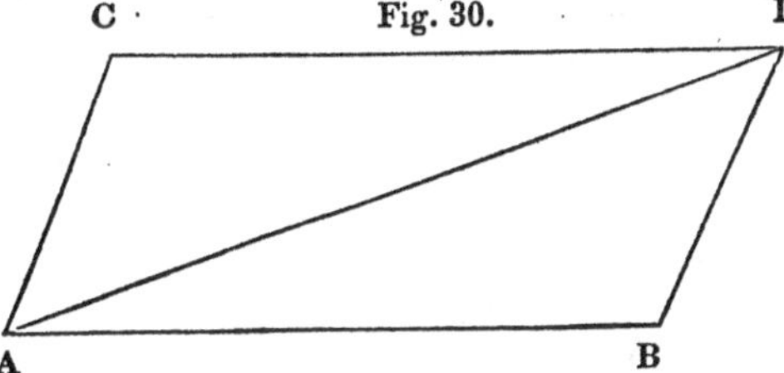

Since AD (Fig. 30) is equivalent to AB and AC, a body placed at A and urged by AB and AC in one direction, and by DA in the opposite direction, would remain at rest. But these three forces correspond in magnitude and direction with the three sides of the triangle ACD.

56. *If a body be kept at rest by three forces, those three forces will be represented by the three sides of a triangle formed by lines drawn in their respective directions.*

For suppose a body be kept at rest by three forces, and that AC, CD (Fig. 30) represent the quantities and directions of two of those forces, then the compound force arising from those two forces will be represented by the line AD; a third force, therefore, represented in quantity and direction by the line DA, equal and opposite to AD, must exactly counterbalance AD and keep it at rest. Whenever, therefore, a body is kept at rest by three forces, if a triangle be drawn, whose sides are respectively in the directions of those forces, those sides will represent the quantity and direction of the several forces thus acting upon the body.

57. *The proportion of the three forces which keep a body at rest will be represented by the three sides of any triangle, drawn parallel or perpendicular to the sides of the triangle which are in the directions of the forces.*

For, let the triangle ABC be that whose sides are drawn in the direction of the three forces, then the triangle $\alpha\beta\gamma$ (whose sides are *parallel* to AB, BC, CA,) and the triangle *abc* (whose sides are *perpendicular* to AB, BC, CA,) being each of them similar*

* Since the sides of the triangle $\alpha\beta\gamma$ are respectively parallel to the sides of the

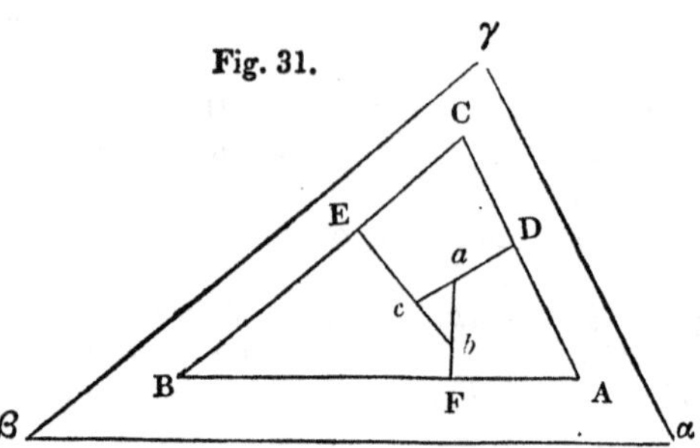

Fig. 31.

to the triangle ABC, must have their sides $\alpha\beta$, $\beta\gamma$, $\gamma\alpha$, or *ab*, *bc*, *ca*, respectively proportional to the three sides AB, BC, CA, which represent the quantity and direction of the forces acting upon a body.

58. *Any one of the three forces which keep a body at rest, is as the sine of the angle included between the other two.*

For, (Fig. 31,) AB : BC : : sin. BCA : sin. BAC ; BC : CA : : sin. BAC : sin. ABC ; and CA : AB : : sin. ABC : sin. BCA.

Conversely, *if a body be acted upon by three forces, each of which varies as the sine of the angle included between the directions of the other two, it will remain at rest,* since the sines are as the sides opposite to them, and when the forces are proportional to these sides, the body will remain at rest by Art. 55.

59. *A body will be kept at rest if it be acted upon by any number of forces, which are represented in quantity and direction by the sides of a polygon taken in order.*

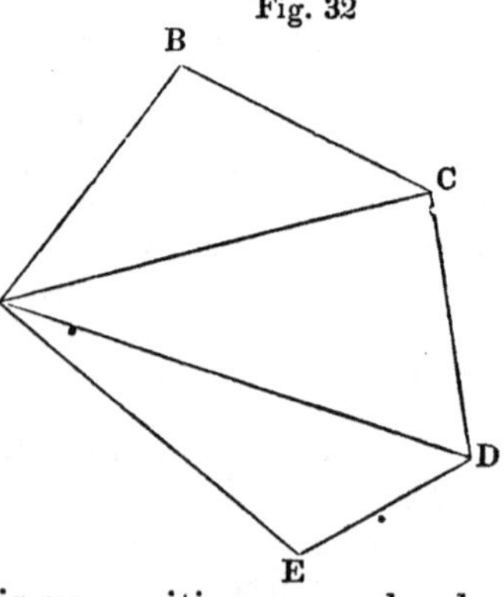

Fig. 32

For, let a body be acted upon by any number of forces represented by the sides AB, BC, CD, DE of the polygon ABCDE ; then (by Art. 44,) these forces compounded together will be represented in quantity and direction by the remaining side AE ; if, therefore, at the same time that the body is acted upon by the forces AB, BC, CD, DE, it is also acted upon by another force represented in quantity and direction by EA, (equal and opposite to AE,) it will remain at rest. The converse of this proposition may also be

triangle ABC, it is evident that the angles $a\beta\gamma$, $\beta\gamma a$, $\gamma a\beta$, are respectively equal to ABC, BCA, CAB. With respect to the triangle *abc* ; since the angles at D, E, F, are *right*-angles, we have D*a*F+DAF=180°, also D*a*F+*bac*=180°, ∴ DAF=*bac* ; and in the same manner it appears that EBF=*abc*, and ECD=*acb*. (See Legendre's Geometry, III, 209.)

established by the same mode of reasoning as that made use of in Art. 56, viz.: *If a body be kept at rest by any number of forces, those forces will be represented, in quantity and direction, by the sides of a polygon formed by the intersection of lines drawn in the direction in which the forces respectively act.*

1. A body is acted upon by two forces a and b, which are at right angles to each other: It is required to find the magnitude and direction of a third force, which shall keep the body at rest.

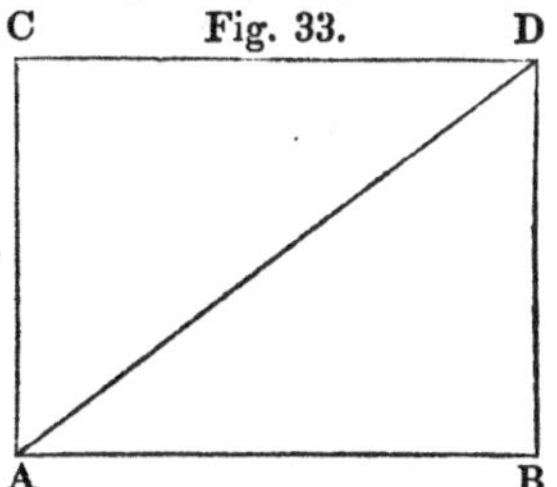

Fig. 33.

Let AC$=a$, AB$=b$, CAB$=90°$ } Complete the parallelogram ABDC, and join AD; then the two forces acting upon the body may be represented by AC, CD; consequently DA (the third side of the triangle ACD) will represent the force which shall keep the body at rest. (Art. 55.)

Now DA$=(AC^2+CD^2)^{\frac{1}{2}}=(a^2+b^2)^{\frac{1}{2}}$ for the magnitude of the force; and sine of CAD : sine of CDA (=DAB) :: CD : CA :: $b : a$; ∴ the direction DA of the third force divides CAB into two angles, whose sines are to each other as $a : b$.

2. A body, acted upon by two forces, is kept at rest by a third force (a), whose direction divides the angle contained between the directions of the two former into the given angles A and B: What is the magnitude of those two forces?

Let AC, AB be the two forces, (Fig. 30); complete the parallelogram ABDC, then DA (=a) is the third force; let CAD=A, DAB=B, then ACD=180°−(A+B.) Now,

AC : AD (a) :: sin. CDA or DAB (B) : sin. ACD ($180°-\overline{A+B}$,)
& AB : AD (a) :: sin. ADB or CAD (A) : sin. ACD ($180°-\overline{A+B}$.)

$$\text{Hence AC}=\frac{\sin. B\times a}{\sin. (180°-\overline{A+B})}=\frac{\sin. B\times a}{\sin. \overline{A+B}},$$

$$\text{and AB}=\frac{\sin. A\times a}{\sin. (180°-\overline{A+B})}=\frac{\sin. A\times a}{\sin. \overline{A+B}}.$$

60. It appears by Art. 58, that any one of the three forces which keep a body at rest is proportional to the sine of the angle included between the directions of the two others. An important consequence of this truth is, that, of three forces that keep a body at rest, the two components and the resultant may severally be represented by the sine of the angle included between the directions of the two others; viz. the resultant by the sine of the angle comprehended between the directions of the two

components, and each of the components by the sine of the angle comprehended between the resultant and the other component. Hence are derived the principles of the composition and resolution of PARALLEL FORCES.

Thus, let a body at A be kept at rest by three forces represented in quantity and direction by the sides of the triangle ADC, AD and DC being the components and AC the resultant. Complete the parallelogram ABCD; produce the lines AB, AC, AD indefinitely, and with any radius describe the arc EFG, and from the points E, G, draw the sines EH, EI, GK.

From what has been said it will be seen, that the force AB may be represented by the sine EI, the force AD by GK, and the

Fig. 34.

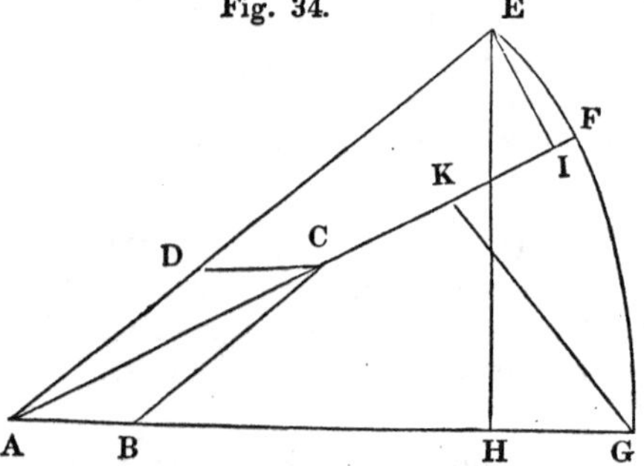

force AC by EH. Suppose now that the lines AE, AF, AG, approach toward parallelism by making the center A continually recede from the arc EFG; then that arc will continually approach toward a straight line, while the sines will approach toward a coincidence with it, until finally, when the lines AE, AF, AG, become parallel, the sine EH will cross the parallels at right angles, and the sines EI and GK will form parts of the same straight line with EH. Hence, when the two forces AB and AD become parallel, their resultant AC forms another parallel with them both; and since the resultant is represented by the sine EH, which, when these lines become parallel, equals the two sines EI and GK, we hence derive the following THEOREMS.

I. *The resultant of two parallel forces is in a direction constituting another parallel, and is equal to their sum.*

II. *If a straight line be drawn perpendicular to the directions of these three forces, (viz. the two components and their resultant,) each of the components will be represented by the part of the perpendicular contained between the directions of the two others.*

61. We have thus far considered the two parallel forces as acting the same way: when they are directed toward opposite parts, the investigation is the same with that in the last article, and the conclusion the same, except that the resultant is equal to the difference of the two components. A great number of par

allel forces may be compounded into a single force equivalent to them all, by proceeding as in Art. 44; that is, by first finding the resultant of two forces, and a new resultant for that resultant and one of the remaining forces, and so on to the last; and any single force may be resolved into any number of parallel forces, by a method the reverse of this.

62. In estimating the effects produced by the composition and resolution of forces, we have hitherto considered them as acting in the same plane; we proceed now to the solution of the problem, by means of which we are enabled to determine the motion of a body resulting from the operation of any number of forces acting IN DIFFERENT PLANES.

All the forces which can possibly act upon a body, may be resolved into equivalent forces acting in the direction of THREE STRAIGHT LINES OR AXES, *at right angles to each other.*

Let AK, AG be two straight lines drawn at right angles to each other in the same plane, and let AL be drawn at right angles to that plane, and, consequently, at right angles to each of the lines AK, AG.* Suppose AB to represent the quantity and direction of a force acting upon a body at A; let fall the perpendicular BP upon the plane passing through AK, AG; join AP, and complete the parallelogram APBC. From P draw PD parallel to AK, and PE parallel to AG. Since AB is the diagonal of the parallelogram APBC, the force represented by AB is resolved into two others AC, AP, equivalent to it; and since AP is the diagonal of the parallelogram ADPE, the two AD.

Fig. 35.

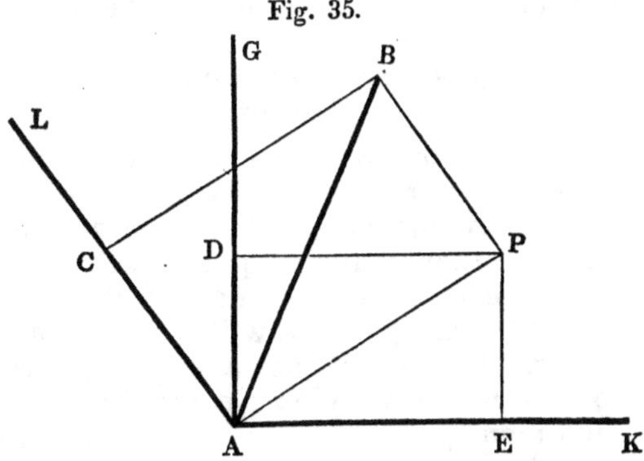

AE are equivalent to AP. Hence the given force AB is resolved into three others, AC, AD, AE, in the direction of the three straight lines AL, AG, AK, which are at right angles to each other, and issue from the point A.

Produce LA to *l*, (Fig. 36,) GA to *g*, and KA to *k*; so that the

* Euc. 2, Sup. Def. I.

three lines (or axes) L*l*, G*g*, K*k*, shall cut each other at right angles in the point A; then it is evident that the directions of all the forces which can possibly act upon a body at the point A, will fall within one or other of the eight solid angles formed by the intersection of three planes cutting each other at right angles, and passing through the axes L*l*, G*g*; L*l*, K*k*; G*g*, K*k*, respectively; and from what has just been shown, each of those forces may be resolved into *three* others, in the directions,*

AL, AG, AK;
AL, AG, A*k*;
AL, A*g*, A*k*;
AL, A*g*, AK;
or A*l*, AG, AK;
A*l*, AG, A*k*;
A*l*, A*g*, A*k*;
A*l*, A*g*, AK;

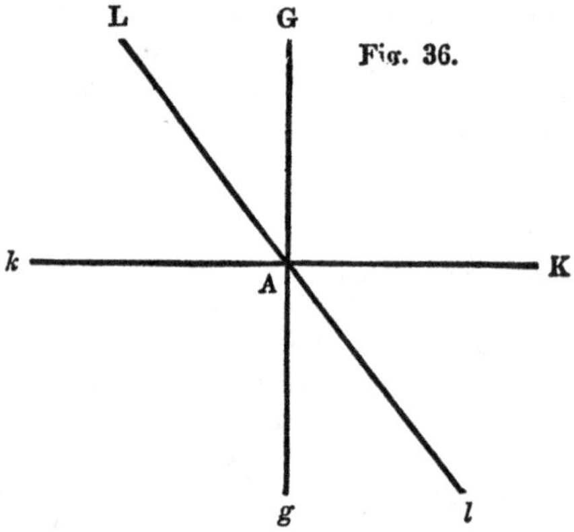

Fig. 36.

according to the solid angle in which it is included. Thus, then, all the forces which can possibly act upon a body at the point A, may be resolved into others acting along the three axes L*l*, G*g*, K*k*; for the forces acting in the directions AL, A*l*; AG, A*g*; AK, A*k*, respectively, are merely forces acting in *opposite* directions in the same straight line.

Cor. If the sum of the opposite forces in the direction of each axis be equal to one another, the body will be at rest.

63. Miscellaneous Questions on the Composition and Resolution of Motion.

1. Three men, equal in strength, undertake to pull down the steeple of an ancient church. They fasten three ropes to a ring near the top, and, standing at equal distances from the circular base of the steeple, they pull at equal angles of 30° to each other. The ropes severally make an angle of 40° with the perpendicular axis of the steeple. Now if a single force of 500 lbs. were applied at right angles at the same point, it would be just sufficient to overturn the steeple: Required the force actually exerted by each man?† *Ans.* 284.717 *lbs.*

2. A body at the equator moves, by the diurnal revolution of the earth, about 1000 miles, and in lat. 40° about 766 miles per

* It will be observed that the first four angles lie *above* and the last four *below* the plane that passes through AK and AG.

† This problem requires no resolution in *different planes*. As the given force acts at right angles to the axis of the steeple, the three forces may be considered as first acting in the same horizontal plane, and their resultant determined. This force increased in the ratio of radius to the sine of 40°, gives the answer.

hour. Now, were a wind to blow from the equator, commencing with a course directly north, and blowing with a uniform velocity of 60 miles per hour, in what DIRECTION would it blow when it reached the latitude of 40°, supposing it still to retain the easterly motion it had in common with other bodies at the equator? *Ans.* N. 75° 37′ E.

3. If a man were taken up at the latitude of 40°, and at the same instant set down at the equator, in what DIRECTION and with what velocity would he move on the equator?

Ans. He would move directly westward, at the rate of 234 *miles per hour.*

4. A ferry boat crosses a river $\frac{3}{4}$ of a mile broad in 45 minutes, the current running all the way at the rate of 3 miles an hour: At what ANGLE with the direct course must the boat head up the stream in order to move perpendicularly across?

Ans. 71° 34′.

5. The same things being given, in what RATIO is the force required to move the boat INCREASED, in consequence of the current? *Ans. It is increased* 3.162 *times.*

6. I shot an eagle that was flying directly over my head. On account of its inertia, it retained some motion in a horizontal direction, and therefore fell, at the end of 4 seconds, 60 feet from the place where I stood: Required the nature of the CURVE which the bird described in its fall?

Ans. The curve is a PARABOLA, *of which the* equation* is $P \times 257\frac{1}{3} = 3600$; *and consequently the* parameter $(P) = 13.99$ (See Art. 48.)

CHAPTER IV.

OF THE CENTER OF GRAVITY.

64. THE *center of gravity of a body is that point about which, if supported, all the parts of a body (acted upon only by the force of gravity) balance each other in any position.*

In order to ascertain this point, it will be necessary to resolve a body into its constituent parts, and then to find two lines, about each of which (if supported) these parts will balance each other in all positions; the common intersection of those two lines will be the center of gravity required. In bodies of a regular form and uniform texture this is very easily effected, but the difficulty increases as the nature or shape of the body becomes more complex. We shall begin with showing the method of

* Bridge's Conic Sec. Prop. 8.—Cor.

finding the center of gravity of a body, or of a *system* of bodies, in a few familiar instances.

METHOD OF FINDING THE CENTER OF GRAVITY OF A BODY OR SYSTEM OF BODIES.

65. *In regular plane figures, such as squares, parallelograms circles, polygons inscribed in circles, &c., the center of gravity is the same as the center of the figures.*

Fig. 37.

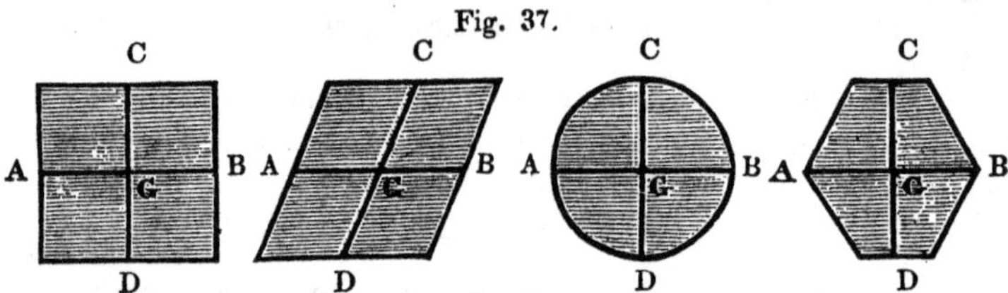

Let the annexed figures represent thin laminæ of matter of a uniform density, and let them be divided into two equal parts by the straight lines AB, CD. Conceive now each of those laminæ to be resolved into lines of particles equal and parallel to AB, there will then be the same quantity of mattter similarly disposed on each side of AB; if, therefore, AB be supported, the parts ACB, ADB will balance themselves about it; the center of gravity will consequently be in the line AB. For the same reason, because all lines drawn parallel to AB are bisected by CD, the center of gravity will also be in the line CD; it must therefore be in their common intersection G. In the same manner it might be shown, that the center of gravity of regular solids, such as the cube, parallelopiped, cylinder, sphere, &c., is the same with the center of magnitude. For each of these solids might be divided into two equal and similar parts by planes passing through it in three different directions; the intersection of two of these planes would be a right line, and the intersection of that line with the third plane would be the center of gravity of the solid. Let the three first figures in the preceding page represent respectively the section of a cube, parallelopiped, and sphere, cut through their middle; then may the line CD represent the intersection of a plane at right angles to ACBD, and AB the intersection of a third plane cutting these solids in a similar manner; the point G will therefore be the center of gravity. In the cylinder it is evident that the center of gravity will be in the point which bisects its axis.

66. *When a body is supported by a prop placed under its center of gravity, the pressure is the same whether the whole quantity of matter is uniformly diffused through the space occupied by the body, or whether it is all concentrated in that center of gravity.*

Suppose now A and B (Fig. 38) to be two equal particles of matter connected together by the inflexible rod AB void of gravity; bisect AB in G, then G will be the common center of gravity of A and B; for it is evident, that if G be supported, the

Fig. 38.

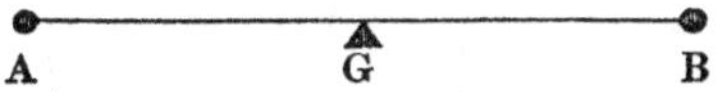

A G B

two particles will balance themselves about it. The pressure upon G will be equal to the weight of the particles A and B, and this pressure does not at all depend upon the length of the line AB; it will therefore be the same whether the particles be placed at A and B, or a particle equal to A+B be placed at G. The same may be said with respect to the particles A, B, C, D, &c., (in Fig. 39,) which are disposed uniformly along the inflexible

Fig. 39.

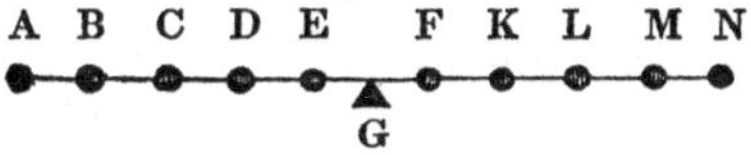

A B C D E F K L M N

G

rod AN void of gravity; viz. that the pressure of A and N is the same as if A+N were placed at G; of B and M the same as if B+M were placed at G; and that the whole pressure of the particles A, B, C, D, &c. is the same as if A+B+C+D+E+F, &c. were placed at G.*

This reasoning might be extended to the lines of particles composing the laminæ in Art. 65, for the particles A, B, C, D, &c. (Fig. 39,) may be increased in number till they become contiguous to each other, and the effect is the same whether we consider them as connected together by an inflexible rod void of gravity, or as actually united together by the power of cohesion. Supposing CD therefore to be supported, (see figures in Art. 65,) the pressure upon it will be the same as if all the particles contained in the lines parallel to AB were incumbent upon it; and supposing the point G only to be supported, the pressure will be the same as if the particles thus collected in CD were incumbent upon it; the pressure of the lamina ACBD upon the center of gravity, is therefore the same as if all the matter contained in it were incumbent upon G. The same mode of demonstration might be applied to the laminæ composing the regular solid bodies in Art. 65.

67. *Two weights or pressures acting at the extremities of an inflexible rod void of gravity, will be in equilibrio about a given point, when their distances from that point are to each other inversely as those weights or pressures.*

Let ABCD, CDEF, (Fig. 40,) represent the sections of two

* We shall come to the same conclusion by considering A and B, &c. as parallel forces, and G as their resultant; then, by Art. 60, G will be equal to their sum.

cylinders of uniform density and of the same diameter, whose axes are MN, NO; bisect MN in G and NO in g, then will G, g, be their centers of gravity. Let ABCD be suspended from the

Fig. 40.

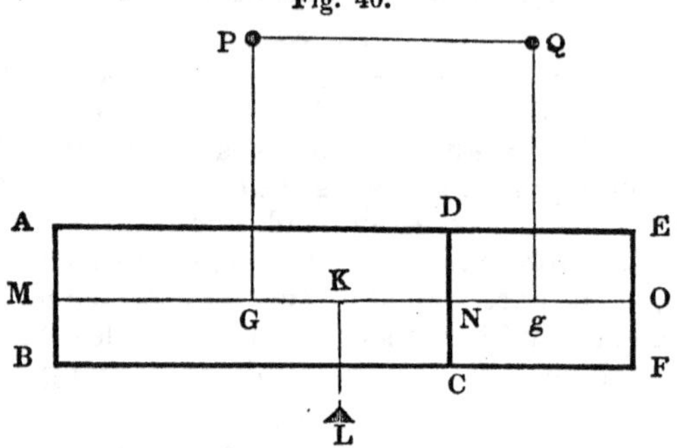

hook P, by the string PG attached to its center of gravity; and let CDEF be suspended in the same manner from the hook Q, by the string Qg; let them also be so placed that their ends may be contiguous to each other. Then will ABCD balance itself about G, and CDEF about g, so that the two axes NM, NO, (after the cylinders are suspended,) will lie in the same straight line; and the pressures upon G, g, will be the same as if the whole weights of the cylinders were collected respectively in those points. Suppose now the two ends which are contiguous to each other, to be firmly cemented together, so that the two cylinders should become one mass; this will not at all affect the pressures upon G, g, but will merely serve to connect those two points together in such a manner, that the pressures upon them may be considered as acting at the extremities of an inflexible rod Gg void of gravity. Bisect the axis MO of the whole cylinder ABFE in the point K, and K will be its center of gravity; let the prop KL be placed under K, and let the two strings PG, Qg, by which it is suspended, be removed, and the cylinder will then balance itself about the point K; or in other words, the two pressures acting at G, g, will be in equilibrio about that point. It only remains, therefore, to find the relation of KG to Kg; now $MK=\frac{1}{2}MO$, and $MG=\frac{1}{2}MN$, $\therefore MK-MG$ (or KG)$=\frac{1}{2}(MO-MN)$ $=\frac{1}{2}NO$; again, $OK=\frac{1}{2}MO$, and $Og=\frac{1}{2}NO$, $\therefore OK-Og$ (or Kg) $=\frac{1}{2}(MO-NO)=\frac{1}{2}MN$; hence $KG : Kg :: \frac{1}{2}NO : \frac{1}{2}MN :: NO : MN ::$ cylinder CDEF : cylinder ABCD. But the pressure upon G (P) is equivalent to the weight of the cylinder ABCD, and the pressure upon g (p) to that of the cylinder CDEF;

$$\therefore KG : Kg :: p : P, \text{ or } P : p :: Kg : KG.$$

68. This furnishes us with the method of finding the common center of gravity of any number of bodies whatever, connected together by inflexible rods void of gravity.

Let A, B, C, D, &c. be the bodies, and let the centers of gravity of A and B be connected together by the inflexible line AB.

Take A : B :: BG : AG,* or A+B : B :: BG+AG (AB) : AG, then will the bodies A and B balance themselves about G, (Art. 67,) and consequently G will be their common center of gravity, (by Art. 64;) and the three first terms of the above proportion being known, the distance of G from A is thus found.

Fig. 41.

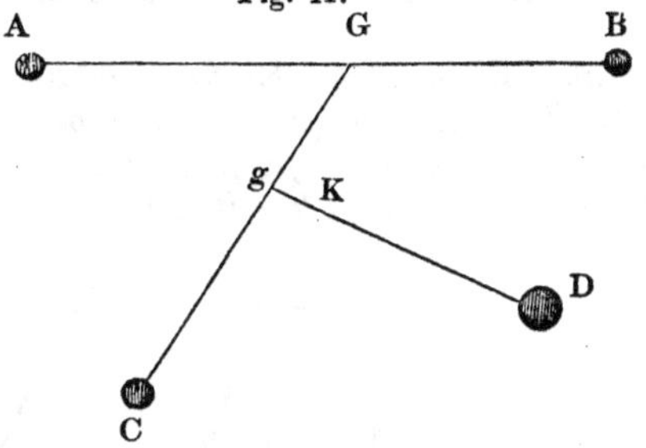

Next, let the center of gravity of C be connected with G by the inflexible line CG, then for the reason assigned in Art. 66, the pressure upon G will be the same as if a body equal to A+B were placed at G; take, therefore,

A+B : C :: C*g* : G*g*, or A+B+C : C :: C*g*+G*g* (CG) : G*g*, then *g* will be the center of gravity of A+B and C, and consequently the common center of gravity of the three bodies A, B, C.

Again, let the center of gravity of D be connected with *g* by the inflexible line D*g*, then the pressure upon *g* will be the same as if A+B+C were placed at *g*. Take, therefore,

A+B+C : D :: DK : *g*K, or, A+B+C+D : D :: DK+*g*K (D*g*) : *g*K, then K will be the center of gravity of A+B+C and D, and consequently the common center of gravity of the four bodies A, B, C, D; moreover, the pressure upon K will be the same as if A+B+C+D were placed at K; and thus we might proceed for any number of bodies.

It is evident that the foregoing demonstration does not at all depend upon the number or weight of the bodies, or their distance from each other; it rests merely on the supposition that their centers of gravity are connected together by inflexible rods void of gravity. It may therefore be applied to any number of particles of matter situated either in the same or in different planes, and placed at all possible distances from each other. Increase the number of these particles till they become contiguous to each other, and for the imaginary line void of gravity substitute the power of cohesion, then the system of bodies, A, B, C, D, &c. may represent an irregular mass of compact matter, not unlike such as are to be met with in the works of nature or of art; and although it may be difficult to find the actual center of

* Since A : B :: BG : AG, by multiplying extremes and means A×AG=B×BG; i. e., when two bodies are in equilibrio, *the product of one of the bodies into its distance from the center of gravity, is equal to the product of the other body into its distance from the same center.* These quantities, A×AG, and B×BG, therefore, express the respective forces by which A and B counteract each other's effects in their tendency to motion round G.

gravity of such a mass, yet the latter part of our proposition still remains true; viz. that if this mass be supported, its pressure downward will be the same as if the whole quantity of matter contained in it were concentrated in its center of gravity.

69. *Whatever be the form or dimensions of a body upon a plane parallel to the horizon, it will remain at rest, if the line drawn from its center of gravity perpendicular to the horizon, (called* THE LINE OF DIRECTION,) *falls within its base.*

Fig. 42.

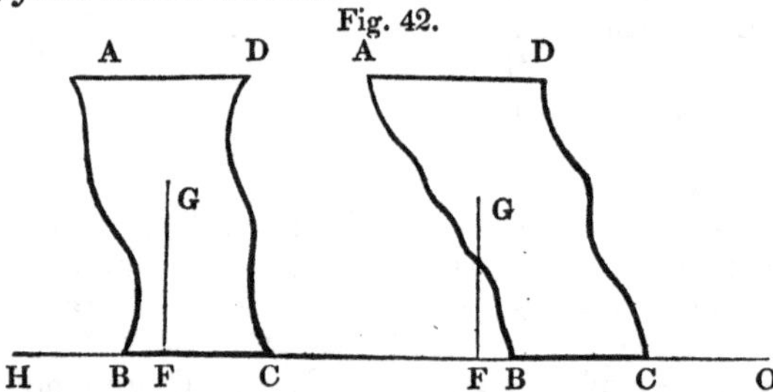

For let ABCD (Fig. 42) represent the section of a body passing through its center of gravity G, and draw GF perpendicular to HO, the plane upon which it stands; then, since the tendency of the body to descend is the same as if its whole weight were concentrated in G, it will rest or fall according as G is supported or not; or according as F falls within or without the base BC; moreover, the stability of the body will depend upon the distance at which the point F falls within the base.

70. *If a body be suspended freely from any point, it will not rest till the line which joins the center of gravity and the point of suspension, is perpendicular to the horizon.*

Fig. 43.

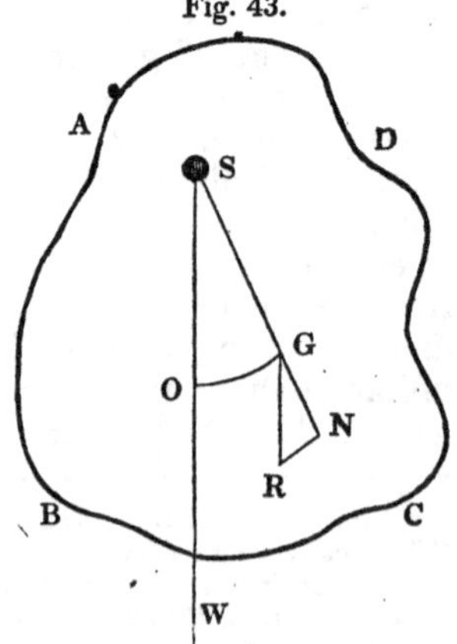

For let ABCD represent the section of a body as before, G its center of gravity, S the point of suspension; join SG, and draw SOW perpendicular to the horizon; produce SG to N, and draw GR parallel to SW; then, since the weight of the body may be considered as collected in G, its tendency to motion will be along the line GR. Let GR therefore represent this tendency, which resolve into GN in the direction SG, and RN perpendicular to it; the part GN is counteracted by the reaction from the point of suspension S, and NR is employed in producing motion in the direction of the circular arc GO; G therefore (and consequently the body) will not remain at rest till NR vanishes, i. e. till the angle NGR (=OSG) vanishes, or SG coincides with SO.

Hence it follows, that if a body be suspended successively by different points, and perpendiculars to the horizon be drawn through the points of suspension, the center of gravity will lie in each of these perpendiculars, and consequently in the point of their intersection.

We proceed to apply the principles just now investigated to the solution of a few practical examples.

71. *In a* TRIANGLE, *if a line be drawn from one of the angles bisecting the opposite side, the center of gravity of the triangle is in that line at the distance of $\frac{1}{3}$ of its length from the base.**

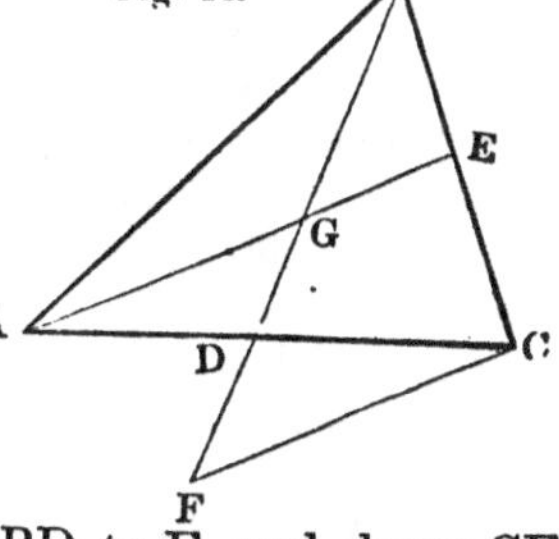

Bisect the side AC in D, and join BD, which will bisect all lines drawn parallel to AC; consequently, if BD be supported, the parts ABD, DBC of the triangle ABC will balance themselves on each side of it; hence the center of gravity is in the line BD. Bisect the side BC in E, and join AE; then, for the same reason as before, the center of gravity will be somewhere in the line AE; it must therefore be at their common intersection G. Produce now BD to F, and draw CF parallel to EA; then since BE=EC, BG will be equal to GF; but the two triangles AGD, DFC, have one side and two angles equal; ∴ GD=DF, and consequently GF (or BG)=2GD; hence BG=$\frac{2}{3}$BD, and GD=$\frac{1}{3}$BD.

72. *In a* TRAPEZOID, *the center of gravity is in the line that bisects the two opposite sides.*

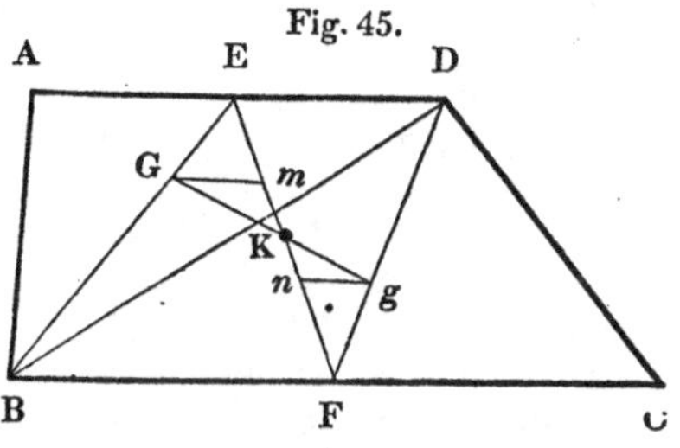

Let ABCD (Fig. 45) be a trapezoid, and bisect AD, BC, in E, F, and join EF; then since EF bisects AD, BC, it will bisect all lines drawn parallel to BC,† and, consequently, the center of gravity of the trapezoid is in the line EF. Join BE, BD, DF, and take GE=$\frac{1}{3}$BE, gF=$\frac{1}{3}$DF; then G is the center of gravity of the triangle ABD, (Art. 71,)

* In finding the centers of gravity of plane figures, a lamina of matter of uniform density, in the shape of those figures, is of course understood.

† For if BA, FE, CD be produced, they will meet in the same point, which will be the vertex of a triangle whose base is BC; and since EF bisects the base BC, it will bisect all lines drawn parallel to it.

and g the center of gravity of the triangle BDC. Join Gg; then, conceiving the triangles ABD, BDC to be collected in G,g, their common center of gravity must be in the line Gg; i. e. the center of gravity of the trapezoid ABCD must be in the line Gg; it is also in the line EF; consequently it is in K, the intersection of EF and Gg. Draw Gm, gn, parallel to AD or BC; then since EG=$\frac{1}{3}$BE, Em must be equal to $\frac{1}{3}$EF; and for the same reason Fn=$\frac{1}{3}$EF; ∴ Em=mn=nF. Now K being the common center of gravity of the triangles ABD, BDC,

$$GK : Kg :: BDC : ABD :: BC : AD.$$
$$GK : Kg :: Km : Kn$$
$$\therefore Km : Kn :: BC : AD.$$
$$Km : Km+Kn :: BC : BC+AD.$$

And since $Km+Kn=mn=Em$

$$\therefore Km+Em : mn :: 2BC+AD : BC+AD$$
$$EK : mn :: 2BC+AD : BC+AD \quad (1)$$
$$Km+Kn : Kn :: BC+AD \,.\, AD.$$

And since $mn=n$F, and Kn+nF=FK

$$\therefore mn : Kn+nF :: BC+AD : BC+2AD$$
$$mn : FK :: BC+AD : BC+2AD \quad (2)$$

uniting (1) and (2)

$$EK : FK :: 2BC+AD : 2AD+BC.$$

If, therefore, the line EF be divided in the ratio of the two last terms of this proportion, (formed of the known sides of the trapezoid,) it will give the center of gravity.

When AD=0, then the figure becomes a triangle, and EK : FK :: 2BC : BC; that is, FK=$\frac{1}{3}$EF, as was found by a different process in Art. 71.

When AD=BC, the figure becomes a parallelogram, and EK : FK :: 3BC : 3BC; consequently, the center of gravity is in the center of the figure, as was shown in Art. 65.

73. *The center of gravity of a* POLYGON *may be found by dividing the polygon into triangles, and finding the common center of gravity of these.*

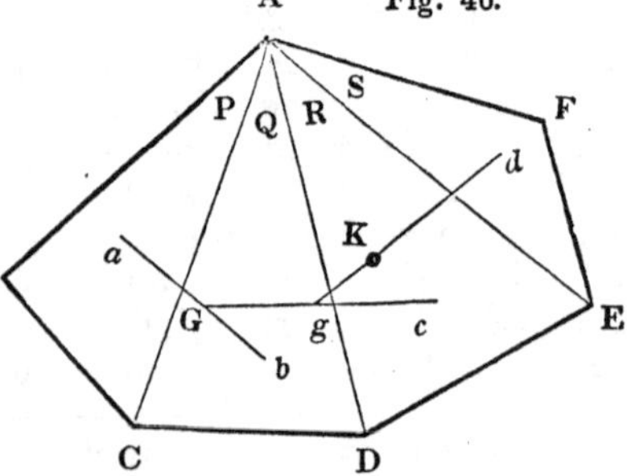

Let ABCDEF be an irregular polygon, divided into triangles whose areas are represented by P, Q, R, S, and whose centers of gravity are respectively a, b, c, d. Conceive these triangles to be collected in the points a, b, c, d; join ab, and take bG : aG :: P : Q, then G will be the center of gravity of the figure ABCD. Join Gc, and take cg : Gg :: P+Q : R; then g will be the center of gravity of the figure ABCDE. Let g and d be joined, and make

$dK : gK :: P+Q+R : S$, then K will be the center of gravity of the whole polygon; and so we might proceed, whatever be the number of sides.

If it were required to find the center of gravity of the *perimeter* of the polygon; then bisect the sides in the points a, b, c, &c., (Fig. 47,) and (since the center of gravity of a right line is in its middle point) a, b, c, &c., will be the centers of gravity of the sides AB, BC, CD, &c., respectively. Join ab and take $bG : aG :: AB : BC$; then G would be the center of gravity of that part of the perimeter represented by ABC. Again, join Gc, and take $cg : Gg :: AB+BC : CD$, then g is the center of gravity of such part of the perimeter as is represented by ABCD; and so we might proceed till we had found the center of gravity of the whole perimeter.

Fig. 47.

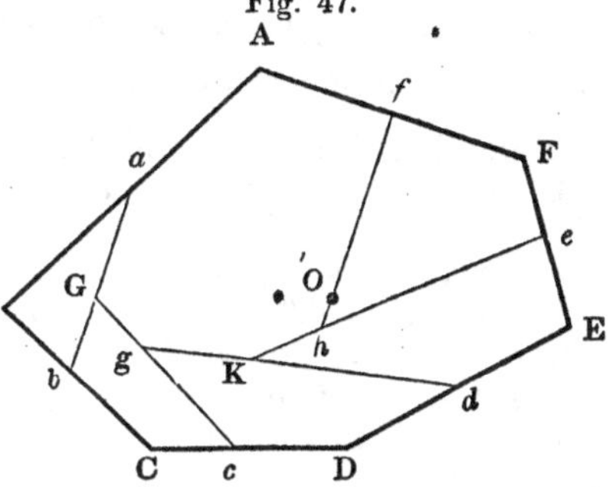

74. *The distance from any assumed point of the common center of gravity, of any number of bodies which have their centers of gravity in a right line passing through that point, is equal to the sum of the products arising from multiplying each body into its distance from the assumed point, divided by the sum of the bodies.*

Let the bodies A, B, C, D, be so placed, that the line OD may pass through their respective centers of gravity; it is required to find the distance of their common center of gravity from any point O, in the line OD.

Fig. 48.

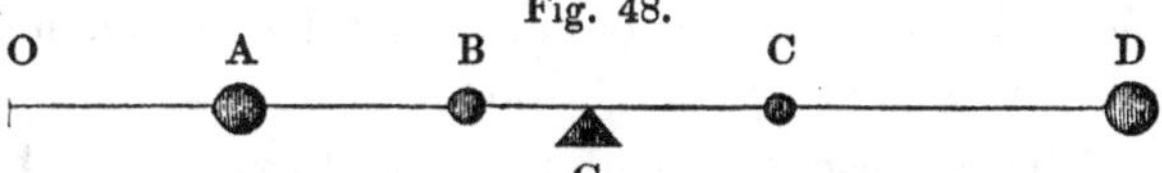

Suppose OD to be an inflexible line void of gravity, and let G be the common center of gravity of the bodies; then, if G be supported, the effort of each body to produce motion round G would be measured by the product of its weight into its distance from G, (Art. 67;) i. e. the effort of $A=A\times AG$; of $B=B\times BG$, &c.; and as the bodies are supposed to be in equilibrio about G, the sums of their efforts on each side of G must be equal to each other, or

$$A\times AG+B\times BG=C\times CG+D\times DG, \text{ i. e.}$$

$$A\times(OG-OA)+B\times(OG-OB)=C\times(OC-OG)+D\times(OD-OG)$$

$$\therefore A\times OG+B\times OG+C\times OG+D\times OG=A\times OA+B\times OB+C\times OC +D\times OD$$

$$\text{Hence } OG = \frac{A \times OA + B \times OB + C \times OC + D \times OD}{A+B+C+D}.$$

75. Miscellaneous Examples.

1. Three bodies, A, B, C, weighing respectively 3, 2, and 1 pounds, have their centers of gravity joined by the lines AB, BC, CA; of which AB=5 feet, BC=4, CA=2: What is the distance of their common center of gravity from the body C?

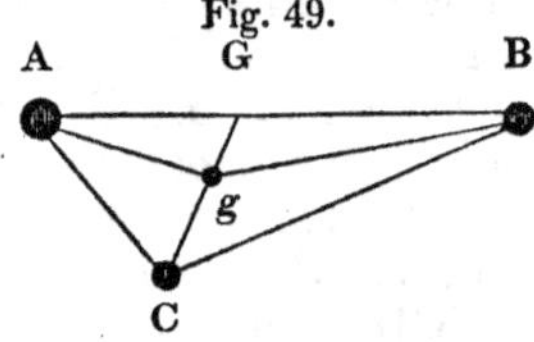

Fig. 49.

Since the three sides of the triangle ABC are 5, 4, 2, the three angles A, B, C, will be found, by the rules of trigonometry, to be respectively 49° 27½′, 22° 20′, and 108° 12½′ Let G be the center of gravity of the bodies A and B, and *g* the common center of gravity of the three bodies found as in Art. 67; then, since AB=5, A=3, B=2, and A : B :: BG : AG, AG will be equal to 2 feet, and BG to 3 feet; hence in the triangle GAC there is given AC=2, AG=2, and the angle CAG=49° 27½′, ∴ each of the angles AGC, ACG=65° 16¼′, from which CG is found to be equal to 1.673.

$$\text{But } Cg : Gg :: A+B : C,$$

$$\therefore Cg : Cg+Gg\ (=CG=1.673) :: A+B\ (5) : A+B+C\ (6);$$

$$\text{hence } Cg = \frac{1.673 \times 5}{6} = 1.394 \text{ feet.}$$

2. A cylindrical tower, consisting of uniform materials closely cemented together, is 20 feet high, and the diameter of its base is four feet: How far may it deviate from its perpendicular position, before it is in danger of falling?

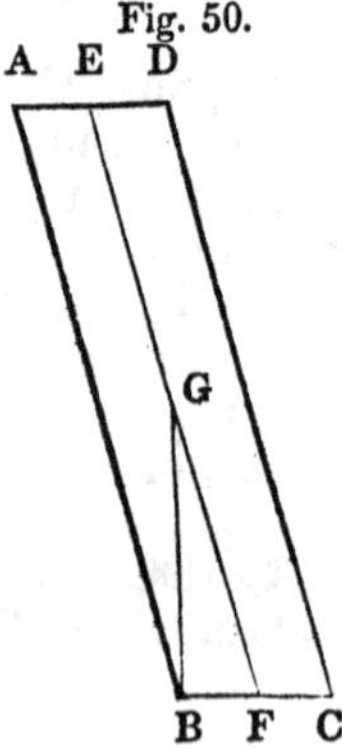

Fig. 50.

Let ABCD represent a section of the tower passing through its axis EF, and let G be its center of gravity. Suppose it to be so much inclined, that the perpendicular line GB, let fall from G, falls upon the edge of its base BC; then GF (10) : BF (2) :: radius (1) : cos. GFB; ∴ cos. $GFB = \frac{2}{10} = .200 = \cos. 78° 27′$. An angle of 78° 27′ is therefore the limit of its inclination, before it is in danger of falling. (Art. 69.) If the angle GFB is less than 78° 27′ then the perpendicular GB falls without the base, and the tower cannot sustain itself.

3. A piece of timber of uniform density and prismatic form, a section of which perpendicular to its sides, and passing through its center of gravity G is represented by the square ABCD, is placed upon an inclined plane: It is required to show when it will have a tendency to roll, and when to slide down the plane.

Draw GF perpendicular to the plane, and GK perpendicular to its base, and let PLN be greater and *p*LN less than half a right angle. In the former case, since ELK is greater than 45°, LEK or GEF will be less than 45°; ∴ the angle EGF is greater than the angle GEF and consequently EF is greater than GF or BF; hence the body has always a tendency to fall over in the direction GE, and will therefore roll down the plane PL. In the latter case the angle EGF is less than GEF, ∴ EF is less than GF or BF; the whole weight of the body, therefore, presses upon the plane *p*L. Let GE represent this weight, which resolve into two, GF, FE; GF will represent the reaction of the plane upon the body, and FE will represent a force which tends to make the body slide down the plane. Hence it appears, that the body will have a tendency either to roll or slide, according as the angle of the plane's inclination is greater or less than 45°.

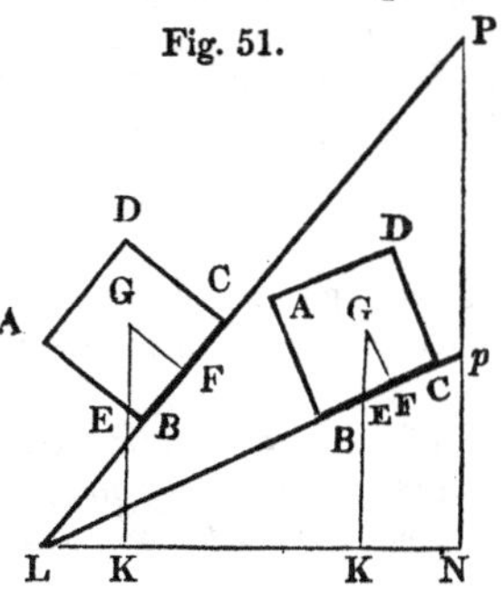

In considering the circumstances under which a body would slide or roll down an inclined plane, it should be observed, that if the surfaces of the body and the plane be perfectly smooth, no rolling will take place, whatever be the angle of inclination of the plane. To give a body a tendency to rotary motion about its center of gravity (G), it is evident that there must be some mutual action between the surface of the body and the surface of the plane, (such as that, for instance, which arises from friction, or the unrolling of a rope;) if there be not some such action as this, all the parts of the body being equally accelerated, the body will, under all circumstances, slide down the plane.

76. Questions on the Center of Gravity.

1. If three equal bodies be placed at the angles of any triangle; show that the common center of gravity of those bodies is in the same point with the center of gravity of the triangle.

2. Four bodies A, B, C, D, weighing respectively 2, 3, 6, and 8 pounds, are placed with their centers of gravity in a right line, at the distance of 3, 5, 7, and 9 feet from a given point: What is the distance of their common center of gravity from that given point; and between which two of the bodies does it lie?

Ans. Between C *and* D; *and its distance from the given point* $7\frac{2}{19}$ *feet.*

3. The bodies A, B, C, weighing respectively 5, 3, and 12 pounds, are so placed, that AB=8 feet, AC=4 feet, and the angle BAC is a right angle: What is the distance of their common center of gravity from the body C? *Ans.* 2 *feet.*

4. Supposing the height of the cylinder in Exam. 2, (Art. 75,) to be only twice the diameter of its base: What is the limit of its angle of inclination before it is in danger of falling?

Ans. 60°.

EFFECT PRODUCED UPON THE COMMON CENTER OF GRAVITY OF A SYSTEM OF BODIES, WHEN SOME OR ALL OF THEM ARE ACTUALLY IN MOTION.

77. *If two bodies approach to or recede from each other, with velocities inversely proportional to their weights, their common center of gravity will remain at rest.*

Fig. 52.

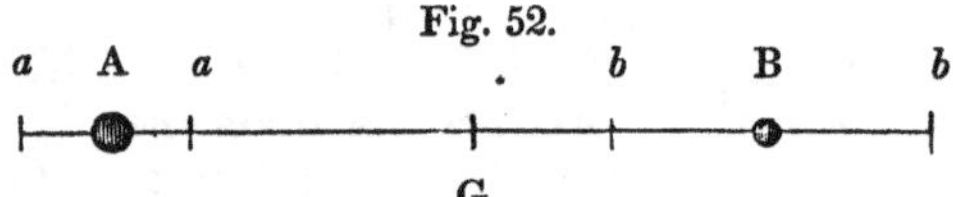

Let A and B (Fig. 52) be two unequal bodies; then if they approach to or recede from each other with velocities inversely proportional to their weights, (in which case their momenta will be equal by Art. 14,) their common center of gravity G will remain at rest. For take A*a* : B*b* :: B : A, and suppose A to move through A*a* while B moves through B*b*, then (since V∝ S when T is given) velocity of A : velocity of B :: A*a* : B*b* :: B : A, (Art. 67;) hence we have

AG : BG :: B : A, and A*a* : B*b* :: B : A;

∴ * AG±A*a* : BG±B*b* :: B : A, i. e. *a*G : *b*G :: B : A;

from which it appears that G is their common center of gravity when the bodies are arrived at *a* and *b*, i. e. the center of gravity has remained at rest while the bodies have approached to or receded from each other through the spaces A*a*, B*b*.

78. *When one body moves uniformly, describing any figure around another at rest, the center of gravity of the two bodies describes a similar figure around the central body.*

Let A (Fig. 53) remain at rest, while B moves uniformly along the sides BC, CD, DE, of the polygon BCDE. When the body arrives at C, join AC, and take

AK : KC :: B : A, or AK : AC :: B : A+B,

then K will be the place of the center of gravity, (Art. 68.) When the body arrives at D, E, join also AD, AE, and divide them in the points L, M, in the ratio of B : A, then will L, M be

* Algebra, 388.

Fig. 53.

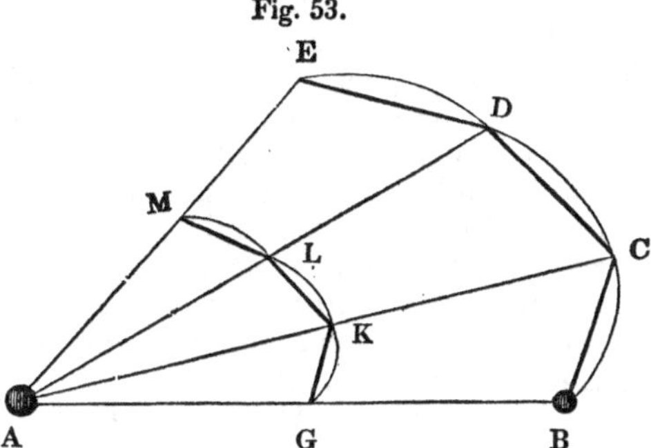

the position of the center of gravity at the end of those respective times. Let GK, KL, LM be now joined; and since
AG : AB : : B : A+B, and AK : AC : : B : A+B,

$\therefore$ AK : AC : : AG : AB;

hence GK is parallel to BC, and the triangle AGK similar to the triangle ABC. In the same manner it may be proved that the triangles AKL, ALM are respectively similar to ACD, ADE, and the whole figure AGKLM to the polygon ABCDE. While the body B therefore moves uniformly along the sides of the polygon BCDE, the common center of gravity G describes with a uniform motion a similar polygon GKLM; and since, from the nature of similar figures,*

GK+KL+LM : BC+CD+DE : : (AG : AB. i. e.) B : A+B,

the velocity of the center of gravity will be to the velocity of the body B as B to A+B. (Art. 12.) Suppose now the number of the sides of the polygon BCDE to be increased without limit, so that it may be considered as assuming the form of a curve, then shall we come to this general conclusion, that, while the body B proceeds uniformly along the perimeter of the figure BCDE, whether rectilinear or curvilinear, the center of gravity G will describe with a uniform motion a similar figure GKLM, with a velocity which is to that of B, as B is to A+B.

79. *When a system of bodies are in motion, their common center of gravity will move in the same manner as if a body equal to the sum of the bodies were placed in that point, and the same motions were communicated to it as are communicated to the bodies separately.*

Let us take the case of three bodies, A, B, C, moving with uniform velocities, in equal successive parts of time, through the spaces A*a*, B*b*, C*c*. Let G be the position of the common center of gravity of the three bodies, and *g* that of B and C, before they begin to move; then (Art. 68) G*g* : A*g* : : A : A+B+C. While A moves from A to *a*, B+C may be considered as at rest in *g*, $\therefore$ (Art. 78,) the common center of gravity (G) will in the same

* Algebra, 388.

time describe GK parallel to Aa, and

$$GK : Aa :: Gg : Ag :: A : A+B+C.$$

When A is arrived at a, join aC and BK; produce BK till it meets aC in m, then m will be the center of gravity of A and C;* join mb, then while B moves from B to b, the common center of gravity will describe KL parallel to Bb, and

$$KL : Bb :: mK : mB :: B : A+B+C.$$

When B is arrived at b, join ab, CL; produce CL till it meets ab in n, then n will be the center of gravity of A and B; join nc, then while C moves from C to c, the common center of gravity will describe LM parallel to Cc, and

$$LM : Cc :: nL : nC :: C : A+B+C.$$

Fig 54.

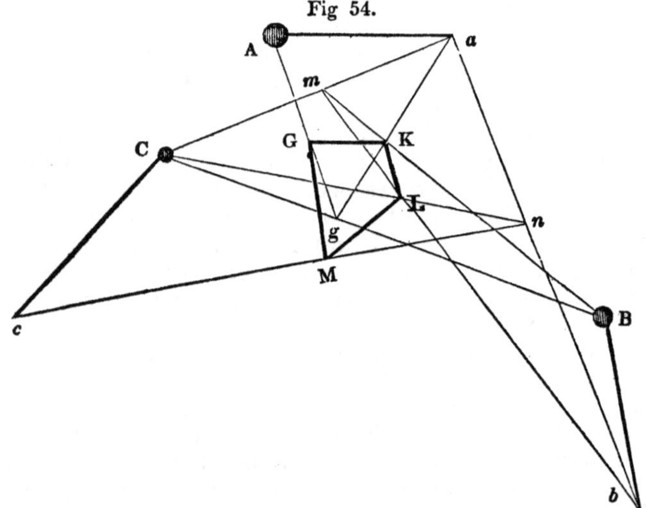

While the bodies A, B, C, therefore, in equal successive parts of time, move uniformly through the spaces Aa, Bb, Cc, their common center of gravity will in the same time describe the polygon GKLM, whose sides GK, KL, LM, are respectively parallel to Aa, Bb, Cc, and bear to them the ratio of A, B, and C to A+B+C.

80. If, instead of moving in successive intervals of time, the three bodies A, B, C, were all to begin to move at the same instant, and describe the lines Aa, Bb, Cc, cotemporaneously; let us then consider what effect would be produced upon their common center of gravity. Now since GK, Aa, are described in the same time, calling the velocity of the common center of

* For, when A moves to a, the center of gravity of A and C is somewhere in the line Ca. But when A moves to a, the center of gravity of the three bodies moves to K; therefore the center of gravity of A and C must also be in the line BK produced, since it must be such a point that A and C when placed there shall balance B. It must therefore be in the intersection of Ca and BK, or at m.

gravity v, and that of the body A, V, then v : V :: GK : Aa :: A : A+B+C ; hence A×V=(A+B+C)×v ; i. e. the momentum of A (Art. 14) is equal to the momentum of a body equal to A+B+C moving with the velocity v; the same force which causes the body A to move over Aa, would in the same time cause a body equal to A+B+C to move over GK. For the same reason, the forces which impel B and C over Bb, Cc, are such as would in the same time cause a body=A+B+C to move over KL, LM. Hence it appears that the motion of the common center of gravity along the sides of the polygon GKLM, is analogous to the motion of a body equal to A+B+C acted upon by three forces which would carry it over GK, KL, LM in the same time that they would carry the bodies A, B, C, over the spaces Aa, Bb, Cc, respectively. But a body acted upon at once by these forces would (Art. 45) describe the other side GM of the polygon GKLM in the same time that it would describe either of the sides GK, KL, LM, when the forces act separately; if the bodies A, B, C, therefore, move cotemporaneously, their common center of gravity will describe the line GM, while the bodies themselves describe the three lines Aa, Bb, Cc,* and the same reasoning is applicable to any number of bodies.

81. Hence, in the first place, if the bodies which compose a system move uniformly in right lines, then their common center of gravity will either remain at rest, or will move uniformly in a right line; for if a body equal to the sum of the bodies were placed in that center, and then acted upon by the same forces which cause the bodies to move separately in right lines, it would either remain at rest, (viz. when the forces counteract each other,) or would describe uniformly the remaining side of a polygon, whose other sides represent the quantity and direction of the several forces acting upon it. In the second place, the common center of gravity of the system will not be affected by the mutual action of the bodies upon each other; for action and reaction being equal, the effect produced upon the common center of gravity by such mutual action, will only be that of two equal and opposite forces acting upon a body equal to the sum of the bodies placed in that center; which would evidently not disturb its state, either of motion or quiescence. Lastly, if the motion of the bodies in these right lines were to cease, and they were left to the mutual attraction of each other, then their common center of gravity would remain at rest, and the bodies would approach each other, in lines drawn to it from their re-

* We have here supposed the bodies A, B, C, to have their centers of gravity in the same plane; in which case it is evident that the motion of their common center of gravity will be in the same plane. If the motion of the bodies be in different planes, then the value of each line GK, KL, LM, might be found as before; but as they will then lie in different planes, the resulting quantity GM must be ascertained according to the principles laid down in Arts. 43 and 44.

spective centers of gravity, and all collect together in that common center.

82. Examples.

1. Let two equal bodies A and B, move from the point D, with the same uniform velocity, along the sides DE, DF, of the isosceles triangle DEF, whose angle EDF=120°: It is required to compare the velocity of their common center of gravity with that of either of the bodies A or B.

Fig. 55.

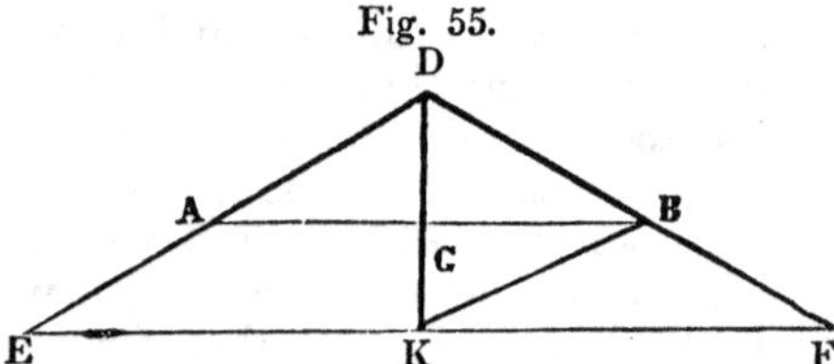

When the bodies are arrived at the points A, B, (Fig. 55,) join AB; and since the bodies move with the same uniform velocity, DA=DB; ∴ DA : DB :: DE : DF, and AB is parallel to EF. Again, because A=B, the center of gravity G will bisect AB; hence, while the bodies move uniformly along DE, DF, the center of gravity will move through the line DK, which bisects AB, EF at right angles. Now EDF=120°, ∴ EDK=60°, and DEK =30°; but since DE, DK are described in the same time, velocity of A : velocity of G :: DE : DK :: rad. : sin. 30° :: 2 : 1; the center of gravity, therefore, moves with half the velocity of either of the bodies A and B.

2. Let the two bodies A and B be placed at the extremity A of the diameter of the circle ADF, and then let B describe the circle ADF while A remains at rest in the point A: In what manner will their common center of gravity move?

Fig. 56.

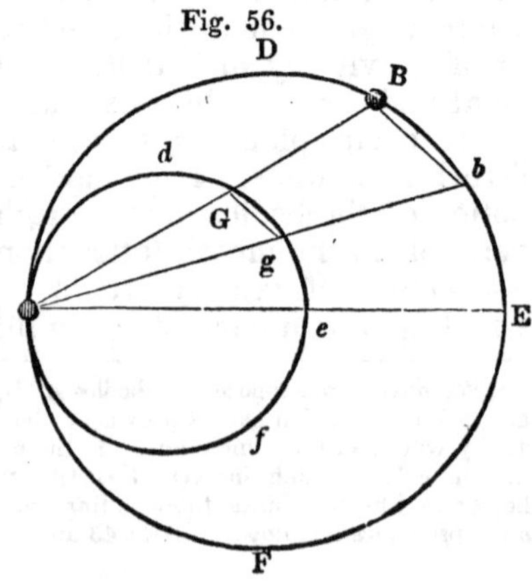

Let B, *b*, (Fig. 56,) be any two positions of the body B, and G, *g*, the corresponding positions of the common center of gravity of A and B. Join B*b*, G*g*; then by Art. 78, G*g* is parallel to B*b*, ∴ AG*g*=AB*b*, hence AD*b*, A*dg*, are similar segments of circles; and when B has described the semicircle ADE, the center of gravity will have described the semicircle A*de*; and so for the semicircles on the other side of AE. Now A*e* : *e*E :: B : A,

$\therefore$ Ae : AE :: B : A+B; hence while B describes the circle ADF, the common center of gravity of A and B will describe the circle A*df*, whose diameter : diameter of ADF :: B : A+B. This conclusion, indeed, follows immediately from the reasoning in Art. 78, for it was there shown that the whole figure described by the common center of gravity, is similar to that which the moving body B describes.

3. Two bodies A and B, begin to move in opposite directions at the same instant from the extremity D of the diameter DE of the circle DAEB, and continue to move on with the same uniform velocity till they meet in E; they pass each other at E, and then continue to move on till they arrive at the point D, whence they set off: What is the course of the common center of gravity during this revolution of the two bodies?

Fig. 57.

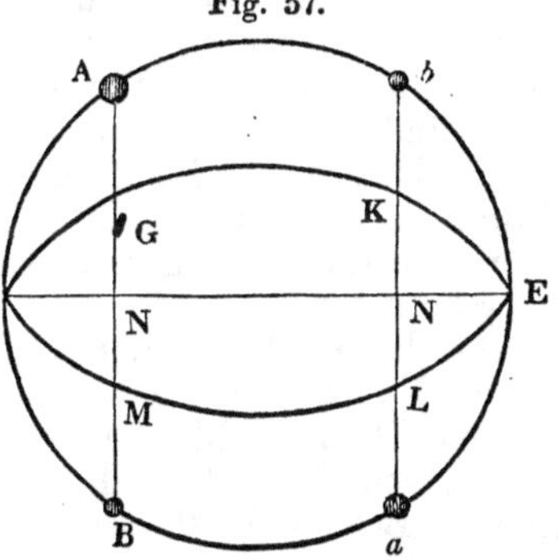

Suppose the bodies arrived at the position AB, (Fig. 57,) then since DA=DB, the line AB will be bisected by DE in N. Let G be the common center of gravity of A and B, then A+B : B :: AB : AG, $\therefore$ $\frac{1}{2}$(A+B) : B :: $\frac{1}{2}$AB (or AN) : AG; hence AN : AG :: A+B : 2B, and $\therefore$ AN : AN−AG (GN) :: A+B : A+B−2B (A−B); i. e. AN : GN in the given ratio of A+B : A−B; consequently while the bodies A and B describe respectively the semicircles DAE, DBE, their common center of gravity describes the semi-ellipse DGKE. In the same manner it may be proved that while A and B describe the semicircles EBD EAD, their common center of gravity would describe a semi-ellipse ELMD, equal and similar to DGKE. While the bodies A and B therefore perform their respective revolutions, their common center of gravity will describe the ellipse DGEM, whose major axis : minor axis :: (AN : GN ::) A+B : A−B.

Cor. If the bodies be equal, A−B=0, and the ellipse becomes a straight line. Indeed it is evident that in this case the common center of gravity would move in a line which always bisects AB, *ab*, i. e. in the diameter DE.

4. Three bodies A, B, C, at the same instant begin to move uniformly from the three angles of a given triangle, and in the same time change places in the direction ABC: How will their common center of gravity be affected by this motion of the bodies?

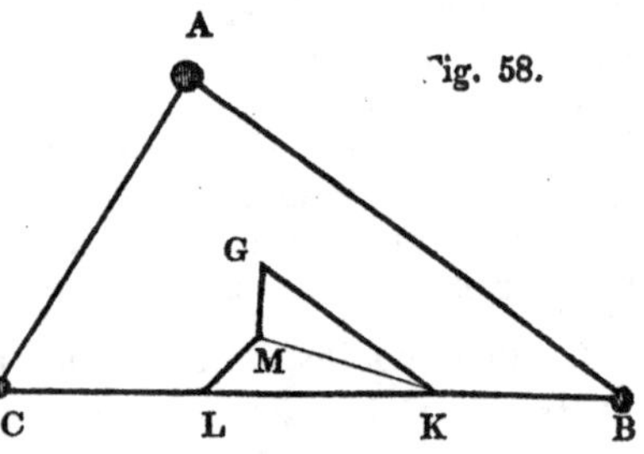

Let G (Fig. 58,) be their common center of gravity, and suppose the bodies first to move in succession. While A moves from A to B, their common center of gravity will (by Art. 78) describe GK parallel to AB, and GK : AB :: A : A+B+C, $\therefore \mathrm{GK}=\frac{\mathrm{A}\times\mathrm{AB}}{\mathrm{A+B+C}}$. While B moves from B to C, the center of gravity will describe $\mathrm{KL}=\frac{\mathrm{B}\times\mathrm{BC}}{\mathrm{A+B+C}}$; and if L M be drawn parallel to CA, and equal to $\frac{\mathrm{C}\times\mathrm{AC}}{\mathrm{A+B+C}}$, while C moves from C to A, the center of gravity will describe LM. Suppose now the bodies to move cotemporaneously, then their common center of gravity will describe GM, (the remaining side of the polygon GKLM,) while the bodies change places in the direction ABC. (Art. 44.)

To find the actual value of GM, we have KL, LM, and the angle KLM (=ACB) given, from which MK and MKL may be found; but GKM=GKL (or ABC)−MKL; in the triangle GKM there are therefore given GK, KM, and GKM, from which GM may be determined.

Cor. If A=B=C, then GK=$\frac{1}{3}$AB, KL=$\frac{1}{3}$BC, and LM=$\frac{1}{3}$AC, ∴ GK, KL, LM, are to each other as AB, BC, AC; and since the angles GKL, KLM are respectively equal to ABC, BCA, the three lines GK, KL, LM will form a triangle similar to the triangle ABC. GM therefore in this case is equal to 0, and the body remains at rest. This also follows from the general theorem in Art. 79; for the common centre of gravity, being under the same circumstances as a body acted upon by three forces which are to each other as the three sides of a triangle taken in order, will, by Art. 55, remain at rest.

83. *The distance of the common center of gravity of any number of bodies or particles of matter from a plane given in position, is equal to the sum of the products arising from multiplying each body into its distance from the plane, divided by the sum of the bodies.*

Let p, p', p'', (Fig. 59,) be any number of small bodies or particles of matter, and ABCD a plane placed in any position with respect to them. Join pp', and let g be the common center of gravity of p and p'; draw px, gk, $p'x'$ at right angles to the plane ABCD, and consequently parallel to each other; join xx', and since the points p, g, p' are in a straight line, the points x, k, x' will also be in a straight line, and therefore xx will pass through k. Join gp'', and let G be the common center of gravity of p, p', p''; draw GK, $p''x''$ perpendicular to the plane; and through g draw mn parallel to xx' meeting px produced in n.

Fig. 59.

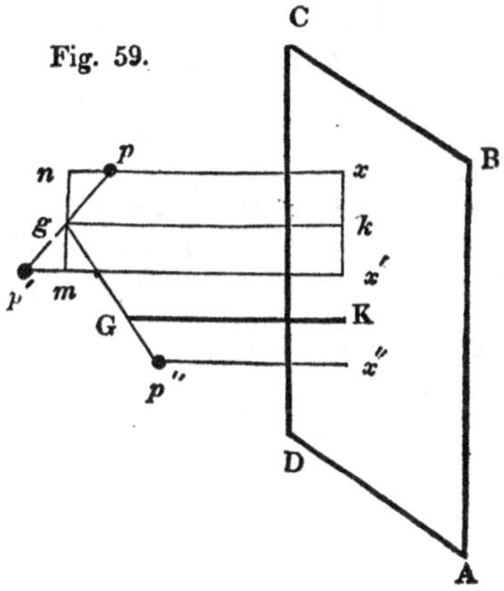

Now $p : p' :: p'g : pg ::$ (by sim. triangles) $p'm : pn$;

$$\therefore p \times pn = p' \times p'm,$$

or $p \times (nx - px) = p' \times (p'x' - mx')$;
but $nx = gk = mx'$,

$$\therefore p \times (gk - px) = p' \times (p'x' - gk)$$

and $(p+p') \times gk = p \times px + p' \times p'x$

$$\therefore gk = \frac{p \times px + p' \times p'x'}{p+p'};$$

for the same reason,
if $p+p'$ is placed at g, we have

$$GK = \frac{(p+p') \times gk + p'' \times p''x''}{(p+p') + p''}$$
$$= \frac{p \times px + p' \times p'x' + p'' \times p''x''}{p+p'+p''};$$

and thus we might proceed, whatever be the number of particles.

CHAPTER V.

OF THE COLLISION OF BODIES.

84. Bodies are divided into elastic and inelastic. Elastic bodies are such as, when compressed, restore themselves to their former state. Inelastic bodies are such as do not thus restore themselves. Thus, sponge, wool, cotton, and India rubber, are more or less elastic; and air, which restores itself with a force equal to that which compresses it, is perfectly elastic. But lead and clay are inelastic bodies, since, when they impinge upon one another, they do not rebound. Ivory, glass, and steel, are among the most elastic substances known. If we suspend two ivory balls by strings of the same length, and let them fall upon one another, (as in Fig. 4, page 29,) we may render the compression which they undergo on meeting apparent, by dotting the points of contact with ink; after impact, these dots will be enlarged in a circular space around the original point. Experiments on this subject are supposed to be made with two spheres or balls of the same density, moving uniformly in the line which joins their centers of gravity.

85. *When one* INELASTIC *body strikes upon another at rest, or moving with less velocity in the same direction, the two bodies move on together as one mass, with a velocity equal to the sum of the momenta divided by the sum of the bodies.*

Thus, let A, B represent the two bodies, and a, b, their respective velocities; then Aa will be the momentum of A, and Bb that of B. (Art. 14.) The sum of their momenta is $Aa+Bb$. Let v be the common velocity after impact; then $(A+B)\times v$=the momentum of the mass. Then $Aa+Bb=(A+B)\times v$; $\therefore v=\frac{Aa+Bb}{A+B}$.

If B is at rest, then the common velocity equals the momentun of A divided by the sum of the bodies; for then Bb becomes 0, and $v=\frac{Aa}{A+B}$.

The velocity lost by A equals the product of B into the DIFFERENCE of their velocities, divided by the sum of the bodies; and that gained by B, equals the product of A into the DIFFERENCE of the velocities divided by the sum of the bodies. For, the velocity lost by $A=a-v=a-\frac{Aa+Bb}{A+B}=\frac{B(a-b)}{A+B}$. The velocity gained by $B=v-b=\frac{Aa+Bb}{A+B}-b=\frac{A(a-b)}{A+B}$.

When B is at rest, these expressions become $\frac{Ba}{A+B}$ and $\frac{Aa}{A+B}$

86. *When the bodies move in* OPPOSITE *directions, the common velocity after impact equals the difference of their momenta divided by the sum of the bodies.*

The momentum of the mass after impact is the difference of their momenta before impact. Hence $Aa-Bb=(A+B)\times v$, $\therefore$ $v=\frac{Aa-Bb}{A+B}$. The velocity lost by A equals the product of B into the SUM of the velocities divided by the sum of the bodies; and that gained by B equals the product of A into the SUM of the velocities, divided by the sum of the bodies.

For, the velocity lost by $A=a-v=a-\frac{Aa-Bb}{A+B}=\frac{B(a+b)}{A+B}$.

The velocity gained by B (in the direction of A)$=\frac{A(a+b)}{A+B}$.

When the two bodies are equal, and meet with equal velocities, the expression $v=\frac{Aa-Bb}{A+B}$ becomes $v=0$, and both bodies remain at rest. Since, in this case, $Aa=Bb$ $\therefore$ $A:B::b:a$; therefore, conversely, *when bodies move before impact with velocities inversely proportional to their quantities of matter, they will be at rest after impact.* The same conclusion may be drawn from the consideration that, in this case, the bodies would meet with equal momenta. (See Art. 14.)

87. Examples for Inelastic Bodies.

1. A, weighing 3 oz., and moving 10 feet per second, overtakes B, weighing 2 oz., and moving 3 feet per second: What is the common velocity after impact? *Ans.* $7\frac{1}{5}$ *feet per second.*

2. A weight of 7 oz., moving 11 feet per second, strikes upon another at rest weighing 15 oz.: Required the velocity after impact? *Ans.* $3\frac{1}{2}$ *feet per second.*

3. A weighs 4 and B 2 pounds; they meet in opposite directions, A with a velocity of 9, and B with one of 5 feet per second: What is the common velocity after impact?

Ans. $4\frac{1}{3}$ *feet per second.*

4. A=7 pounds, B=4 pounds; they move in the same direction, with velocities of 9 and 2 feet per second: Required the velocity lost by A and gained by B? *Ans.* A $2\frac{6}{11}$, B $4\frac{5}{11}$.

5. A body moving 7 feet per second, meets another moving 3 feet per second, and thus loses half its momentum: What are the relative magnitudes of the two bodies?

Ans. A : B : : 13 : 7.

6. A weighs 6 pounds and B 5; B is moving 7 feet per second, in the same direction as A; by collision B's velocity is doubled: What was A's velocity before impact?

Ans. $19\frac{5}{6}$ *feet per second.*

88. *In the collision of* ELASTIC *bodies, the velocity lost by the one and gained by the other, is twice that which it would have been, had the bodies been inelastic.*

According to the definition of elasticity, the body restores itself with a force equal to that which compresses it; consequently as much momentum is exerted in the restitution as in the compression. In a given body, therefore, the velocity of restitution is equal to that of compression. Suppose, for example, that a ball of lead A, strikes upon another B; then what B gains A loses by reaction, and both bodies move on together; but when a ball of ivory (supposed perfectly elastic) impinges on another, it not only loses the momentum which it at first imparted to B, but the latter, in restoring itself after compression, exerts a force equal to that of reaction, and therefore destroys as much more of the motion of A. Again A, while receiving this second impulse from B, reacts with an equal force, and thus doubles the effect of its impulse upon B.

Distinguishing the corresponding elastic body by an accent, since, the direction being the same, the velocity lost by A=

$\frac{B(a-b)}{A+B}$; ∴ that lost by $A'=\frac{2B(a-b)}{A+B}$ ∴ the velocity of A′ after impact $=a-\frac{2B(a-b)}{A+B}=\frac{(A-B)a+2Bb}{A+B}$.

So the velocity of B′ after impact $=\frac{(B-A)b+2Aa}{A+B}$.

When the directions are opposite,

The velocity lost by $A=\frac{B(a+b)}{A+B}$. By $A'=\frac{2B(a+b)}{A+B}$. Hence, velocity of A′ after impact $=a-\frac{2B(a+b)}{A+B}=\frac{(A-B)a-2Bb}{A+B}$.

And velocity of B′ after impact $=\frac{(A-B)b+2Aa}{A+B}$.

89. *When equal elastic bodies impinge upon one another, each moves after impact with the previous velocity of the other body.*

For if B=A, then A−B or B−A are each equal to 0; ∴ when the bodies move in the same direction before impact, the velocity of A after impact $=\frac{2Bb}{A+B}=\frac{2Bb}{2B}=b$; and the velocity of B after impact $=\frac{2Aa}{A+B}=\frac{2Aa}{2A}=a$. If they move before impact in opposite directions, then the velocity of A after impact $=\frac{-2Bb}{A+B}=\frac{-2Bb}{2B}=-b$, and the velocity of B after impact $=\frac{2Aa}{A+B}=\frac{2Aa}{2A}=a$. Hence, in all cases when the bodies are equal, they move after impact with interchanged velocities; that is, when the directions are the same, A moves on, after impact, with the velocity of B, and B moves on with the velocity of A; and when the directions are opposite, A returns with the velocity of B, and B returns with the original velocity of A.

90. *When equal elastic bodies, moving with equal velocities, in opposite directions, meet, each is reflected back with its original velocity.*

For, by Art. 89, the velocity of A after impact $=-b$, and that of $B=a$; and since $a=b$, each returns with its previous velocity.

If B rests before impact, then $b=0$; ∴ the V_r of A after impact $=\frac{(A-B)a}{A+B}$, and velocity of B after impact $=\frac{2Aa}{A+B}$. If A be greater than B, then $\frac{(A-B)a}{A+B}$ is positive, ∴ A moves after impact in the same direction as it did before with $V_r=\frac{(A-B)a}{A+B}$, and B precedes it with a velocity $=\frac{2Aa}{A+B}$ (which is greater than a.) If A be less than B, then $\frac{(A-B)a}{A+B}$ is negative, ∴ A is reflected back by its impact upon B with a velocity $=\frac{(A-B)a}{A+B}$; and B moves forward in A's original direction with a velocity $=\frac{2Aa}{A+B}$ (which is less than a.)

91. *If one elastic body strikes on another equal to it at rest, the first body will be brought to a state of rest, while the second will move on with the velocity of the first.*

If A be equal to B, then $\frac{(A-B)a}{A+B}$ (which is the velocity of A after impact) $=0$, and $\frac{2Aa}{A+B}$ (which is the velocity of B after impact) $=\frac{2Aa}{2A}=a$; i. e. if B rests before impact, then A will rest after impact; and B will move forward in A's direction with A's velocity before impact.

Let there be a row of equal elastic balls A, B, C, &c. . . X, (Fig. 60,) placed contiguous to each other; then (by Art. 89,) if A is moved from its position and made to impinge upon B, it will rest after impact, and B will have a tendency to move on with A's velocity; after the impact of B upon C, it will remain at rest, and C have a tendency to move on with A's velocity; after the impact of C upon D, it will remain at rest, and D will have a tendency to move on with the same velocity; and so the motion will be propagated through the whole row, and the last body X will move forward with the velocity of A, all the others remaining at rest.

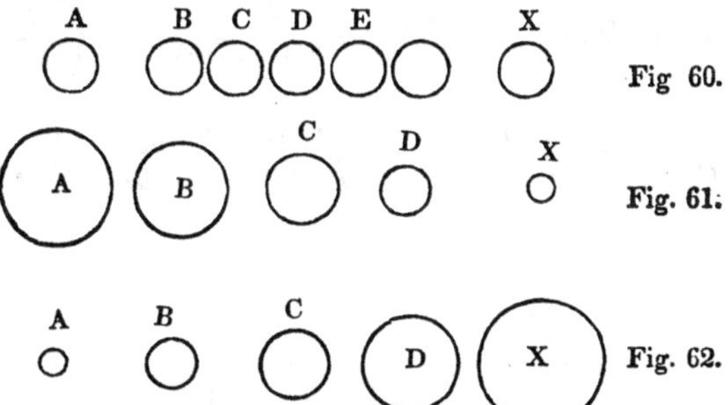

Fig 60.
Fig. 61.
Fig. 62.

If the bodies decrease in magnitude, (Fig. 61,) then, since A is greater than B, (by Art. 90,) the velocity communicated to B will be greater than that of A; and the velocity communicated from B to C greater than that of B, &c.; so that the last body will move forward in the direction of A's motion with a velocity much greater than that of A, and the other bodies will follow it in such a manner, that the velocity of each succeeding body shall be greater than that of the preceding. On the contrary, if the bodies increase in magnitude, (Fig. 62,) since A is less than B, (by Art. 90,) the velocity communicated to B will be less than that of A, and A will be reflected back by B; for the same reason the velocity communicated from B to C will be less than that of B, and B will be reflected back by C; so that in this case all the bodies will move backward except the last, and that will move forward in the direction of A's original motion, but with a velocity much less than that of A.

92. *When a row of bodies are in geometrical progression, and the first impinges on the second, and motion is thus propagated through the series, the velocity of the first is to the velocity of the last, as* $1 : \left(\frac{2}{1+r}\right)^{n-1}$.

Let the series be A, Ar, Ar^2. . . . Ar^{n-1}.

By Art. 90, when A impinges on B at rest the velocity communicated to B is $\frac{2Aa}{A+B}=\frac{2Aa}{A+Ar}=\frac{2a}{1+r}=b$.

Again, the velocity imparted to C is $\frac{2Bb}{B+C}=\frac{2Ar}{Ar+Ar^2}\times\frac{2a}{1+r}=\frac{2^2a}{(1+r)^2}$. Hence the successive velocities are $a, \frac{2a}{1+r}, \frac{2^2a}{(1+r)^2}$, &c. from which it appears that any term in the series is found by multiplying the original velocity by 2, raised to a power one less than the number of terms, and divided by $1+r$ raised to the same power. Consequently, the last term is $\frac{2^{n-1}a}{(1+r)^{n-1}}$. Hence,

V, of the first : V, of the last :: $a : \frac{2^{n-1}a}{(1+r)^{n-1}} :: 1 : \left(\frac{2}{1+r}\right)^{n-1}$.

93. *When a perfectly elastic body impinges on a perfectly smooth plane, it makes the angle of reflexion equal to the angle of incidence.*

Fig. 63.

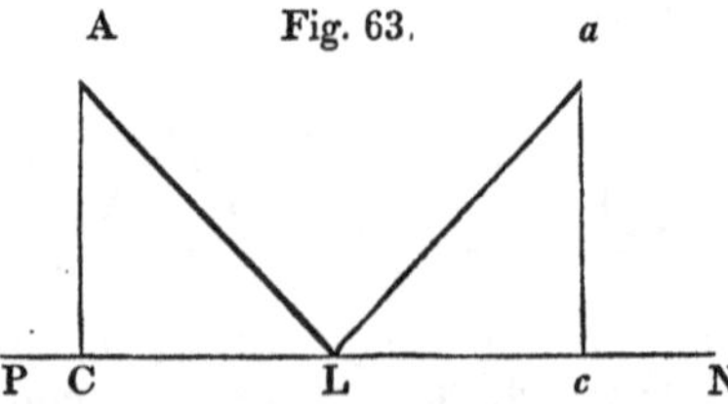

If a perfectly elastic body impinges perpendicularly upon

perfectly smooth plane, then, since the force of restitution is equal to the force of compression, it will ascend from the plane with the same velocity as that with which it impinged upon it. But if moving uniformly, it impinges upon the plane PN, (Fig. 63,) in the oblique direction AL, then resolve AL into two AC, CL, of which, as in the former instance, the perpendicular part AC will not be destroyed, but will represent the velocity with which the body ascends from the plane; and CL will represent the velocity it has in the direction of the plane, the same as before. Take therefore L*c*=LC, and from *c* draw *ca* at right angles to L*c* and equal to CA, and join L*a*; then L*a* will represent the direction and velocity of the body after impact. But since L*c*, *ca* are equal to LC, CA, and the angles L*ca*, LCA are right angles, L*a* will be equal to LA, and the angle *a*L*c*, to the angle ALC; hence the body will move after impact with the same velocity which it had before impact, and in a direction making the angle of reflexion equal to the angle of incidence.

94. Examples for Elastic Bodies.

1. A weighing 10 lbs. and moving 8 feet per second, impinges on B weighing 6 lbs. and moving in the same direction, 5 feet per second: What are the velocities of A and B after impact?
Ans. A's=$5\frac{3}{4}$. B's=$8\frac{3}{4}$.

2. A : B :: 4 : 3; directions the same; velocities 5 : 4: What is the ratio of their velocities after impact? *Ans.* 29 : 36.

3. A weighing 4 lbs., velocity 6, meets B weighing 8 lbs., velocity 4: Required their respective directions and velocities after collision?
Ans. A *is reflected back with a velocity of* $7\frac{1}{3}$, *and* B *with a velocity of* $2\frac{2}{3}$.

4. A and B move in opposite directions; A equals 4B, and $b=2a$: How do the bodies move after collision?
Ans. A *returns with* $\frac{1}{5}$, B *with* $1\frac{2}{5}$ *its original velocity.*

5. There are ten bodies whose magnitudes increase geometrically by the constant ratio 3, and the first impinges on the second with the velocity of 5 feet per second: Required the motion of the last body?
Ans. The last body would move with the velocity of $\frac{5}{512}$ *feet per second.*

CHAPTER VI.

OF THE LEVER.

95. In the preceding chapters, the motion of bodies has been supposed to arise either from collision, or from the immediate action of one or more forces. We now proceed to consider the effects produced, when these forces are made to act by the intervention of other bodies. These intermediate bodies are called *Machines*; and by means of them the effect of a given force may be increased or diminished in any given ratio. Machines are divided into *simple* and *compound.*

96. The *simple* machines, or what are commonly called the MECHANICAL POWERS, are six in number; viz. 1. The *Lever;* 2 The *Wheel and Axle;* 3. The *Pulley;* 4. The *Inclined Plane;* 5 The *Screw;* 6. The *Wedge.* In philosophical strictness, the number of simple machines may be reduced to three; viz. the *lever*, the *inclined plane*, and the *cords* or *ropes* which connect the power and weight with the different parts of the machine; for the mechanism of the wheel and axle, and of the pulley, merely combines the principle of the lever with the tension of cords; the properties of the screw depend entirely on those of the lever and the inclined plane; and the case of the wedge, so far as it is capable of mathematical demonstration, is very analogous to that of a body sustained between two inclined planes. *Compound* machines are formed from the combination of two or more simple ones. But it is not the object of this treatise to enter upon a full description of the nature and use of compound machinery; our intention is rather to explain, upon mathematical principles, the general theory of mechanical action.

97. *The* LEVER *is an inflexible bar or rod, some point of which being supported, the rod itself is movable freely about that point as a center of motion.*

This center of motion is called the FULCRUM or PROP. When two forces act on one another by means of any machine, that which gives motion is called the POWER, that which receives it, the WEIGHT.

98. In treating of the Mechanical Powers, the first inquiry is, *What are the conditions of an equilibrium?* that is, When do the power and weight exactly balance each other? This point being ascertained, any addition to the power puts the weight in motion. The investigation first proceeds on the supposition that the action of the mechanical powers is not impeded by their own

weight, or by friction and resistance, a suitable allowance being afterward made for the various impediments.

We shall begin with estimating the relation between the forces acting upon the arms of a straight lever, which, of all the mechanical powers, is the most simple.

99. *If any two forces, acting in the same plane, and perpendicular to the extremities of a straight lever, be in equilibrio, they will be to each other inversely as the lengths of the arms upon which they respectively act.*

Fig. 64.

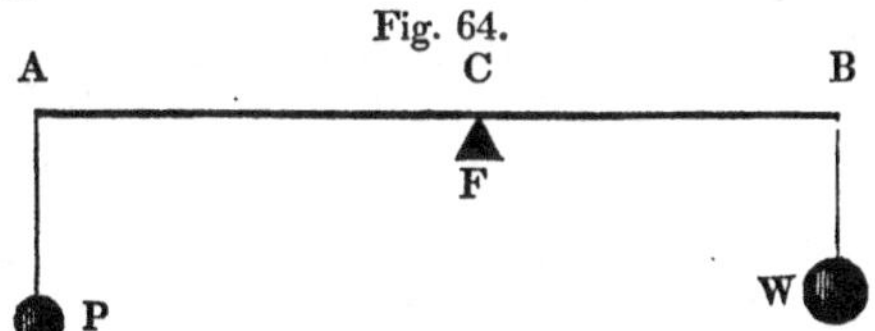

Let ACB (Fig. 64) be a straight lever, supported by a prop or fulcrum F, and movable about the point C as its fulcrum. From the extremities of its arms, CA, CB, let two weights, P, W, be suspended; and suppose them to be in equilibro about C, the lever itself remaining in a horizontal position. In the present instance, let us also suppose that the lever AB, and the cords AP, BW, by which the weights are suspended, are entirely void of gravity; in which case it is evident that the equilibrium of the bodies does not at all depend upon the length of the cords AP, BW; and as (Art. 70) the centers of gravity of the bodies P, W. are in the direction of the lines AP, BW, the effect will be the same whether the bodies are suspended by the strings AP, BW; or whether they are placed with their centers of gravity in the points A, B, respectively.* In this latter case, the point C becomes the center of gravity of the weights P, W, and consequently P : W : : BC : AC, (Art. 67.)

But it is evidently quite immaterial to the truth of the foregoing demonstration, whether the equilibrium of the lever is produced by the force of gravity of the two weights, P, W, or by the action of any other forces in the directions AP, BW.

100. *The effect of any forces to turn the lever about the center of motion, is measured by the product arising from multiplying each force into the distance at which it acts from the fulcrum.*

For if the magnitudes of the forces acting at A and B are represented by P and W respectively, then (since P : W :: BC : AC) $P \times AC = W \times BC$; $\therefore P \times AC$ represents the effect of P, and $W \times BC$ represents the effect of W, to turn the lever round C.

* For since the equilibrium does not at all depend upon the length of the lines AP BW, we may suppose those lines to vanish; in which case the centers of gravity of P, W, may be considered as coinciding with the extremities A, B, of the lever.

101. *Any number of weights will keep each other in equilibrio upon the arms of a straight lever, when the sums of the products arising from multiplying each weight by its distance from the fulcrum are equal on the two sides of that center.*

Fig. 65.

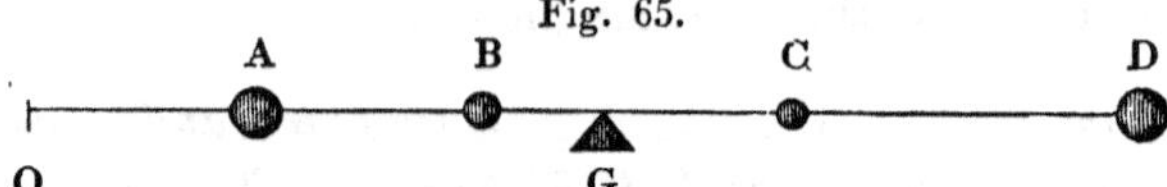

Let AD (Fig. 65,) represent a straight lever whose fulcrum is G, and let the bodies or weights A, B, C, D, be placed upon its arms AG, DG, at different distances from G; then the effort of A to turn the lever about G being represented by $A \times AG$, of B by $B \times BG$, of C by $C \times CG$, &c., the whole effect upon the arm AG will be represented by $A \times AG + B \times BG$, and upon the arm DG by $C \times CG + D \times DG$; there will consequently be an equilibrium when $A \times AG + B \times BG = C \times CG + D \times DG$.

102. Levers are divided into three different orders, according to the position of the power and weight with respect to the fulcrum. I. In a lever of the first order, the fulcrum is between the power and weight, as in the preceding instance; and here the pressure on the fulcrum is equal to the sum of the weights. II. In a lever of the second order, the weight is placed between the power and the fulcrum, as in the annexed figure, where the weight W is supported by the power P acting upward in the direction AP. In this case also there is an equilibrium, when the power and weight are inversely as the arms on which they respectively act; for the effort of the weight W to turn the lever about C is measured by $W \times BC$ (Art. 68, Note;) the effort of the power P (acting in the direction PA) to turn the lever about C,

Fig. 66.

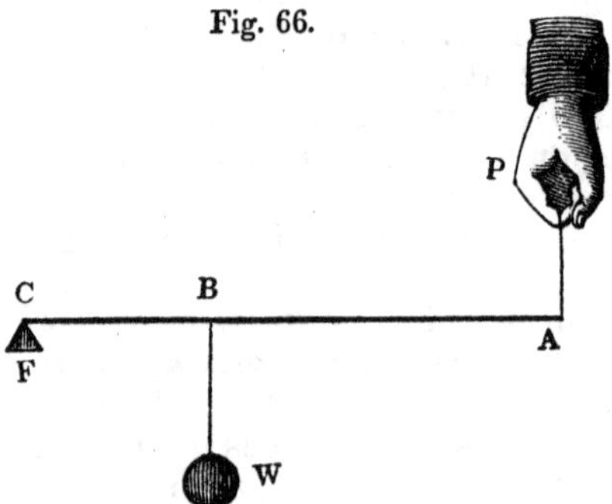

i. e. to sustain W, is measured by $P \times AC$; when there is an equilibrium, therefore, $P \times AC$ must be equal to $W \times BC$, or $P : W :: BC : AC$, as before. Therefore, P is less than W; and the pressure upon the fulcrum (P and W acting in opposite directions) is equal to W—P; for the pressure at A is the same as would be exerted on a fulcrum at that point, in which case the pressure on both points C and A would equal the whole weight W; therefore, the pressure on C equals W—P. III. In a lever of the third order, the power acts between the weight and the fulcrum; but the equilibrium is produced on the same principle as before; for an equilibrium will take place when the opposite forces P and W are equal; which will be when $P\times AC=W\times BC$; or when

Fig. 67.

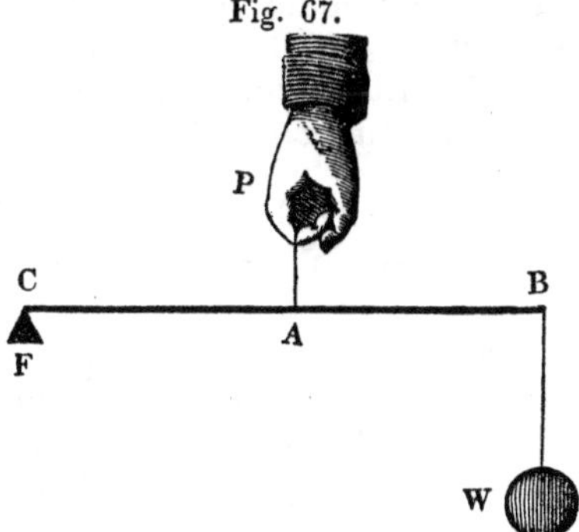

P : W :: BC : AC. Since BC is greater than AC, P is greater than W, and the pressure upward from the fulcrum is represented by P—W.* Hence we have the following general principle applicable to the three orders of levers,

When the forces act PERPENDICULARLY TO THE ARMS OF A STRAIGHT LEVER, *an equilibrium is produced, if the power is to the weight as the distance of the weight from the prop is to the distance of the power from the prop.*

In the second kind of lever the weight is greater than the power; in the third kind, less.

When a weight is sustained between two props, the part sustained by each prop is inversely as the distance of the weight from it

Fig. 68.

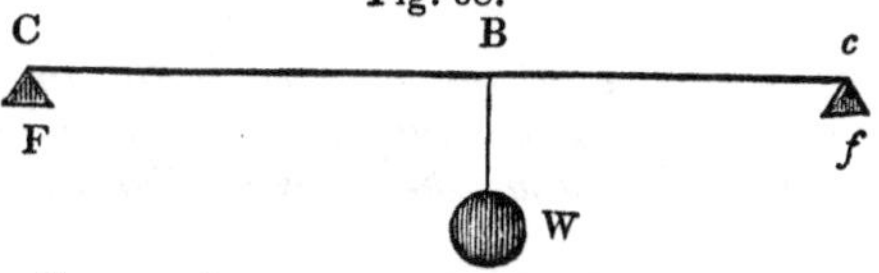

For suppose C, *c* to be successively the centers of motion, then

Press. on fulcrum *f* : weight W :: BC : C*c*; for same reason,

Weight W : press. on fulcrum F :: C*c* : B*c*; ∴

Press. on fulcrum *f* : press. on fulcrum F :: BC : B*c*; and as the whole weight is sustained by the two props, it is divided between them in the ratio of BC : B*c*.

* In this third order of levers, although the lever is supposed to move freely round the center of motion C, it is yet necessary to consider it as firmly connected with the prop at that point.

103. Let us next estimate the relation of the forces which keep a lever in equilibrio, when its own weight is taken into consideration. Since this weight may be considered as collected in the center of gravity of the lever, (Art. 66,) its effective force is equal to the weight, multiplied into the distance of its center of gravity from the fulcrum. (Art. 100.) Suppose the lever to be of a cylindrical or prismatic form, and that its weight$=w$, then,

In a lever of the first order, (Fig. 64,) since the center of gravity of the lever is in the middle point, and the distance of this point from the fulcrum $=\frac{1}{2}$ (AC$-$BC), the effect of the weight of the lever $=\frac{1}{2}w$ (AC$-$BC), and is exerted in the direction of the longer arm AC. In the case of an equilibrium, therefore, we have,

$$\text{P}\times\text{AC}+\tfrac{1}{2}w(\text{AC}-\text{BC})=\text{W}\times\text{BC},$$

$$\text{or P}=\frac{\text{W}\times\text{BC}-\frac{1}{2}w(\text{AC}-\text{BC})}{\text{AC}}.$$

In a lever of the second order, (Fig. 66,) the whole weight of the lever operates in conjunction with W, and the distance of the center of gravity from the fulcrum in this case $=\frac{1}{2}$AC, $\therefore$

$$\text{P}\times\text{AC}=\text{W}\times\text{BC}+\tfrac{1}{2}w\times\text{AC},$$

$$\text{or P}=\frac{\text{W}\times\text{BC}}{\text{AC}}+\tfrac{1}{2}w.$$

In a lever of the third order, (Fig. 67,) the whole weight of the lever operates in conjunction with W, and the distance of the center of gravity from the fulcrum $=\frac{1}{2}$BC, $\therefore$

$$\text{P}\times\text{AC}=\text{W}\times\text{BC}+\tfrac{1}{2}w\times\text{BC},$$

$$\text{or P}=\frac{(\text{W}+\frac{1}{2}w)\times\text{BC}}{\text{AC}}.$$

We have thus far confined our attention to the case in which the lever is supposed to be straight, and the forces to be applied at right angles to it; we now propose to take a more general view of the properties of the lever, whatever be its shape, or the directions of its forces.

104. *Two forces acting at the extremities of the arms of* ANY *lever will be in equilibrio, when they are to each other inversely as the perpendiculars let fall upon the lines of direction in which they respectively act.**

Let ACB (Fig. 69) be any lever whose fulcrum is C; and let two forces, P, p, act in the directions AP, Bp, upon the extremities of its arms CA, CB. Produce PA, pB, to M, N, and let fall the perpendiculars CM, CN; with the longer perpendicular CN and center C describe the circular arc ND, and join CD. Let DP represent the magnitude of the force acting upon the lever at A in direction AP, and let it be resolved into two others, viz.

* This theorem evidently embraces the proposition in Art. 99.

Fig. 69.

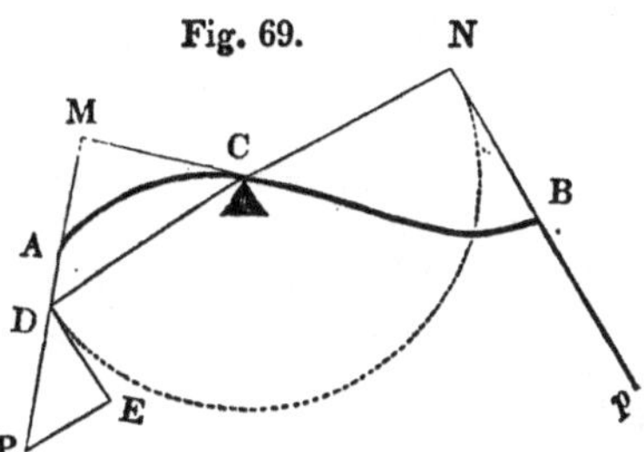

DE perpendicular and EP parallel to the radius CD; then DE only is effectual to produce motion round the center C, the part EP being exerted merely to produce pressure upon the fulcrum in direction CD. Supposing, therefore, the lever to be perfectly inflexible, this effort of P to produce motion round C would be counteracted by a force equal to DE, applied perpendicularly at N, in direction NB. Let then the power p be equal to that part of P which is represented by DE, and (as it is indifferent in what point of the line of direction this power acts) conceive it to act at N. In this case the forces P, p will be in equilibrio; i. e. when P : p :: PD : DE, the lever will be kept in equilibrio about the center of motion C; but by similar triangles

PD : DE :: CD (CN) : CM, ∴ P : p :: CN : CM.

105. Produce PA, pB (Fig. 70) till they meet in S; join CS, and draw CO parallel to pS, in which case the angle OCS=CSN. Now if CS is made radius, CM becomes the sine of CSO, and CN the sine of CSN or OCS; but as SO : OC :: sin. OCS(CN) : sin. CSO(CM;) hence P : p :: (CN : CM ::) SO : OC. The two sides SO, OC of the triangle SOC represent, therefore, the relative magnitude and direction of the two forces P, p; the third

Fig. 70.

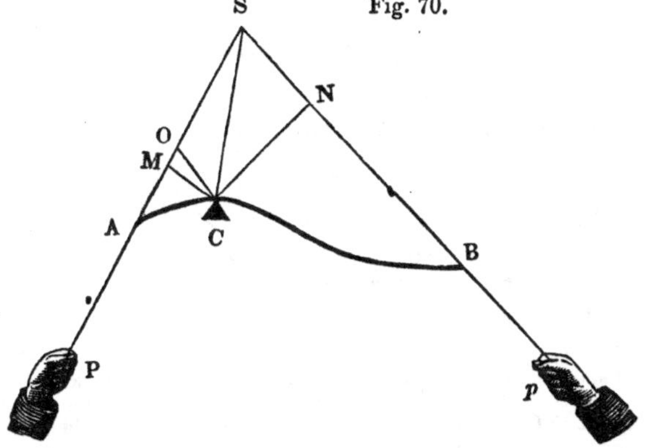

side SC will consequently represent a force equivalent to them both, (Art. 43;) and as this compound force acts directly toward

C, the pressure upon the fulcrum will be represented, in quantity and direction, by the line SC.

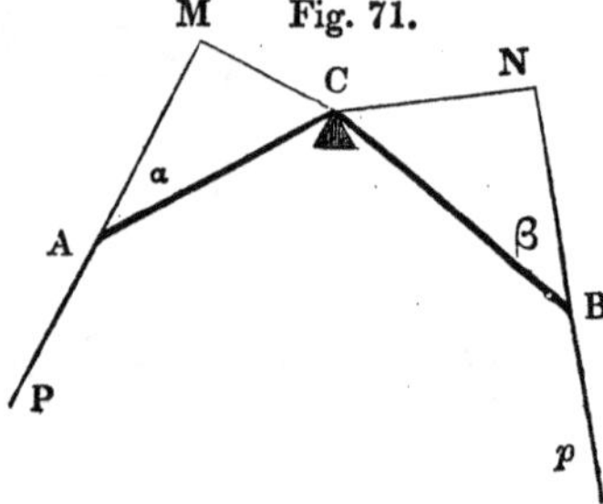

Fig. 71.

106. If each arm of the lever is straight, (Fig. 71,) but the two arms are inclined to each other in the given angle ACB, let CAM=α, CBN=β, and rad.=1; then

AC : CM :: rad. (1) : sin. α,

∴ CM=AC × sin. α, and

BC : CN :: rad. (1) : sin. β,

∴ CN=BC × sin. β. Hence

P : p :: BC × sin. β : AC × sin. α.

If sin. α=sin. β, then P : p :: BC : AC, or P, p, are to each other inversely as the arms of the lever upon which they respectively act; which shows that the same law of equilibrium obtains in the bent as in the straight lever, when the forces act at equal angles.

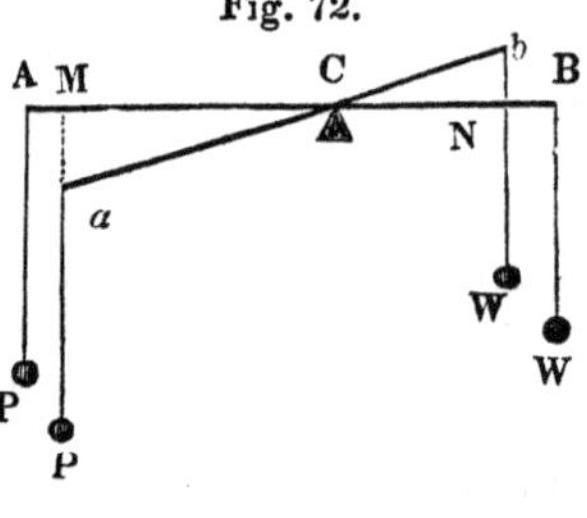

Fig. 72.

107. If the lever is straight, (Fig. 72,) and the forces act parallel to each other, then sin. α=sin. β,

∴ P : W :: BC : AC,

as in Art. 106; and this will be the case whatever be the position of the lever; if therefore P and W are in equilibrio when the lever is in the horizontal position ACB, they will also be in equilibrio when it is in any other position aCb; i. e. the lever thus acted upon will rest in any position.*

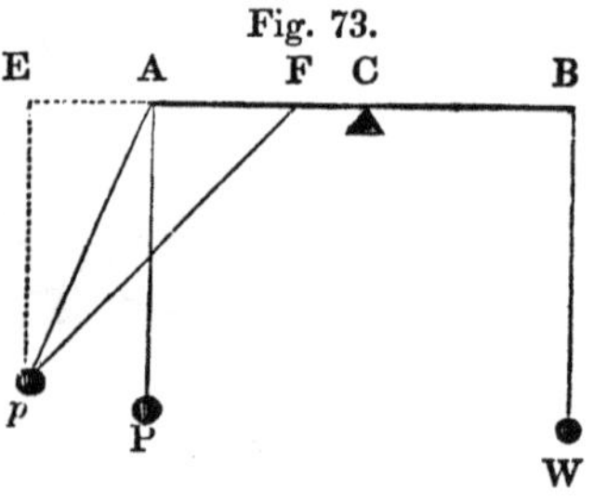

Fig. 73.

108. In the common balance or scales, the arms AC, CB, (Fig. 73,) are equal to each other; ∴ when there is an equilibrium, P=W. But this equilibrium will be destroyed, if either P or W is removed from its perpendicular position. Suppose, for instance, a person placed in the scale P is balanced by the weight W, but by pushing in the

* This also appears from Art. 104; for produce Pa to M, then, when the lever is in position aCb, P : W :: CN : CM :: (by sim. triangles) Cb : Ca :: (since the arms of the lever are invariable) CB : CA. In thus asserting that the lever will rest in any position, it is of course taken for granted, that the common center of gravity of P, W, and the lever, coincides with the center of motion; for it is evident, from the principles laid down in Chap. IV, that the lever will only rest when that center is supported

oblique direction pF against the arm CA, the scale is protruded into the position pA; then draw pE parallel to PA, and produce CA to meet it in E; and at the instant the scales arrive at the position pA, the power will act at the perpendicular distance CE from the center of motion; its effect, therefore, (by Art. 100,) to turn the lever about C, will be measured by $P \times CE$, i. e. its effect will exceed that of W in the proportion of CE : CA, and consequently the scale in which the person is will preponderate, and the equilibrium be destroyed.

109. *In the* COMPOUND LEVER, *the opposite forces are in equilibrio, when the power is to the weight, as the product of all the arms on the side of the weight, is to the product of all the arms on the side of the power.*

In a combination of levers connected with each other in the manner represented in the annexed figure, (Fig. 74,) there will be

Fig. 74.

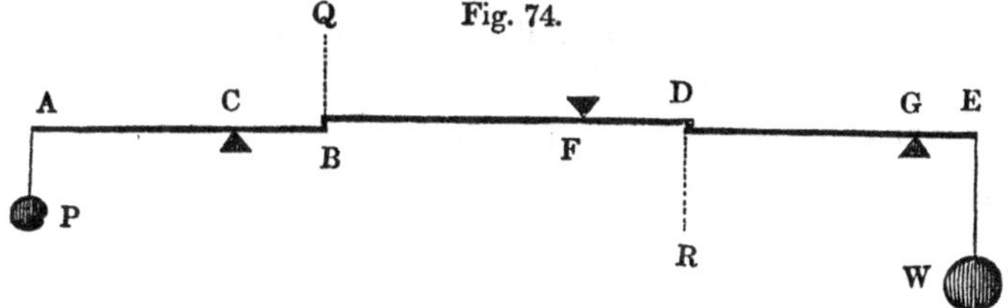

an equilibrium when $P : W :: BC \times DF \times EG : AC \times BF \times DG$. For suppose the equilibrium to exist, and that the forces which act at B, D, are represented by Q, R, respectively, then

$$P : Q :: BC : AC \text{ or } Q = \frac{P \times AC}{BC},$$

$$Q : R :: DF : BF,$$

$$R : W :: EG : DG \text{ or } R = \frac{W \times EG}{DG};$$

$$\therefore P : W :: BC \times DF \times EG : AC \times BF \times DG. \qquad \text{(A.)}$$

Also, the pressure on $C = P + Q = P + \frac{P \times AC}{BC} = \frac{P \times AB}{BC}$,

" " $F = Q + R = \frac{P \times AC}{BC} + \frac{W \times EG}{DG}$,

" " $G = R + W = \frac{W \times EG}{DG} + W = \frac{W \times DE}{DG}$.

We have here supposed the forces to act perpendicularly to the extremities of the several levers; if they acted obliquely, or if the arms of the levers were inclined to each other, then for these arms must be substituted, in the proportion marked (A,) the perpendiculars let fall from the centers of motion, C, F, G, upon the lines of direction in which the forces act. We now proceed to illustrate the foregoing theory by a few plain examples.

110. Examples.

1. At the extremities of a straight lever, whose length is 24 inches, are placed two weights of 5 and 7 pounds: At what point must the fulcrum be placed so that these weights shall balance each other, the weight of the lever not being taken into the account?

This is the case of the lever of the first order, in which (Fig. 64) P=5, W=7, and AB=24; and when there is an equilibrium, P (5) : W (7) :: BC : AC; ∴ 12 : 7 :: BC+AC (AB=24) : AC=$\frac{168}{12}$=14; hence BC=24−14=10 inches.

2. At the extremity of a lever of the second order, there acts a power which is of itself able to sustain only a weight of 15 pounds; but when acting under this mechanical advantage, it is able to sustain a weight of 100 pounds, placed five feet from it: What is the length of the lever?

Referring to Fig. 66, 100 : 15 :: AC : AC−5, ∴ 100 : 85 :: AC : 5, ∴ AC=$\frac{500}{85}$=5$\frac{15}{17}$ feet.

3. A body suspended at the extremities of a balance whose arms are unequal, weighs p pounds at one end, and q pounds at the other: What is its real weight?

A balance of this kind is called a false balance, because when the body is suspended at the extremity of the longer arm, the weight which balances it, is above, and when suspended at the extremity of the shorter arm, is below the true weight. But the true weight of the body is easily found by the following operation; viz. Let x=the true weight, and a the arm of the lever upon which p is suspended to balance it; and b the arm upon which q is suspended to balance it; then $x:p::a:b$, or $x=\frac{pa}{b}$, and $x:q::b:a$ or $x=\frac{qb}{a}$; multiply these two equations together and we have $x^2=pq$ or $x=\sqrt{pq}$; i. e. the true weight of the body is a mean proportional between the apparent weights thus obtained. Hence to find the weight of a body by a false balance, we have this

Rule.—*Take the weight of the body in each scale; multiply together the two weights thus found, and take the square root of the product.*

4. To explain the construction of the steelyard.

A steelyard is a lever of the first order, having two unequa.

arms BC, CD, (Fig. 75;) a given weight P is movable along the longer arm CD, so as to sustain a weight of a variable magnitude suspended from the extremity of a shorter arm CB.

Fig. 75.

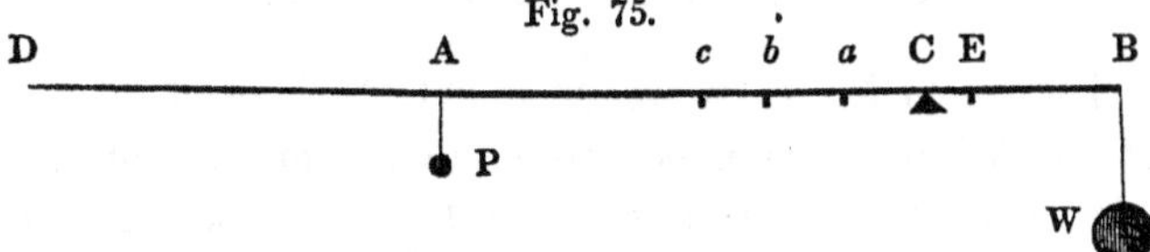

1. Suppose the weight of the arms not to be taken into consideration, then P : W :: BC : AC (Art. 99,) ∴ W×BC=P×AC, and as P and BC are given, W ∝ AC. Hence if C*a*, *ab*, *bc*, &c., be taken equal to each other, (or C*b*=2C*a*, C*c*=3C*a*, &c.,) then if P balances one pound when placed at *a*, it will balance two pounds at *b*, three pounds at *c*, &c.

2. Let us next suppose the steelyard to have weight, and that the excess of the weight of the longer arm CD above that of the shorter CB, is such that the movable weight P, when placed at E, would keep the arms in equilibrio; in which case this excess would be measured by P×CE; if therefore a weight W, placed at B, be in equilibrio with the weight P, placed at A, we should have W×BC=P×AC+P×CE=P (AC+CE) =P×AE; consequently, since P and BC are given, W ∝ AE. The construction of the steelyard, therefore, would be the same as in the former instance, except that the graduation must begin from E, instead of from C.

If the longer arm be divided into equal parts, to indicate the number of pounds, ounces, &c., which are contained in the variable weight W, the magnitude of the divisions may be found. For W×BC=P×AE; ∴ $\frac{AE}{W}=\frac{BC}{P}$, and as AE is the extent of the graduated arm corresponding to W, $\frac{AE}{W}$ or $\frac{BC}{P}$ will be the length of a division corresponding to 1 pound or 1 ounce, &c.

Hence, when only BC and P are given, the magnitude of each division will be equal to $\frac{BC}{P}$.

5. ACB is a cylindrical straight lever whose weight is (*w*) at whose extremity A, a given weight (P) is suspended: It is required to determine the position of the fulcrum C, so that P may be in equilibrio with the longer arm BC.

Fig. 76.

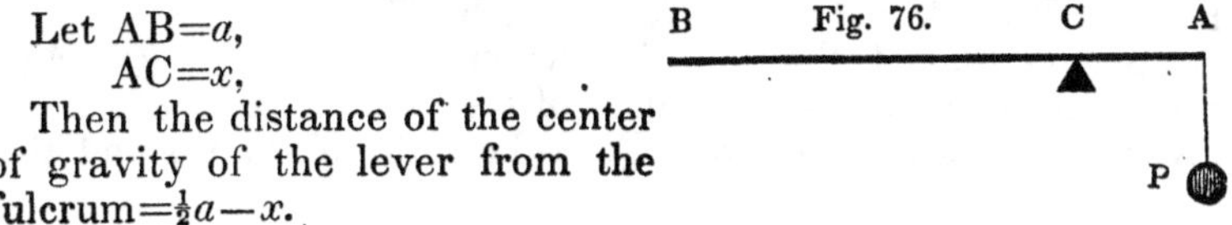

Let AB=*a*,
AC=*x*,

Then the distance of the center of gravity of the lever from the fulcrum=$\frac{1}{2}a-x$.

Hence, from the principles laid down in Art. 103, when there s an equilibrium, we have

$$P\times x=w(\tfrac{1}{2}a-x),\text{ or }2Px+2wx=wa,\text{ and }x=\frac{wa}{2(P+w)}.$$

Cor. If $P=w$, then $x=\frac{1}{4}a$, or $AC=\frac{1}{4}AB$, the length of the lever.

6. P and W are suspended from the extremities of the arms of the bent lever ABC, (whose weight is not taken into consideration:) It is required to find the angle of inclination (ACB,) so that when there is an equilibrium, AC shall be parallel to the horizon.

Fig. 77.

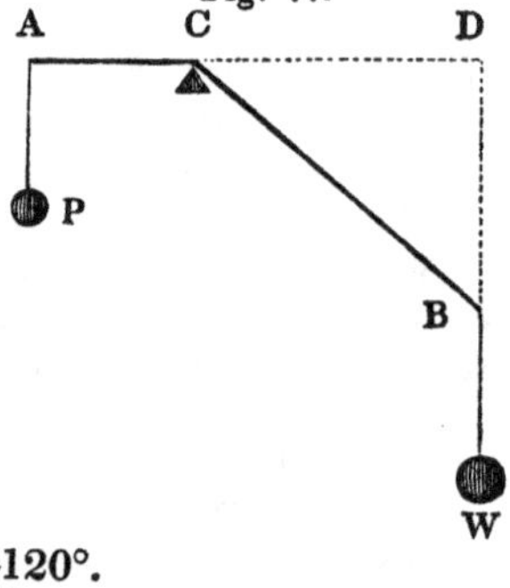

By Art. 106,
P : W :: BC × sin. CBD : AC × sin. PAC; but sin. PAC = rad. = 1, and sin. CBD = cos. BCD; ∴ P : W :: BC × cos. BCD : AC × 1, hence cos. $BCD=\frac{P\times AC}{W\times BC}$, from which BCD and consequently ACB is known.

Cor. 1. If P=W, then cos. $BCD=\frac{AC}{BC}$;

and if BC=2AC, then $\frac{AC}{BC}=\frac{1}{2}$;

∴ cos. BCD=$\frac{1}{2}$=cos. 60°; hence ACB=120°.

Cor. 2. If AC=BC, then cos. $BCD=\frac{P}{W}$; and if W=2P, then $\frac{P}{W}=\frac{1}{2}$; ∴ cos. BCD=$\frac{1}{2}$=cos. 60°; hence, in this case also, ACB =120°.

7. From the extremities of the arms CA, CB, of a bent lever, the weights P, W, are suspended: It is required to determine the position of the lever when these weights are in equilibrio.

Through C (Fig. 78,) draw MN parallel to the horizon, and produce PA, WB to meet it in M and N; by Art. 104, the lever will be in equilibrio, when

Fig. 78.

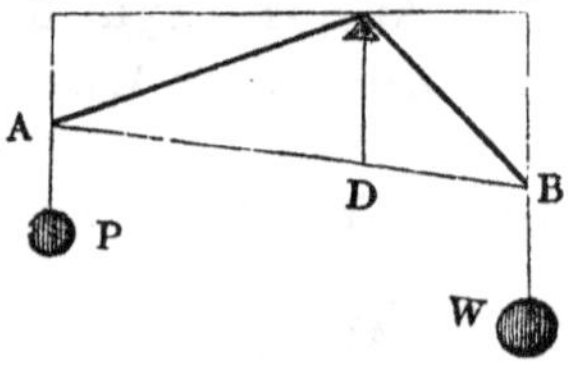

P : W :: CN : CM.

Join AB, and draw CD parallel to MP or NW; CD will cut AB in the same ratio that it does MN,* i. e.

DB : DA :: CN : CM;

hence DB : DA :: P : W; from which it appears that if the line AB, which joins the extremities of the arms of the lever, be divided in the point D, in the ratio of P : W, and that point be brought immediately under the center of motion, it will give the position of the lever when P and W are in equilibrio.

111. Questions upon the Principles of the Lever.

1. At one extremity of a straight lever whose length is 7 feet, a weight of 10 pounds is suspended; at the distance of 5 feet from the point of suspension a fulcrum is placed: What weight must be suspended from the other extremity of the lever, to keep it in equilibrio? *Ans.* 25 *pounds.*

2. A lever of the second order is 25 feet long: At what distance from the fulcrum must a weight of 125 pounds be placed, so that it may be supported by a power able to sustain 60 pounds, acting at the extremity of the lever? *Ans.* 12 *feet.*

3. A cylindrical straight lever is 14 feet long, and weighs 6 lbs. 5 oz.; its longer arm is 9, and its shorter 5 feet; at the extremity of its shorter arm a weight of 15 lbs. 2 oz. is suspended: What weight must be placed at the extremity of the longer arm to keep it in equilibrio? (See Art. 103.) *Ans.* 7 *pounds.*

4. A body weighs 11 pounds at one end of a false balance, and 17 lbs. 3 oz. at the other: What is its real weight?

Ans. 13 *lbs. and* 12 *oz.*

5. A and B are of the same height, and sustain upon their shoulders a weight of 150 pounds, placed on a pole 9$\frac{1}{3}$ feet long; the weight is placed 6$\frac{2}{3}$ feet from A: What is the weight sustained by each person? (See Art. 102.)

Ans. A *sustains* 42$\frac{6}{7}$ *pounds, and* B *sustains* 107$\frac{1}{7}$ *pounds.*

6. The longer arm of a steelyard is 2 feet 2 inches in length, and the shorter 2$\frac{2}{3}$ inches; and its apparatus of hooks, &c., is so contrived, that a weight of two pounds placed upon the longer arm, at the distance of 10 inches from the center of motion, will balance 8 pounds placed at the extremity of the shorter arm; the movable weight (of 2 pounds) cannot conveniently be placed nearer to the fulcrum than $\frac{2}{3}$ of an inch: What must be the graduation of the steelyard that it may weigh ounces, and what will be the greatest and least weights that can be ascertained by it?

Ans. The graduation is to 12*ths of an inch; and it will weigh from* 1 *to* 20 *pounds.*

7. The arms of a straight lever are to each other as 7 : 9, and it is acted upon obliquely by two forces; the force (P) applied at the extremity of the longer arm, is inclined to it at an angle of 50°, and (p) at the shorter at an angle of 80°: What is the proportion between the forces, when the lever is in equilibrio? (See Art. 106.) *Ans.* $P=p$.

8. The arms of a bent lever are equal, and P : W :: 1 : 2: What must be the inclination of the arms to each other, that the arm from whose extremity P is suspended may be parallel to the horizon? *Ans.* 120°.

9. In a combination of levers connected together in the manner represented in Fig. 74, the three shorter arms (BC, DF, EG,) are respectively 2, 5, and 3 feet; the three longer arms (AC, BF, DG,) are 13, 14, and 15 feet; the weight (P) suspended from A is 5 pounds: What weight will it sustain at E?

Ans. 455 *pounds*

CHAPTER VII.

OF THE WHEEL AND AXLE; AND THE PULLEY.

WHEEL AND AXLE.

112. In order to explain the manner in which the wheel and axle operate upon each other, suppose DE (Fig. 79,) to be a cylindric roller supported upon the props LH, MQ, and movable about the axis LM. Let two straight inflexible rods AG, BC be inserted into this cylinder, in a direction perpendicular to the axis,

Fig. 79.

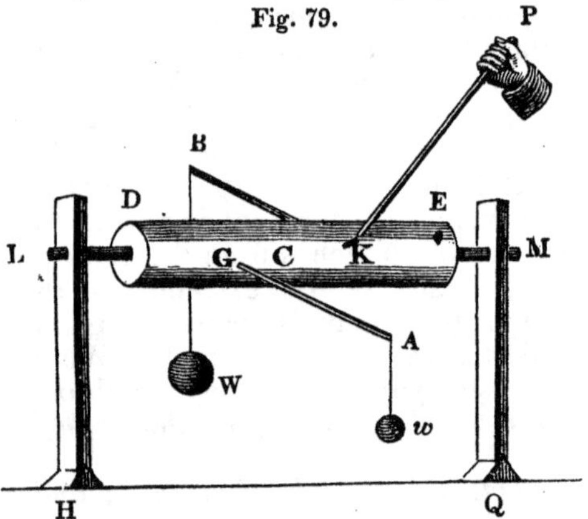

but parallel to each other and the horizon; let there be another rod PK perpendicular to the axis, but making any angle with the plane passing through BC or AG, and the axis. From the extremities of the rods BC, AG, let the weights W, w be suspended; then (Art. 100) W×BC will represent the effect of W, and w× AG the effect of w, to turn the roller about the axis LM; and supposing the rods AG, BC, and the roller DE, to be perfectly rigid and inflexible, it is evident that these two weights will counteract each other's effects, in the same manner as if they were acting at the extremities of the arms of a straight lever. When

W and w therefore are in equilibrio, W×BC will be equal to w ×AG, or w : W :: BC : AG; and this will be the case whatever be the length of the rod BC. Suppose that rod to be equal to the radius of the roller, then the string BW will become a tangent to the roller, and the foregoing proportion becomes w : W :: the radius of the roller : AG. Let us next suppose the weight W to be kept in equilibrio by a power P acting at right angles to the extremity of the rod PK; then may P and W be considered as acting at the extremities of the arms PK, BC of the bent lever; and since they act at right angles to those arms, (Art. 106) P : W :: BC : PK;* and (when BW acts as a tangent to the roller) P · W :: radius of the roller : PK.

113. *In the wheel and axle an equilibrium is produced when the power acting at the circumference of the wheel : weight sustained upon the axle :: radius of the axle : the radius of the wheel.*†

Fig. 80.

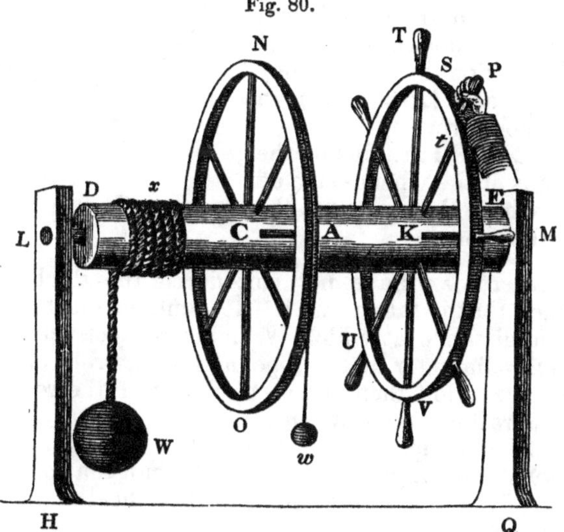

For let the weight W (Fig. 80) (which is suspended by a rope going round the axle DE) be kept in equilibrio, either by another weight (w) suspended from a rope going freely round the wheel NAO, or by a power P acting at right angles to the handles TS, Pt, &c. of the wheel StV, and let the planes of these wheels be at right angles to the axis LM of the machine; then, in the for-

* For in this case sin. α=sin. β.

† Let R=radius of the wheel, r=radius of the axle, then P : W :: r : R, ∴P×R=W×r; if W and r be given, and P and R variable, then $P \propto \frac{1}{R}$; i. e. to sustain a given weight upon a given axle, the power must be increased as the radius of the wheel is diminished; and vice versa.

mer case, the weight W may be considered as sustained by a weight (w) acting at right angles to the extremity of the arm AC of a straight lever, and in the latter, by a power (P) acting at right angles to the extremity of the arm tK of a bent lever, the weight itself in each case acting at the distance of the radius of the axle from the center of motion; hence, (by Art. 112,)

w or P : W : : radius of the axle : AC or tK.

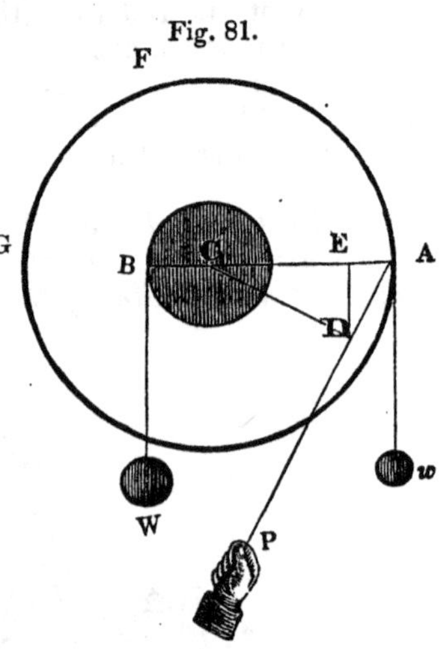

Fig. 81.

114. If the power does not act at right angles to the radius of the wheel, but in some oblique direction, as AP (in the annexed figure, which represents a section of the wheel and axle,) then let fall CD perpendicular to AP. By the property of the lever, P and W are to each other inversely as the perpendiculars let fall from the center of motion upon the lines of direction in which they respectively act, (Art. 104;) in this case, therefore,
P : W : : CB : CD : : radius of the axle : radius of the wheel × sine of the angle which P makes with the radius of the wheel.

115. Hitherto we have not considered the thickness of the rope; when that is taken into the account, we must add the half of it to the distance at which W and w respectively act.* Let therefore $2t$=diameter of the rope, and let R=rad. of the wheel, r=rad. of the axle; then if the thickness of the rope be taken into consideration, we have $w : W :: r+t : R+t$; and since in this case the same quantity (t) is added to each term of the ratio $r : R$, w must bear a greater ratio to W than that of $r : R$, or of the radius of the axle to the radius of the wheel.†

116. In a combination of wheels, such as is represented in figure 82, where a power (P) acts upon the winch or handle PQ,‡ which turns the wheel A, which acts upon the wheel

* For W, w, act in the direction of the axis of the rope, and this axis is evidently removed from the circumference of the wheel or axle by half the thickness of the rope.

† In this article we have considered the ropes which go round the wheel and axle to be of the same thickness, and that the rope coils round the axle but once. But suppose the thickness of the rope to which W is appended to be 2T, that of w to be $2t$, and the rope to coil round the axle any number of times denoted by n; then it is evident, that for each coil of the rope after the first, W will be further removed from the circumference of the axle by the whole thickness (2T) of the rope; the most general form therefore under which the relation of w : W can be exhibited, when the thickness of the rope is taken into consideration, is $w : W :: r+(2n-1)T : R+t$.

‡ In this case the effect will evidently be the same as if the power acted at the circumference of a wheel whose radius is PQ.

Fig. 82.

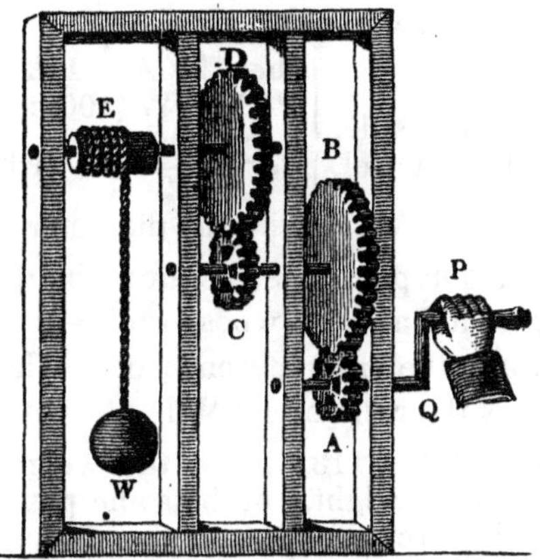

B, from which the motion is propagated through the wheels C and D to the axle E, about which the rope that sustains the weight (W) is wound, let the force exerted by the wheel A upon the wheel B$=p$, by C upon D$=q$; then, supposing P and W to be in equilibrio, we have,

$$P : p :: \text{rad. of wheel A } (r) : PQ,$$
$$p : q :: \text{rad. of wheel C } (r') : \text{rad. of wheel B } (R),$$
$$q : W :: \text{rad. of axle } (a) : \text{rad. of wheel D } (R');$$
$$\therefore P : W :: a \times r \times r' : PQ \times R \times R';^*$$

where the demonstration goes upon the same principle as that of a combination of levers, in Art. 109.

117. Instead of the power being applied to the handle or winch PQ, (Fig. 82,) suppose it to be applied at the circumference of a wheel whose radius is (R;) let the radius of the axle$=r$, the radii of the small wheels $=r'$, r'', &c., and of the larger ones $=R'$, R'', &c.; then, whatever be the number of such wheels, the proportion expressing the relation between P and W, when there is an equilibrium, will be $P : W :: r \times r' \times r''$ &c. : $R \times R' \times R''$ &c. :: product of the radii of all the smaller wheels (or axles) : product of the radii of all the larger ones.

If the radii of the large wheels are equal to each other, and also those of the small wheels, then $P : W :: r^n : R^n$, where n equals the number of wheels or axles.

118. Examples.

1. What must be the diameter of a wheel by which a weight of 100 pounds suspended by a rope going round an axle, whose

* Algebra, Art. 392

radius is six inches, may be kept in equilibrio by a power acting upon it equivalent to 12 pounds.

Let r=radius of the axle, x= do. do. wheel, { then, by Art. 113, P (12) : W (100) : : r (6) : x, $\therefore x=\frac{600}{12}$=50 inches=4 Ft. 2 In, and the diameter 8 4 }

2. A weight of 500 pounds is sustained by a rope of one inch diameter, going round an axle whose radius is 8 inches; and the power acts close to the circumference of a wheel whose radius is 4 feet: What is the ratio of P : W?

This is a case of Art. 113, where the weight is not kept in equilibrio by another weight, but by some power acting upon a handle close to the circumference of the wheel, (as P acts upon the wheel StV in Fig. 80,) and since t disappears in the 4th term of the proportion, w (or P) : W : : $r+t$: R$+t$, which becomes P : W : : $r+t$: R. In the present instance, W=500, r=8 inches, $t=\frac{1}{2}$ inch, R=48 inches, $\therefore$ P : 500 : : $8\frac{1}{2}$: 48; or $P=\frac{4250}{48}$=88.54 pounds.*

3. In Fig. 82, PQ=1 foot; the radii of the wheels A, C, are each 4 inches; the radii of the wheels B, D are each 15 inches; and the radius of the axle E is 3 inches: What power must be applied to P to support a weight of 600 pounds?

By Art. 116, P : W (600) : : $a\times r\times r'$: PQ$\times$R$\times$R$'$,
: : 3$\times$4$\times$4 : 12$\times$15$\times$15,
: : 4 : 225, $\therefore P=\frac{2400}{225}=10\frac{2}{3}$ lbs.

119. Questions upon the Principles of the Wheel and Axle.

1. A power of 14 pounds acts upon a wheel whose diameter is 9 feet: What weight will keep it in equilibrio, supposing the rope which supports that weight to be wound round an axle whose diameter is 7 inches? *Ans.* 216 *pounds.*

2. A power of 4 pounds keeps in equilibrio a weight of 176 pounds, by means of a wheel whose diameter is 11 feet: What is the diameter of the axle? *Ans.* 3 *inches.*

3. A power (P) acting by means of a rope going over a wheel

* If the thickness of the rope be not considered, then P : W (500) : : r (8) : R (48,) $\therefore P=\frac{4000}{48}$=83.33; it makes a difference, therefore, of 5.21 lbs., whether this thickness be or be not taken into the account.

whose diameter is 7 feet 11 inches, supports a weight of 528 pounds; the diameter of the axle is 9 inches, and the rope by which P and W are suspended is two inches thick: What must be the magnitude of P, supposing the thickness of the rope to be taken into consideration? *Ans.* P=59.876 *pounds.*

4. Four wheels, A, B, C, D, whose diameters are 5, 6, 10, and 2 feet respectively, are put in motion by a power of 15 pounds applied at the circumference of the wheel A; these wheels act upon each other by means of three smaller wheels, the diameter of each of which is 10 inches; the last wheel D turns an axle whose diameter is 4 inches: What weight may be sustained by a rope going over this axle? *Ans.* 46,656 *pounds*

THE PULLEY.

120. A pulley is a small grooved wheel movable about a pivot, the pivot itself being at the same time either fixed or movable. The principle upon which a weight is sustained by means of a pulley or system of pulleys, is very simple, and will be readily understood from the following investigation.

In the single fixed pulley A, (Fig 83,) about which the weight W is sustained by the power P acting on a string WAP passing along the groove in the circumference, no mechanical advantage is gained; for since the rope passes freely round the pulley, it is evident that the tension on each side of it must be the same, and consequently that the power must be equal to the weight which it sustains. The only advantage attending a pulley of this kind

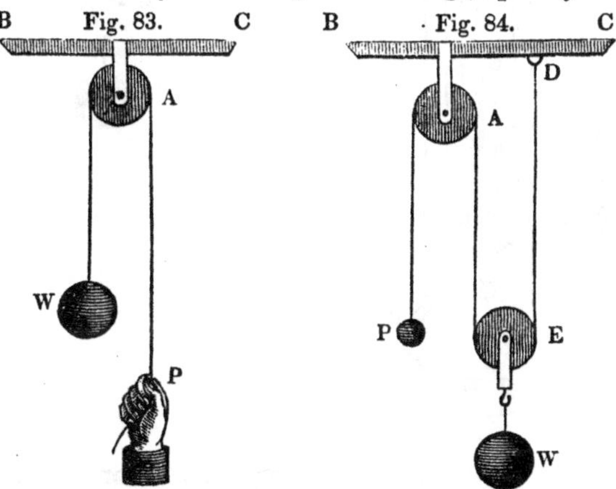

is, that a given power may be made to sustain or put in motion a given weight in a more convenient manner, by altering at pleasure the direction in which the power acts. The pressure upon the pivot or axis of the pulley A is evidently equal to P+W.

121. But if a weight W (Fig. 84) be sustained by a power P acting on a string going over a movable pulley E as well as the fixed one A, then it is evident that this weight is sustained by two strings AE, DE; and as it is suspended from the center of the pulley E, these ropes must act at equal distances from that center; consequently, each string must sustain half the weight. But it is evident, that whatever is the weight sustained by the string AE, the same must be sustained by the power P, which acts upon a string going freely over the fixed pulley A; hence, when there is an equilibrium, $P=\frac{1}{2}W$, or $W=2P$, $\therefore P : W :: 1 : 2$. With respect to the pressure upon the hook D, it is $\frac{1}{2}W$ or P, and upon the axis of the pulley A it is equal to $P+\frac{1}{2}W=2P$.

122. The same principle applies to the system of pulleys, in which the same string goes round all the pulleys, as in Fig. 85. For it is evident that the weight W is supported by all the strings at the lower block; if therefore the whole number of these strings be (n,) each string must support $\frac{1}{n}$th part of the weight. But when

Fig. 85.

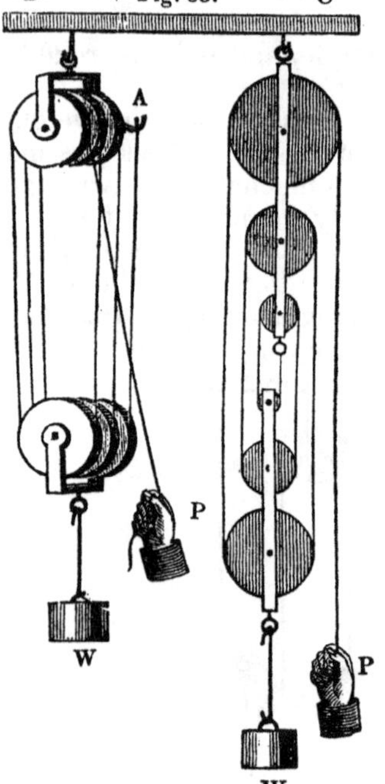

Fig. 86.

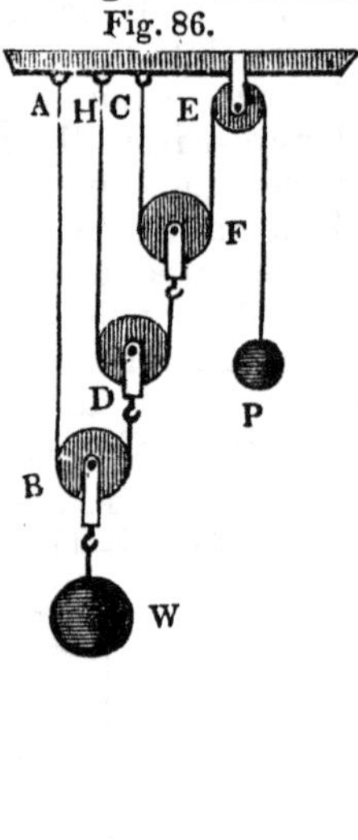

there is an equilibrium, whatever be the tension upon each of these strings which support the weight, the same will be the tension on the string upon which the power P acts; hence $P=\frac{1}{n}W$, or $W=nP$, $\therefore P : W :: 1 : n$* (where n=number of strings at the

* If two blocks of pulleys of this kind (in which m and n are respectively the

lower block, or twice the number of movable pulleys.) The pressure upon the hook B or C is evidently equal to $P+W=P+nP=(n+1)P$.

123. When the same string does not go round all the pulleys, but each pulley (Fig. 86) has a separate string CFE, HDF, ABD, &c., going round it, and fastened to the hooks A, H, C, &c., then the relation between P and W must be estimated by a different method. Thus (since the string CFE goes over a single movable pulley) by Art. 121, P : weight sustained by pulley F :: 1 : 2; and weight sustained by F : weight sustained by D :: 1 : 2, weight sustained by D : weight sustained by B, i. e. W :: 1 : 2, $\therefore$ P : W :: 1 : $2\times2\times2$, &c., :: 1 : 2^n, or $W=2^nP$ (if n be the number of movable pulleys.) In this system of pulleys, the pressure upon the hook A $=\frac{1}{2}W=\frac{1}{2}\times2^nP=2^{n-1}P$; upon the hook H ($=\frac{1}{2}$ pressure upon A) $=\frac{1}{2}\times2^{n-1}P=2^{n-2}P$, &c.; and the pressure upon pulley $E=2P$.

Fig. 87.

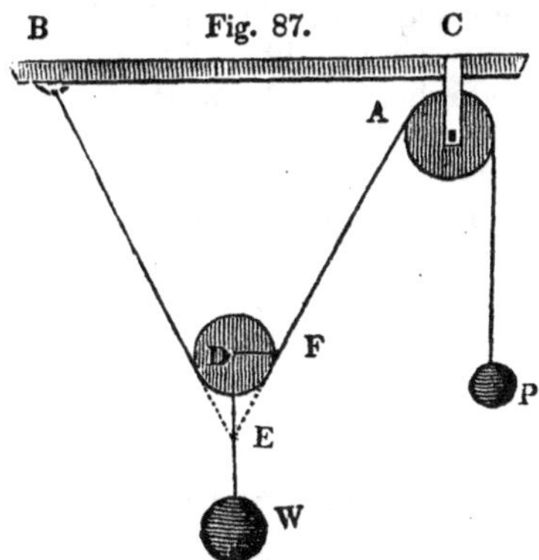

124. Hitherto we have considered the strings as acting parallel to each other; but suppose the power P (Fig. 87,) to act upon the weight W by a string going over the movable pulley D in an oblique direction; then produce the string AF to E, and draw DF at right angles to DE, (D being the center of the pulley.) Let FE represent the magnitude of the power acting in the direction EF, which resolve into ED, DF; then ED is that part of it which is efficacious in supporting the weight W; and since the string BD supports the same weight as the string AF, the whole weight sustained by the string BFA will be represented by 2DE; hence,

$$P : W :: EF : 2DE :: \text{rad.} : 2 \cos. DEF; \text{ or,}$$

number of strings) were combined together, so that the effect (E) produced by the first block should act as power upon the second, then P : E :: 1 : m, and E : W :: 1 : n; $\therefore$ P · W :: 1 : mn.

the power is to the weight, as radius to twice the cosine of the angle of inclination of the direction of the power to that of the weight.

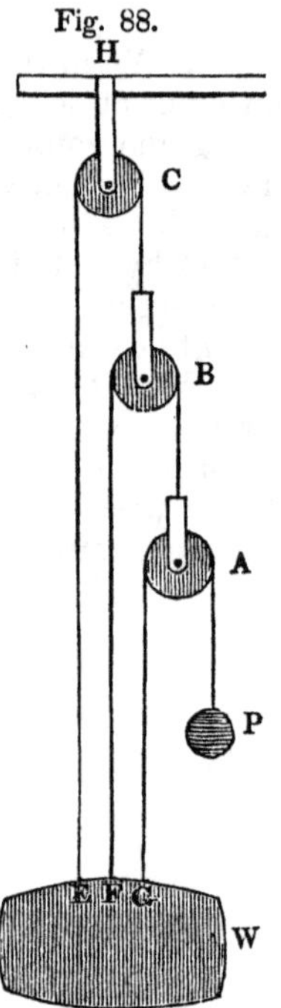

Fig. 88.

125. There is another mode of combining pulleys together, which we have not yet noticed; viz. when each string is fixed into the weight, as in Fig. 88. In this case, supposing there is an equilibrium, and the power P acts upon a string going freely over the pulley A, then it is evident that the pressure upon that pulley will be equal to 2P, ∴ the string BA supports a part of the weight equal to 2P. For the same reason, since the string FBA goes freely over the pulley B, the string CB supports 4P, &c.; hence the portions of weight supported by the strings AG, BF, CE, &c. are P, 2P, 4P, &c. respectively, and consequently $W=P+2P+4P \ldots . 2^{n-1}P$ (where n=number of strings attached to the weight) $= P(1+2+4 \ldots . 2^{n-1})=P(2^n-1)$,*

$$\therefore P : W :: 1 : 2^n-1.$$

The pressure upon the support at H is evidently equal to

$$P+W=P+(2^n-1)P=2^nP.$$

126. Examples.

1. A weight of 56 pounds is kept in equilibrio by a power of 7 pounds, by means of a system of pulleys, in which the same string goes round every pulley: What is the number of movable pulleys?

Let n=the number of strings at the lower block of pulleys, then, by Art. 122, $P(7) : W(56) :: 1 : n=\frac{56}{7}=8$=twice the number of movable pulleys, ∴ the number required is 4.

2. In the system of pulleys described in Art. 123, find the general relation between P, W, and n.

In this system, $W=2^nP$, $\therefore P=\frac{W}{2^n}$;

* The sum of a geometric series, whose first term is 1, common ratio 2, and number of terms n, is 2^n-1. (Alg. 442.)

$$\text{and } 2^n=\frac{W}{P},$$

$$\therefore n\,.\,\log.\ 2=\log.\ W-\log.\ P,$$

$$\text{or } n=\frac{\log.\ W-\log.\ P}{\log.\ 2.}.$$

From which it appears, that if any two of the three quantities P, W, and n be given, the third may be found.

3. Find the general relation between P, W, and n, in the sys tem of pulleys described in Art. 125; and also the number of pulleys necessary for a power of 3 pounds, to support a weight of 381 pounds.

$$\text{Here } W=P(2^n-1),\ \therefore P=\frac{W}{2^n-1};$$

$$\text{also, } 2^n-1=\frac{W}{P},\ \therefore 2^n=\frac{W}{P}+1=\frac{W+P}{P};$$

hence $n\times\log.\ 2=\log.\ (W+P)-\log.\ P$, or $n=\dfrac{\log.\ (W+P)-\log.\ P}{\log.\ 2.}$

If W=381, and P=3; then $n=\dfrac{\log. 384-\log. 3}{\log.\ 2}=\dfrac{2.107210}{0.301030}=7.$

127. Questions on the Principles of the Pulley.

1. By means of a system of pulleys, of which five are movable, and in which the same string goes round all the pulleys, what power will be necessary to sustain a weight of 165 lbs?
Ans. $16\frac{1}{2}$ *lbs.*

2. A weight is sustained by a power attached to a rope going over one movable pulley, (as in Fig. 87,) the direction of the rope making an angle of 60° with a vertical line passing through the center of the pulley: What is the relation between P and W?
Ans. P=W.

3. A weight of 240 lbs. is sustained by a power equivalent to $7\frac{1}{2}$ lbs. by means of the system of pulleys described in Art. 123: What is the number of pulleys? *Ans.* 5 *pulleys.*

4. What power will be necessary to sustain a weight of 2387 lbs. in a system of 10 pulleys, constructed according to Fig. 88, where the strings are all fastened to the weight? *Ans.* $2\frac{1}{3}$ *lbs*

CHAPTER VIII.

OF THE INCLINED PLANE, THE SCREW, AND THE WEDGE.

128. This chapter will comprehend the three remaining mechanical powers; viz. the *Inclined Plane*, the *Screw*, and the

Wedge; beginning with the *Inclined Plane,* being that upon which the properties of the *Screw* more immediately depend.

THE INCLINED PLANE.

129. *In the inclined plane, an equilibrium is produced, when the power is to the weight, as the sine of the inclination of the plane is to the sine of the angle, which the direction of the power makes with a perpendicular to the plane, at the point where the weight rests upon it.*

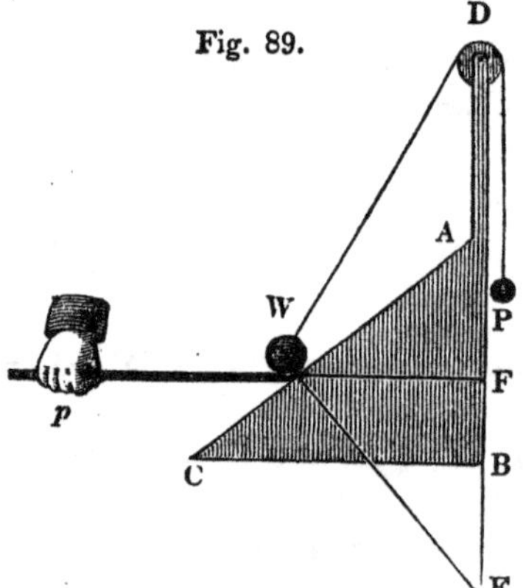

Fig. 89.

Let AC (Fig. 89,) be an inclined plane, whose length is AC, height AB, and base BC; and suppose the weight W to be kept in equilibrio by any other weight (or power) P acting freely over a pulley fixed at D. Draw* WE at right angles to AC, meeting AB (produced if necessary) in E; the weight W may be considered as kept at rest by three forces, viz. the action of the power in the direction WD, its own weight (or gravity) in direction DE, and the reaction of the plane in direction EW; ∴ (by Art. 56,) these three forces are to each other as the three sides of the triangle DWE, in the directions of which they respectively act. Hence,

P : W :: WD : DE :: sin. WED or ACB : sin. DWE.

By the third law of motion, the pressure of W upon the plane must be equal to the reaction of the plane upon W; if, therefore, EW represents that reaction, WE will represent the pressure upon the plane; hence,

P : press. on the plane :: WD : WE :: sin. WED or ACB : sin. WDE;
W : press. on the plane :: DE : WE :: sin. DWE : sin. WDE.

130. *In the inclined plane, when the power acts* PARALLEL TO THE PLANE, 1. *The power is to the weight as the height of the plane to its length;* 2. *The power is to the pressure on the plane, as the height of the plane to its base;* 3. *The weight is to the pressure on the plane as the length of the plane to its base.*

If the power acts parallel to the plane, then WD may be considered as coinciding with WA, and the power, the weight, and the pressure, will be respectively represented by the three sides WA, AE, WE, of the triangle AWE; hence,

* This figure is to be considered as a section of the plane passing through the center of gravity of the weight; and if the weight be not large, that center of gravity may be considered as placed in the angular point (W) of the triangle DWE.

P : W :: WA : AE :: AB : AC :: height : length.
P : press. on the plane :: WA : WE :: AB : BC :: height : base.
W : press. on the plane :: AE : WE :: AC : BC :: length : base.

131. *In the inclined plane, when the power acts parallel to the* BASE *of the plane,* 1. *The power is to the weight, as the height of the plane to its base;* 2. *The power is to the pressure on the plane as the height of the plane to its length:* 3. *The weight is to the pressure on the plane, as the base of the plane to its length.*

If the power acts parallel to the base of the plane, (i. e. if the weight W be sustained upon the plane by a force acting in the direction pW, and pushing horizontally against the plane,) then produce pW to F; and when there is an equilibrium, the power, the weight, and the pressure will be respectively represented by the three sides WF, FE, WE, of the triangle WFE; therefore,
P : W :: WF : FE :: AB : BC :: height : base.
P : press. on the plane :: WF : WE :: AB : AC :: height : length.
W : press. on the plane :: FE : WE :: BC : AC :: base : length.

132. *The least power will be required to raise or sustain a given weight upon a given inclined plane, when the direction in which that power acts is parallel to the plane; and, conversely, the greatest weight will also be raised or sustained by a given power upon a given inclined plane, when the direction in which the power acts is parallel to the plane.*

Let α=angle of inclination of the plane, β=angle which the direction of the power makes with a perpendicular to the plane at the point where the weight rests upon it; then, by Art. 129, P : W :: sin. α : sin. β; $\therefore \mathrm{P}=\frac{\mathrm{W}\times\sin.\alpha}{\sin.\beta}$. Suppose W and sin. α to be given, then P varies as $\frac{1}{\sin.\beta}$, and will consequently be the least when sin. β is the greatest, i. e. when the angle DWE becomes a right angle, or P acts parallel to the plane.

Again, $\mathrm{W}=\frac{\mathrm{P}\times\sin.\beta}{\sin.\alpha}$; if, therefore, P and sin. α be given, then W $\propto$ sin. β, and will consequently be greatest when sin. β is greatest.

133. *The pressure on a given inclined plane, with a given power, is greatest when the power acts parallel to the base of the plane.*

Let γ=WDE; then, by Art. 129, P : pressure upon the plane :: sin. α : sin. γ, $\therefore$ pressure upon the plane$=\frac{\mathrm{P}\times\sin.\gamma}{\sin.\alpha}$; if P and sin. α be given, then the pressure upon the plane $\propto$ sin. γ, and will consequently be greatest when sin. γ is greatest, i. e. when

WDE becomes a right angle, or the power acts parallel to the base of the plane.*

134. When a weight W is sustained by another weight P going over a fixed pulley D, (Fig. 90,) since the angle DWE varies at every point of the plane, it is evident that there is but one point of the plane where a given power will sustain a given weight; and that point may be thus determined. Take BG : BC : : P : W, and with center B and radius BG, describe a circular arc cutting AC in F; join BF, and draw DKL perpendicular to BF, or to BF produced: then the point W, where DL cuts

Fig. 90.

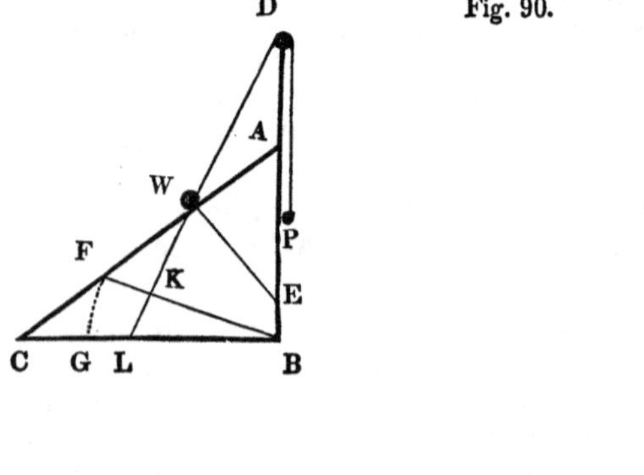

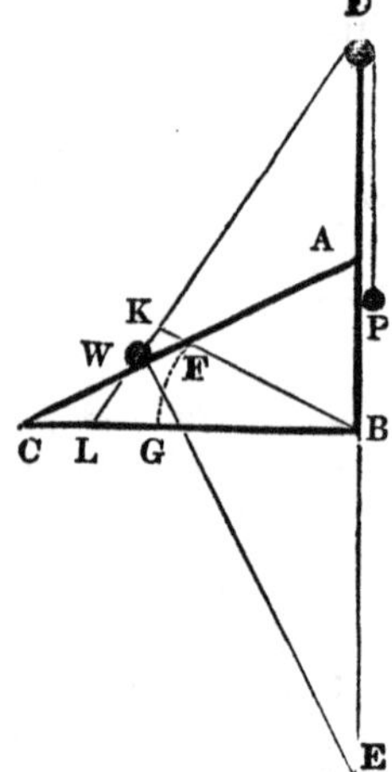

AC, will be the point required; for the triangles DWE, BFC being similar,† we have

WD : DE : : BF or BG : BC;
but P : W : : BG : BC,
∴ P : W : : WD : DE; hence,

by Art. 129, the relation between the sides of the triangle WDE is such as to give the position W when there is an equilibrium between P and W.

But when the power acts parallel to the plane or parallel to the base of the plane, the ratio of P : W is constant‡ through every part of the plane; if therefore a given power be in equilibrio

* In this case, since WE is greater than FE, it appears that the pressure is greater than W; but in this there is no inconsistency, for it is evident that when a weight is sustained upon an inclined plane by means of a force acting in direction *p*W, part of the pressure arises from the power as well as from the weight, and therefore the whole pressure may be greater than the weight.

† Draw WE at right angles to AC, as before, then AEW=ACB; and in the right-angled triangles BKL, DBL, KLB is common, ∴ KBL=KDB; hence CFB =DWE, and consequently the triangle CFB is similar to the triangle DWE.

‡ For (in one case) P : W : : height of the plane : length of the plane; and in the other, P : W : : height of plane : base of plane. (Art. 130, 131.)

with a given weight at any one point of a plane, the same power would also be in equilibrio with the weight, when placed at any other point of that plane. We now proceed to give a few examples for illustration.

135. Examples.

1. A person is just able to sustain by his strength a weight of 200 pounds. What weight would he be able to sustain on an inclined plane whose elevation is 50°, by means of a rope going round it, and fixed to the top of the plane in the manner represented in the annexed figure?

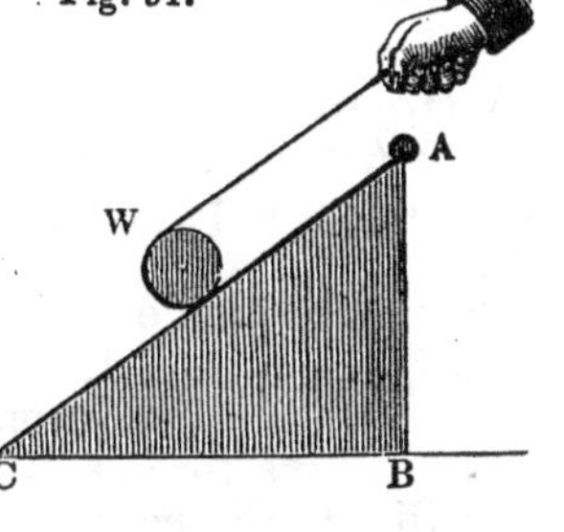

Fig. 91.

In this case, (Fig. 91,) the power which supports the weight acts parallel to the plane, ∴ by Art. 130, P : W :: AB : AC :: sin. ACB (50°) : radius :: 7660 : 10000, or $W=\frac{P\times 10000}{7660}$; this is the weight supported by the rope PWA; but since that rope is fixed at A, each part PW, WA, of that rope supports half* the weight; hence, if the force exerted by the rope PWA=P, the force acting in the direction WP=½P; calling that force (p,) then $p=\frac{1}{2}P$, or $P=2p$; substitute this for P and we have

$$W=\frac{2p\times 10000}{7660}=(\text{if } p=200)\frac{400000}{766}=522.19 \text{ pounds};$$

if therefore a person by his natural strength is able to lift a weight of 200 pounds, acting under the circumstances here represented, he will be able to sustain a weight of 522.19 pounds.

2. Upon an inclined plane, whose length is 20 feet, and elevation 30°, a weight of 3 pounds is sustained by a power of 2 pounds, acting over a pulley fixed at the distance of 10 feet from the top of the plane, (in the manner represented in Fig. 90:) It is required to find the distance of W from the top of the plane, when there is an equilibrium.

Since AC=20 feet, and ACB=30°, AB (=sin. 30°) will be equal to 10 feet, and consequently $BC=\sqrt{(AC^2-BA^2)}=\sqrt{300}$ =17.32 feet; constructing the figure, therefore, as in Art. 134, we have BG : BC(17.32) :: P(2) : W (3), ∴ BG or $BF=\frac{2\times 17.32}{3}$ =11.55 feet; hence, in the triangle BFC we have BC=17.32,

* The case being similar to that of a weight supported by two parallel strings going over a pulley.

BF=11.55, FCB=30°, from which the angles FBC, CFB are found to be respectively 18° 34′ and 131° 26′; but the triangle DWE is similar to the triangle BFC, ∴WED=30°, WDE=18° 34′, and DWE=131° 26′.

Again, since DWE is 131° 26′, and AWE a right angle, DWA must be 41° 26′; hence, in the triangle DWA, we have AD=10 feet, WDA=18° 34′, and DWA=41° 26′, from which AW is found to be 4.812 feet, which gives the distance of W from the top of the plane when P is in equilibrio with W.

3. A body is sustained upon an inclined plane, first by a power acting parallel to the plane, and afterward by a power acting parallel to the base of the plane. Compare the pressures upon the plane in these two different cases.

By Art. 130, when the power acts parallel to the plane,
W : press. on the plane (P) : : length : base.

By Art. 131, when the power acts parallel to the base,
Press. on the plane (p) : W : : length (L) : base (B ;) ∴
p : P : : $L^2 : B^2$.

Thus, in a plane whose elevation is 60°, (and whose length is consequently double* of its base,) it makes a difference of 4 : 1 as to the pressure upon the plane, whether a body is sustained upon it by a force acting parallel to the plane, or by one acting parallel to the base of the plane.

4. Two weights P, W, resting upon the inclined planes AC, AD, (Fig. 92,) whose common height is AB, keep each other in equilibrio by means of a string going over a pulley fixed at A. Compare the two weights.

Fig. 92.

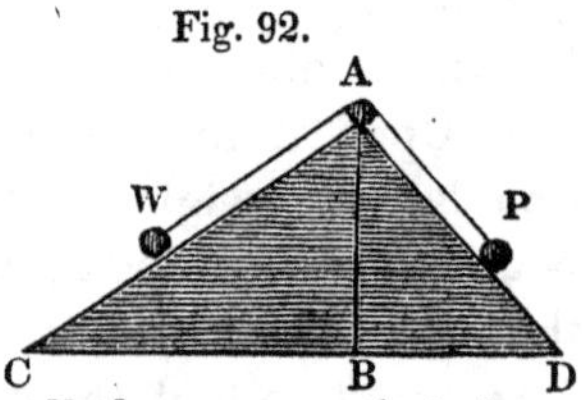

Since the string passes freely over the pulley at A, and the two weights are at rest, it is evident that the tension of the string WAP must be everywhere the same, i. e. whatever power is exerted at A to sustain W on the plane AC, the same is exerted at that point to sustain P upon the plane AD; call that power (p,) then, since the power in each case may be considered as acting parallel to the plane, we have, by Art. 130.

$$p : W : : AB : AC, \text{ and } P : p : : AD : AB;$$

∴ P : W : : AD : AC : : plane upon which P rests : plane upon which W rests. Hence, two weights resting on two inclined planes which meet, (forming a ridge,) will balance each other, when they are to one another as the lengths of the planes on which they respectively rest.

* In Fig. 91, if ACB=60°, then CAB=30°; ∴ BC=½ radius=½AC, or AC=2BC

5. A body is supported between two inclined planes of given elevations: Compare the pressure upon the planes.

Let NHL, (Fig. 93,) represent a perpendicular section of the body passing through its two points of contact H, L, with the planes; and from those points draw HF, LF at right angles to the planes DC, AC respectively. From their intersection F, draw FG perpendicular to the horizon, and let it represent the weight of the body. Through G draw GO parallel to LF; then the three sides of the triangle GOF will be in the direction of the three forces which keep the body at rest upon the plane AC, viz. GO will represent the reaction of the plane AC; OF the reaction of the plane DC; and FG the weight of the body; and in the same manner it may be shown that the three sides of the same triangle will represent the three forces which keep the body at rest upon the plane DC. Through G draw MGK parallel to the horizon; then since the three sides of the triangle GOF are perpendicular to the three sides of the triangle MCK, MCK must be similar to GOF, (see note, p. 55;) and since the weight of the body, the pressure upon the plane DC, and the pressure upon the plane AC, are respectively represented by the three sides FG, FO, OG of the triangle GOF, they will also be represented by the three sides MK, MC, CK, of the triangle MCK. Hence,

Fig. 93.

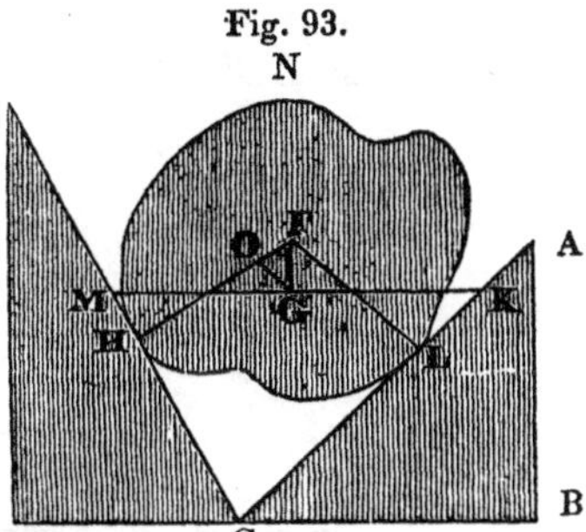

Pr. on DC : Pr. on AC :: MC : CK

:: sin. MKC or ACB : sin. CMK or DCE.

∴ Pr. on DC : Pr. on AC :: sin. ACB : sin. DCE.

Thus suppose DCE=60°, ACB=30°, then pressure on DC : pressure on AC :: sin. 30° : sin. 60° :: $\frac{1}{2} : \frac{\sqrt{3}}{2}$ :: 1 : $\sqrt{3}$. Hence, when a weight is supported between two inclined planes, the pressures on the planes are reciprocally as the sines of the angles of inclination of the planes.

THE SCREW.

166. *The screw is a spiral thread or groove, winding round a cylinder, so as to cut all the lines drawn on its surface parallel to its axis, at the same angle. The spiral may be either on the convex or the concave surface of the cylinder, and the screw is denominated accordingly, the external or the internal screw.*

The distance between the two contiguous threads of a screw.

corresponds to the height of an inclined plane, and the circum ference of the cylinder corresponds to the base of the same plane; hence the forces necessary to produce an equilibrium in the screw, are the same as in the inclined plane. Thus let the inclined plane ABC (Fig. 94) be wrapped round a cylinder, the

Fig. 94.

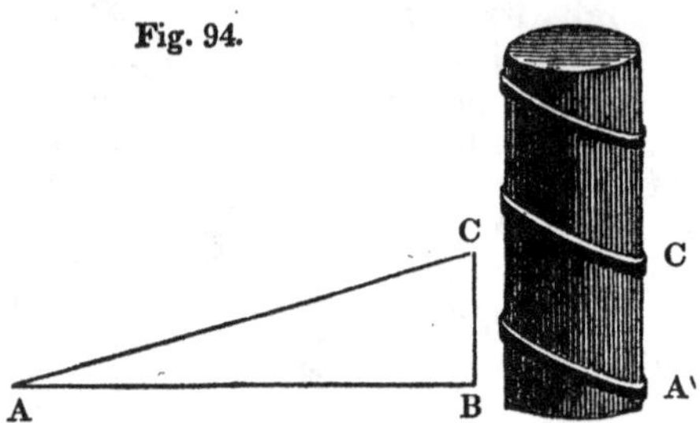

circumference of whose base is equal to the line AB; then the point A being placed on A′, the point B will come round to A′, and the point C will fall on C′, and the line AC will trace out the thread of the screw on the surface of the cylinder as far as C′, and may be continued in the same manner. By Art. 131, when the power acts parallel to the base of the plane, an equilibrium is produced when the power is to the weight as the height of the plane to its base; or, applied to the screw, an equilibrium is produced, when the power is to the weight, as the distance between two contiguous threads is to the circumference of the base.

Fig. 95

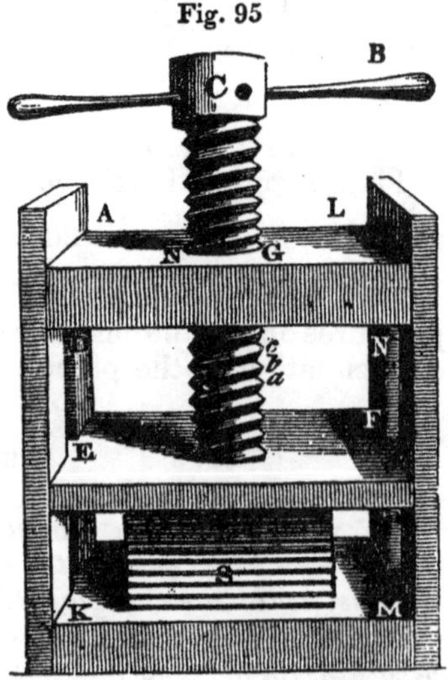

137. Let the external and internal screws be fitted to each other in the manner represented in Fig. 95, and let the external screw be turned round by a power applied to the lever BC, (acting parallel to the base of the cylinder,) while the internal screw remains fixed; then it is evident, from the manner in which the two screws act upon each other, that while the lever BC makes one revolution, the external screw will be elevated or depressed through one of the spaces *ab*, *bc*, according to the direction in which it is turned. When the screw is depressed it drives before it the board EF, which moves in the grooves DE, NF, by which means a pressure is created

upon any substance (S) placed between that board and the fixed board KM.

138. Let us now estimate the quantity of this pressure, by finding the relation which it bears to the power which produces it. To do this, it will be necessary, in the first place, to consider the force which would be generated in the elevation of the screw; which force may be estimated by showing separately what part of it arises from the action of the spirals of the screw upon each other, and what from the action of the lever. As the machine turns round, each point of the external screw acts upon the corresponding one of the internal screw, with a force analogous to that by which a body is sustained upon an inclined plane when the power acts parallel to the base of the plane;* the whole force therefore of the screw will be of the same kind, and (by Art. 131) will bear to the weight which it could support, the ratio of the distance between two spirals to the circumference of the cylinder. This would be the case, supposing the force to act close to the surface of the cylinder; when it acts therefore at the extremity of the lever BC, it will be increased in the proportion of the length of the lever to the radius of the cylinder.

139. *In the screw, an equilibrium is produced when the power is to the weight, as the distance between two contiguous threads is to the circumference of the circle described by one revolution of the power.*

Fig. 96.

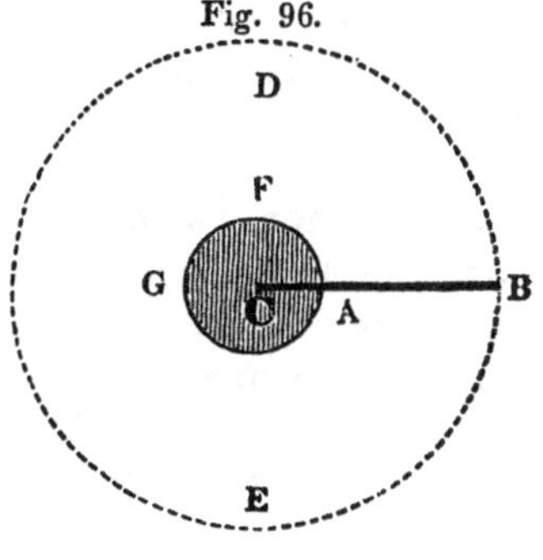

Let d=the distance between two spirals of the screw, which is cut upon the cylinder of which AFG (Fig. 96) is the section, a=length of the lever (CB), π=3.1415, &c., then the circumference (BDE) of the circle described by the extremity of the lever $= 2\pi a$. Let P = the power acting at the extremity of the lever; p = the power acting at the surface of the cylinder, W=the weight which is kept in equilibrio by P, then, by Art. 138,

$$p : W :: d : \text{circumference AFG},$$
$$P : p :: CA : CB :: \text{circumf. AFG} : \text{circumf. BDE};$$
$$\therefore P : W :: d : \text{circumf. BDE} = (2\pi a).$$
$$\therefore 2\pi a P = Wd, \therefore P = \frac{Wd}{2\pi a},\ W = \frac{2\pi a P}{d},\ d = \frac{2\pi a P}{W},\ \text{and } a = \frac{Wd}{2P\pi};$$

* Instead of moving a body up an inclined plane, we here move the plane itself against a resistance, which is overcome in the same manner as that of a body, and which may therefore be properly considered as a weight.

hence, if any three of the four quantities W, P, a, d, be given, the fourth may be found.*

140. We have thus estimated the magnitude of the weight which might be sustained by a given power applied to the elevation of the screw; but this machine is oftener used for the purpose of creating a pressure than for raising a weight; and it is evident, that whatever force is exerted by the screw to sustain a weight when it is turned in one direction, will also be exerted to create a pressure downward, when it is turned in an opposite direction. Whether, therefore, the screw be applied to raising a weight, or creating a pressure, the power necessary to sustain the weight or produce the pressure, will always bear to that weight or pressure the ratio of the distance between any two spirals of the screw, to the circumference of the circle which the power describes.

141. Examples.

1. A screw, the distance between whose spirals is one inch, is turned horizontally by a lever whose length is 2 feet, reckoning from the axis of the screw: What weight could be sustained or pressure produced by it, when a power of 30 pounds acts at the extremity of the lever?

By Art. 139, $W=\frac{2\pi aP}{d}=\frac{2\times3.1415\times24\times30}{1}=4523.76$ pounds; i. e. a power of 30 pounds applied to a machine of this kind would be sufficient to sustain a weight, or create a pressure, equivalent to about 4523 pounds, or somewhat more than two tons.

2. A person who could just lift a weight of 60 pounds found himself able, by means of a lever 3 feet long, acting as a handle to a screw, to sustain a ton weight: What was the distance between the spirals of the screw?

By Art. 139, $d=\frac{2\pi aP}{W}=\frac{2\times3.1415\times3\times60}{2240}=.504$ feet = about 6 inches.

3. In Fig. 97, the screw AB, which is turned by a power P acting upon the handle PQ, turns at the same time the wheel C,

* From this proportion it appears that the relation P : W depends entirely upon the distance between the spirals and the circumference which the power describes, whatever be the thickness of the cylinder upon which the screw is cut: and since $P\left(=\frac{Wd}{2\pi a}\right)\propto\frac{Wd}{a}$, when W is given, $P\propto\frac{d}{a}$; i. e. the power necessary to sustain a given weight varies directly as the distance between the spirals, and inversely as the length of the lever.

in such a manner as to cause it to draw up the weight W, by a rope wound round the axle D: This is called the endless screw, and it is required to find the ratio of P : W.

Supposing all the parts of this machine to be nicely adjusted, it is a very powerful one; inasmuch as it combines the energy of the screw with the multiplying power of the wheel and axle.

Fig. 97.

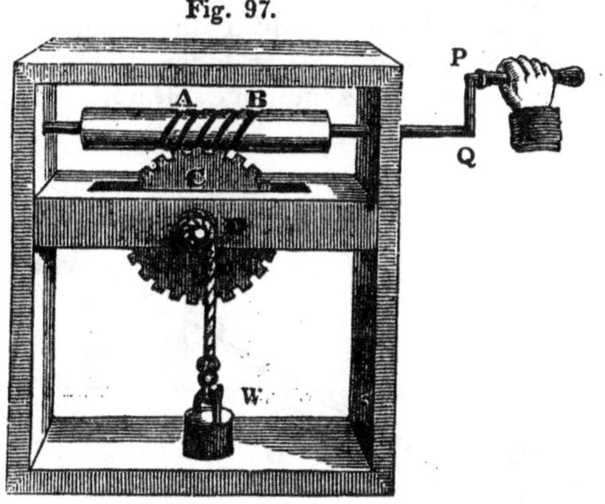

To estimate the effect, therefore, let PQ$=a$, the distance between two spirals of the screw$=d$, radius of the wheel=R, radius of the axle$=r$, $\pi=3.14159$, and Q=the force exerted by the screw upon the wheel; then, by Art. 139, $P : Q :: d : 2\pi a$,

and, by Art. 113, $Q : W :: r : R$,

$\therefore P : W :: dr : 2\pi a R$.

Hence $P=\frac{Wdr}{2\pi a R}$, and $W=\frac{2P\pi a R}{dr}$; let $d=1$ inch, $a=12$ inches, $r=4$ inches, $R=18$ inches, and $P=30$ pounds, then $W=\frac{40715}{4}$ pounds $=4.54$ tons; so that, by means of this machine, a power of 30 pounds would be sufficient to keep in equilibrio a weight of about $4\frac{1}{2}$ tons.

142. Questions on the Principles of the Inclined Plane and Screw.

1. If a man can draw a weight of 125 pounds up the side of a perpendicular wall, 20 feet high, what weight will he be able to raise along a smooth plank 44 feet long, laid sloping from the top of the wall? *Ans.* 275 *pounds.*

2. Suppose that a horse is able to draw a weight of 440 pounds out of a well, (by a rope passing over a fixed pulley, which allows the horse to draw in a horizontal direction;) what

weight will the same animal draw up a railway having a slope of five degrees, no allowance being made for friction?

Ans. 5048.5 *pounds.*

3. A lever five feet long is fixed at right angles in a screw whose threads are one inch asunder, so that the lever turns just once round in raising or depressing the screw one inch. If then this lever be urged by a force of 65 pounds, with what force will the screw press? *Ans.* 24504.35 *pounds.*

4. A shipwright wishing to haul a ship upon the stocks, employed a machine, combining the lever, the screw, the wheel and axle, the pulley, and the inclined plane, as represented in Fig. 98.

Fig. 98.

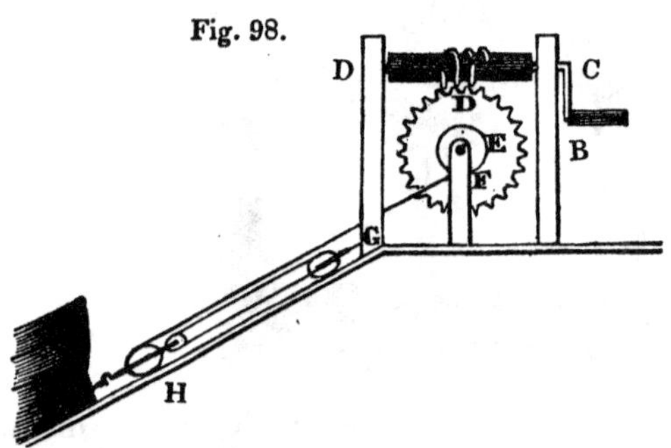

The handle of the winch BC =18 inches.
The distance of the threads on CD=1 inch.
The radius of the wheel ED =2 feet.
The radius of the axle EF =6 inches.
G is a fixed, and H a movable pulley, the number of strings being =4.
Inclination of the plane =30°.

Allowing a man to turn on the handle at B, with a force equal to 100 pounds, how much force could he exert on the ship?

Ans. 361911.168 *pounds, or more than* 161½ *tons.*

THE WEDGE.

143. All those instruments which are used for the separation of the parts of bodies, such as knives, axes, coulters, and chisels, come under the general denomination of the wedge; but these instruments are made of such variety of shapes, and forces are applied to them in such various ways, that of all the mechanical powers, the wedge is that whose properties are least capable of being brought to mathematical calculation. In the particular case where the wedge is of the form of a triangular prism, and the resistance upon its sides can be considered as forces acting in given directions, the relation between those resistances and

the power which counteracts them, may be estimated in the following manner.

144. *In the wedge, an equilibrium is produced when that part of the power, which, when resolved, acts perpendicularly to the back of the wedge, is equal to the sum of those parts of the resistances which also act perpendicularly to the back; these resistances being to each other inversely as their respective distances from the line of direction in which the resultant of the power acts.*

Fig. 99.

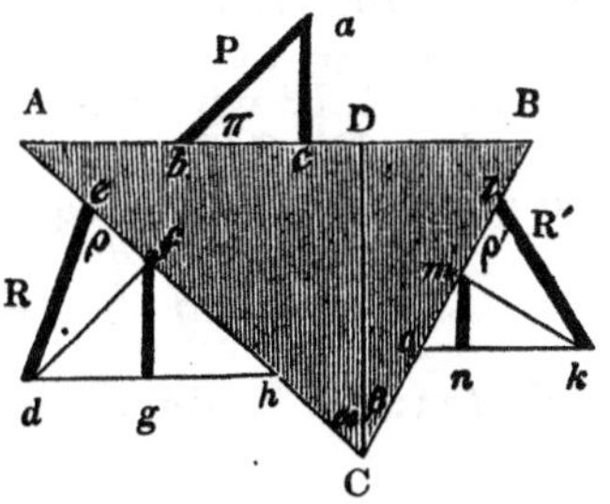

Let ABC (Fig. 99) represent a section of the wedge perpendicular to the axis of the prism, and suppose its sides AC, BC, to be perfectly smooth. Let a power P, whose magnitude and direction is represented by *ab*, act upon AB, the back of the wedge, and let it be counteracted by two resistances R, R′, (which are represented in quantity and direction by the lines *de*, *kl*,) acting upon the sides AC, BC. Resolve *ab* into *ac* perpendicular, and *bc* parallel to the back of the wedge, and let *de* also be resolved into *df* perpendicular and *ef* parallel to the side AC; and since the side AC is perfectly smooth, *df* only is effectual to stop the progress of the wedge; resolve *df* again into *dg* parallel, and *gf* perpendicular to the back of the wedge, then *fg* is the only part of the resistance R which is directly opposed to that part of the power (viz. *ac*) which acts perpendicularly to the back of the wedge. Let the resistance R′ be resolved in the same manner, and let *mn* be that part of it which is directly opposed to *ac*; then, supposing every part of the wedge to be perfectly hard and inflexible, it is evident that in the case of an equilibrium between P and R+R′, $ac=fg+mn$.

But in ascertaining this relation between P and R+R′ when they are in equilibrio, it should be recollected that the point *c* cannot be arbitrarily assumed; for there evidently will be a tendency to *vibratory* or *rotary* motion, unless the two resistances balance themselves about that point. To determine the point *c* so that this tendency to vibratory motion shall be prevented, produce *gf*, *nm*, (Fig. 100,) to *k* and *o*, then it is evident that the resistances *fg*, *mn* will only balance themselves about *c*, when, according to Theorem II, of parallel motion, (Art. 60,) $fg : mn :: co : ck$; that is, when the effective parts of the resist-

Fig. 100.

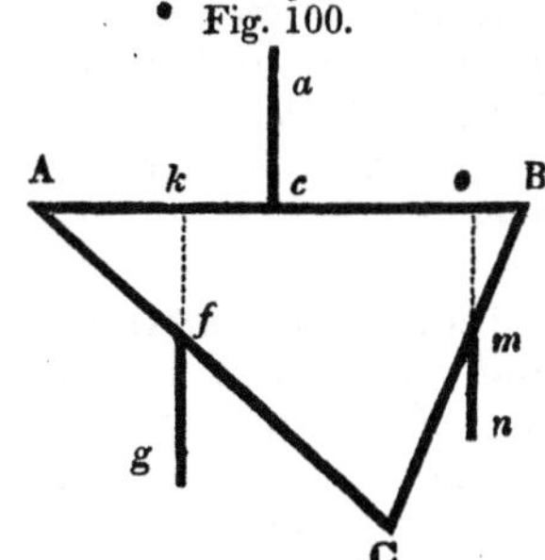

ances are to each other inversely as their respective distances from the line of direction in which the resultant of the power acts.

145. But for the purpose of computing the actual power of the wedge in particular cases, we must find an expression for the power in known terms, arising from the shape of the wedge, and the conditions of the resistances. Therefore produce *dg* to *h*, (Fig. 99,) and let a perpendicular CD, be drawn from the vertical angle to the back of the wedge. Let α, β denote the two vertical angles, π the angle made by the power with the back of the wedge, and ϱ and ϱ' the respective angles made by the resistances R, R′ with the sides. Then, since in the right-angled triangles *dfh*, CAD, the alternate angles *dhf*, CAD, are equal, the angle *fdg* must be equal to the angle ACD (α.) Now

$$de(\mathrm{R}) : df :: \mathrm{rad.} : \sin.\ def, \text{ or } \sin.\ \varrho,$$
$$\text{and } df : fg :: \mathrm{rad.} : \sin.\ fdg, \text{ or } \sin.\ \alpha,$$

$$\therefore \mathrm{R} : fg :: \overline{\mathrm{rad.}}|^2 : \sin.\ \varrho \times \sin.\ \alpha, \text{ or } fg = \frac{\mathrm{R} \times \sin.\ \varrho \times \sin.\ \alpha}{\overline{\mathrm{rad.}}|^2}.$$

For the same reason, $$mn = \frac{\mathrm{R}' \times \sin.\ \varrho' \times \sin.\ \beta}{\overline{\mathrm{rad.}}|^2}.$$

Again, $ab(\mathrm{P}) : ac :: \mathrm{rad.} : \sin.\ \pi, \therefore ac = \dfrac{\mathrm{P} \times \sin.\ \pi}{\mathrm{rad.}}$.

But when there is an equilibrium, $ac = fg + mn$; hence

$$\frac{\mathrm{P} \times \sin.\ \pi}{\mathrm{rad.}} = \frac{\mathrm{R} \times \sin.\ \varrho \times \sin.\ \alpha + \mathrm{R}' \times \sin.\ \varrho' \times \sin.\ \beta}{\overline{\mathrm{rad.}}|^2},$$

$$\text{or } \mathrm{P} = \frac{\mathrm{R} \times \sin.\ \varrho \times \sin.\ \alpha + \mathrm{R}' \times \sin.\ \varrho' \times \sin.\ \beta}{\mathrm{rad.} \times \sin.\ \pi}.$$

This formula enables us to compute the power of a wedge, (or its ratio to the weight,) when we have given the angles made by a perpendicular drawn from the vertical angle to the back of the wedge, and likewise the angles at which the power and resistances respectively act, whatever these angles may be: but when (as is frequently the case) the power acts perpendicularly to the back, and the resistances perpendicularly to the sides, the formula becomes much simpler. For then, sin. π, sin. ϱ, and sin. ϱ', each becomes equal to radius, and the general formula becomes $\mathrm{P} = \mathrm{R} \times \sin.\ \alpha + \mathrm{R}' \times \sin.\ \beta$.

146. In the particular case when the directions of the power and resistances meet in the same point, as in Fig. 101, *the power is to the weight as the back of the wedge to the sum of the sides;* for then it is evident, that this equilibrium is produced under the same circumstances as that of a body kept *at rest* by three forces whose directions meet in that point; but (by Art. 57)

those three forces would be to each other as the three sides of a triangle perpendicular to their respective directions; P, R and R′, will therefore be to each other as the three sides of the triangle ABC; i. e.

Fig. 101.

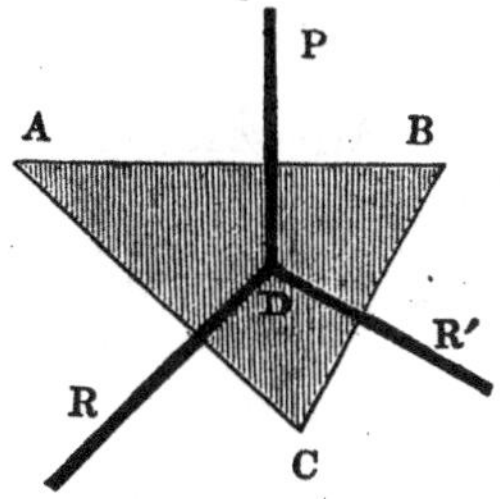

$P:R::AB:AC$; $P:R'::AB:BC$

$\therefore P:R+R'::AB:AC+BC$.*

If the wedge is isosceles, and two equal resistances act perpendicularly to the sides, while the power acts perpendicularly to the back, then $P:W::\frac{1}{2}AB:AC$;

For, $P:(2R)\ W::AB:2AC$

$\therefore P:W::\frac{1}{2}AB:AC$; that is,

If the wedge is *isosceles, the power is to the weight as half the back of the wedge is to one of its sides.*

Hence, the thinner the wedge, that is, the longer the sides of the wedge in proportion to its thickness, the greater is its power of overcoming resistance, with a given blow; a well known property of sharp instruments, which are referred to the wedge

147. Examples.

1. A wedge, whose sides are perfectly polished planes, is driven into the trunk of a tree, until that part of the pressure on each side, which acts perpendicularly to the back, is 1500 pounds: What force must be applied at the back to prevent its recoil?

Here $P=R+R'=2R=3000$ pounds.

2. Two equal resistances (R, R′) acting at angles of 60° and 30° upon the sides of a perfectly smooth wedge, are kept in equilibrio by a power acting perpendicularly to its back; the angle which a perpendicular (from the vertical angle of the wedge to the back) makes with the side upon which R acts is 45°, and with that upon which R′ acts 30°: Required the ratio of $P:R+R'$ or 2R.

This is a case of Art. 145, where $R'=R$, sin. π=rad.$=1$, $\therefore P=R(\sin.\varrho\times\sin.\alpha+\sin.\varrho'\times\sin.\beta;)$ but,

$$\left.\begin{array}{l}\sin.\varrho=\sin.60^\circ=\frac{\sqrt{3}}{2}\\ \sin.\alpha=\sin.45^\circ=\frac{1}{\sqrt{2}}\\ \sin.\varrho'=\sin.30^\circ=\frac{1}{2}\\ \sin.\beta=\sin.30^\circ=\frac{1}{2}\end{array}\right\}\begin{array}{l}\therefore P=R\left(\frac{\sqrt{3}}{2\sqrt{2}}+\frac{1}{4},\right)\\ =\frac{R(\sqrt{6}+1)}{4};\text{ hence}\\ P:R::\sqrt{6}+1:4,\text{ and }P:2R::\sqrt{6}+1:8.\end{array}$$

* Algebra, Art. 388.

148. Questions on the Principles of the Weige.

1. An isosceles wedge, whose acute angle was 5°, was inserted in a cleft of a rock. The pressure exerted perpendicularly on each side was equal to 6500 pounds: What force applied at the back of the wedge would overcome the resistance?

Ans. 567.33 *pounds.*

2. In a wedge of the form of a triangular prism, the power and resistances act perpendicularly to the back. The pressures upon the two sides of the wedge, whose lengths are 5 and 7 inches, are estimated at 950 and 300 pounds respectively, acting at the center of each side. The length of the back of the wedge is 5 inches: What power must be exerted perpendicularly to the back of the wedge, to produce an equilibrium?

Ans. P=1250 *pounds.*

3. A force of 1600 pounds, acting at an angle of 75° upon the back of a wedge, is just sufficient to balance two resistances, which are to one another as 5 : 7, and whose directions make angles with their respective sides of 80° and 40°. The angle which a perpendicular from the vertical angle upon the back, makes with the side, upon which is exerted the greater resistance, is 10°, and that which it makes with the other side is 4°: What is the amount of the resistances? *Ans.* 16487.76 *pounds.*

General Principle, applicable to all the Mechanical Powers.

149. The most simple view which can be taken of the mechanical powers, is by a comparison of the respective VELOCITIES OF THE POWER AND THE WEIGHT. In order clearly to understand this subject, it must be recollected,

That an equilibrium implies the action of opposite and equal FORCES;

That the measure of a force is its MOMENTUM, *and, consequently, that in an equilibrium, the momenta on the opposite sides are equal*,

That momentum is compounded of the quantity of matter and velocity; and hence, that a small body may have as much momentum as a large one, if it moves over as much greater space in the same time, as its quantity of matter is less.

Now let us apply the foregoing principles to the mechanical powers.

When the power and weight are in equilibrio, one has just as much momentum as the other; and therefore the product of the weight into its velocity, equals the product of the power into its velocity; and it will be seen by reviewing the several mechanical powers, that in the theorem by which the law of equilibrium is in each case enunciated, the line into which the power or the weight is multiplied, is proportional to the space over which it moves in a given time, and therefore (by Art. 12) is the measure of its velocity. This doctrine will be clearly comprehended by reviewing each of the mechanical powers separately.*

150. Suppose P and W (Fig. 101′,) to vibrate in equilibrium on the end of a *straight lever*, PCW: they will describe similar arcs Pp, Ww, which will be the measures of their respective velocities; or, V : v :: Pp : Ww :: PC : WC :: W : P, ∴ V : v :: W : P.

Fig. 101′.

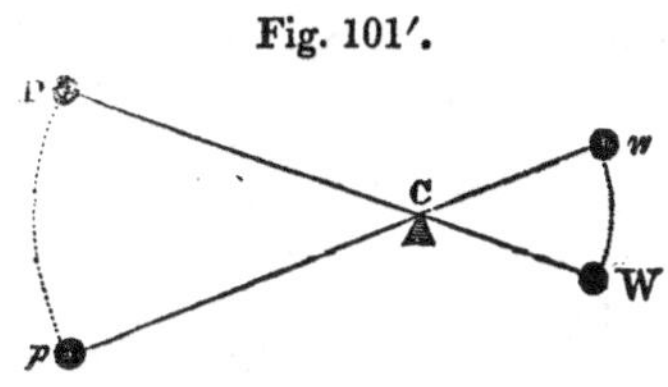

In the *wheel and axle*, while the power descends through a space equal to the circumference of the wheel, the weight ascends through a space equal to the circumference of the axle, ∴ the velocity of the power : the velocity of the weight :: the circumference of the wheel : the circumference of the axle :: radius of the wheel : radius of the axle :: W : P.

In the single *fixed pulley*, the power and weight move through equal spaces in the same time, ∴ the velocity of the power is equal to the velocity of the weight; and in this machine P=W, ∴ velocity of power : velocity of weight :: W : P.

In the system of n pulleys, (Fig. 85,) while the power descends through any space (x,) each of the strings belonging to the block of pulleys to which the weight is appended is shortened by $\frac{1}{n}x$, ∴ the weight ascends through a space equal to $\frac{1}{n}x$ in the same time that P descends through the space x; hence the velocity of the power : the velocity of the weight :: $x : \frac{x}{n} :: n : 1$ :: W : P.

In the system of pulleys (Fig. 86,) while P descends through any space (x,) the pulley F is raised through a space$=\frac{1}{2}x$; the pulley D through a space$=\frac{1}{2}\times\frac{1}{2}x=\frac{1}{4}x$; the pulley B through a space$=\frac{1}{2}\times\frac{1}{4}x=\frac{1}{8}x$; hence (if n=the number of pulleys) velocity of P : velocity of W :: $x : (\frac{1}{2})^n \times x :: 2^n : 1$ :: W : P.

Let ABC (Fig. 102,) be an *inclined plane*, up which the weight W is drawn by a power P acting over a pulley at A, then P : W :: AB : AC. Draw Wb parallel to CB, and ab parallel to AB; then while P descends through a space equal to Wa, W ascends upon the plane through the same space; but the space actually described by W in this time in the direction of gravity, is ba; ∴ (when the velocity of

Fig. 102.

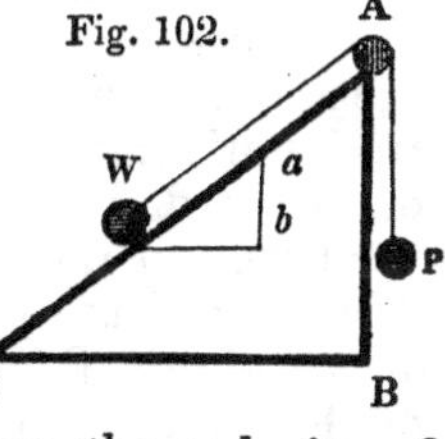

* In estimating the increase of the power necessary to put the machine in motion, the friction of its different parts should be taken into the account. If p=the sum of the impediments arising from friction, and P=the power which would keep the weight (W) in equilibrio, then it is evident, that, before the machine can be put in motion, the power actually employed must exceed P+p.

the power and weight are estimated in the direction in which they respectively act)

velocity of P : velocity of W :: Wa : ab :: AC : AB :: W : P.

In the *screw*, while P describes the circumference of a circle whose radius is BC, (Fig. 96,) the weight is elevated or depressed through a space equal to the distance between two contiguous spirals (d,) ∴ the velocity of P : the velocity of W :: circumference (DBE,) : d :: W : P.

Finally, in the *wedge*, the power of overcoming the resistance is proportioned (Art. 146) to the acuteness of the wedge; and the distance to which the parts are separated, that is, the space over which the weight moves, when compared with the space through which the power (namely, the wedge itself in the direction of the power) moves, is constantly diminished as the acuteness of the wedge is increased.

In the same manner we might trace this relation between the velocity of the power and the velocity of the weight, in machines where the power and weight act obliquely to each other; but as the operation then becomes somewhat more intricate, and as what has been already shown is sufficient to illustrate the truth of our proposition, it seems unnecessary to pursue the investigation any further.

THE ROPE MACHINE.

151. *If a body fixed to two or more ropes, is sustained by powers which act by means of those ropes, this assemblage is called the Funicular or Rope Machine.*

152. *A given weight is in equilibrio with two given powers, which are equal to one another, and which pass over pulleys situated in the same horizontal line, when either power is to the weight as the sine of the angle formed by the directions of the power and the weight, is to the sine of the angle formed by the directions of the two powers.*

Let A and B (Fig. 103,) be two pulleys fixed at a given distance from each other in the same horizontal line, and suppose a cord, PAWBP′, to pass over them, at the extremities of which are suspended the two equal weights P, P′; these two weights being kept in equilibrio by a third weight W.* Draw AE parallel to CB, and produce WC to meet it in E; then the three sides CA, AE, EC, of the tri-

Fig. 103.

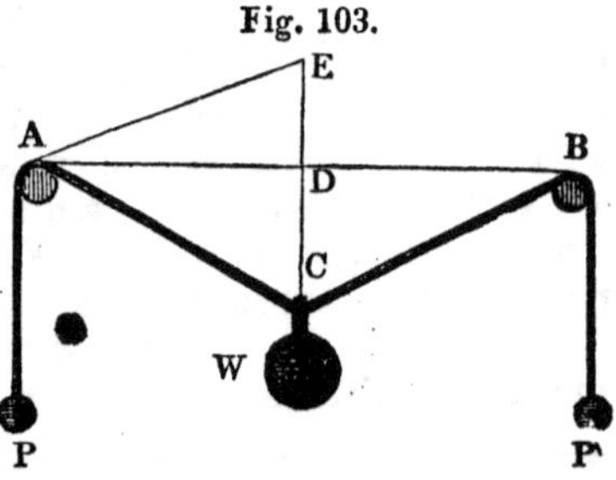

* In this and the following articles we suppose the cords to be without weight, and perfectly inextensible; and the pulleys to be so small, and so adjusted, that the center of gravity of P, P′, W, &c. may all lie in the same vertical plane.

angle ACE, will represent the quantity and direction of the three forces by which the weight W is kept at rest; for these three forces are, 1st, the tension produced by P in direction CA; 2dly, the tension produced by P′ in direction CB or AE; and 3dly, its own gravity in the direction EC; and since P′=P, AE or CB must be equal to AC. Hence

P or P′ : W :: AC or CB : EC :: sin. AEC : sin. ACB.

153. Join AB; then, since AB is parallel and CE perpendicular to the horizon, the angles at D are right angles, and since ACE and ACB are isosceles triangles, the lines AB, CE, bisect each other in D. Hence EC=2CD, and P : W :: AC : (EC or) 2CD; ∴ 2P : W :: 2AC : 2CD :: AC : CD, and $4P^2 : W^2 :: AC^2 : CD^2$; also $4P^2 - W^2 : W^2 :: AC^2 - CD^2\ (AD^2) : CD^2\ \therefore CD^2 = \frac{W^2 \times AD^2}{4P^2 - W^2}$, or $CD = \frac{W \times AD}{\sqrt{4P^2 - W^2}}$; which gives the position of W when the equilibrium takes place, for P, W, and AD ($\frac{1}{2}$AB) are known quantities.

Cor. 1. If $P = \frac{1}{2}W$, or $2P = W$, then $4P^2 = W^2$, and $4P^2 - W^2 = 0$; if W be greater than 2P, then $4P^2 - W^2$ is negative; in the former case, therefore, the value of CD becomes infinite, and in the latter impossible; which shows that if W be equal to or greater than 2P, no equilibrium can take place.

Cor. 2. If W and AD be given, CD will vary inversely as $\sqrt{4P^2 - W^2}$; and if P be indefinitely increased, W^2 may be neglected, and CD will vary inversely, as $\sqrt{4P^2} = 2P$; that is, CD will diminish as the sum of the forces is increased, but can never become nothing until the sum of the forces becomes infinite. Suppose, for example, the line ACB to be a rope whose weight is equal to W; then *it would require an infinite force to draw it into a straight line, by powers applied at its extremities.*

CHAPTER IX.

OF THE MOTION OF BODIES UPON INCLINED PLANES; AND THE DOCTRINE OF THE PENDULUM.

154. In the second chapter we investigated the relation which takes place between the space described, the velocity acquired, and the time of its motion, when a body ascends or descends perpendicularly near the earth's surface. The object of this chapter is, to ascertain that relation when the bodies ascend or descend upon planes inclined to the horizon.

MOTION ON INCLINED PLANES.

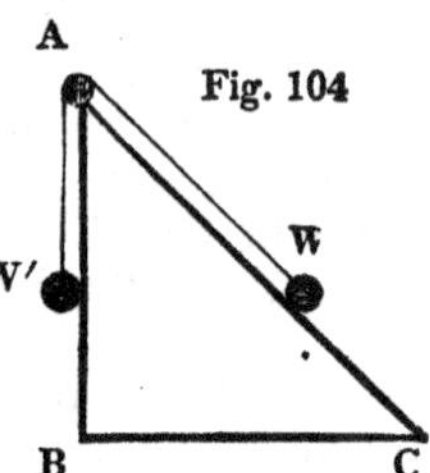

154. In Art. 130, it was shown, that if a weight W be sustained upon an inclined plane AC, (Fig. 104,) by another weight W′ acting upon it in a different direction parallel to the plane, then W′ : W :: AB : AC. Suppose now the string WAW′ to be cut in two, then it is evident that the weight W would descend down the plane with a force which bears to its own weight the ratio of AB : AC; and since (by Art. 134) this force is constant through every part of the plane, the body thus descending may be considered as acted upon in every point of its descent by a constant force, which bears to the force of gravity the given ratio of the height of the plane H to the length of the plane L; call this force F, and let the force of gravity be represented by unity, then $F : 1 :: H : L$, and $F = \frac{H}{L}$;* i. e. the force which accelerates the motion of a body down an inclined plane is such a part of the force of gravity as may be represented by the fraction $\frac{H}{L}$; this force, therefore, differs not from the force of gravity in kind, but in degree; the effects produced by it must consequently be analogous to the effects produced by gravity.

156. In order to estimate these effects, we have only to consider, that if a body be acted upon by different constant forces for the same time, the velocity generated will evidently be proportional to the intensity of those forces; and that if it be acted upon by the same force for different times, the velocity will be proportional to the time for which the forces act; from which it follows, that if a body be acted upon by different constant forces for different times, the whole velocity generated will be as the force and time conjointly.† Suppose now that the force of gravity is represented by unity, that $m = 16\frac{1}{12}$ feet, and that V = the velocity acquired in the time T while a body describes the space S acted upon by some other constant force F, then, from what has just been shown, V : the velocity acquired by gravity in $1'' = 2m$ (Art. 34) :: $F \times T : 1 \times 1$, $\therefore V = 2mFT$, and $T = \frac{V}{2mF}$. Again, since $V \propto F \times T$, $T \times V$ must vary as $F \times T^2$;‡ but $S \propto T \times V$ (Art. 29,) $\therefore S \propto F \times T^2$; hence S : the space described by gravity

* Since $F : 1 :: AB : AC :: \sin. ACB : \text{rad.}$, $\therefore F = \frac{\sin. ACB}{\text{rad.}} \propto \sin. ACB$; therefore, the force which accelerates a body down an inclined plane varies as the sine of the angle of the plane's elevation.

† Algebra, 420.

‡ Algebra, 413

in 1″ (m) :: $F\times T^2 : 1\times 1^2$, ∴ $S=mFT^2$, and $T=\left(\frac{S}{mF}\right)^{\frac{1}{2}}$. Lastly, since $V \propto F\times T$, V^2 must vary as $F\times T\times V$, or as $F\times S$,* ∴ V^2 : square of the velocity acquired by gravity in 1″ ($4m^2$) :: $F\times S : 1\times$ space described by gravity in 1″ :: $F\times S : 1\times m$; hence $V^2=4mFS$, $V=2\sqrt{mFS}$, and $S=\frac{V^2}{4mF}$.

157. Let us now apply these expressions to the case before us, i. e. let a body descend down an inclined plane whose height is (H) and length (L ;) then $F=\frac{H}{L}$ (Art. 155 ;) and if the body descends from rest,

Space (S) described in time $T=\frac{H}{L}\times mT^2$, and $T=\left(\frac{L\times S}{mH}\right)^{\frac{1}{2}}$

Velocity (V) acquired in time $T=\frac{H}{L}\times 2mT$, and $T=\frac{L\times V}{2mH}$.

V^y acquired through space (S)$=2\left(\frac{H}{L}\times mS\right)^{\frac{1}{2}}$, and $S=\frac{L\times V^2}{4mH}$

158. Since $S=\frac{H}{L}\times mT^2$, and H, L, m are given, ∴ $S \propto T^2$, that is, the space described varies as the square of the time when a body falls from rest down an inclined plane, as well as when it descends freely by the force of gravity; the spaces described from rest in equal successive portions of time will therefore be as the odd numbers 1, 3, 5, 7, &c.; and if the body be projected upward with the velocity acquired in falling through any space upon the plane, it will ascend to the point from which it fell, the spaces described in equal successive portions of time being as the numbers, 1, 3, 5, 7, &c. taken in the inverted order. If, moreover, at any point of its descent, it moves forward with the velocity acquired continued uniformly, it will describe twice the space in the same time as that in which it has fallen to acquire the velocity; and if it be projected downward or upward with the velocity (V,) and moves for the time (T,) the space described in that time will be equal to $T\times V\pm\frac{H}{L}\times mT^2$. All this follows from the law of acceleration and retardation of bodies moving upon inclined planes, being the same as that which regulates the motion of bodies descending or ascending freely by the force of gravity. (See Arts. 31, 32, and 33.)

---

* Algebra, 418, cor.

159. The expressions contained in Art. 157, apply to the case of a body descending from rest through any part of an inclined plane whose height is (H) and length (L.) If the body falls through the whole length, then S=L, $\therefore$ the velocity acquired in falling down the whole length of the plane $=2\left(\frac{H}{L}\times m S\right)^{\frac{1}{2}}=2\sqrt{mH}$=(by Art. 34) the velocity acquired by descending freely through the height H; $T=\left(\frac{L\times S}{mH}\right)^{\frac{1}{2}}=\frac{L}{\sqrt{mH}}$; but the time of falling freely down $H=\left(\frac{H}{m}\right)^{\frac{1}{2}}$, $\therefore$ the time of describing the whole length of the plane : the time of falling freely down its height $::\frac{L}{\sqrt{mH}}:\left(\frac{H}{m}\right)^{\frac{1}{2}}::L:H::$ length of the plane : height of the plane.

Since $T=\frac{L}{\sqrt{mH}}$, $V=2\sqrt{mH}$, and m is a given quantity, T varies as $\frac{L}{\sqrt{H}}$, and V as $\sqrt{H}$; i. e. the time of describing any inclined plane varies as its length directly, and as the square root of its height inversely; and the velocity acquired varies as the square root of the height, whatever is the length of the plane.

Hence we deduce the following THEOREMS.

I. *The* VELOCITY *acquired in falling down an inclined plane, is the same as that acquired by falling freely through the perpendicular height of the plane.*

II. *The* TIME *of describing the whole length of an inclined plane is to the time of falling freely through its height, as the length of the plane to its height.*

III. *The time of describing any inclined plane* VARIES AS *its length directly, and as the square root of its height inversely.*

IV. *The velocity acquired in describing any inclined plane* VARIES AS *the square root of its height, whatever be the length of the plane.**

160. EXAMPLES.

1. How far will a body descend from rest in 4″, upon an inclined plane whose length is 400 feet, and height 300 feet?

Ans. 193 feet.

* The expressions deduced in this section are true only when the body slides down a perfectly smooth plane; for in this case it is evident that every particle of the body is equally accelerated, and therefore whatever is proved of any one point of it will apply equally to all; but if the body in its fall has a rotary motion communicated to it, then it is evident that all the points of it will not be equally accelerated.

2. How long would a body be in falling down 100 feet of a plane, whose length is 150 feet and height 60 feet?

Ans. 3.94 seconds.

3. The length of an inclined plane is 60 feet, and its elevation 30°: What velocity would a body acquire in falling from rest down it for 2″? Ans. $32\frac{1}{6}$ feet in 1″.

4. The height of a plane : length of a plane :: 7 : 15: How long would a body be in falling down it, to acquire a velocity of 20 feet per second? Ans. 1.33 seconds.

5. H : L :: 5 : 14: What space must a body fall through, to acquire a velocity of 10 feet per second? Ans. 4.35 feet.

6. H : L :: 25 : 90: What velocity would a body acquire in falling down 70 feet? Ans. 35.37 feet in 1″.

7. The length of an inclined plane is 100 feet, and its elevation 60°: How long would a body be in falling down it, and what velocity would it acquire at the end of its fall?

Ans. T=2.68 seconds; V=74.64 feet in 1″.

8. A body is projected up an inclined plane whose length is 10 times its height, with a velocity of 30 feet in 1″: In what time will its velocity be destroyed?

Ans. The time in which a body would fall down an inclined plane of this elevation to acquire a velocity of 30 feet per second=9.32″.

9. A body is projected up an inclined plane, whose height is $\frac{1}{6}$ of its length, with a velocity of 50 feet per second: Find its place, and velocity after 6″ are elapsed?

Ans. S=$203\frac{1}{2}$ from the bottom of the plane; V=$17\frac{5}{6}$ feet in 1″.

10. A body falls from rest down the inclined plane AC (Fig. 105:) Compare the times of describing the first and last halves of it.

Bisect AC in D, and draw DE parallel to CB; by Art. 159, the time down AC : the time down AD :: $\frac{AC}{\sqrt{AB}} : \frac{AD}{\sqrt{AE}}$ :: (by similar triangles) $\frac{AC}{\sqrt{AC}} : \frac{AD}{\sqrt{AD}} :: \sqrt{AC} : \sqrt{AD} :: \sqrt{2} : 1$. Hence,

Fig. 105.

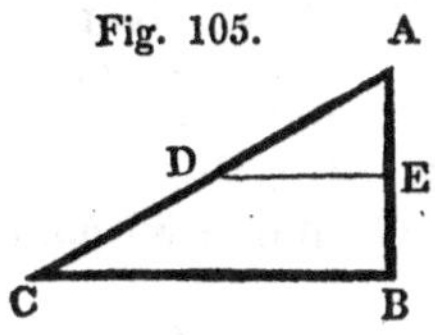

Fig. 106.

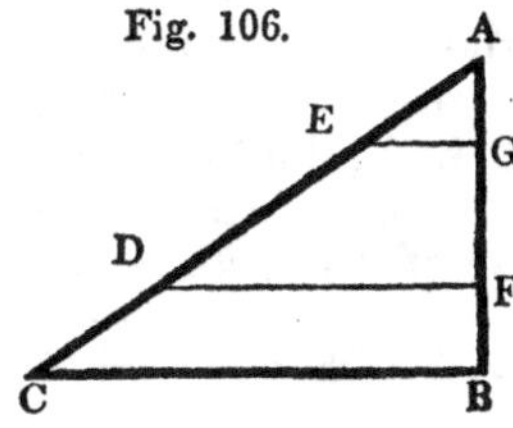

17

(1.) The times down different parts of the same inclined plane (when the body falls from rest from the top of the plane) are to each other as the square roots of the lengths of those parts.

(2.) The time down AC—the time down AD (i. e. the time down DC) : the time down AD :: $\sqrt{2}-1:1$.

11. To mark out upon a plane AC (Fig. 106,) a part ED which shall be equal to the height AB, and which a body (falling down AC) would describe in the same time as one falling freely through AB.

$$\left.\begin{array}{r}\text{Let AC}=a\\ \text{AB or ED}=b\\ \text{AE}=x\\ \text{then AD}=b+x\end{array}\right\}\quad \begin{array}{l} TAD^{*}:TAE::\sqrt{\text{AD}}:\sqrt{\text{AE}}\\ \qquad ::\sqrt{b+x}:\sqrt{x},\end{array}$$

$$\therefore TAD-TAE \text{ (i. e. } TDE):TAE::\sqrt{b+x}-\sqrt{x}:\sqrt{x},$$

but $$TAE:TAC::\sqrt{\text{AE}}:\sqrt{\text{AC}}::\sqrt{x}:\sqrt{a},$$

$$\text{and by Art. 159, } TAC:TAB::a:b::\sqrt{a}:\frac{b}{\sqrt{a}};$$

$$\dagger\ TDE:TAB::\sqrt{b+x}-\sqrt{x}:\frac{b}{\sqrt{a}}.$$

$$\text{But } TDE=TAB,\ \therefore\ \sqrt{b+x}-\sqrt{x}=\frac{b}{\sqrt{a}};$$

$$\text{which equation solved gives } x=\frac{\overline{a-b}|^2}{4a} \text{ or AE}=\frac{\overline{\text{AC}-\text{AB}}|^2}{4\text{AC}}.$$

161. Questions on the Inclined Plane.

1. The length of an inclined plane is 480 feet, and the height 210; a body falls from rest from the top of the plane: What space will it have fallen through in 6″; what time will it be in falling through 450⅓ feet; and what velocity will it have acquired, when it has arrived within 124.7 feet of the bottom of the plane?

Ans. S=253$\frac{5}{16}$ *feet;* T=8 *seconds;* V=100 *feet in* 1″.

2. A body has been falling for 15″ down an inclined plane whose length is 2½ times its height: What velocity will it have acquired at the end of its fall? *Ans.* V=193 *feet in* 1″.

3. The elevation of a plane is 30°; a body in falling from the top to the bottom of it, acquires a velocity of 579 feet in 1″: What is the length of the plane? *Ans.* L=10422 *feet.*

4. A car broke loose from the top of an iron railway eleven miles in length, which was uniformly inclined to the horizon at an angle of 1 degree. Supposing the car to move without re-

* *TAD* means the time down AD, and so of the rest. † Alg. 393.

sistance, in what time would it reach the lower end of the railway, and what velocity would it acquire?

Ans. T=7′ 34″.88; V=255.36 *feet per second.*

5. At Alpnach, in Switzerland, is a celebrated slide for conveying timber trees from Mount Pilatus to Lake Luzerne, whence they are transported down the Rhine. The slide consists of an inclined plane formed of logs in the shape of a trough, into which the trees are launched, and down which they descend by the force of gravity. It is 8 miles in length, and is inclined to the horizon on an average, at an angle of 3° 14′: In what time will a tree descend from the top to the bottom of this plane, no allowance being made for friction? *Ans.* 3′ 35″.8

6. Trees descending the slide sometimes "bolt out" of the trough, and occasion great destruction: With what force would a tree weighing 1500 pounds, leaping out of the slide at the end of 7 miles, strike upon an obstacle, as for example, a standing tree? *Ans. With a force equal to* 549318.2 *lbs.*

7. Two inclined planes have a common height of 75 feet; the elevation of one of them is 50°, and of the other 20°: With what velocity must a body be projected from the bottom of the former, that it may just rise to the top of the latter; and what will be the whole time of its ascending and descending through the two planes?* *Ans.* V=69.46 *feet in* 1″; T=9.133 *seconds.*

8. How long will a body be in falling down the last half of a plane, whose height is 1 mile, and angle of elevation 1 minute?

Ans. T=5 *hours,* 4 *minutes,* 3 *seconds.*

9. The length of a plane is 250 feet, and height 150: Mark out upon it a part equal to the height which a body in falling down it describes, while another body would descend freely through the height.

Ans. It begins to describe this part when it has fallen through 10 *feet from the top of the plane.*

MOTION OF BODIES DOWN DIFFERENT SYSTEMS OF INCLINED PLANES.

162. It has already been shown, that when a body descends down an inclined plane whose length is L, and height H, the velocity acquired varies as $\sqrt{H}$, and the time of description as $\frac{L}{\sqrt{H}}$; let us now apply these expressions to finding the relation between the times and velocities of bodies falling down different systems of inclined planes.

163. Let AC, AD, AE, (Fig. 107,) &c., be a system of inclined planes having the same height, AB; then since the velocities ac-

* The planes are placed as in Fig. 92; a body is projected from D with a velocity just sufficient to carry it to A, and then falls from rest down the plane AC.

quired by bodies falling down these planes are as $\sqrt{AB}$, and the times of description as $\frac{AC}{\sqrt{AB}}, \frac{AD}{\sqrt{AB}}, \frac{AE}{\sqrt{AB}}$, &c. (i. e. as AC, AD AE, &c.) it is evident that bodies falling down a system of planes of this kind would acquire at the end of their fall the same velocity,* and that the times of description would be as their respective lengths.

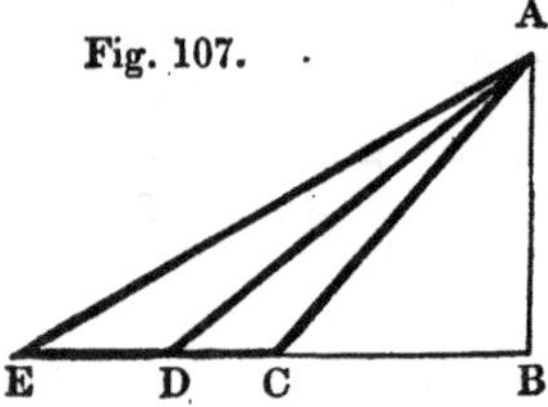

Fig. 107.

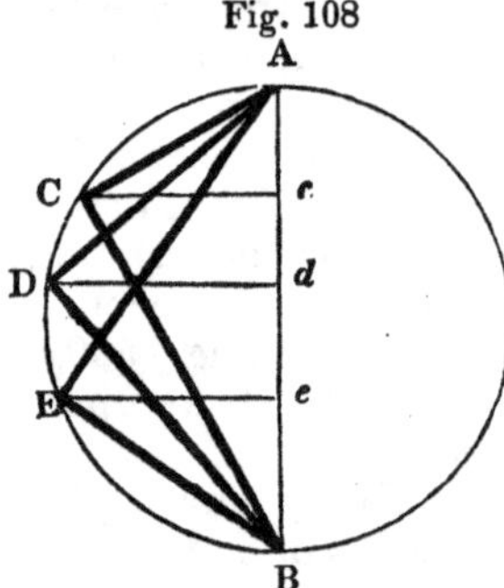

Fig. 108

164. *If chords be drawn in a circle from the extremity of that diameter which is perpendicular to the horizon, the* VELOCITIES *which bodies acquire in falling through them, are proportional to their lengths ; and the* TIMES *of describing these chords are all equal to one another, and are severally equal to the time of describing the diameter.*

Let the diameter AB of the circle ACB (Fig. 108,) be perpendicular to the horizon ; draw the chords AC, AD, AE, and CB, DB, EB ; draw also C*c*, D*d*, E*e*, &c., parallel to the horizon ; then the velocities acquired by bodies falling down the former system of chords are as $\sqrt{Ac}$, $\sqrt{Ad}$, $\sqrt{Ae}$, and down the latter as $\sqrt{cB}$, $\sqrt{dB}$, $\sqrt{eB}$, but $Ac=\frac{AC^2}{AB}$,† $Ad=\frac{AD^2}{AB}$; and $cB=\frac{CB^2}{AB}$, $dB=\frac{DB^2}{AB}$; hence $\sqrt{Ac}$, $\sqrt{Ad}$, $\sqrt{Ae}$, vary as AC, AD, AE, and $\sqrt{cB}$, $\sqrt{dB}$, $\sqrt{eB}$, vary as CB, DB, EB. The times of describing AC, AD, AE, and CB, DB, EB, are as $\frac{AC}{\sqrt{Ac}}, \frac{AD}{\sqrt{Ad}}, \frac{AE}{\sqrt{Ae}}$, and $\frac{CB}{\sqrt{cB}}$ $\frac{DB}{\sqrt{dB}}, \frac{EB}{\sqrt{eB}}$; but each of these quantities is equal to $\sqrt{AB}$.‡

* And by Art. 159, this velocity is equal to the velocity which a body would acquire by falling freely through the height (AB) of the plane.

† For (Euc. 8. 6.) $Ac : AC :: AC : AB$, or $Ac=\frac{AC^2}{AB}$.

‡ For since $Ac : AC :: AC : AB$, $\frac{AC^2}{Ac}=AB$, $\therefore \frac{AC}{\sqrt{Ac}}=\sqrt{AB}$; and so of the rest

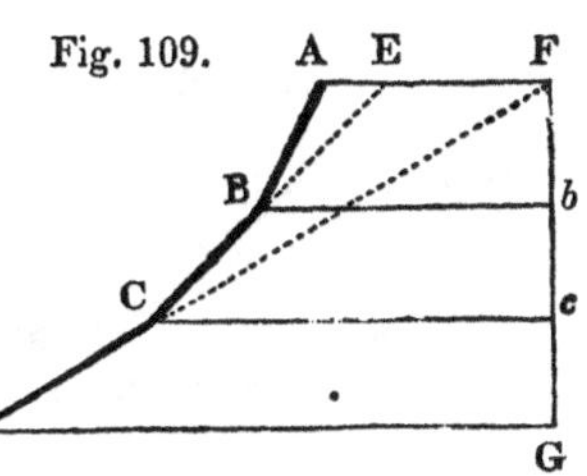

Fig. 109.

165. Suppose now that a body falls down a system of planes AB, BC, CD, (Fig. 109,) inclined to each other, and that no velocity is lost in falling from one plane to the other. Draw AF parallel to the horizon, and produce DC, CB to meet it in F, E; through B, C, draw B*b*, C*c*, parallel to AF, and let fall FG at right angles to the horizon. By Art. 163, the velocity down AB=velocity down EB=velocity down F*b*; ∴ the velocity down AB+BC=velocity down EB+BC (or EC)=velocity down FC=velocity down F*c*; and reasoning in the same manner, the velocity down AB+BC+CD=velocity down FD=velocity down FG; i. e. the whole velocity acquired by a body falling down successive planes, is equal to the velocity which a body would acquire in falling freely through their joint height FG.

Fig. 110.

166. If the number of planes be increased infinitely, the figure will become a *curve*, as in Fig. 110; and hence the velocity acquired in descending through any perfectly smooth curved surface, is the same as that acquired by falling through the perpendicular height of the curve.

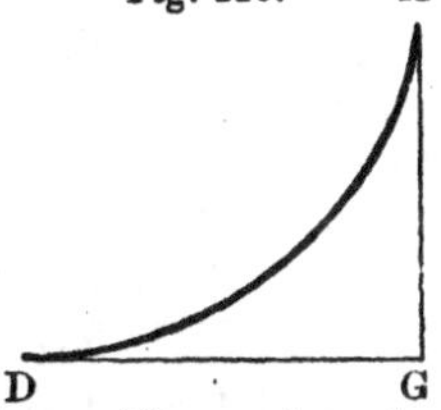

Hence, in general, *a body, by descending from a certain height to the same horizontal line, will acquire the same velocity, whether the descent be made perpendicularly or obliquely, over an inclined plane, or over many successive inclined planes, or over a curve surface.**

167. *The times and velocities of bodies falling down planes similarly inclined to the horizon, are to each other both as the square roots of the lengths, and as the square roots of the heights of the planes*

Fig. 111.

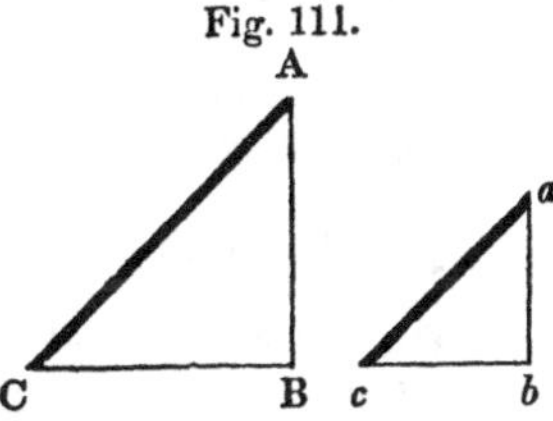

Let AC, *ac*, (Fig. 111,) be two planes similarly inclined to the horizon, then AC : *ac* :: AB : *ab*, and $\sqrt{AC} : \sqrt{ac} :: \sqrt{AB} : \sqrt{ab}$.

Now the time down AC : the time down *ac* :: $\frac{AC}{\sqrt{AB}} : \frac{ac}{\sqrt{ab}}$ ::

* Cavallo.

$\frac{AC}{\sqrt{AC}} : \frac{ac}{\sqrt{ac}} :: \sqrt{AC} : \sqrt{ac}$, or $\sqrt{AB} : \sqrt{ab}$; and the velocity acquired in falling down AC : the velocity down ac :: $\sqrt{AB} : \sqrt{ab}$, or $\sqrt{AC} : \sqrt{ac}$.

168. *The times of descent down* SIMILAR SYSTEMS *of inclined planes are as the square roots of the lengths of the planes.**

Fig. 112.

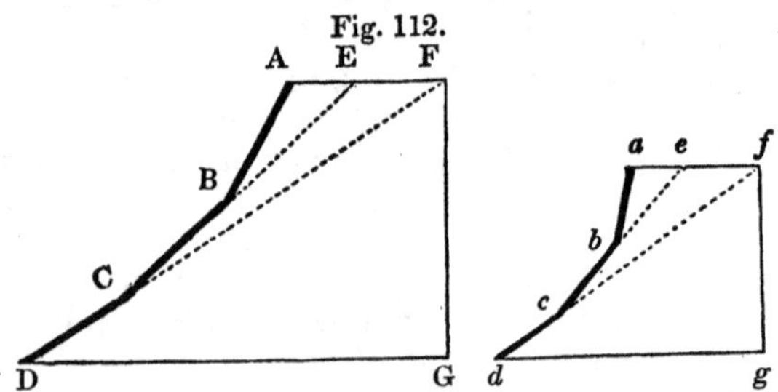

Let there be two systems of planes, AB, BC, CD, *ab*, *bc*, *cd*, (Fig. 112,) similar, and similarly situated with respect to the horizon, and complete the figures as in Art. 164; then, supposing no velocity to be lost in passing from one plane to the other, and using the notation employed on page 130, we have, by Art. 167,

$$T\textit{AB} : T\textit{ab} :: \sqrt{AB} : \sqrt{ab}. \qquad (X.)$$

$$T\textit{EC} : T\textit{ec} :: \sqrt{EC} : \sqrt{ec} :: \sqrt{AB} : \sqrt{ab},\dagger$$

$$T\textit{EB} : T\textit{eb} :: \sqrt{EB} : \sqrt{eb} :: \sqrt{AB} : \sqrt{ab};$$

Hence,

$$T\textit{EC}-T\textit{EB} \text{ (or } T\textit{BC}) : T\textit{ec}-T\textit{eb} \text{ (or } T\textit{bc}) :: \sqrt{AB} : \sqrt{ab}. \qquad (Y.)$$

In the same manner it may be shown, that

$$T\textit{CD} : T\textit{cd} :: \sqrt{AB} : \sqrt{ab}. \qquad (Z.)$$

From the proportions marked (X,) (Y,) (Z,) therefore, we have

$$T\,(\textit{AB}+\textit{BC}+\textit{CD}) : T\,(ab+bc+cd) :: \sqrt{AB} : \sqrt{ab}$$
$$:: \sqrt{(AB+BC+CD)} : \sqrt{(ab+bc+cd)}.\ddagger$$

169. *The times of descending through* SIMILAR CURVES, *similarly situated with respect to the horizon, are as the square roots of the lengths of those curves.*

If the number of planes AB, BC, CD, &c., *ab*, *bc*, *cd*, &c., be in-

* The velocity down AB+BC+CD : velocity down $ab+bc+cd$ :: $\sqrt{FG} : \sqrt{fg}$ :: $\sqrt{FD} : \sqrt{fd}$ :: $\sqrt{AB} : \sqrt{ab}$ :: (AB+BC+CD) : $\sqrt{(ab+bc+cd)}$. The velocities as well as the times are therefore as the square roots of the lengths of the planes.

† For by similar triangles, EB : eb :: AB : ab; and from similar planes BC : bc :: AB : ab; ∴EB+BC (or EC) : $eb+bc$ (ec) :: AB : ab; and $\sqrt{EC} : \sqrt{ec}$ :: $\sqrt{AB} : \sqrt{ab}$. In the same manner it may be shown that FD : fd :: AB : ab, or $\sqrt{FD} : \sqrt{fd}$ :: $\sqrt{AB} : \sqrt{ab}$.

‡ For since AB : ab :: BC : bc :: CD : cd, AB+BC+CD : $ab+bc+cd$ :: AB . ab; or $\sqrt{AB} : \sqrt{ab}$ :: $\sqrt{(AB+BC+CD)} : (ab+bc+cd)$.

creased, and their lengths and their inclinations to each other be diminished ad infinitum, then the polygons ABCD, *abcd*, become similar curves, in falling down which no velocity is lost.

Suppose the curves to be circular arcs; then, since similar circular arcs are to each other as the radii of the circles to which they belong, the times of descending through these arcs will be to each other as the square roots of their radii.

THE PENDULUM.

170. Definitions.—A *Pendulum* is a body suspended by a right line from any point, and moving freely about that point as a center. The point about which the pendulum revolves, is called the *center of suspension.* The *vibration* of a pendulum, is its motion from a state of rest at the highest point on one side, to the highest point on the other side.* The *center of oscillation* of a pendulum, is such a point, that, were all the matter of the pendulum collected in it, the quantity of motion (or momentum) would be equal to the sum of the momenta of all the parts taken separately.

Thus, (Fig. 113,) the parts of the pendulum about *b* move faster than those about *a*, and consequently have more momentum; but there is a point about which the momenta balance each other, and therefore in the investigations relating to the pendulum, all the parts of which it consists may be considered as concentrated in that point.

Fig. 113.

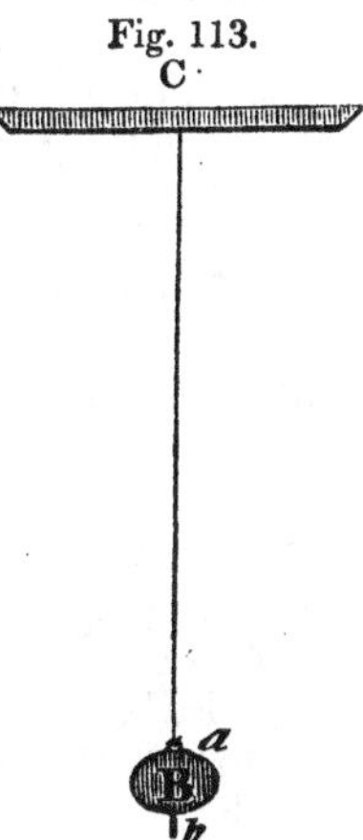

The center of oscillation is below the center of gravity; for, since the parts more remote from the center of suspension have more velocity than the parts that are nearer to it, the quantity of matter below the center of oscillation must be less than the quantity of matter above it.

171. In order to understand the doctrine of the pendulum, it is necessary to become acquainted with a few of the leading properties of the curve called the Cycloid.†

---

* In these investigations, as in those of the Mechanical Powers, pendulums are supposed to move without any resistance from the air or from friction. The conclusions, therefore, will be accurately true only when applied to vibrations performed in a perfect vacuum, round a perfectly smooth axis of suspension.

† The learner will remark that the mode of reasoning is this: it is first proved that, were a pendulum to vibrate in a cycloid, all its vibrations, whether performed in larger or in smaller arcs, would be equal. It is then shown that pendulums vibrating in small *circular* arcs are subject, very nearly, to the same law. The calculus affords an easier method than Geometry of investigating the properties of this as well as of

A Cycloid is the curve described by a point in the circumference of a circle rolling in a straight line on a plane.

Fig. 114.

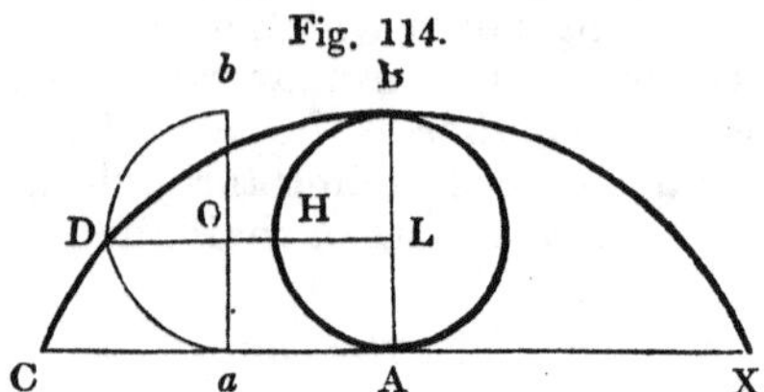

Let the circle AHB (Fig. 114) make one revolution upon the line CAX, equal to its circumference; the curve line CDBX, traced out by that point of the circle which was in contact with C when the circle began to revolve, is called a Cycloid. If CX be bisected in A, and AB be drawn at right angles to it, it is evident from the manner in which the curve is generated, that it will have similar branches on both sides of AB, and that its vertex B will be so placed as to make its *axis* AB equal to the diameter of the generating circle. Its properties, as applied to the vibration of the pendulum, are the following.

172. The *cycloidal ordinate* DH *equals the circular arc* BH.—For, let bDa (Fig. 114) be the position of the circle when the generating point is at D; draw the diameter ba parallel to BA, and from D draw DHL parallel to CA; then the arc Da=arc HA, ∴ sin. DO=sin. HL, and consequently DH=OL; but from the mode in which the cycloid is generated, Ca=arc Da, and CA= semicircle BHA; hence DH=OL=aA=CA−Ca=semicircle BHA−arc HA=arc BH.

173. *A tangent to the cycloid at any point,* E, (Fig. 115,) *is parallel to the corresponding chord* BK *of the generating circle.*—Draw DHL indefinitely near to EKM; join BK, and produce it to k; let fall Ho at right angles to Kk. The indefinitely small triangle HKk is similar to the triangle KRB formed by the tangents (KR, BR) to the circle at the points K, B, and is consequently isosceles; ∴ KH=Hk. Now by Art. 172, arc BKH=DH, ∴ BKH−KH (=arc BK=EK) =DH−Hk=Dk; but since EK and Dk are equal and parallel, ED and Kk must also be equal and parallel; and as the tangent at the point E may be considered as coinciding with ED, it must therefore be parallel to the chord BK.

Fig. 115.

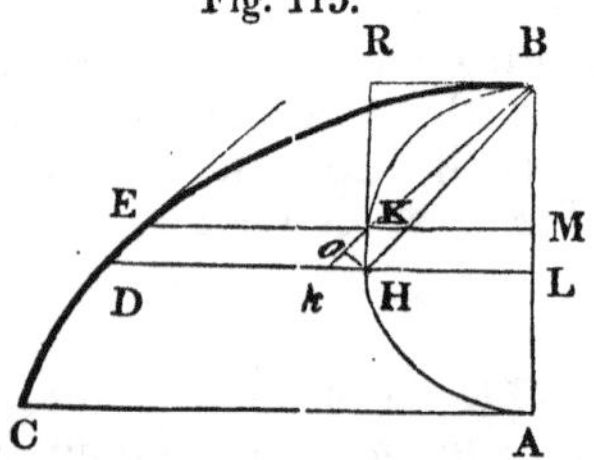

other curves; but the geometrical method is retained in this treatise, to render the study intelligible to such as are not acquainted with the Calculus.

174. *The cycloidal arc* BE *is equal to twice the corresponding chord* BK *of the generating circle.*—For since the triangle KHk is isosceles, Ho bisects the base Kk, ∴ Kk or ED=2Ko; and since Ho may be considered as a small circular arc described with radius BH, Ko=Bo−BK=BH−BK; hence ED and Ko are contemporaneous increments of the cycloidal arc BE and the chord BK; and as the arc and chord begin together from the point B, and the former increases by ED or Kk while the latter increases by Ko=½ED, the arc BE must be equal to twice the chord BK: consequently, the whole arc BC=twice the diameter AB.

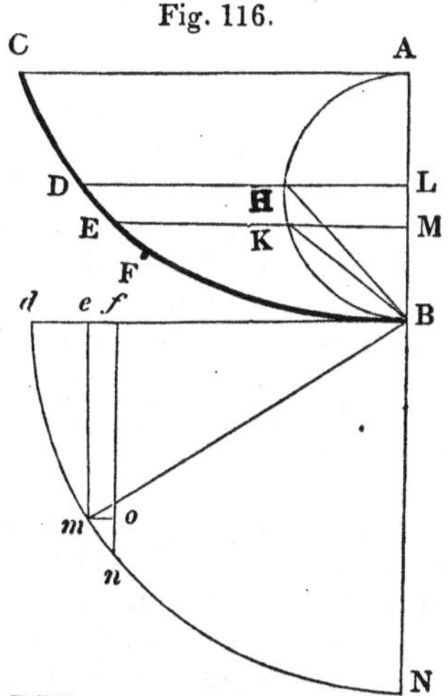

Fig. 116.

175. AHB (Fig. 116) is a circle, whose diameter AB is perpendicular to the horizon, and CDB a cycloidal arc. Now, by Art. 159, if a body begins to descend from any point D, its velocity at the point E will be the same as the velocity at the point M of a body falling freely through the perpendicular height LM; and its velocity at every other point during its descent through the cycloidal arc DEB will be the same as the velocity of the body falling freely at every other corresponding point of the line LMB. By Art. 167, therefore, the velocity V of the body thus descending along the cycloidal arc DE will vary as $\sqrt{(\text{LM})}$ $\propto \sqrt{(\text{BL}-\text{BM})} \propto \sqrt{(\text{AB}\times\text{BL}-\text{AB}\times\text{BM})} \propto \sqrt{(\text{HB}^2-\text{KB}^2)} \propto \sqrt{(\text{DB}^2-\text{EB}^2.)}$ Hence, let Bd be drawn parallel to AC and equal to BD, and upon Bd describe the quadrant of a circle dmN; take Be equal to BE, and draw em at right angles to Bd; then will $em^2=\text{B}m^2-\text{B}e^2=\text{B}d^2-\text{B}e^2=\text{DB}^2-\text{EB}^2$, and consequently V, which varies as $\sqrt{(\text{DB}^2-\text{EB}^2,)}$ will vary as em, the sine of the arc dm, whose versed sine is de or DE, the space fallen through.

176. Let EF be an indefinitely small part of the cycloidal arc, and make ef equal to it; draw fn at right angles to Bd, and mo parallel to it. Since EF is very small, it may be considered as described with the velocity V at E continued uniformly, and therefore the time of describing EF $\left(\text{since } \text{T}=\frac{\text{S}}{\text{V}}\right)$ will be represented by $\frac{\text{EF}}{\text{V}}$ or $\frac{ef}{em}$. Now since the sine em represents the velocity at any point E, the whole velocity acquired in falling down DB will be represented by the radius BN or Bm; if, therefore, a body were to describe the quadrantal arc dmN with the velo-

city at the lowest point continued uniformly, the time of describing any small part (mn) of it would be represented by $\frac{mn}{Bm}$. But by similar triangles, Bme, mno, Bm : em : : mn : mo or ef, $\therefore \frac{ef}{em} = \frac{mn}{Bm}$; hence the time of describing the small cycloidal arc EF is equal to the time of a body's moving through the corresponding small circular arc mn, with the velocity in B continued uniformly; *the whole time of descent therefore through* DEB, *will be equal to the time of a body's describing the quadrantal arc dm*N *with the velocity at* B *continued uniformly.*

177. Now the velocity at the lowest point B of the cycloid is equal to the velocity acquired in falling down the chord HB, and by Art. 31, with this velocity continued uniformly it would describe 2HB=BD=Bd in the same time. But by Art. 164, the time of falling down the chord HB=the time of falling down the diameter AB or the axis of the cycloid; hence the time of descending down the cycloidal arc DB, and the time of falling freely through the axis AB of the cycloid, are to each other as the times of describing the arc dmN and the straight line Bd with the same uniform velocity, i. e. *as the quadrant of the circle to its radius.**

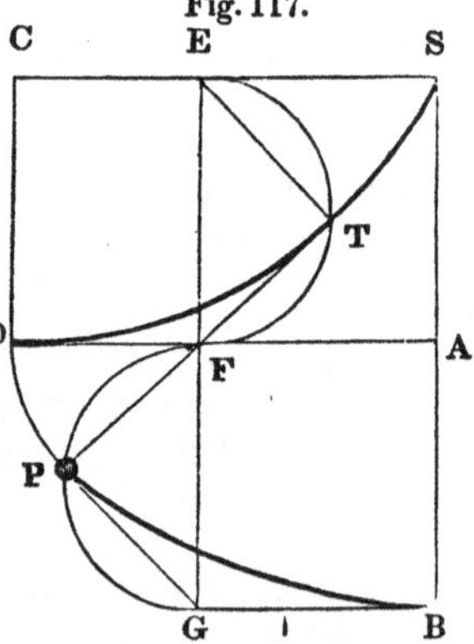

178. Take any line SC, (Fig. 117,) and draw SA at right angles to it; make SC : SA : : semi-circumference of a circle : its diameter; and complete the parallelogram SCDA. Produce SA to B, making AB=SA; upon SC, AD, describe two semi-cycloids SD, DB, the vertex of the former of which is at D, and the latter at B; then if a body be suspended from a point S by a string whose length is equal to the cycloid SD, and begins to descend from D, its place will always be in the semi-cycloid DB, the part TP of the string being always at right angles to the cycloidal arc DB. For through any point F, draw EFG perpendicular to SC, and through B draw BG parallel to SC; then EG=SB; on EF, FG, describe the two semicircles ETF, FPG, and draw the chords TF, FP, the former of which (Art. 173) is a tangent to the cycloid SD in T. Now SE=arc ET, and SC=ETF, $\therefore$ CE(=DF)=arc TF=arc FP; hence the angles TEF, FGP are equal, and consequently the triangles TEF, FGP similar

* In uniform motions, when the velocities are equal, the times are as the spaces (Art. 12.)

and equal to each other; ∴ TF, FP, are in the same straight line; moreover, TP=2TF=(Art. 174) the cycloidal arc TD; if, therefore, the string TP be always equal to the cycloidal arc TD, i. e. if the whole string STP be equal to the semi-cycloid SD, P will always be found in the cycloid DB; and since the angle TPG is a right angle, and PG (Art. 173) is a tangent to the cycloid, therefore TP is always at right angles to the curve.

179. *A pendulum may be made to vibrate in a cycloid, by causing the flexible rod or string to apply itself, as it moves, to the sides of two semi-cycloids.*

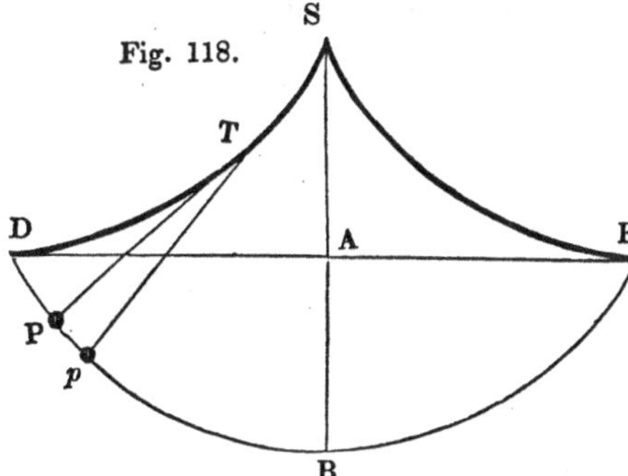

Fig. 118.

Instead of supposing the body to descend down a curve in the form of a cycloid, the effect will evidently be the same with respect to the acceleration of the body, whether it be kept in a curve of this form by the reaction of the surface in a direction perpendicular to it, or by the tension of a string acting in the same direction. Now a small body P, (Fig. 118,) suspended from the point S by a string STP of the same length with either of the semi-cycloids SD, SE, and made to vibrate between them, (the string gradually unwinding from the semi-cycloid SD as it descends to the lowest point B, and winding round the semi-cycloid SE as it ascends to the highest point E,) will always be found in a cycloid similar and equal to the two semi-cycloids SD, SE; and as the string TP is always at right angles to the curve, it will be under precisely the same circumstances as the body descending down the curve. Having descended to the lowest point B, it will then ascend through an arc BE, equal to BD, in the same time in which it descended down DB; so that the whole time of one vibration will be twice the time of descending down DB; hence, (Art. 177,) *the time of one vibration will be to the time of a body's falling freely down half the length of the pendulum, as the circumference of a circle to its diameter.*

180. Since (Art. 179) the time of a vibration is to the time down the axis in the same given ratio, whatever be the point

from which the pendulum begins its oscillations, *all the vibrations of a pendulum of the same length are equal to each other.* But from the difficulty of constructing plates of the exact form SD, SE, and from other causes, the cycloidal pendulum is of little or no practical utility. In a geometrical point of view, however, this mode of comparing the time of a vibration with the time of falling down a space equal to half the length of the string, is of considerable importance; for the cycloid at the lowest point B may evidently be considered as a circular arc described with the radius SB; if therefore a body be suspended by a string whose length is SB, and vibrates in a circular arc only to a very short distance on each side of the point B, *the time of the vibration of a pendulum in small* CIRCULAR *arcs, is to the time down half the length of the pendulum, as the circumference of a circle to its diameter; and therefore, within moderate limits, the time will be the same, whether the arc of vibration be larger or smaller.*

181. *The times of vibration of pendulums of different lengths are to each other as the square roots of the lengths.*

Let L=length of the string or thin inflexible rod by which a small body is suspended, $\pi=3.14159$, &c. $m=16\frac{1}{12}$ feet; then, (Art. 34) the time of a body's falling down half the length of the string $=\left(\frac{S}{m}\right)^{\frac{1}{2}}=\left(\frac{L}{2m}\right)^{\frac{1}{2}}$. Hence the time of a vibration (T) $:\left(\frac{L}{2m}\right)^{\frac{1}{2}}::\pi:1$, or $T=\left(\frac{\pi^2 L}{2m}\right)^{\frac{1}{2}}$, which varies as $\sqrt{L}$. Let T=1: then $\left(\frac{\pi^2 L}{2m}\right)^{\frac{1}{2}}=1$, or $\pi^2 L=2m$, and $L=\frac{2m}{\pi^2}=\frac{32.1666}{9.87}=3.259$ feet$=$ 39.11 inches; if the space fallen through from rest by gravity in 1″, therefore, be $16\frac{1}{12}$ feet, the length of a pendulum which vibrates seconds will be 39.11 inches.

Let x=the space fallen through by gravity in 1″, then $T=\left(\frac{\pi^2 L}{2x}\right)^{\frac{1}{2}}$ and $T^2=\frac{\pi^2 L}{2x}$, or $x=\frac{\pi^2 L}{2T^2}$; if therefore the length of a pendulum which vibrates in T″ is given, the space fallen through by gravity in 1″ may be found; thus, let the length of the pendulum which vibrates seconds=39.2 inches, then

$x=\frac{\pi^2 L}{2}=\frac{9.87\times 39.2}{2}=193.452$ inches, or 16.121 feet in 1″.

182. *The times of vibration of pendulums of different lengths acted upon by* DIFFERENT ACCELERATIVE FORCES, *will vary as the square roots of the lengths directly, and as the square roots of the forces inversely.*

If the accelerative force be not given, then (Art. 156) the time of falling down half the length of the string $=\left(\frac{S}{mF}\right)^{\frac{1}{2}}=\left(\frac{L}{2mF}\right)^{\frac{1}{2}}$.

$\therefore T:\left(\frac{L}{2mF}\right)^{\frac{1}{2}}::\pi:1$, or $T=\left(\frac{\pi^2 L}{2mF}\right)^{\frac{1}{2}}$, which varies as $\left(\frac{L}{F}\right)^{\frac{1}{2}}$...

The times of vibrations of pendulums of the SAME LENGTH *vary inversely as the square roots of the accelerative forces.*

If L be given, then $T \propto \frac{1}{\sqrt{F}}$.

The lengths of pendulums vibrating in the SAME TIME *vary as the forces which accelerate them*

If T be given, then $L \propto F$.

183. *The* NUMBER *of vibrations performed in a given time by pendulums of different lengths, acted upon by different accelerative forces, are directly as the square roots of the forces, and inversely as the square roots of the lengths.*

Let n=number of vibrations performed in any given time, and T=time of one vibration; then n will vary inversely as $T=\left(\frac{\pi^2 L}{2mF}\right)^{\frac{1}{2}} \propto \left(\frac{L}{F}\right)^{\frac{1}{2}} \therefore n \propto \left(\frac{F}{L}\right)^{\frac{1}{2}}$.

If therefore the lengths be given, the number of oscillations will be directly as the square roots of the forces; and if the forces be given, the number of oscillations will be inversely as the square roots of the lengths.

184. Examples.

1. What is the time of an oscillation of a pendulum whose length is 10 feet; and what must be the length of a pendulum which shall oscillate ten times a minute?

The length of a pendulum which oscillates in 1″ is 39.11 inches; and (Art. 181) the times of vibration of pendulums of different lengths, are to each other as the square roots of their lengths; hence $1'' : T :: \sqrt{39.11} : \sqrt{120} :: 625 : 1095$, $T=\frac{1095}{625}=1\frac{470}{625}$ seconds.

Again, let $T=6''$=time of one oscillation of a pendulum which makes ten in a minute, then $6'' : 1'' :: \sqrt{L} : \sqrt{39.11}$, $\therefore 36 : 1 :: L : 39.11$, or $L=39.11\times 36$ inches$=117\frac{1}{3}$ feet.

2. Compare the times of vibration T, t, of two pendulums whose lengths are L, l, when carried to the distances D, d, above the Earth's surface.

Let r=the radius of the Earth, then since the force of gravity varies inversely as the square of the distance from the Earth's center, the force which accelerates the pendulum whose length is (L) : the force which accelerates the pendulum whose length is (l) :: $\frac{1}{\overline{r+D}|^2}$

$: \frac{1}{\overline{r+d}|^2} :: \overline{r+d}|^2 : \overline{r+D}|^2$; but by Art. 182, $T \propto \left(\frac{L}{F}\right)^{\frac{1}{2}}$.

$\therefore T : t :: \left(\frac{L}{\overline{r+d}|^2}\right)^{\frac{1}{2}} : \left(\frac{l}{\overline{r+D}|^2}\right)^{\frac{1}{2}} :: \overline{r+D}.\sqrt{L} : \overline{r+d}.\sqrt{l}$.

If $L=l$, then $T : t :: r+D : r+d$; i. e. the times of vibration of the same pendulum, when carried to different heights above the Earth's surface, are to each other as the distances of those heights from its center.

3. If a pendulum at the Earth's surface vibrates (m) times in T″; how must its length be altered so that it may vibrate (n) times in T″?

Let L=the length of the pendulum which vibrates (m) times in T″, and $L+x$=the length of that which vibrates (n) times in T″; then by Art. 183, (since F is given) $m : n :: \frac{1}{\sqrt{L}} : \frac{1}{\sqrt{(L+x)}}$,

$\therefore m^2 : n^2 :: L+x : L$, or $n^2L+n^2x=m^2L$, and $x=\frac{(m^2-n^2)L}{n^2}$.

Let $m=n\pm y$, where y is very small with respect to n; then $\frac{(m^2-n^2)L}{n^2}=\frac{(\pm 2ny+y^2)L}{n^2}$=(rejecting y^2 as very small with respect to $2ny$) $\frac{\pm 2Ly}{n}$, and consequently $L+x=L\frac{\pm 2Ly}{n}$, which furnishes us with a convenient theorem for ascertaining the quantity, by which the pendulum of a clock must be lengthened or shortened according as it gains or loses a few seconds or minutes in a day.*

* Suppose, for instance, that a clock gains 3 minutes in a day, i. e. instead of performing $24\times60\times60$ or 86400 vibrations in a day, it performs 86400+180 or 86580 vibrations in that time; then n=86400 and y=180, $\therefore \frac{2Ly}{n}=\frac{360L}{86400}=\frac{L}{240}$; and although this acceleration of the pendulum indicates that its length is a little less than 39.11 inches, yet it is evident that in finding the value of $\frac{L}{240}$, L may be assumed equal to 39.11 inches without any material error; hence $\frac{L}{240}$ or $\frac{39.11}{240}$ (=.16 of an inch) is the quantity by which the pendulum must be lengthened to make it vibrate seconds. In the example here given, m (=86580) is greater than n, and therefore y is positive; if the pendulum loses a certain number of seconds in a day, then m is less than n, and consequently y is negative; the value of $\frac{2Ly}{n}$ must in this case be subtracted

4. A pendulum, which vibrated seconds and kept true time at the Earth's surface, was carried to the top of a mountain, and there lost (t) seconds in an hour: What was the height of the mountain?

Let r=radius of the earth, x=height of the mountain, T=the number of seconds in an hour or the number of vibrations in an hour at the Earth's surface, then T$-t$ will be the number of vibrations in an hour at the top of the mountain. By Art. 183, when the length of the pendulum is given, the number of vibrations in a given time are directly as the square roots of the forces which act upon the pendulum; hence, since the forces are inversely as the squares of the distances from the center of the earth

$$T : T-t :: \left(\frac{1}{r^2}\right)^{\frac{1}{2}} : \left(\frac{1}{(r+x)^2}\right)^{\frac{1}{2}} :: \frac{1}{r} : \frac{1}{r+x}, \therefore$$

$$t : T-t :: x : r, \therefore x = \frac{tr}{T-t}.*$$

185. Questions on the Pendulum.

1. The lengths of pendulums vibrating seconds at St. Thomas near the equator, at New York, at London, and at Spitzbergen, in lat. 80°, have been ascertained by very accurate experiments to be as stated below: Required the space through which a body would fall in one second at those places respectively? (Art. 181.)

	Length of Pend.	Space in 1 sec.
St. Thomas,	39.02074	192.56 inches.
New York,	39.10168	192.96
London,	39.13860	193.14
Spitzbergen,	39.21469	193.52.†

2. What is the length of the seconds pendulum at New Haven, where a body falls from rest $16\frac{1}{12}$ feet the first second?

Ans. 39.110, *or* $39\frac{1}{10}$ *inches nearly.*

---

from L, or the pendulum must be shortened. The theorem alluded to in the 3d example, therefore, gives this rule: Multiply twice the length of the pendulum by the number of seconds gained or lost, and divide the result by the number of seconds in a day; the quotient will give the number of inches or parts of an inch by which the pendulum is to be lengthened or shortened.

* If t be very small with respect to T, then x may be considered as equal to $\frac{tr}{T}$ without any material error. Now T=3600″, if therefore t be equal to 1, 2, 3, &c. seconds, then x will be equal to $\frac{r}{3600}$, $\frac{2r}{3600}$, $\frac{3r}{3600}$, &c.; and supposing r to be equal to 4000 miles, the heights of mountains upon which a pendulum, that vibrates seconds at the Earth's surface, loses 1, 2, 3, &c. seconds in an hour, will be $1\frac{1}{9}$, $2\frac{2}{9}$, $3\frac{3}{9}$, &c. miles respectively.

† Hence it appears that the space through which a body falls from a state of rest in one second, is less at the equator than at the latitude of 80° by $\frac{96}{100}$ inch, or nearly 1 inch.

3. The length of the seconds pendulum being $39\frac{1}{10}$ inches, what are the lengths of pendulums vibrating $\frac{1}{2}$ and $\frac{1}{4}$ seconds; also in 2 seconds; and how long must a pendulum be to vibrate once an hour?

Ans. For half sec. 9.775 inch.$=\frac{1}{4}$ the length of the seconds pend.
quarter " 2.444 " $=\frac{1}{16}$ " "
two " 13.03 feet $=4$ times " "
one hour 7997.7 miles $=$diameter of the earth nearly.*

4. A pendulum which vibrated seconds at the level of the sea, was found to vibrate but 3597 times in an hour, on the top of a neighboring mountain: Required the height of the mountain?

Ans. $3\frac{1}{3}$ *miles.*

185′. Central Forces.

Central Force is that power or energy with which a body moving in any curve around a center is made to approach to, or recede from that center. In the former case it is called the *centripetal*, and in the latter the *centrifugal*, force.

Fig. 118′.

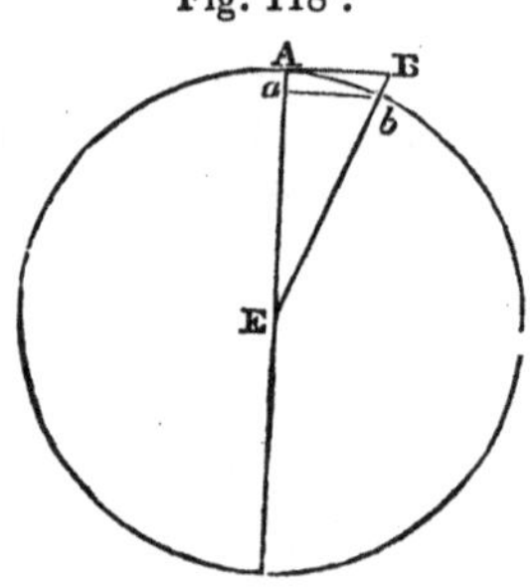

(1.) When a body revolves in a *circle* the centripetal and centrifugal forces are *equal to each other.* In Fig. 118′, B*b* represents the force by which a body revolving in a circle around the center E, is carried toward the center by the centripetal force; but since, were its motion not thus constrained it would have receded from the center through the space B*b*, consequently, this line also represents the centrifugal force. Hence the two forces are equal to each other.

(2.) The central force (either the centripetal or the centrifugal) of a body moving uniformly in a circle, is *as the square of the velocity divided by the radius of the circle.* Let r be the radius of the circle, v the velocity of the body, and A*b* the arc described in an indefinitely short time t. Then (since by our hypothesis the motion is considered as uniform) the space $Ab=vt$. But A*b* being a very small arc, may be taken as equal to its chord, which is a mean proportional between the diameter $2r$ and the versed sine A*a*, ($=$B*b*). Hence, $Aa\times 2r=Ab^2$, and $Aa=\frac{Ab^2}{2r}=\frac{v^2t^2}{2r}$, which, when the time is given, varies as $\frac{v^2}{r}$, $\therefore F \propto \frac{v^2}{r}$.

(3.) Again, the central force is as the *radius of the circle divided by the square of the time.* Let t equal the time of describing the

* The diameter of the earth is 7912 miles.

whole circumference$=2\pi r$; then $2\pi r=tv$, and $v=\frac{2\pi r}{t}$, $\propto\frac{r}{t}$ and $\therefore v^2\propto\frac{r^2}{t^2}$. But $F\propto\frac{v^2}{r}\propto\frac{r^2}{t^2 r}\propto\frac{r}{t^2}$, $\therefore F\propto\frac{r}{t^2}$.

CHAPTER X.

OF PROJECTILES.

186. *A body thrown into the air at any angle with the horizon, is called a projectile.*

The doctrine of Projectiles proceeds on the supposition that the force of gravity acts uniformly, and that bodies move without resistance from the air, neither of which suppositions is strictly true; but, for distances so small as those usually involved in these inquiries, the variation in the force of gravity is not material, and for all practical purposes the effect of the resistance of the air is separately computed, and allowed for.

187. *If a body be projected in any direction, not perpendicular to the horizon, it will describe a parabola.*

This proposition has already been demonstrated in Art. 48.

The most concise and comprehensive method of treating this subject, is first, to investigate by means of Analytical Trigonometry, a *general formula*, showing the relations between the time, velocity, range, and angle of elevation of a projectile, however these may vary among themselves; and secondly, to deduce from this, formulæ of greater simplicity that serve for particular cases.

188. In order to investigate the *general formula*,* let A (Fig. 119,) be the point of projection, AB or AB′ the plane on which the body is projected, passing through A. AB also denotes the *range*, or distance to which the body is thrown. Let AC be drawn parallel, and BCD perpendicular to the horizon; let the angle of elevation CAD$=a$, the angle of depression of the plane CAB$=b$, the velocity of projection$=v$, the time of flight$=t$, the range AB$=r$, and the space fallen through by gravity in one second or $16\frac{1}{12}$ feet$=m$.

Then, by the laws of uniform motion, at the end of the time t, if gravity did not act, the body would be found in the point D,

* See *Encyc. Metropolitana*, Art. Mechanics, p. 105, from which a large part of this chapter is taken.

while, by the laws of falling bodies, it would, in the same time pass through the perpendicular DB, consequently,

$AD=tv$; and $DB=mt^2$.

Fig. 119.

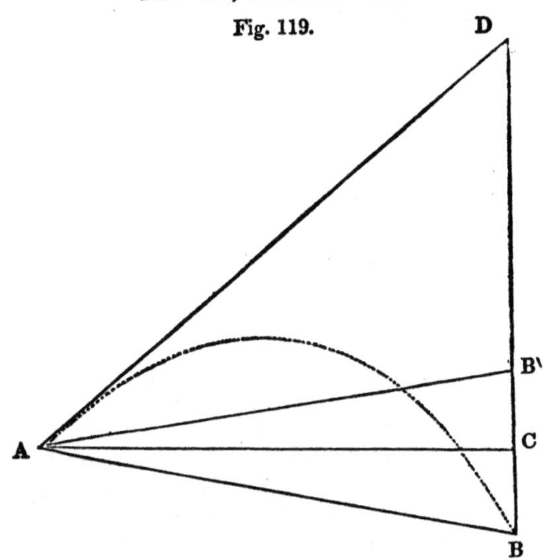

In the right-angled triangles ABC and ADC, the angle B is the complement of b, and the angle D is the complement of a; and, since the sides are as the sines of the opposite angles,

$$\cos.\ b : \sin.\ (a \pm b)^* :: tv : tv\frac{\sin.\ (a\pm b)}{\cos.\ b}=mt^2.$$

$$\text{Or, } \frac{mt}{v}=\frac{\sin.(a\pm b)}{\cos.\ b}. \qquad (1.)$$

$$\text{Again, } \cos.\ a : \sin.\ (a\pm b) :: r : \frac{r\sin.\ (a\pm b)}{\cos.\ a}=mt^2.$$

$$\text{Or, } \frac{mt^2}{r}=\frac{\sin.\ (a\pm b)}{\cos.\ a} \qquad (2.)$$

$$\text{From equation (1.) } t=\frac{v\sin.\ (a\pm b)}{m\cos.\ b}.$$

$$\text{and } t^2=\frac{v^2\sin.^2\ (a\pm b)}{m^2\cos.^2\ b}.$$

$$\text{From equa. (2.) } t^2=\frac{r\sin.\ (a\pm b)}{m\cos.\ a}.$$

$$\text{Hence, } \frac{r}{v^2}=\frac{\sin.\ (a\pm b)\cos.\ a}{m\cos.^2\ b}. \qquad 3.)$$

From these three equations, all the relations between the time,

---

* *Plus* when the plane descends, *minus* when the plane ascends.

velocity, range, and angle of elevation, are readily determined; so that any two of these four quantities being given, the other two may be found. Thus,

By equation (1.) $v=\frac{mt \cos. b}{\sin. (a\pm b)}$.

By equation (2.) $r=\frac{mt^2 \cos. a}{\sin. (a\pm b)}$.

If the *range* and *elevation* be given to find the *time* and *velocity*,

By equation (2.) $t=\left(\frac{r \sin. (a\pm b)}{m \cos. a}\right)^{\frac{1}{2}}$.

By equation (3.) $v=\left(\frac{rm \cos.^2 b}{\sin.(a\pm b) \cos. a}\right)^{\frac{1}{2}}$.

If the *velocity* and *elevation* be given to find the *time* and *range*

By equation (1.) $t=\frac{v \sin. (a\pm b)}{m \cos. b}$.

By equation (3.) $r=\frac{v^2 \sin. (a\pm b) \cos. a}{m \cos.^2 b}$.

If any two of the above quantities be given to find the *angle of elevation*, then (the inclination of the plane to the horizon (b) being supposed to be known) in order to find the value of a, we must substitute for $\sin. (a\pm b)$, its value involving the sine of each are separately. Now $\sin. (a\pm b)=\sin. a \cos. b\pm\sin. b \cos. a$,* whence either the $\sin. a$ or $\cos. a$ may be obtained.

The value of a and b being known, we shall then find in all the preceding cases, $\sin. (a\pm b) \cos. a$ equal to a known quantity Let this be denoted by C; then $\cos. a \sin. a \cos. b\pm\sin. b \cos.^2 a=C$;

Let $x=\sin. a$, then $\cos. a=(1-x^2)^{\frac{1}{2}}$

$$x(1-x^2)^{\frac{1}{2}}\pm\tan.\dagger\, b(1-x^2)=\frac{C}{\cos. b}.$$

189. As this is a quadratic equation, the solution will give two values of x, which shows, that there are always two different angles of elevation, which equally satisfy the conditions of the problem, except in the case where the two roots of the equation are equal to each other, when only one angle will be found. In this case, as is shown below, $\sin. a=\sin. \frac{1}{2} (90\pm b)$, under which limitation the range will be a maximum, or the greatest possible; and all angles of elevation equally above and below that which gives the maximum range, will give ranges equal to each other. For, by the value of r as determined above,

$$r=v^2\frac{\sin. (a\pm b) \cos. a}{m \cos.^2 b}.$$

---

* Day's Trigonometrical Analysis, Art. 208. † Ib. Art. 216.

If v and the angle b are given, the range will vary as sin. $(a \pm b)$ cos. a. But

$$\sin.\,(a \pm b)\cos.\,a = \sin.\,a \cos.\,a\,\cos.\,b \pm \sin.\,b\,\cos.^2\,a.$$
$$= \tfrac{1}{2}\sin.\,2\,a\,\cos.\,b \pm \sin.\,b(\tfrac{1}{2}+\tfrac{1}{2}\cos.\,2a)^*$$
$$= \tfrac{1}{2}\sin.\,2\,a\,\cos.\,b \pm \tfrac{1}{2}\cos.\,2\,a\,\sin.\,b + \tfrac{1}{2}\sin.\,(\pm b)$$
$$= \tfrac{1}{2}\sin.(2\,a \pm b) + \tfrac{1}{2}\sin.(\pm b)$$

; and since the second part of this expression is constant, the range will vary as sin. $(2a \pm b)$. This quantity will be greatest, when

$$2a \pm b = 90°.$$
$$\text{Then } 2(a \pm b) = 90° \pm b$$
$$\text{And } a \pm b = 45° \pm \tfrac{1}{2}b$$
$$a = 45° \pm \tfrac{1}{2}b \mp b = 45° \pm \tfrac{1}{2}b = \tfrac{1}{2}(90° \pm b)$$
$$\therefore a = \tfrac{1}{2}(90° \pm b)$$
$$\sin.\,a = \sin.\,\tfrac{1}{2}(90° \pm b).$$

Therefore, as above, the range will be a maximum where the sine of the angle of elevation equals the sine of $\frac{1}{2}(90° \pm b)$.

And since all angles equally above and below 90° have the same sine, all angles equally above and below that which gives the maximum have equal ranges. Thus, a cannon ball fired at an angle of 60° (in a vacuum) above a horizontal plane, would reach the plane at the same distance from the point of projection as if fired at an elevation of 30°. When the data of the problem give or require a greater value for sin. $(2a \pm b)$ than 1, that is, than the sine of 90°, it [illegible] the problem, under the proposed conditions, to be impossible.

190. To find the greatest *height* to which the projectile will ascend, we must recollect that a body projected perpendicularly upward, will rise to the same height from which it must have fallen to acquire the velocity of projection. (Art. 30.) Calling the time of rising to the greatest height t', we shall have $2mt'$ =the velocity of descent from gravity; and, v representing the whole velocity of projection in an oblique direction, the perpendicular part will be represented by v sin. a; whence

$$2mt' = v\sin.\,a, \text{ and } t' = \frac{v\sin.\,a}{2m} \text{ and } t'^2 = \frac{v^2\sin.^2\,a}{4m^2}.$$

But the space fallen through in the time $t' = mt'^2 = \dfrac{v^2\,sin.^2\,a}{4m}$.

And the ascent in the same time from projection is,

$t'v\sin.\,a = \dfrac{v^2\sin.^2\,a}{2m}$. Consequently, the difference of these will be the greatest height of the projectile above the point A; that is, $h = \dfrac{v^2\sin.^2\,a}{2m} - \dfrac{v^2\sin.^2\,a}{4m} = \dfrac{v^2\sin.^2\,a}{4m}$, and $\dfrac{4mh}{v^2} = \sin.^2\,a.$ (4.)

If therefore the angle of elevation and velocity are given, the

* Day's Trigonometrical Analysis, Art. 213.

greatest height may be determined; or if the range (r) or the time (t) be given, (the angles being known,) the value of v^2 may first be ascertained by preceding formulæ, and then the height (h) from equation (4.)

191. All the preceding equations become much more simple when the projection is along a *horizontal* plane; for then $b=0$, and sin. $b=0$, and cos. $b=1$; hence,

In eq. $(1.)\frac{mt}{v}\left(=\frac{\sin.(a\pm b)}{\cos. b}\right)=\sin. a$ (1′.)

$$(2.)\frac{mt^2}{r}\left(=\frac{\sin.(a\pm b)}{\cos. a}\right)=\frac{\sin. a}{\cos. a}=\tan. a. \quad (2'.)$$

$$(3.)\frac{r}{v^2}\left(=\frac{\sin.(a\pm b)\cos. a}{m\cos.^2 b}\right)=\frac{\sin. a\cos. a}{m}=\frac{\frac{1}{2}\sin. 2a}{m}. \quad (3'.)$$

Hence from (1′.) $t=\frac{v\times\sin. a}{m}\propto v\times\sin. a.$ (5.)

$$(3'.)\ r=\frac{v^2\times\sin. 2a}{2m}\propto v^2\times\sin. 2a. \quad (6.)$$

$$(4.)\ h=\frac{v^2\times\sin.^2 a}{4m}\propto v^2\times\sin.^2 a. \quad (7.)$$

On a horizontal plane, therefore, (the most usual case,) we have the following THEOREMS.

I. *The* TIME OF FLIGHT *varies as the velocity of projection multiplied by the sine of the angle of elevation.*

II. *The* RANGE* *varies as the square of the velocity of projection. multiplied by the sine of twice the angle of elevation.*

III. *The* GREATEST HEIGHT *varies as the square of the velocity of projection, multiplied by the square of the sine of the angle of elevation.*

Moreover, since the sine of twice 45° equals the sine of 90°, which equals radius, hence, by Theorem II,

IV. *The* RANGE IS GREATEST *when the angle of elevation is* 45°, *and is the same at elevations equally above and below* 45°.†

And since the square of the sine of the angle of elevation must be greatest when the angle is a right angle, therefore, by Theorem III,

V. *A projectile rises to the* GREATEST HEIGHT *when thrown perpendicularly upward.*

Finally, since the sine of the angle of elevation is greatest when the angle is a right angle, therefore, by Theorem I,

VI. *The* TIME OF FLIGHT IS GREATEST, *when the body is thrown perpendicularly upward.*

---

* Sometimes called *random.*

† For the sine of twice any angle below 45° is the same as the sine of twice any angle of the same number of degrees above 45°.

192. QUESTIONS ON PROJECTILES.

1. A body is projected at an angle of 15 degrees with the horizon, with the velocity of 140 feet per second: What is its *range, greatest height*, and *time of flight?*

By (6.) $r=\frac{v^2 \times \sin. 2a}{2m}=\frac{19600\times\frac{1}{2}}{32\frac{1}{6}}=304.663$ feet.

By (4.) $h=\frac{v^2\times\sin.^2 a}{4m}=$ (by logarithms) 20.409 feet.

By (5.) $t=\frac{v\times\sin. a}{m}=2.253$ seconds.

That is, its horizontal range is 304.663 feet; its greatest altitude 20.409 feet; and its time of flight is 2.253 seconds.

2. A body was projected at an angle of 60° with the horizon, and descends to it at the distance of 100 feet: With what velocity was it projected, and what was its greatest altitude and its time of flight?

From (6.) $v=\left(\frac{2mr}{\sin. 2a}\right)^{\frac{1}{2}}=60.94$ feet.

From (4.) $h=\frac{v^2\times\sin.^2 a}{4m}=43.3$ feet.

From (5.) $t=\frac{v\times\sin. a}{m}=3.28$ seconds.

3. I fired an arrow which remained in the air 4 seconds, and fell at the distance of 100 feet: With what angle of elevation was it fired, with what velocity, and how high did it ascend?

By equation (2′.) tan. $a=\frac{mt^2}{r}=2.57333=$tan. 68° 46′.

By equation (6.) $v=\left(\frac{2mr}{\sin. 2a}\right)^{\frac{1}{2}}=69.024$ feet per second.

By equation (4.) $h=\frac{v^2\times\sin.^2 a}{4m}=64.343$ feet.

4. A gun was fired at an elevation of 50°, and the shot struck the ground at the distance of 4898 feet: With what velocity did it leave the gun, and how long was it in the air?

Ans. Velocity, 400 *feet per second.*
Time, 19.05 *seconds.*

5. Random 4898 feet, time of flight 16 seconds: Required the angle of elevation and the velocity of projection?

Ans. El. 40° 3′, V. 400 *feet per sec.*

6. Random 2898 feet, velocity of projection 389.1 feet: What were the elevation, and time of flight?

Ans. El. 19° *or* 71°, T. 7.87 *sec.*

7. Elevation 40°, random 4898: Required the random when the elevation is 29½°? Art. 191. (6.) *Ans.* 4263.

8. Elevation 40° 3′, time of flight 16 seconds: Required the random and velocity of projection?

Ans. R. 4898, V. 400 *feet.*

9. Velocity 510 feet per sec., time of flight 15 seconds, to find the elevation and random. *Ans.* El. 28° 14′, R. 6740.

10. On a side-hill ascending uniformly above a horizontal level at an angle of 10° 20′, a ball was fired at an angle of elevation above the horizon of 34°, and with a velocity of 401 feet per second: What was the range on the hill-side when the gun was directed up the hill, and what when directed downward?

Ans. 3438 *and* 5985 *feet.**

CHAPTER XI.

OF THE STRENGTH OF MATERIALS.

193. THE importance to the architect and the engineer of ascertaining the form and position of the materials which he employs, in order to secure the greatest degree of strength and stability at the least expense, has led mathematicians and writers on mechanics, to devote much attention to this subject. How is the strength of a beam affected by giving to it different shapes and different positions? how must a given quantity of matter be disposed in order that it may have the greatest possible strength? and upon what principles depends the stability of columns, roofs, and arches? these, and many similar inquiries, have been the objects of profound investigation.

The strength of *beams*, or pieces of timber, is the first object of inquiry. STRENGTH is the power to *resist* fracture: STRESS, the power to *produce* fracture.

194. *The strength of a beam resting, horizontally, on its two ends, from a weight on its center, is proportioned to the area of a cross section, multiplied by the depth of its center of gravity.*

Thus, in Fig. 120, if AB represent a stick of timber, resting horizontally on supports at its two ends, and W be a weight placed at the center, and *a b c d* be a cross section, then, (supposing the weight to be sufficient to break the beam,) the fracture will commence at the bottom, and proceed regularly to the

* See Art. 188, formula (3.)

top, ending at W.* Now since the tendency to resist fracture from cohesion, must depend upon the total amount of all the separate forces acting between contiguous particles, it must evidently depend upon the number of the particles, that is, upon the extent of surface, or area of the section.

Fig. 120.

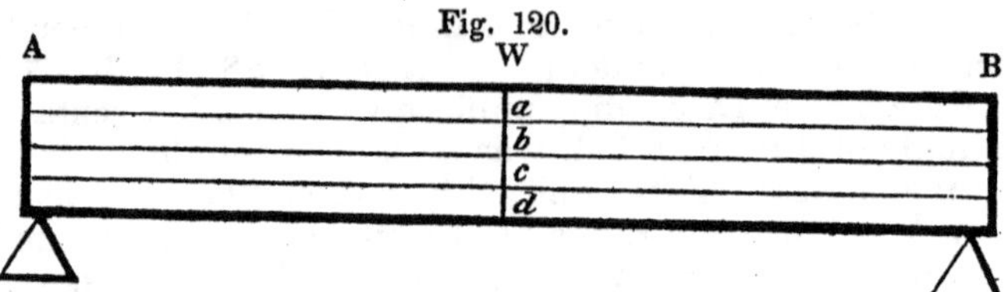

This would be the case were no other mechanical force involved but cohesion; but the *reaction* of the supports at A and B, being equal to the weight, and severally acting at the longer end of a bent lever AW*a*, AW*b*, &c., consequently, the tendency to fracture, from the leverage, will be lessened, as the shorter arm of the lever is increased, while the longer arm remains the same: therefore the strength being inversely as the stress, will be regularly increased as the distance of any lamina from W is increased, and the whole effect will be as the distance of the center of gravity from that point. Hence from both causes, the strength varies as the area of the section multiplied by the depth of the center of gravity.†

Fig. 121.

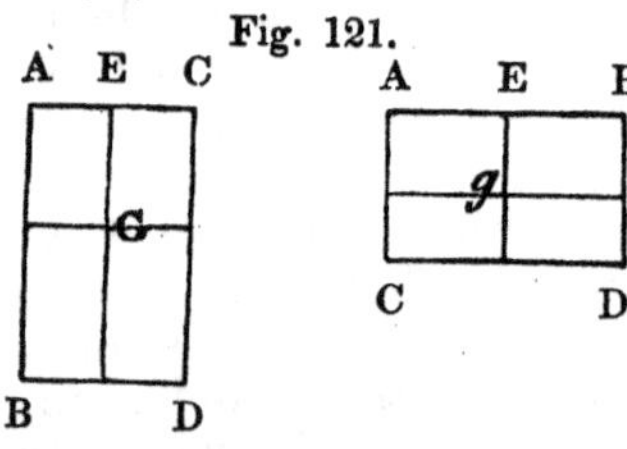

195. This proposition is general, and applies to a number of distinct cases. In *cylindrical* and *square* beams, since the area of the section varies as the square of its diameter, and the distance of the center of gravity from the point E, (Fig. 121,) varies as the diameter, their strength is as the *cubes* of the diameters. In beams of an *oblong* figure, the strength varies as the *breadth* and *square of the depth;* for here the area being as the product of the two sides, and the distance of the center of gravity from

* It is here supposed, according to the views embraced by most writers on Mechanics since the time of Galileo, that the parts of a fractured beam turn about the line where the fracture terminates; but Mr. Barlow, in his essay on the Strength and Stress of Timber, proves by experiment, that the tendency is to turn about a line entirely within the section, the fibres on that side of the line where the fracture begins being extended, and those on the other side compressed. This line he calls the *neutral axis.*

† Day's Algebra. (Art. 408.)

E being equal to half the perpendicular side, and therefore proportioned to that side, the proposition is, that the strength varies as the breadth×depth×depth, or as the breadth into the square of the depth. Hence, the same oblong beam with its narrow side upward, is as much stronger than with its broad side upward, as the depth exceeds the breadth. For the area being the same in both cases, the strengths are proportioned to EG and E*g*, or as AB to AC. Thus if a joist be 10 inches broad and $2\frac{1}{2}$ thick, it will bear four times more weight when laid on its edge than on its side. Hence the modern mode of flooring with very thin, but deep pieces of timber.

196. *In beams of different lengths, resting on two supports, the strength will vary as the area of the section into the depth of the center of gravity, divided by the length into the weight.*

Let L, l, denote the lengths; W, w, the weights; A, a, the areas of the sections; and G, g, the depths of the centers of gravity, of two prismatic beams, resting horizontally on their two ends.

The *stress*, or tendency to *produce* fracture, *from the weight of the beam itself*, will be expressed by $\frac{1}{8}L\times W$, and $\frac{1}{8}l\times w$.* But the tendency to *resist* fracture is denoted by $A\times G$, and $a\times g$. Hence the aggregate strength of the timber will be directly as the latter and inversely as the former. That is,

$$S:s::\frac{A\times G}{\frac{1}{8}L\times W}:\frac{a\times g}{\frac{1}{8}l\times W}::\frac{A\times G}{L\times W}:\frac{a\times g}{l\times w}.$$

197. *A triangular beam is twice as strong when resting on its broad base, as when resting on its edge.*

For, the area being the same in both cases, the strength varies as EG and E*g*, (Fig. 122,) which are as 2 to 1. (Art. 71.)

Fig. 122.

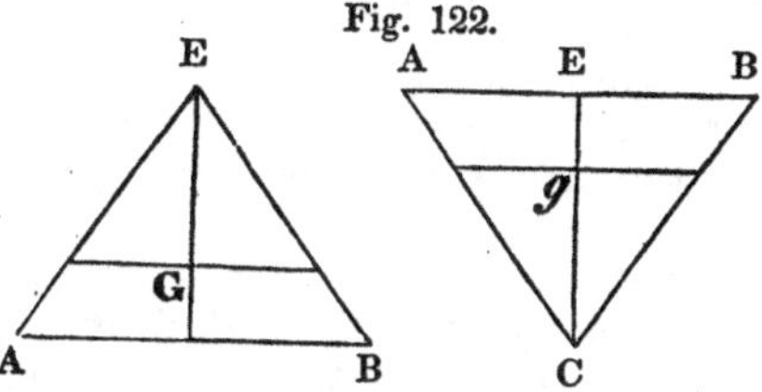

These principles apply not only to beams, but to bars, and similar structures of every sort of matter.

198. *The strength of any bar in the direction of its length, is directly proportional to the area of its transverse section.*

* For the reaction at each support is a force which $=\frac{1}{2}W$, acting upon $\frac{1}{2}L$ at its center of gravity; but the center of gravity of $\frac{1}{2}L$, is at a distance from the prop $=\frac{1}{2}$ of $\frac{1}{2}L=\frac{1}{4}L$; therefore the efficacy of the force $=\frac{1}{4}L\times\frac{1}{2}W=\frac{1}{8}L\times W$. (Fig. 120.)

Here each line of particles, in a longitudinal direction, may be considered as exerting a separate force, and therefore the aggregate force will manifestly depend on their number, and of course on the area of its section, the whole being equal to the sum of all its parts. Hence the various shapes of bars make no difference in their absolute strength, since this depends only on the area of the section, and must obviously be the same when the area is the same, whatever be its figure. A rope, therefore, or a wire, to which a weight is appended, is as likely to break in one place as in another, but when the weight of the rope becomes considerable, and the force is applied perpendicularly, the increase of weight, as its length increases, renders it more liable to break in the upper than in the lower parts.

199. *The lateral strengths of similar beams, are inversely as their lengths or breadths.*

Let D, *d*, represent the diameters of two *cylindrical* beams, or the sides of two beams in the form of *square prisms ;* then,

$S : s :: \frac{D^3}{L \times W} : \frac{d^3}{l \times w}$. In similar beams, $L \propto D$ and $W \propto D^3$

$$\therefore S : s :: \frac{D^3}{D^4} : \frac{d^3}{d^4} :: \frac{1}{D} : \frac{1}{d} :: \frac{1}{L} : \frac{1}{l}.$$

In *oblong* beams,

$S : s :: \frac{B \times D^2}{L \times W} : \frac{b \times d^2}{l \times w}$. But in similar beams, $L \propto B$,

$$\therefore S : s :: \frac{L \times D^2}{L \times D^3} : \frac{l \times d^2}{l \times d^3} :: \frac{1}{D} : \frac{1}{d} :: \frac{1}{L} : \frac{1}{l}.$$

When the beams are of the *same figure*, and their *lateral sections the same*, then, the breadth and depth of one being respectively equal to the breadth and depth of the other,

$S : s :: \frac{1}{L \times W} : \frac{1}{l \times w}$. And since $W \propto L$, $S : s :: \frac{1}{L^2} : \frac{1}{l^2}$.

Hence, half the length of a beam, supported at both ends, will bear four times as great a pressure as the whole beam ; and a prop placed under the center of a beam increases its strength in the same ratio.

Long beams are weak from their own weight ; and the length may be so increased, that they will break from this cause alone. The strength arising from making the beam larger, increases as the *square* of one of the homologous sides, while the weight increases as the *cube*, and therefore preponderates in long beams. To consider, in connection, the several circumstances which affect the strength of timber, it appears that a beam twice as broad as another of the same length, is also twice as strong ; that one twice as deep, the other dimensions remaining the same, is four times as strong ; and that one twice as long as another *similar* beam, has only half the strength.

200. But if besides their own weights, these beams are made to support other weights, W′, w', placed at their middle points, then their tendency to fracture will be increased.

For, since the reaction, arising from the pressure on each support$=\frac{1}{2}$W′, and this force acts at the point of fracture with a leverage equal to $\frac{1}{2}$L, the stress produced by $W'=\frac{1}{2}L\times\frac{1}{2}W'=\frac{1}{4}L\times W'$. But the stress arising from the weight of the beam itself$=\frac{1}{8}L\times W$. Therefore the whole stress$=\frac{1}{8}L\times W+\frac{1}{4}L\times W'=\frac{1}{4}L(\frac{1}{2}W+W')\propto L\,(\frac{1}{2}W+W')$,

$$\therefore S : s :: \frac{A\times G}{L(\frac{1}{2}W+W')} : \frac{a\times g}{l(\frac{1}{2}w+w')};$$

or, in the case of cylinders and square prisms,

$$S : s :: \frac{D^3}{L(\frac{1}{2}W+W')} : \frac{d^3}{l(\frac{1}{2}w+w')}.$$

If the weights of the beams be so small, when compared with the weights supported, as to make it unnecessary to take them into consideration, then

$$S : s :: \frac{A\times G}{L\times W'} : \frac{a\times g}{l\times w'} :: \text{(in cyl. and sq. prisms,)}\ \frac{D^3}{L\times W'} : \frac{d^3}{l\times w'}.$$

201. In order that the foregoing general formulæ may be applied to practice, so as to find the actual strength of bars or beams, it is necessary to have some standard of strength ascertained by experiment, which may be employed as the unit of comparison. For example, it is found by experiment that a small beam of oak, one foot long and one inch square, is able, when supported at both ends, to sustain a weight of 600 pounds; and that a bar of iron of the same dimensions, would sustain in the same circumstances 2190 pounds. The beam weighs half a pound, and the iron three pounds. With these data applied to the foregoing formulæ, we may perform such problems as the following.

1. What weight might be sustained at the middle point of a prismatic beam of oak, whose length is 6 feet, and its end 4 inches square?

Let S=strength of the beam required.

s=strength of a beam whose length is one foot and square end one inch.

W=weight of the larger beam, and w that of the smaller$=\frac{1}{2}$ pound. Let $L=6$, $l=1$, $D=4$, $d=1$. Weight required=W′. Given weight (600 pounds) $=w'$.

Then, the weight of the beams not being taken into the account, $S : s :: \frac{D^3}{L\times W'} : \frac{d^3}{l\times w'} :: \frac{4^3}{6\times W'} : \frac{1^3}{1\times 600}$.

But the strength, at the moment of fracture=0 in both cases, **i. e.** $S=s$; $\therefore \frac{4^3}{6\times W'}=\frac{1^3}{1\times 600}$; whence W′=6400 pounds.

If the weight of the beams be taken into the account, then (Art 200,) $S : s :: \frac{D^3}{L \times \frac{1}{2}W + W'} : \frac{d^3}{l \times \frac{1}{2}w + w'} :: \frac{4^3}{6 \times 24^* + W'} : \frac{1^3}{1 \times \frac{1}{4} + 600}$.

Hence, $\frac{64}{6 \times 24 + W'} = \frac{1}{600\frac{1}{4}}$; and $W' = 6378\frac{2}{3}$ pounds. Ans.

2. What must be the depth of a beam of an oblong prismatic form, whose breadth is 2 inches and length 8 feet, to support a weight of 6400 pounds, its own weight not taken into consideration?

Here, (Art. 200,) $S : s :: \frac{B \times D^2}{L \times W'} : \frac{1^3}{l \times w'} :: \frac{2 \times D^2}{8 \times 6400} : \frac{1}{600}$, $\therefore 2D^2 = 85.333$, or $D = 6.53$ inches. Ans.

3. What weight might be supported at the middle point of a bar of iron 10 feet long, and the side of whose square end is 3 inches, its own weight not being taken into consideration?

Ans. 5913 pounds.

202. *The stress (or tendency to fracture) on any part of a horizontal beam supported at both ends, is proportional to the product of its two distances from the supported ends.*

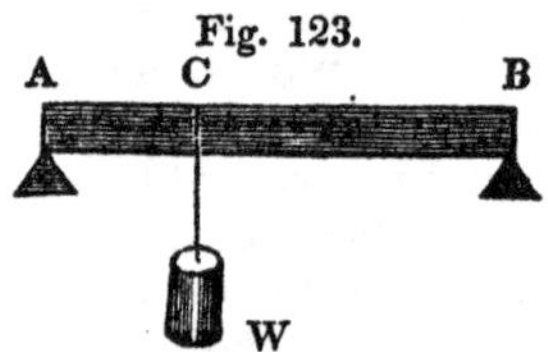

Fig. 123.

The sum of the pressures on A and B, (Fig. 123,) must obviously be equal to the whole weight. But, (Art. 102,)

Pressure at $A = \frac{W \times BC}{AB}$, and pressure at $B = \frac{W \times AC}{AB}$.

But the *reaction* of either point of support is equal to the pressure on that point; and this force acts at C with a *leverage* AC on one side, and BC on the other, so that the stress at $C = \frac{W \times BC}{AB} \times AC$ or $\frac{W \times AC}{AB} \times BC$, either of which expressions = stress at C, and $\propto AC \times BC$. And since this rectangle is greatest when $AC = CB$, and diminishes as these lines become more and more unequal in length, so the tendency of a horizontal bar to break is greatest in the middle, and decreases toward the points of support.

* For $W : w :: L \times D^2 : l \times d^2$, $\therefore W : \frac{1}{2} :: 6 \times 16 : 1 :: 96 : 1$ $\therefore W = 48$.

Hence a beam, in order to be equally strong throughout, must be thickest in the middle, being thinned off toward the ends; and if the sides of such a beam are parallel planes, the figure of the beam must be *elliptical.*

Fig. 124.

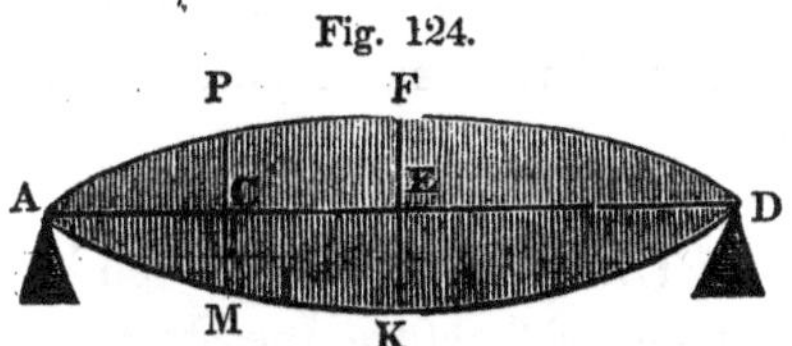

For let the curve APDM, (Fig. 124,) whose axis is AD, represent this longitudinal section, and let a=the thickness or breadth of the beam; then a lateral section of the beam at any point C, will be an oblong, whose breadth is a, depth PM, and the depth of its center of gravity $\frac{1}{2}$PM. Hence, the tendency of the beam to resist fracture at any point C, is as $a\times \text{PM}^2$; but the *stress* at C is as $\text{AC}\times\text{CD}$; therefore

$$\text{the strength at C} \propto \frac{a\times\text{PM}^2}{\text{AC}\times\text{CD}} \propto \frac{\text{PM}^2}{\text{AC}\times\text{CD}};$$

hence if $\text{PM}^2 \propto \text{AC}\times\text{CD}$, the strength will be the same at every point: but in this case the curve APDM is an ellipse, whose transverse is AD, and conjugate FK.

203. The timbers which compose the horizontal part of the frame of a house, being usually rectangular parallelopipeds of uniform dimensions throughout, it is manifest that a considerable portion of the material is wasted; but, in such cases, the attempt to save the material would be attended with paramount disadvantages. When however the material is expensive, or where lightness is important, as in many kinds of machinery, the foregoing principles may be applied with great advantage. A useful application of it is seen in the shape sometimes given to the iron bars of railways, as represented in the following figure.

Fig. 125.

204. Examples.

1. What must be the length of a beam 4 inches square, to support 6400 pounds at its middle point?

Let S=strength of the required beam,

s=strength of a beam 1 foot long, and its end 1 inch square.

W=weight of the larger beam,
w=weight of the smaller beam or $\frac{1}{2}$ lb.
By the question, L=required length, $l=1$, $D=4$, $d=1$.
$W'=6400$ lbs.
w'=weight sustained by the unit=600 lbs.

(1.) If the weight of the beams be not taken into consideration, by Art. 200,

$$S : s :: \frac{D^3}{L\times W'} : \frac{d^3}{l\times w'} :: \frac{4^3}{L\times 6400} : \frac{1^3}{1\times 600}.$$

But, at the moment of fracture, the strength in both cases becomes equal to nothing; and then, S being equal to s, $\frac{64}{L\times 6400} = \frac{1}{1\times 600}$; whence $L=6$ feet.

(2.) If the weight of the beams be taken into account, then, (Art. 200,) $S : s :: \frac{D^3}{L(\frac{1}{2}W+W')} : \frac{d^3}{l(\frac{1}{2}w+w')} :: \frac{4^3}{L(4^*+6400)} : \frac{1^3}{1(\frac{1}{4}+600)}$

$$\therefore \frac{64}{L(4L+6400)} = \frac{1}{\frac{1}{4}+600}.$$ Whence $L=5.98$ feet.

2. What must be the breadth of a beam of an oblong prismatic form, whose depth is 8 inches and length 6 feet, to support a weight of 6400 lbs., its own weight not being taken into the account?

Let B=required breadth.
b=breadth of the beam 1 foot long, and its end one inch square.

Then, (Art. 200,)
$S : s :: \frac{B\times D^2}{L\times W'} : \frac{b^3}{l\times w'} :: \frac{B\times 64}{6\times 6400} : \frac{1^3}{1\times 600}$. Hence at the moment of fracture, when $S=s$, $\frac{B\times 64}{6\times 6400} = \frac{1}{600}$, whence $B=1$ inch. In the two preceding examples, the beams are of the same length, and have equal strengths, each supporting a weight of 6400 lbs. But in beams of equal length, the solid contents are as the areas of the sections. In example 1st, the section$=4^2=16$; and in example 2d, the section$=1\times 8=8$. Hence,

The first beam : second beam :: 16 : 8. Therefore, the oblong beam placed edgeways is as strong as the square one, although it contains only one half as much material.

3. What weight may be supported at the middle of a bar of

* The weights being as the solid contents,
$W : w :: L\times D^2 : l\times d^2 :: L\times 16 : 1\times 1, \therefore W : \frac{1}{2} :: L\times 16 : 1, \therefore \frac{1}{2}W = \times L4$

iron, 10 feet long, and the side of whose square is 3 inches, its own weight being taken into the account?

$$S : s :: \frac{D^3}{L(\frac{1}{2}W+W')} : \frac{d^3}{l(\frac{1}{2}w+w')} :: \frac{3^3}{10(135^*+W')} : \frac{1^3}{1(1\frac{1}{2}+2190)}.$$

Hence, $W'=5782.05$ lbs.

205. The foregoing investigations and examples relate to beams supported at *both* ends: we proceed to the case where the beam is supported at only *one* end.

In similar cylindrical and prismatic beams, supported at one end, *the strength varies inversely either as the diameter or as the length.*

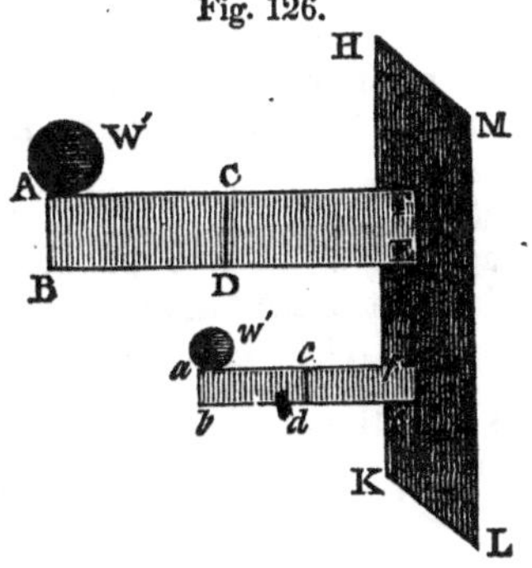

Fig. 126.

Let ABEF, *abef*, (Fig. 126,) represent the longitudinal sections of two prismatic beams fixed horizontally into the wall HKLM; then the tendency of these beams to *resist* fracture at the ends EF, *ef*, where they are inserted into the wall, will be measured in the same manner as in the preceding cases, that is, *by the area of the lateral section into the depth of its center of gravity,* except that in this case, the fracture will begin at the upper points F, *f*, and end at the lower points E, *e*. The tendency to *produce* fracture will be the weight of the beams acting at the distance of their centers of gravity from the ends EF, *ef*. Hence, (Art. 196,)

$S : s :: \frac{A\times G}{\frac{1}{2}L\times W} : \frac{a\times g}{\frac{1}{2}l\times w}$, or if any weights, W', w', are placed at the other ends of the beams, then (since the effects of these weights to produce fracture will be measured by $W'\times L$ and $w'\times l$) we have $S : s :: \frac{A\times G}{L\frac{1}{2}W+W'} : \frac{a\times g}{l\times\frac{1}{2}w+w'}$; and if the weights W, w of the beams are very small when compared with the weights W' w', then $S : s :: \frac{A\times G}{L\times W'} : \frac{a\times g}{l\times w'}$.

Hence, in similar beams, as in Art. 199, $S : s :: \frac{1}{D} : \frac{1}{d}$ or $\frac{1}{L} : \frac{1}{l}$.

Let W, w, represent the weights of the parts ABCD, *abcd*, of the beams, then the tendency of those parts to produce fracture at C, c, will be measured by $\frac{1}{2}AC\times W$, and $\frac{1}{2}ac\times w$; therefore, if

* The bar of iron weighs 270 lbs.

S, s represent the strengths of the beams at the points C, c, then $S : s :: \frac{A\times G}{AC\times\overline{\frac{1}{2}W+W'}} : \frac{a\times g}{ac\times\overline{\frac{1}{2}w+w'}}$; or if W, w, be very small with respect to W′ w', then $S : s :: \frac{A\times G}{AC\times W'} : \frac{a\times g}{ac\times w'}$.

Hence if a given weight W′ be supported at the end of a given beam whose weight is so small as not to be taken into consideration, the strength of that beam to support the weight W′ at any point C, between A and F, will vary as $\frac{A\times G}{AC\times W'}$; or since W′ is constant, as $\frac{A\times G}{AC}$.

A beam supported at one end in the form of an ISOSCELES WEDGE, *or of a* PARABOLA, *is equally strong throughout; or, when a weight is hung at the end, the beam is as liable to break in one place as in another.*

Let the beam be in the form of an *isosceles wedge*, (Fig. 127,) whose flat sides are parallel to the horizon, and whose given depth$=d$; then $A=ED\times d$, and $G=\frac{1}{2}d$, $\therefore A\times G=\frac{1}{2}ED\times d^2$, which varies as ED or EC, which varies as AC. Hence the strength is as $\frac{AC}{AC}$, that is, it is constant.

Fig. 127.

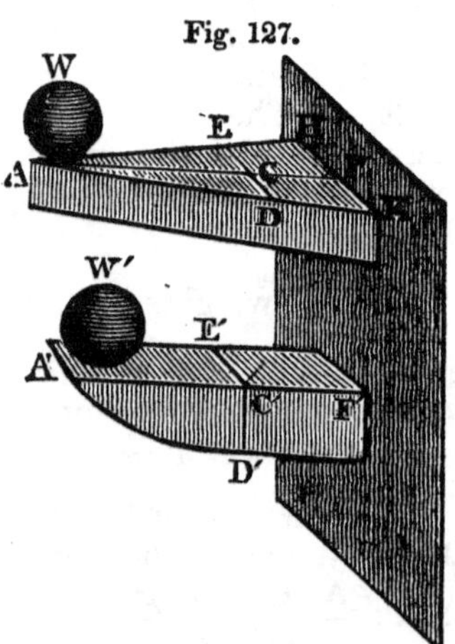

If the sides of the beam be parallel planes, and its longitudinal section a semi-parabola, (see Fig. 127,) as A′ C′ F′ D′; then let d equal the given breadth, and we have $A=d\times C'D'$, $G=\frac{1}{2}C'D'$, $\therefore A\times G=\frac{1}{2}d\times C'D'^2$, which varies as $C'D'^2$. Hence $S \propto \frac{C'D'^2}{A'C'}$. But since A′D′ is a parabola, $A'C' \propto C'D'^2$, $\therefore S \propto \frac{C'D'^2}{C'D'^2}$.

206. Suppose the beams to be cylinders or square prisms, (resting either on one end or on both ends,) whose diameters or sides are D, *d*, then $S : s :: \frac{D^3}{L\times\frac{1}{2}\overline{W+W'}} : \frac{d^3}{l\times\frac{1}{2}\overline{w+w'}}$. Let $w'=0$, then $S : s :: \frac{D^3}{L\times\frac{1}{2}\overline{W+W'}} : \frac{d^3}{\frac{1}{2}l\times w}$, which expresses the relative strengths of two cylindrical beams whose lengths are L, *l*, diameters D, *d*, weights W, *w*, the former of which supports the given weight W′ at the end of it, and the latter supports only its own weight. Let $d=D$;

$$\text{then } S : s :: \frac{1}{L\times\frac{1}{2}\overline{W+W'}} : \frac{1}{\frac{1}{2}l\times w} :: \frac{1}{L\times\frac{1}{2}\overline{W+W'}} : \frac{2L}{W\times l^2};$$

for, since $w : W :: l : L, \therefore w=\frac{W\times l}{L}$. In this case, therefore, the ratio of $S : s$ expresses the relative strengths of two cylindrical beams of the same diameter, one of which supports the given weight W′ at its end, and the other supports only its own weight; and if the beam whose length is L breaks when W′ is placed at the end of it, the beam whose length is *l* will break by its own weight. Hence, let $S=s$, then $\frac{1}{L.\frac{1}{2}\overline{W+W'}}=\frac{2L}{W\times l^2}$

$$\therefore l^2=\frac{L^2(W+2W')}{W} \therefore l=L\left(\frac{W+2W'}{W}\right)^{\frac{1}{2}}$$ =length of the beam of the same diameter that would break by its own weight.

207. Let the beams be *similar* cylinders, then $D^3 : d^3 :: L^3 : l^3$, $\therefore$ (Art. 206,) $S : s :: \frac{L^2}{\frac{1}{2}W+W'} : \frac{l^2}{\frac{1}{2}w} :: \frac{L^2}{\frac{1}{2}W+W'} : \frac{2L^2}{W\times l}$.* And when $S=s, \frac{L^2}{\frac{1}{2}W+W'}=\frac{2L^2}{W\times l} \therefore \frac{1}{2}W+W'=\frac{W\times l}{2L}$, or $l=\frac{L(W+2W')}{W}$. If, therefore, a cylindrical beam whose length is L breaks with the given weight W′ placed at the end of it, a similar cylindrical beam whose length is $\frac{L(W+2W')}{W}$ will break with its own weight.

208. If a horizontal beam be supported at *both ends*, the stress produced by its own weight, W, is measured by $\frac{1}{8}L\times W$, (Art. 196.)

If the beam be supported at *one end* only, the stress is meas-

* The weights of similar cylinders of the same density are as the cubes of their diameters or lengths; therefore, $w : W :: l^3 : L^3 \therefore w=\frac{Wl^3}{L^3}$.

ured by the whole weight applied at the center of gravity, and consequently the stress$=\frac{1}{2}L\times W$.

Therefore a beam supported at both ends has four times the strength of the same beam, supported only at one end. And if a certain beam resting on one end breaks by its own weight, a beam of the same dimensions twice as long will break by its own weight when resting on two supports, the former having just four times the strength it would have if twice as long.

If, however, instead of the weight of the beam itself, this is left out of the account, and a weight W′ be appended, then the stress on the beam when supported at *one end* will be measured by $L\times W'$; while, in the case of the beam supported at *both ends*, (since the weight being at the center is also at the center of gravity of the beam,) the stress is measured as before, by $\frac{1}{4}L\times W'$. (Art. 200.) Therefore, a weight appended at the end of a beam supported only at one end produces four times the stress, as the same weight applied at the center of the beam when supported at both ends.

209. Examples.

1. What must be the length of a beam of oak one inch square, supported at both ends, which is just capable of bearing its own weight?

By Art. 201, a beam of oak 1 foot long and 1 inch square, weighing $\frac{1}{2}$ pound, just supports 600 pounds. And by Art. 206. the expression $l=L\left(\frac{W+2W'}{W}\right)^{\frac{1}{2}}$ denotes that when a beam whose length is L breaks when W′ is placed at the end of it, l is the length of a beam that will break with its own weight; consequently, since here $L=1$, $W=\frac{1}{2}$, and $W'=600$, $l=\left(\frac{\frac{1}{2}+1200}{\frac{1}{2}}\right)^{\frac{1}{2}}=(2401)^{\frac{1}{2}}=49$ feet.

2. What must be the length of a bar of iron 1 inch square, supported at one end, which would break by its own weight?

Here $L=1$ foot, $W=3$ pounds, and since (Art. 208) a beam supported at one end will break with $\frac{1}{4}$ as great a weight as when supported at both ends, $W'=547\frac{1}{2}$ pounds,

$$\therefore l=1\left(\frac{3+2\times547\frac{1}{2}}{3}\right)^{\frac{1}{2}}=19.13.$$

3. Since a bar of iron 1 inch square, and 1 foot long, will support a weight of 2190 pounds, what must be the dimensions and

weight of a similar bar, which will break with its own weight when supported at both ends?

The required beam being *similar* to the given beam, therefore, by Art. 207, its length equals 1461 feet. And L (1) : l (1461) :: D (1 inch) : $d=121\frac{3}{4}$ feet=the side of its square. Again, since the weights are as the cubes of the homologous sides,

w : W(3) :: $l^3(1461)^3$: 1, ∴ the weight=9355605543 pounds.

4. Two beams are of equal length and weight, the first being a square prism whose section is 4 inches square, the second an oblong 8 by 2 inches: How much stronger is the second beam than the first, and how much stronger when laid on the narrow than on the broad side?

Ans. The second beam is TWICE as strong as the first, and FOUR times as strong when laid on the narrow, as on the broad side.

210. On the foregoing principles Dr. Gregory makes the following remarks, most of which were originally suggested by Galileo, to whom we are indebted for the earliest investigation of these propositions. From the preceding deductions (says Gregory) it follows, that greater beams and bars must be in greater danger of breaking than less similar ones; and that, though a less beam may be firm and secure, yet a greater similar one may be made so long as necessarily to break by its own weight. Hence Galileo justly concludes, that what appears very firm, and succeeds well, in models, may be very weak and unstable, or even fall to pieces by its weight, when it comes to be executed in large dimensions, according to the model. From the same principles he argues that there are necessarily limits in the works of nature and art, which they cannot surpass in magnitude; that immensely great ships, palaces, temples, &c., cannot be erected, since their yards, beams, bolts, and other parts of their frame, would fall asunder by their own weight. Were trees of a very enormous magnitude, their branches would, in like manner, fall off. Large animals have not strength in proportion to their size; and if there were any land animals much larger than those we know, they could hardly move, and would be perpetually subjected to the most dangerous accidents. As to marine animals, indeed, the case is different, as the specific gravity of the water sustains those animals in a great measure; and in fact these are known to be sometimes vastly larger than the greatest land animals.* It is (says Galileo) impossible for Nature to give bones to men, horses, or other animals, so formed as to subsist, and successfully to perform their offices, when such animals should be enlarged to immense heights, unless she uses matter much firmer

* Whales in the Northern Regions, are sometimes found sixty feet long, and weighing seventy tons.

and more resisting than she commonly does; or should make bones of a thickness out of all proportion; whence the appearance and figure of the animal must be monstrous. Hence we naturally join the idea of greater strength and force with the grosser proportions, and that of agility with the more delicate ones. The same admirable philosopher likewise remarks, in connection with this subject, that a greater column is in much more danger of being broken by a fall, than a similar small one; that a man is in greater danger from accidents than a child; that an insect can sustain a weight many times greater than itself, whereas a much larger animal, as a horse, could scarcely carry another horse of his own size.*

211. *The lateral strengths of two cylinders, of the same matter, and of equal weight and length, one of which is hollow, and the other solid, are to each other as the diameters of their sections.*

Fig. 128.

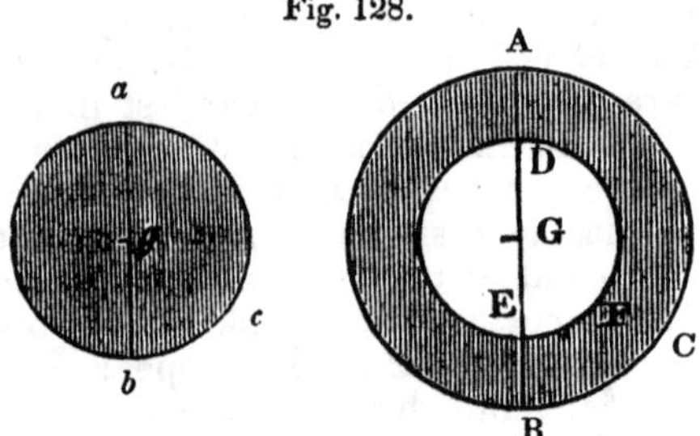

Let ABC, *abc*, (Fig. 128,) represent sections of two cylinders, of equal length, and containing equal quantities of matter, of which ABC is hollow, and *abc* is solid. Then the area of the ring whose breadth is AD, is equal to that of the circle *abc*. But the strengths of these areas are as the areas multiplied by the distances of their centers of gravity from the points of pressure A, *a*, (Art. 194;) or, since the areas are equal, the strengths are as AG : *ag*, that is, as the diameters of their sections.

The strongest form, therefore, in which a given quantity of matter can be disposed, is that of a hollow cylinder; and leaving out of view the diminished rigidity of their structures or fabrics, there would seem to be no limits to the strength which might be given to such a cylinder by increasing its diameter. But the proposition is true only when the sections are perfectly circular; and this condition, connected with the want of rigidity when the annulus becomes very thin, occasions limits to the actual operation of the principle.

212. From this proposition Galileo justly concludes, that Nature in a thousand operations greatly augments the strength of

* Gregory's Mechanics, I, 110.

substances without increasing their weight; as is manifested in the bones of animals, and the feathers of birds, as well as in most tubes or hollow trunks, which, though light, greatly resist any effort to bend them. Thus, (says he,) if a wheat straw, which supports an ear heavier than the whole stalk, were made of the same quantity of matter, but solid, it would bend or break with far greater ease than it now does. And with the same reason, art has observed, and experience confirmed the fact, that a hollow cane, or tube of wood or metal, is much stronger or firmer, than if, while it continues of the same weight and length, it were solid; as it would then, of consequence, be not so thick. For the same reason, lances, when they are required to be both light and strong, are made hollow.*

213. The area $ABC : abc :: AB^2 : ab^2$.

And $DEF : abc :: DE^2 : ab^2$.

$\therefore ABC-DEF : abc :: AB^2-DE^2 : ab^2$.

Or the area of the ring is to the area of the solid section, as $AB^2-DE^2 : ab^2$. If the area of the ring is equal to the solid section, then $AB^2-DE^2=ab^2$, and $ab=\sqrt{(AB^2-DE\ )}$.

What weight could be sustained at the middle point of a cylindrical iron tube 8 feet long, whose diameter is $1\frac{1}{2}$ inches, and thickness $\frac{1}{4}$ of an inch; supposing the tube to be supported at both ends?

The diameter of a solid cylinder of the same length and weight=1.11 inches, and (Art. 200)

$$S : s\dagger :: \frac{A\times G}{L\times W'} : \frac{a\times g}{l\times w'} :: \frac{.532}{8\times W'} : \frac{.3927}{1720}.$$

Therefore $W'=291.3$ lbs.=weight which would be sustained by a solid cylinder containing the same quantity of matter as the tube, and which consequently measures the strength of the cylinder. But, (Art. 211,) putting S for the strength of the tube and S′ for that of the cylinder,

$$S : S' :: \tfrac{3}{2} : 1.11, \therefore S : 291.3 :: \tfrac{3}{2} : 1.11, \therefore S=393.64 \text{ lbs.}$$

* Gregory, I, 112.

† Since a *bar* of iron 1 foot long and 1 inch square, weighs 2190 lbs., (Art. 201,) a *cylinder* of the same dimensions weighs 1720 lbs.

PART II.—PRACTICAL APPLICATIONS OF THE PRINCIPLES OF MECHANICS.

CHAPTER I.

OF THE MECHANICAL PROPERTIES OF MATTER.

214. *Matter* constitutes the great subject of Chemistry; *motion*, that of Mechanical Philosophy. Chemistry inquires, first, whether a given body is simple or compound,—whether it consists of one kind of matter, or of two or more different kinds of matter united in one body; and, secondly, what are the peculiar properties of each individual body. Mechanical Philosophy, on the other hand, takes cognizance of those properties of matter only which belong either to all bodies whatsoever, or at least to extensive classes of bodies. The changes it contemplates are those which appertain to the quantity, position, figure, or motion of bodies, while it leaves to Chemistry all those changes which alter the constitution of bodies, transforming them into something of a different nature from what they were before.

215. The leading mechanical properties of matter are Divisibility, Porosity, Compressibility, Elasticity, Indestructibility, and Attraction.

Divisibility.—Matter is susceptible of mechanical division beyond any known limits. It is said that a grain of *musk* is capable of perfuming for several years a chamber twelve feet square, without sustaining any sensible diminution of its volume or weight. Such a chamber contains nearly 3,000,000 cubic inches, and as the odor of the musk pervades every part of the room, a certain number of particles are contained in each cubic inch. The air of the room may be, in the mean time, changed many thousand times, and a new supply of odorous particles furnished to each successive portion of air. Hence the number of particles diffused in the time supposed exceeds all computation, and yet the weight of the material is not sensibly diminished. The thickness of a *soap-bubble*, according to Newton, the moment before it bursts, is only the four-millionth part of an inch. The thread of a silk-worm is a perfectly smooth cylinder whose diameter is nearly the two thousandth part of an inch, and yet the spider's web is vastly more attenuated.

216. Porosity.—In many bodies the pores, or vacant spaces, are easily distinguishable by the naked eye, as in the case of sponge, wood, and most kinds of stones. Many substances which do not exhibit pores to the naked eye, still betray them to the microscope. Metals do not usually, when pure, appear porous, even under the microscope, but still such a structure may be detected by mechanical means. Thus if a hollow ball of gold be filled with water, plugged close, and compressed in a vise, the water will exude through the metal. By means of this structure, in animals and vegetables, air and water circulate freely through them, aiding the functions of life, as the sap in trees. A cross section, or thin slice of wood, viewed with the microscope, shows that the pores occupy usually a much greater space than the solid matter of the wood. Indeed, the solid partitions between the larger cells are themselves seen, by powerful magnifiers, to be full of pores. The surface of the body of a middle-sized man has been estimated to contain more than 20,000,000 pores,* the skin being perforated with a thousand holes to every inch. Wood consists of bundles of fibres of different degrees of fineness, usually aggregated together so loosely, that a free circulation of water is easily maintained between them. Glass is the only solid known which appears, as far as experiments have gone, to be absolutely impermeable to all fluids.†

217. Compressibility.—All bodies yield more or less to external pressure, undergoing a diminution of volume proportional, in each case, to the force applied. Aeriform bodies, as common air, yield readily to any compressing force, the diminution of volume being always exactly proportional to that force. Solids, also, as wood and stone, are compressed, in different degrees, under heavy weights. A cork immersed two hundred feet in the sea, is so much compressed as to become heavier than water and to sink; and a pint bottle of fresh water, corked closely, and sunk to a great depth in the ocean, will, when drawn up, be found to be filled with salt water. This remarkable fact is explained by supposing that the cork has been so much contracted in bulk as to admit the salt water, which being heavier than the fresh, displaced it and occupied the bottle. As the cork sustained an equal pressure on all sides, it would not be removed out of its place; and, as the bottle was drawn up, and the pressure was diminished, the cork would regain its original dimensions. Hard mineral substances, as blocks of granite, indicate some contraction of volume when subjected to the pressure of high and massive walls. Liquids resist compression much more than either air or solid bodies. Still, under enormous weights, it may be rendered sensible, as will be more fully explained hereafter.

* Leslie, Nat. Phil. I, 18.

† Pouillet, El. Phys. t. I, 29.

218. ELASTICITY.—Bodies are said to be *perfectly* elastic when they restore themselves to their original dimensions when released, and with a force equal to that with which they were compressed. *Air* and all gases are of this class; and even *liquids*, as water, are found to conform to the same law, and, in this sense therefore, they must also be regarded as perfectly elastic substances.* *Metals*, indeed, have the same property, a double extension or compression, requiring twice the force; triple, three times the force, and so on. The elasticity of wood is exemplified in the cross-bow, and that of a mineral substance peculiarly in mica. The elasticity of *torsion*, or the force by which a wire when twisted endeavors to resume its natural state, is employed as the most delicate test and measure of force known, the force of torsion being always proportional to the angle through which the body has been twisted.

219. INDESTRUCTIBILITY.—*Matter is wholly indestructible.* In all the changes we see going on in bodies around us, not a particle of matter is lost; it merely changes its form; nor is there any reason to believe that there is now a particle of matter either more or less than there was at the creation of the world. When we boil water and it passes into the invisible state of steam, this, on cooling, returns again to the state of water without the least loss; when we burn wood, the solid matter of which it is composed, passes into different forms, some into smoke, some into different kinds of airs, or gases, some into steam, and some remains behind in the state of ashes. If we should collect all these various products, and weigh them, we should find the amount of their united weights the same as that of the body from which they were produced, so that no portion is lost. Each of the substances into which the wood was resolved, is employed in the economy of nature to construct other bodies, and may finally re-appear in its original form. In the same manner, the bodies of animals, when they die, decay and seem to perish; but the matter of which they are composed merely passes into new forms of existence, and re-appears in the structure of vegetables or other animals.

220. ATTRACTION.—This property of matter produces or governs a large part of the phenomena of the natural world. By it all matter tends toward all other matter, and by it the particles of matter unite, forming innumerable compounds. It is chiefly this property in different degrees which constitutes the STRENGTH OF MATERIALS,—a subject which has been already considered *theoretically*, (Arts. 193, 213,) but which it will be useful now to consider *practically*, in its relation to the arts.

* Mosely's Illustrations, Sec. 32.

221. The strength of substances *in the direction of the length* has been determined, experimentally, by suspending small cylinders or wires of each material vertically, and applying weights at the bottom until they broke. Of all substances, that which sustains the greatest load is *iron.* Different materials, before rupture, increase in length in different degrees. Bars of the best wrought iron are elongated about .000082 for a load of one ton to the square inch. Iron in the form of wires admits of a greater extension. A bundle, or cable of small wires, will, under similar circumstances, be elongated .000091; and becomes more extensible in proportion as the wires are smaller. Bar iron will bear to be extended .000714 without injury; and several kinds of wood, as oak, pine, and fir, will bear an elongation three times as great. Iron wire, on account of its extraordinary tenacity, has within a few years been most successfully applied to the construction of suspension bridges. Iron wire $\frac{1}{30}$ inch in diameter, has been found by experiment capable of bearing a load of 60 tons to the square inch without breaking. In one instance, indeed, the load sustained was 90 tons. The Menai bridge, in Wales, one of the most celebrated works of the age, is supported by iron wire cables. Its span between the points of suspension is 560 feet, its height above highwater mark 100 feet, and the roadway 30 feet in breadth. Its weight is over four millions of pounds, (2000 tons,) but the whole is suspended by four lines of strong iron cables by perpendicular rods 5 feet apart. Russia *bar iron* has a tenacity of 27 tons to the square inch. The best *cast iron* is about one third as strong, having a tenacity of about 9 tons to the inch. The tenacity of platina wire is nearly as great as that of bar iron. The comparative strength of several substances much used in the arts, is thus stated by Mosely: 7 rods of mahogany, taken together; 5 of pine, oak, or beach; 3 of box or of cast iron; 2 of gold; 1½ of silver or copper, have respectively the same tenacity as 1 corresponding rod of wrought iron; or as a rod whose section is $\frac{5}{12}$ made of steel or fine wire cable.

222. As the materials used in building are liable to give way by the superincumbent weight, it is often important to know the relative power of different materials to resist forces tending to *crush* them. Numerous experiments on this subject have been instituted by taking small blocks of similar shape and size, and loading them respectively with weights until they were crushed. It appears that cast iron is best adapted of all the materials in common use to sustain such pressures; that bar iron is not more than half as strong; granite, one sixth; Italian marble, one seventh; free-stone, one tenth; brick work, still less. Wooden columns have comparatively little power of resisting a force tending to crush them in the direction of their fibres. Short columns, however, bear stronger pressures than long ones, the

strength diminishing in a geometrical, while the height is increased in an arithmetical ratio. Slight changes of form are found sometimes greatly to affect the strength of a column. Thus merely *rounding* the ends of a perpendicular column makes its strength only one third that of a column whose extremities are flat. If we take three columns, equal in all respects, and round both ends of the first, one end of the second, and leave both ends of the third flat, their respective strengths will be as the numbers 1, 2, 3. The shape of a column has great influence on its strength. A cylindrical column is weakest in the middle; and it is found that the strength of a column of cast iron, containing a given weight of metal, whether it be solid or hollow, is much greater when it is cast in the form of a double cone, that is, with its greatest thickness in the middle of its height, and tapering to its extremities, than when cast in any other form. In lofty stone columns, however, their own weight may constitute a large part of the load to be sustained, and hence it is the practice of architects to make the swell below the center. In some of the Grecian temples, it was one third from the base; and this rule is now frequently adopted.

223. In practice, materials cannot safely be subjected to constant strains or thrusts approaching to those which produce rupture. They are liable to various occasional and accidental pressures; and others of a permanent kind, resulting from the settling of the pile. It is therefore regarded as not entirely safe to load any structure of stone more than one sixth the amount of the pressure that crushes it. Iron, cast or wrought, may be loaded to one fourth that amount.

The view taken by Barlow of the mode in which a horizontal beam undergoes compression, leads to results somewhat different from those investigated in Articles 193, &c. When a beam is bent in the middle, the fibres on the upper side undergo compression, while those on the under side undergo extension, as in Fig. 129,

Fig. 129.

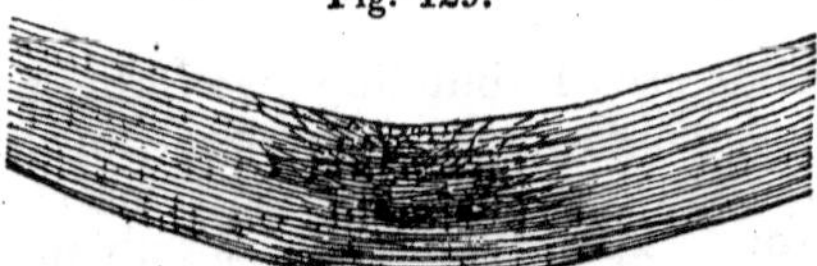

and between the two is a line that sustains neither, which is called the neutral axis. Since, throughout its neutral axis, the strength of the beam is not at all called into action, this will not be impaired by boring a hole through the beam in the direction of that axis. What constitutes the strength of a beam is its resistance to extension on the lower, and to compression on the upper side. These act as antagonist forces, and if either of them yields, the beam is broken Hence the strength of the beam is not impaired by sawing it through the upper side as low as to the neutral axis. This extends to about five eighths of the depth.*

224. Although wood has not intrinsically the strength of iron or stone, yet its lightness in some measure compensates for this, so that large structures of wood have sometimes great power of resistance to external forces. Pine is only one fifteenth as heavy as cast iron, while it has more than half the tenacity. Sixteen bars of it would weigh only the same as one bar of wrought iron, while they would have three times the strength. Many large structures, when constructed of heavy materials, are weak from their own weight. Iron roofs have been known to fall in by their own pressure. Trees often resist the action of external forces which overturn works of art apparently of much more stable materials; and nothing is more deserving the attention of the architect than the rules which nature has observed, both in the selection and distribution of the materials of which trees are constructed. The tapering form of their trunks; the increased diameter and density of their bases; the buttresses that frequently, in a large tree especially, support the trunk on every side; the comparative lightness of the extended top; and the universal symmetry of form that pervades the entire structure; these qualities, severally and collectively, add to the strength of trees, and fit them to encounter the most violent winds. Indeed, mechanical writers, when they have descended to a minute investigation of the structure of trees, have found the most refined use made of such mechanical principles as tend both to the greatest strength and greatest economy of material. Thus, in a cylinder of wood, like the trunk of a tree, the neutral axis is near the center; hence the resistance to compression on one side and to that extension on the other, act in opposition to each other, when a tree is bent by the wind, and thus the trunk is prevented from breaking.

CHAPTER II.

GENERAL OBSERVATIONS ON MOTION.

225. Motion and rest are accidental states of bodies, nor is a body naturally prone to one state more than to the other. If it is found at rest, it is because it is kept in equilibrium by opposite

* Mosely.

and equal forces; and if it is found in motion, it is because it has been put in motion by some force extrinsic to itself. The resistances to motion which exist near the surface of the earth, particularly gravity, create a seeming tendency to a state of rest; but, in reality, rest is no more the natural state of bodies than motion is.

226. Motion is distinguished into absolute and relative. *Absolute motion*, is a change of place in space with respect to any fixed point: *Relative motion*, is a change of place in bodies with respect to each other. A body may be at the same time in a state of absolute motion, and of relative rest. Thus, all the different articles contained in a ship under sail, have a motion in common with the ship, but may be at rest with respect to each other. When a man walks toward the stern of a ship at the same rate as that of the ship, he is in motion with respect to the ship, but at rest with respect to the earth. When a balloon, carried along by the wind, attains the same velocity as the wind, it is relatively at rest, and appears to the aeronaut to be in a perfect calm, although it may be actually moving one hundred miles an hour. Since the earth in its annual revolution around the sun, is moving eastward at the rate of nineteen miles, or 100,000 feet per second, were a cannon ball, at a certain time of day, fired eastward at the rate of 2000 feet per second, the only effect would be to add 2000 feet to the velocity which the ball had before in common with the earth: and were it fired westward, the effect would be merely to stop 2000 out of 100,000 parts of its previous motion, while the cannon would proceed onward, leaving it behind.* Did not the atmosphere partake of the diurnal motion of the earth, but were it to remain at rest with respect to this motion, the progress of any place to the east, would cause a relative motion of the air, or a wind, westward, which would blow with a violence far surpassing that of the most terrible hurricanes.†

Indeed, we cannot be sure that we have ever seen a body absolutely at rest. In our stillest moments, we are revolving with the earth on its axis; we are accompanying the earth in its annual revolution from west to east around the sun; and are perhaps attending the solar system around a common center of motion.‡

227. *Apparent motion*, as distinguished from relative, is that in which the body that seems to be moving is quiescent, and the motion is owing to a real motion in the spectator. Thus, the

* Robinson's Mechanical Phil. I, 31.

† Winds are in fact frequently produced by this cause, viz. by their having a relative velocity different from that of the part of the earth over which they blow.

‡ Young's Natural Phil. I, 19.

backward motion of the trees to one riding rapidly, the receding of the shore to one who is sailing from it with a fair wind, and the diurnal motions of the heavenly bodies from east to west, in consequence of the revolution of the spectator in the opposite direction; these are severally examples of apparent motion. It is often a very difficult problem to deduce the real from the apparent motion. While a planet, as Venus, is revolving about the sun in an orbit nearly circular, its motions, as seen from the earth, are extremely irregular; and to make all these irregularities consistent with the real motion, has been a perplexing problem in astronomy. We can sometimes decide that a given motion is real, because we observe a *cause* in operation, which is competent to produce it. The impulse of the wind, or the direction of the current, will satisfactorily account for a ship's receding from a given object, while no cause appears why the object should recede from the ship; the revolution of the earth on its axis is a cause competent to explain the apparent revolution of the heavens, while we can find no cause for their actual revolution. The *effects* also of a given motion, enable us to decide whether it is real or apparent. Thus a constant tendency to move in a straight line is characteristic of real motion.*

228. The Laws of motion have been already recited in Chapter I, and concise illustrations of them were added in that place. It was necessary to proceed thus far at the beginning of this work, since these constitute the fundamental principles of mechanics. By their great comprehensiveness, they furnish the most convenient classification of the various phenomena of motion, and it will therefore be useful to resume the consideration of them. They are very remarkable examples of a happy generalization; but their very comprehensiveness renders them difficult to be understood by the young learner; nor can they be thoroughly mastered in all their relations, until after considerable proficiency is made in the science of Mechanics. These laws indeed are the chief foundation of Newton's Principia.†

229. First law.—*A body continues always in a state of rest, or of uniform rectilinear motion, until, by some external force, it is made to change its state.*—This law contains the doctrine of inertia, expressed in four particulars. First, that unless put in motion by some external force, a body always remains at rest; secondly, that when once in motion, it continues always in motion, unless stopped by some force; thirdly, that this motion, arising from inertia, is uniform; and, fourthly, that this motion is in right lines. The proofs by which this and the other laws of motion are established, have been already stated. (Art. 22.)

* Wood's Mechanics, p. 22.

† Young's Nat. Phil. I, 26

It is our present object to make the application of these laws to various phenomena of nature and art.

230. And first, with respect to bodies *at rest*. The operation of this principle is seen, when a horse starts suddenly forward, and his rider is thrown backward. "When we desire a person, with suspected disease of the brain, to shake his head, and tell whether he feels pain, we are doing nearly the same as if we touched the naked brain with the finger to find the tender part, for the inertia of the brain, when the skull is moved, causes a momentary pressure between it and the skull, almost equivalent for our purpose to such a touch."* In consequence of the inertia of matter, before a body can be brought to the required velocity, this velocity must be impressed upon every particle of matter it contains. Hence, the more numerous its particles, the greater its inertia, which is therefore proportioned to the quantity of matter. But the weight also is proportioned to the quantity of matter, and therefore the inertia is proportioned to the weight. Yet it must be carefully distinguished from weight, having in fact nothing in common with it, except that both are proportioned to the quantity of matter, and of course to each other. But were we to strike with a hammer upon the top of a body falling toward the earth, the resistance from inertia would be the same as if the body were struck with the same force on the side; or in whatever direction the blow were applied, a similar resistance would be felt. This seems little else than what we commonly understand by the reaction of a body; but we conceive this reaction itself to depend upon an inherent property in matter, to which we give the name of Inertia. Inertia is the *cause* and reaction the *effect*. A vast weight may be moved on a horizontal railway by a comparatively small force, provided it can be got in motion with the required velocity. In transporting large quantities (eighty tons, for instance) of coal, the weight is distributed into a number of different cars, connected together by a loose chain, in order that the inertia of the several parts may be overcome successively.†

231. In consequence of the inertia of matter, the motion applied to a body does not instantly pervade the mass. In order to this, motion must be applied gradually, especially if the body is large; for if it is applied suddenly, it is frequently all expended on a part only of the mass, the cohesion is overcome, and the body is broken. This explanation accounts for several familiar facts. When a team starts suddenly with a heavy load, the effort is either wholly ineffectual, or some part of the har-

* Arnott's El. Phys. p. 50.
† See account of the Hatton Railway in "Strickland's Reports."

ness or tackling gives way. If we draw a heavy weight by a slender string, a slow and steady pull will move the weight, when a sudden twitch would break the string without starting the mass. The same principle applies to bodies already in motion. Thus, when a horse in a carriage starts suddenly forward, he may break loose as well when the carriage was previously in motion, as when it was at rest. The inertia of a body is in fact the same whether the body is in motion or at rest, opposing the same resistance to its moving with increased velocity, as to its beginning to move from a state of rest. Several singular phenomena result from the same cause, showing that time is necessary in order that motion communicated by impulse, may pervade an entire mass. A pistol ball fired through a pane of glass, frequently makes a smooth, well-defined hole, and does not fracture the other parts of the glass. Here, the momentum of the ball is communicated to the particles of glass immediately before it. Had the impulse been gradual, the same motion would have diffused itself over the whole pane, and every part would have felt the shock. A ball fired through a board delicately suspended, causes no vibrations in the board. A cannon ball, having very great velocity, passes through a ship's side, and leaves but little mark, while one with less speed, splinters and breaks the wood to a considerable distance around. A near shot thus often injures a ship less than one from a greater distance.* A soft substance, as clay or tallow, may be fired through a plank,—the body, by its great momentum, forcing its way through the plank, before the motion has had time to diffuse itself through the contiguous parts. The whole momentum being concentrated upon the part immediately before the body, the cohesion of that part is destroyed.

232. *Secondly*, let us consider the effects of inertia as it respects bodies *in motion*. All bodies in contact with each other acquire a common motion; as, for example, a horse and his rider, a ferry-boat and its passengers, a ship and every thing within it, the earth and all things on its surface. Whenever either of these bodies stops suddenly, the movable bodies connected with it are thrown forward. Were the revolution of the earth on its axis to be suddenly arrested, the most dreadful consequences would ensue; every thing movable on its surface, as water, rocks, cities, and animals, not receiving instantaneously this backward impulse, would fly off eastward in promiscuous ruin. Were the diurnal motion of the earth, however, very gradually diminished, until it finally ceased, so that time should be afforded to communicate the loss by slow degrees to the bodies on its surface, no such effects would take place. If a passenger leaps from a carriage in rapid motion, he will fall in

---

* Arnott's El. Phys. p. 104.

the direction in which the carriage is moving at the moment his feet meet the ground; because his body, on quitting the vehicle, retains, by its inertia, the motion which it had in common with it. When he reaches the ground, this motion is destroyed by the resistance of the ground to the feet, but it is retained in the upper and heavier part of the body, so that the same effect is produced as though the feet had been tripped.

233. Although, on account of the numerous impediments to motion which exist on the surface of the earth, bodies are unable to maintain for any considerable time the motion they have acquired, yet we see the first law of motion, so far as it respects the tendency of bodies to persevere in motion, fully confirmed in the continued and unaltered revolution of the heavenly bodies. These are impelled by no renewed forces, but revolve from age to age in an undeviating course, simply because they meet with no impediments.

234. *Thirdly,* bodies, in consequence of their inertia, have a tendency to move over equal spaces in equal times, that is, to move uniformly. In a ball rolled on ice, in a pendulum continuing to vibrate after the moving force is withdrawn, and in numerous cases similar to these, we observe in nature and art this tendency to uniform motion; but in all these cases, the motion is not absolutely uniform, but more or less retarded by the resistances encountered. A much nearer approximation to the truth is obtained by means of a piece of apparatus called *Atwood's Machine.* Its construction, omitting some parts not essential to the principle, is as follows. The triangular base and upright pillars (which are usually of mahogany) constitute the *frame,* which is surmounted by a horizontal table or plate of wood AB, Fig. 130, perforated with several holes. C is a vertical wheel, which, by a contrivance called *friction wheels,* (not represented in the figure,) is made to revolve with the least possible resistance from friction. D and E are two weights exactly equal, and connected by a slender string passing over the wheel C. FG is a perpendicular scale graduated into inches from top to bottom, extending from 0 to 60 or 70, according to the height of the machine. H is a movable ring which slides up and down on the scale, and K is a brass plate

Fig. 130.

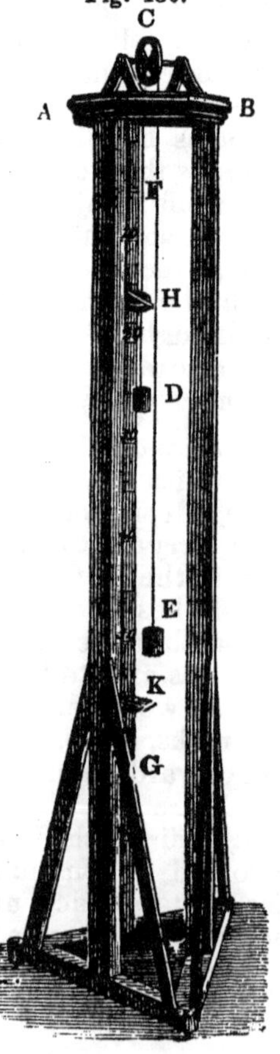

sliding in the same manner.* A great variety of the principles of motion may be established by means of this apparatus, but we are at present concerned only with the method of showing, that a body when once put in motion continues, by its inertia, to move uniformly, after the moving force is withdrawn. It is obvious that the weights D and E balance each other, and consequently, that the power of gravity is entirely removed from D, so that it is at liberty to obey the full and exclusive influence of any force that may be applied to it. If therefore an impulse be given, by the finger, for example, to D, when at the top of the scale, it ought, in conformity to the law under consideration, to move uniformly along down the scale, passing over the same number of inches in each successive second. Such appears to be the fact. But in order to give still greater precision to the experiment, a small brass bar is laid on D, which communicates motion to it, accelerating its progress until it comes to the brass ring H, where the bar lodges, and the weight D, after it leaves the ring, passes accurately over the same number of inches on the scale in each successive second.

235. *Fourthly*, moving bodies have a constant tendency to proceed in right lines. In nature there occur indeed but few examples of rectilinear motion, but almost every moving body describes a curve. Thus, the heavenly bodies move in ellipses ; projectiles describe parabolas ; or if their direction is so altered by a resisting medium, as the atmosphere, that their path is no longer a parabola, it is still changed to some other curve ; and a ship sailing across the ocean, describes a curvilinear path on the surface of the earth. The waving of trees and plants, the courses of rivers, the spouting of fluids, the motions of winds and waves, are likewise more or less curvilinear. Bodies falling toward the earth by gravity, present almost the only examples we observe in nature of a motion purely rectilinear. But notwithstanding the deviation from a right line, observable in actual motions, yet we find there is always some extraneous cause in operation, which accounts for such deviation.

236. In consequence of this tendency of moving bodies to proceed in right lines, when a body revolves in a curve, around some center of motion, it constantly tends to fly off in a straight line, which is a tangent to its orbit. This is called the *centrifugal force.* (Art. 185'.) A stone from a sling, water escaping from the periphery of a revolving wheel, and water receding from the center of a tumbler or pail when the vessel is whirled, are familiar instances of the tendency of bodies when revolving in circles to fly off in straight lines. The action of the centrifu-

* There are frequently connected with the machine a pendulum, and such parts of a clock as are necessary for beating seconds.

gal force may be studied more advantageously by means of an apparatus called the Whirling Tables. These consist of two small circular tables, to which is communicated a horizontal revolution around their centers. Bodies laid on the tables in different ways are made to participate in their rotary motions, and thus the laws of the centrifugal force may be observed. By means of this apparatus the following propositions are illustrated.

237. The centrifugal force of bodies revolving in a given circle, is proportional to their *densities or specific gravities.* If quicksilver, water, and cork, be whirled together in a suspended pail or glass vessel, these bodies will arrange themselves in the order of their specific gravities, so that the cork will be at the least, and the quicksilver at the greatest distance from the center of motion.

238. When bodies revolve in the same circle with different velocities, the centrifugal forces are as the *squares of the velocities.* By doubling the velocity of a revolving body, its centrifugal force is quadrupled. Millstones, revolving horizontally, communicate their circular motion to the corn that is introduced between them near the center. The corn, by the centrifugal force which it gradually acquires, recedes from the center and passes out at the circumference. If too great velocity is given to millstones, they sometimes burst with violence. A horse in swift motion, on suddenly turning a corner, throws his rider; and a carriage turning swiftly is overset on the same principle. In feats of horsemanship, when the equestrian rides rapidly round a small ring, he inclines his body inward in different degrees, according to the velocity with which he is moving, and thus counteracts his tendency to fall outward by the centrifugal force.*

239. When bodies revolve in different circles, in the same time, the centrifugal forces are as the *radii of the circles.* Hence, when spherical bodies revolve on their axes, the equatorial parts, being farthest from their centers of motion, and consequently moving faster, have a proportionally greater centrifugal force. If the revolving body is soft so as to yield, it is elevated in the equatorial and depressed in the polar parts. Thus a mass of clay revolving on a potter's wheel, swells out in the central parts and becomes flattened at the two ends. The earth itself, by its figure, which is an oblate spheroid, indicates the operation of this principle, its equatorial exceeding its polar diameter by 26 miles; and the planet Saturn, which has a far more rapid revolution on

---

* Arnott's El. Phys. p. 62.

its axis, indicates the same modification of its figure in a still higher degree, being strikingly elevated at the equator and depressed at the poles.

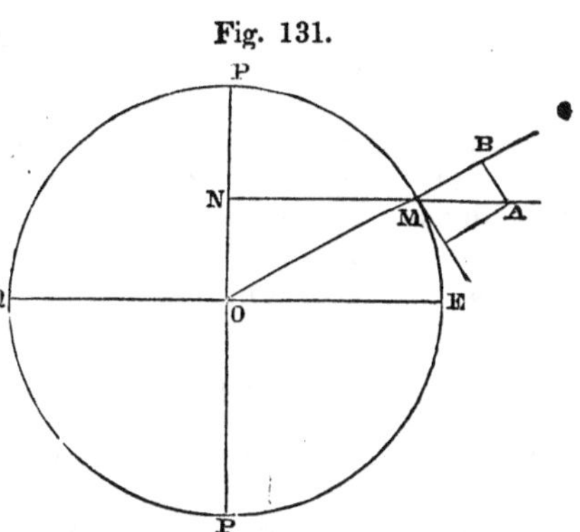

Fig. 131.

240. The centrifugal force is continually opposed to the action of gravity, so that, except at the pole, where this force becomes nothing, the weight of a body is diminished by it in the ratio of *the square of the cosine of the latitude.* For, let EPP represent the earth, EQ the equator, PP the earth's axis, and MN the radius of a parallel of latitude, which equals the cosine of the latitude, since EM is the latitude, MP its complement, and MN the sine of MP. Now, on account of the different velocities of two bodies at E and M respectively, their centrifugal forces would be as EO to MN, or as radius to the cosine of the latitude. But, at the equator, the centrifugal force being exerted in the direction of OE is directly opposed to gravity, while at M the force is exerted in the direction of NM, and is therefore not directly opposed to gravity. Produce MN to A, and let MA represent the centrifugal force at M, and upon OM produced let fall the perpendicular AB; then MB will represent the part of the force which acts in opposition to gravity. But since the angle AMB is equal to the latitude, (being equal to MOE,) therefore, MA is to MB as radius to the cosine of the latitude. Hence, the centrifugal force is diminished twice in the ratio of the cosine of the latitude, and consequently is as the square of the cosine of the latitude. At the equator, the centrifugal force is $\frac{1}{289}$ of the force of gravity, and of course it so much diminishes the weight of bodies; and since this force varies as the square of the velocity, were the earth to revolve with 17 times its present velocity (the square of which is 289) the centrifugal force would be equal to that of gravity, and bodies would lose all their weight; and were the velocity greater than this, they would fly off in the direction of tangents to the surface.*

241. We have hitherto considered the doctrine of central forces only in respect to bodies moving in *circular* orbits; but Newton extended the theory to all possible curves, and established the

* At the equator, the loss of weight by the centrifugal force is $\frac{1}{289}$; and by increased distance from the center of the earth, $\frac{1}{590}$; hence the entire loss is $\frac{1}{194}$.

general laws of motion in every sort of curve. Observing that we can always make a circle pass through three points taken in any curve, and the nearer the points are to each other, the more nearly the circumference of the circle will coincide with the curve. If the three points be taken infinitely near each other, the circle will form what is called the *osculating circle*. We can then suppose that at each point of the curve the body is moving in the osculating circle; and, consequently, its centrifugal force is measured by the square of the velocity divided by the radius of the osculating circle, (Art. 185′.) As the radius changes at each point of the supposed curve, the centrifugal force also is continually changing, while in a circular orbit it remains always the same.*

242. The consideration of centrifugal force proves, that if a body be observed to move in a curvilinear path, some efficient cause must exist which prevents it from flying off, and which compels it to revolve round the center. Thus the bodies of the solar system are constantly impelled or drawn toward the sun by a force which we denominate gravity. If this force did not act constantly, they would resume their motion in the right line in which they were originally projected, when they were first launched into space, and continue moving in it forever.

243. Second Law.—*Motion, or change of motion, is proportional to the force impressed, and is produced in the right line in which that force acts.*

First, motion is proportional to the force impressed. This is very satisfactorily shown by means of Atwood's machine, (Art. 223, Fig. 130.) When the box D is loaded with small bars of different weights, (the bars being left on the ring H, as in Art. 223,) the box descends along the scale, in consequence of the motion given it by the bars, with velocities exactly proportional to the weights of the bars respectively.

Secondly, motion is in the *direction* of the force impressed. Notwithstanding the diversity of motions to which every terrestrial body is constantly subject, the effect of any force to produce motion is the same, when the spectator has the same motion as the body, as though the body were absolutely at rest. In other words, all motions are compounded so as not to disturb each other; each remaining, relatively, the same as if there were no others.† Since, for example, by the diurnal motion of the earth, places toward the equator move faster than those toward the poles, if the foregoing principle were not true, the same forces would produce different quantities of motion in different latitudes; and a body struck in a direction north or south, would not move in that direction, but would deviate to the east or west.

* Pontecoulant, El. Astron. p. 502.

† Whewell's Mechanics, p. 231.

A pendulum also would vibrate differently according as it moved in a north and south, or in an east and west direction, whereas not the slightest difference of time can now be detected. If we are in a ship, moving equally, any force which we can exert will produce the same motion relatively to the vessel, whether it be or be not in the direction of the vessel's motion. If we stand on the deck, supposed to be level, and roll a body along it, the same effort will produce the same velocity along the deck whether the motion be from head to stern, or from stern to head, or across the vessel. Also, a body dropped from the top of the mast will not be left behind by the motion of the ship, but will fall along the mast as it would if the mast were at rest, and will reach the foot of it in the same time. If a body be thrown perpendicularly upward, it will rise directly over the hand and fall perpendicularly upon it again; and if it be thrown in any other direction, the path and motion relatively to the person who throws it will be the same as if he were at rest.*

244. It may seem, at first view, more questionable whether, as is asserted in Art. 20, *the smallest force is capable of moving the largest body*. Agreeably to this doctrine, a blow with a hammer upon the earth ought to move it, and that it would do so may be inferred from the following reasons.

(1.) We can conceive the earth to be divided into parts so small, that the blow would produce upon one of them even a sensible motion. Then it would produce on two of the parts half as much velocity; and upon all the parts together a velocity as much less than upon one, as their number was greater than unity. This velocity might be appreciable in numbers, although too small to be observed by the senses.

(2.) Very heavy weights may be actually put in motion by small forces. Leslie asserts that a ship of any burden in calm weather and smooth water, may be gradually pulled along even by the exertions of a boy.†

(3.) The repetition of very small blows finally produces sensible effects upon large bodies. The wearing away of stone by the dropping of water, the abrasion of marble images by the kisses of pilgrims, and, especially, the demolition of the strongest fortresses by repeated blows of the battering ram, are examples of powerful effects produced by small impulses, each of which must have contributed its share, since the addition of any number of nothings is nothing still.

245. Third Law.—*When bodies act upon each other, action and reaction are equal, and in opposite directions.*

* Robison's Mech. Phil. by Brewster, I, 42.
† Leslie, El. Nat. Phil. I, 30.

The doctrine of action and reaction has been fully investigated and explained in the former part of this work. All we propose to do at present is to add a few familiar and practical illustrations.

246. If I strike one hand upon the other at rest, I perceive no difference in the sensations experienced by each. The resistance to the hand which gives the blow is equal to the impulse given. A boatman presses against the bank with his oar, and receives motion in the opposite direction, which being communicated to the boat, makes it recede from the shore. He strikes the water, the reaction of which at every impulse, carries the boat forward in the opposite direction. An infirm old man presses the ground with his staff, and thus, by lightening the pressure on his lower limbs, makes his arms perform a part of the labor of walking. A bird beats the air with his wings, and by giving a blow whose reaction is more than sufficient to balance the weight of his body, rises with the difference. When the wings are small and slender, as those of the humming-bird, and disproportioned to the weight of the body, the defect is compensated by more frequent blows, giving nimble motions, suited to their short but swift excursions, while the long wings of the eagle are equally fitted, by their less rapid but more effectual blows, for his distant journeys through the skies. Hence, propelling and rowing a boat, flying, and swimming, are processes analogous to each other, depending on the principle of reaction.

247. If a man stands in a boat and pulls upon a rope which is fastened to a post on the shore, the force of the man is expended on the post in one direction, and the post, by its reaction, draws the man in the opposite direction, namely, toward the shore. (See p. 32, Ex. 12.) Call the man A, and let another man B take the place of the post. If B pulls with a force just equal to that of A, he will do nothing more than what the post did before, and therefore the two men together will bring the boat ashore no sooner than A would have done alone in the former case. If A pulls with more force than B, he pulls B toward him, and the reaction, or the force which carries the boat ashore, is the same as before, namely, the force of B. If B were to pull with more force than A, he would pull A out of the boat, were not A attached firmly to the boat, in which case the velocity of the boat would be augmented. By attentively considering this and all analogous cases, we shall perceive, that whenever two bodies act against each other, they give and receive equal momenta, and the momenta being in opposite directions, it follows, that bodies do not alter the quantity of motion they have, estimated in a given direction, by their mutual action on each other. This principle is well explained in Emerson's Mechanics, as follows:

*The sum of the motions of any two bodies in any one line of direction, toward the same part, cannot be changed, by any action of the bodies upon each other; whatever force these actions are caused by, or the bodies exert among themselves.** Hence it follows, that the sum of the motions of all the bodies in the world, estimated in one and the same line of direction, and always the same way, is eternally and invariably the same. Whatever motion, therefore, one body receives toward another, whether it is drawn toward it by attraction, or by a rope, or by any other method, precisely the same quantity of motion it imparts to the other body in the opposite direction.† If a man in a boat pulls at a rope attached to another boat of equal size, the boats will move toward each other with equal velocities; but a man in a boat pulling a rope attached to a large ship seems only to move the boat, but he really moves the ship a little, although its velocity is as much less than that of the boat, as its weight is greater. A pound of lead and the earth attract each other with equal force, and the two bodies approach each other with equal momenta.‡

248. Since momentum is proportioned to the joint product of the velocity and quantity of matter, a great momentum may be obtained either by giving a slow motion to a great mass, or a swift motion to a small body. A striking illustration of this is afforded by problem 9th, page 32, where on the supposition that a grain of light moving with its usual velocity, were to impinge directly against a mass of ice floating at its ordinary slow rate, the grain of light would be competent to stop about 44½ tons of ice. Islands of ice move with such vast momentum, that they instantly demolish the largest ship of war if it comes in their way.

249. If a body in motion strikes a body at rest, the striking body must sustain as great a shock from the collision as if it had been at rest, and struck by the other body with the same force. For the loss of force which it sustains in one direction is an effect of the same kind as if, being at rest, it had received as much force in the opposite direction. If a man walking rapidly, or running, encounters another standing still, he suffers as much from the collision as the man against whom he strikes. When two bodies moving in opposite directions meet, each body sustains as great a shock as if, being at rest, it had been struck by

* Emerson's Mechanics, 4to, p. 17.

† Quantitas motus quæ colligitur capiendo summam motuum factorum ad eandem partem, et differentiam factorum ad contrarias, non mutatur ab actione corporum inter se. (Principia, Lex III, cor. 3.)

‡ The pound of lead does indeed attract the earth only half as much as two pounds would do; nor does it *receive* from the earth but half as much; the power of attracting and of being attracted is the same.

the other body with the united forces of the two. For this reason, two persons walking in opposite directions, receive from their encounter a more violent shock than might be expected. If they be of nearly equal weight, and one be walking at the rate of three and the other four miles an hour, each sustains the same shock as if he had been at rest, and struck by the other running at the rate of seven miles an hour.*

This principle accounts for the destructive effects arising from ships running foul of each other at sea. If two ships of 500 tons burden encounter each other, sailing ten knots an hour, each sustains the shock which being at rest it would receive from a vessel of 1000 tons burden sailing at the same rate. It is a mistake to suppose that when a large and a small body meet, the small body suffers a greater shock than the large one. The shock which they sustain must be the same; but the large body may be better able to bear it. When the fist of a pugilist strikes the body of his antagonist, it sustains as great a shock as it gives; but the part being more fitted to endure the blow, the injury and pain are inflicted on his opponent. This is not the case however when fist meets fist. Then the parts in collision are equally sensitive and vulnerable, and the effect is aggravated by both having approached each other with great force. The effect of the blow is the same as if one fist, being held at rest were struck by the other with the combined force of both.†

250. The question may be asked, Why are the effects so much more severe when we fall from an eminence upon a naked rock than upon a bed of down? In both instances our fall is arrested, and we sustain a contrary and equal reaction; yet in the one case we might suffer hardly any injury, while in the other we should be bruised to death. The reason of the difference is this: when we fall on a bed of down, the resistance is applied gradually; when we fall on a rock, it is applied instantaneously. We do not strike the bed with the same force that we do the rock; we move along with the bed, and of course do not lose our motion at once, and we receive in the opposite direction merely what we lose. A violent blow, if equally diffused over the human body, may be sustained without injury. Thus, if an anvil be laid on the breast, a man may receive on it a heavy blow from a great hammer with impunity.‡

251. There are many instances where action and reaction mutually destroy each other and no motion results. Thus, when a child stands in a boat and pulls by a rope attached to the stern, he labors in vain to make the boat advance. Dr. Arnott tells us of a man who attached a large bellows to the hinder part of his

* Lardner's Mech., p. 47 † Ib. p. 48. ‡ Arnott's El. Phys. p. 104

boat, with the view of manufacturing a breeze for himself, being ignorant that the reaction would carry the boat backward as much as the impulse of the artificial wind carried it forward.* A force which begins and ends within a machine has no power to move it.†

252. Variable motion.—When a moving body is subjected to the energy of a force which acts on it without interruption, but in a different manner at each instant, the motion is called in general, *variable motion.*‡ We have instances of variable motions in the unbending of springs, in the action of the wind on the sails of a ship, and in the action of gunpowder on a ball while it is passing through the barrel of the gun. In each of these cases, the velocity of the moving body is constantly augmented, yet the degree of augmentation is diminishing until it finally ceases.

253. When a moving body receives, each successive instant, the same increase of velocity, it is said to be *uniformly accelerated.* If a small wheel were revolving without resistance, and at the end of every second I should apply a given impulse, the wheel would be uniformly accelerated; for, by its own inertia, it would retain all its previous motion, and, by the second law of motion, the repetition of the same force, at equal intervals, would increase its velocity at a uniform rate. If the intervals at which this force was repeated, were indefinitely diminished, the same kind of effect would take place; and the same would evidently be the case, were the force to operate without cessation. Such a force is that of Gravity.

254. It has already been shown, in articles 4 and 7, that Gravity is a quality which belongs alike to all matter in proportion to its quantity; and that at different distances from the center of the earth, it varies inversely as the square of the distance. The manner in which this force decreases as the distance increases will be seen at one view by the following table, beginning with the distance of the surface from the center.

* Arnott's El. Phys. p. 107.

† It is common in elementary works on Mechanics, to find under the head of "reaction," a class of phenomena which evidently belong to a cause distinct from that of the mutual action of bodies. For example, a little steam carriage is sometimes exhibited, from which a jet of steam issues, and the carriage moves in the opposite direction. This, it is said, is owing to the reaction of the air upon the steam, being supposed analogous to the flying of a bird, which beats the air with its wings, and is borne along by its reaction; but the motion of the carriage in the foregoing experiment, is owing to a very different cause. Before the jet was opened, the steam pressed equally on all sides of the vessel; as soon as the opening is made on one side, the pressure is removed from that side, but remains on the opposite side, and therefore gives motion to the vessel in that direction.

‡ Gregory's Mech., I, 81.

Distance,	1	2	3	4	5	10	20	60
Attraction,	1	$\frac{1}{4}$	$\frac{1}{9}$	$\frac{1}{16}$	$\frac{1}{25}$	$\frac{1}{100}$	$\frac{1}{400}$	$\frac{1}{3600}$

Hence it appears that a body placed 20 times as far from the center of the earth as the surface is from the center, is attracted only $\frac{1}{400}$th part as much; and at the distance of 60 times the radius of the earth the same force is diminished 3600 times.* At this distance therefore it would take 60 seconds, or one minute, for a body to fall through the space it falls at the surface of the earth in one second; that is, through $16\frac{1}{12}$ feet. But all distances within a few hundred feet of the earth bear so small a ratio to the earth's radius, that the force of gravity may be considered as the same unvarying force, in relation both to the weights of bodies and to the velocities with which they fall. (See Art. 8.)

255. It is not alone by the direct fall of bodies that the gravitation of the earth is manifested. The curvilinear motion of bodies projected in directions different from the perpendicular, is a combination of the effects of the uniform velocity that has been given to the projectile by the impulse which it has received, and the accelerated or retarded velocity which it receives from the earth's attraction. Suppose a body placed at any point P, (Fig. 132,) above the surface of the earth, and let PA be the direction of the earth's center. If the body were allowed to move without receiving any impulse, it would descend to the earth in the direction PA with an accelerated motion. But suppose that at the moment of its departure from P, it receives an impulse in the direction PB, which would carry it to B in the time the body would fall from P to A; then, by the composition of motion, (Art. 41,) the body must at the end of that time be found in the line BD, parallel to PA. If the motion in the direction of PA were uniform, the body P would in this case move in the straight line PD. But this is not the case. The velocity of the body in the direction PA is at first so small as to produce very little deflection of its motion from the line PB. As the velocity, however, increases, this deflection increases, so that it moves from P to D in a curve, which is convex toward PB. The greater the velocity of the projectile in the direction

Fig. 132.

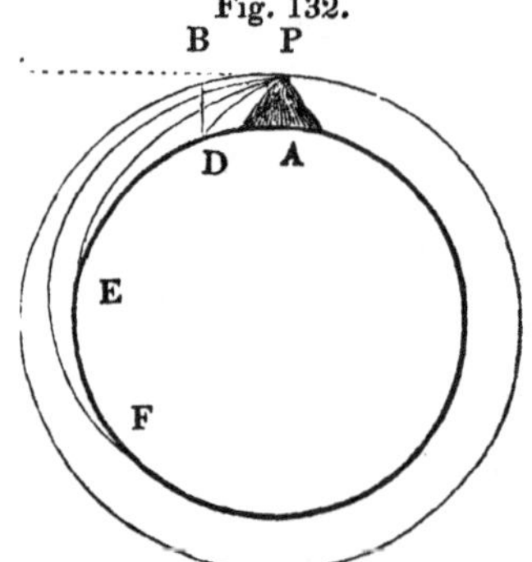

* This last is nearly the distance of the moon from the earth; and it is found by calculation that the moon is actually drawn toward the earth, away from the straight line in which she tends to move, by exactly this force.

PB, the greater sweep the curve will take. Thus it will successively take the forms PD, PE, PF, &c., and that velocity can be computed,* which (setting aside the resistance of the air) would cause the projectile to go completely round the earth, and return to the point P from which it departed. In this case the body P would continue to revolve round the earth like the moon.

256. Hence it is obvious, that the phenomenon of the revolution of the moon round the earth, is nothing more than the combined effects of the earth's attraction, and the tendency it has to move forward in a straight line which is a tangent to its orbit. And were any of the heavenly bodies to explode, we may conceive that the fragments would proceed in a rectilineal direction until, approaching, severally, within the sphere of influence of some large body, whose attraction would combine with their projectile force, they would forever afterward continue to revolve around that body, as the satellites revolve around their primaries.†

257. But the attraction of gravitation is manifested by comparatively small masses of matter. The effect of a high mountain is perceptible upon a plumb line, causing it to deviate sensibly from a perpendicular, so that the same star in the zenith would change its apparent place when viewed on opposite sides of the mountain. This was observed by two French astronomers, near Mount Chimborazo in South America, as early as the year 1738; and the experiment was repeated in 1772 with all possible accuracy, by Dr. Maskelyne, astronomer royal of Great Britain, at the base of the mountain Schehallien, in the eastern part of Scotland.‡ Mr. Cavendish, a distinguished English philosopher of the last century, rendered sensible even the attraction of a sphere of lead, by bringing it near a small bullet, suspended from one arm of an exceedingly delicate balance. The sphere when brought near the bullet disturbed its equilibrium.

258. By gravity, bodies are directed toward the center of the earth. We are not to infer from this fact that there is any peculiar force, (like that of a large magnet, for example,) residing at the center, but merely that the effect of the earth, taken as a whole, is the same as though its matter were condensed into the center. The line of attraction passes through the center because

* A cannon ball shot horizontally from the top of a lofty mountain, would go three or four miles. If there were no atmosphere to resist its motion, the same original velocity would carry it thirty or forty miles before it fell; and if it could be dispatched with about ten times the velocity of a cannon shot, the centrifugal force would exactly balance the force of gravity, and the ball would go quite round the earth. (Arnott, El. Phys. p. 91.)

† This has actually been supposed of the four new planets, Ceres, Pallas, Juno, and Vesta.

‡ Ed. Encyclopedia, III, 76.

such a line is the *resultant* of the separate attractions of all the particles composing a sphere. Thus if AB be a line passing through the center of the sphere, and A be any body, and *a* and *b* two particles of matter in the sphere equally distant from A, it is evident that their combined actions would be expressed by the diagonal which would coincide in direction with AB. The same is true of any other particles taken equally distant from A in opposite hemispheres. At different parts of the earth, therefore, the directions of falling bodies are not *parallel*, but form converging lines. But on account of the great magnitude of the earth, two places 100 feet distant will not vary *one second*, and when a mile asunder, they will not differ *one minute* from perfect parallelism.* For all the purposes of machinery, therefore, as well as for experiment, the direction of the lines of gravity may be considered as parallel.

Fig. 133.

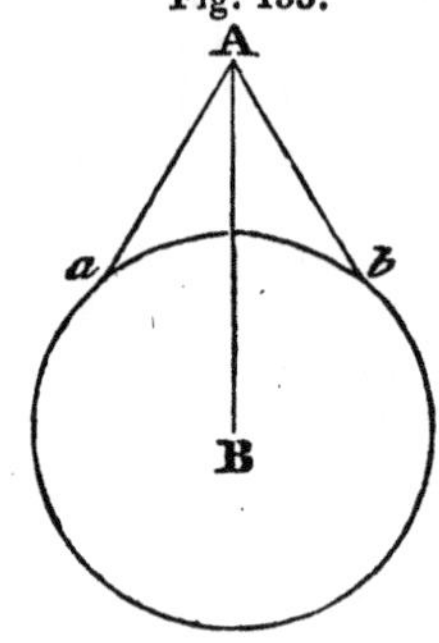

259. Since bodies in falling toward the earth are uniformly accelerated, the *velocity* acquired must be proportioned to the time the body has been falling: at the end of ten seconds it has acquired ten times the velocity which it had at the end of one second. And in Art. 29, it has been shown that the *spaces* described are proportioned to the squares of the times; so that the space described during 100 seconds, is not merely 100 times as great as that described in one second, but it is the square of 100, or 10,000 times as great. This conclusion was arrived at mathematically, long before it was established by actual experiment. There were two difficulties which stood in the way of such a verification, viz., the little time occupied in descending through such perpendicular heights as the experimenter can command, and the resistance of the air, which, when the velocity becomes great, acts as a powerfully retarding force. We can rarely command a perpendicular eminence of more than four hundred feet, and yet the time occupied in the whole descent is only about five seconds, a period too short to enable us to mark distinctly the respective rates at which the successive intervals are described. Atwood's machine (Fig. 130, Art. 231) affords the means of obviating both these difficulties, and verifying the laws of falling bodies with great accuracy.

260. The object of the machine, so far as respects experiments on falling bodies, is to render the descent of bodies so *gradual* that the relation between the times and spaces can be accurately

* Gregory's Mech. I. 193.

observed By recurrence to the figure, and to the descriptions given in Art. 231, we shall readily see how this object may be accomplished. The weights D and E each equal $31\frac{1}{2}$ ounces, and of course the quantity of matter in both is 63 ounces. Now, since one of these rises as the other descends, the force of gravity retards the one as much as it accelerates the other, and they are in effect the same as though they were entirely destitute of gravity. If a small weight, as one ounce, were let fall *freely* from the top of the machine, it would fall through so small a space almost in an instant, and we should be unable to mark the rate at which it would pass over the successive portions of the graduated scale FG; but if it be laid on the weight D, it must carry D along with it; that is, it must make D descend and E ascend, and therefore the motion belonging to one ounce, will be distributed throughout 64 ounces, and its velocity will be retarded in the same ratio. Consequently the weight D will descend only $\frac{1}{64}$th part as fast as a body falling freely; and as a body falling freely descends in one second about 16 feet or 192 inches, the weight D will descend $\frac{192}{64}=3$ inches in one second. The comparative progress of this weight, and of a body falling freely for several successive seconds, will be seen in the following table.

Time,	1	2	3	4	5	6
Body falling freely, in feet,	$16\frac{1}{12}$	$64\frac{1}{3}$	$144\frac{3}{4}$	$257\frac{1}{3}$	$402\frac{1}{12}$	579
Do. in Atwood's Machine, in inches.	3	12	27	48	75	108

Hence it appears that in six seconds, while a body would fall freely through 579 feet, it would in the same time descend only nine feet in Atwood's Machine. But the latter is a uniformly accelerated velocity, and subject to the same laws as the former, and it may therefore be employed to investigate the laws of falling bodies. The results correspond remarkably with theory, so that when the instrument is well constructed, and managed skilfully, the descending weight clicks upon the stage or brass plate K, at the very instant required.

261. We see in nature the law of acceleration of falling bodies indicated, by the impetuosity with which bodies fall from any considerable height upon the earth. Meteoric stones, falling from the sky, sometimes bury themselves deep in the ground. Aeronauts that have fallen from balloons have been dashed in pieces.*

* Any liquid falling from a reservoir, forms a descending mass or stream, of which the bulk diminishes from above downward, in the same proportion in which the velocity increases. This truth is well exemplified by the pouring out of molasses or thick sirup: if the height of the fall be considerable, the bulky mass which first escapes, is reduced, before it reaches the bottom, to a small thread; but the thread is moving with proportionally greater speed, for it fills the receiving vessel with great rapidity. The same truth is exhibited on a grand scale in the Falls of Niagara, where the broad river is seen first bending over the precipice, a vast slow

It is, however, a rare occurrence to see a body falling from any great height *perpendicularly:* most instances of accelerated motion which come under our observation, are in bodies falling down *inclined planes*, where the same law of acceleration prevails. (Art. 158.) A fragment of rock descending from the side of a mountain, has its speed augmented as it goes, until its momentum becomes irresistible, and large trees are prostrated before it.

262. A very remarkable example of the acceleration of bodies descending down inclined planes, occurs at the *Slide of Alpnach* in Switzerland. On Mount Pilatus, near Lake Luzerne, is a valuable growth of fir-trees, which, on account of the inaccessible nature of the mountain, had remained for ages undisturbed, until within a few years, a German engineer contrived to construct a trough in the form of an inclined plane, by which these trees are made to descend by their own weight, through a space of eight or nine miles, from the side of the mountain to the margin of the lake. Although the average declivity is no more than about one foot in seventeen, and the route often circuitous and sometimes horizontal, yet so great is the acceleration, that a tree descends the whole distance in the short space of six minutes. To a spectator standing by the side of the trough, at first is heard, on the approach of the tree, a roaring noise, becoming louder and louder; the tree comes in sight at the distance of half a mile, and in an instant afterward shoots past with the noise of thunder and the rapidity of lightning. When a tree happens to "bolt" from the trough, it cuts the standing trees quite off.* (See p. 131, Prob. 5 and 6.)

263. Composition and Resolution of Motion.—*Simple* motion is that which arises from the action of a *single* force; *Compound* motion is that which is produced by *several* forces acting in different directions. Strictly speaking, we have no example of a simple motion, since in the absolute motion of all bodies, their own proper motion is combined with that of the earth in its diurnal and annual revolutions, and we know not with how many others. (Art. 226.) In an enlarged sense, therefore, all motions are compound. But in the foregoing distinctions we have reference only to relative motions, as those which take place among bodies on the earth. In accordance with the second law of motion, (Art. 243,) a force striking upon a body in motion, will produce the same change of motion as though the body had been at rest when the force struck it. This may at first view appear inconsistent with experience, especially in regard to opposite mo-

---

moving mass, then becoming a thinner and thinner sheet; until it flashes into the deep below, almost with the velocity of lightning. (Arnott's El. Phys. 79.)

* Playfair's Works, I, 96.

tions. Let us, therefore, consider the principle in its application to several different cases. Conceive the ice of a frozen river to be first stationary, and afterwards to float down with the current. Standing on the bank I roll a ball directly across the river. Will it pass in the same direction in both cases? It will not; for in the first case it will pass across *perpendicularly* to the banks, and in the second case it will go across *diagonally*. But now let me stand upon the ice and roll the ball. Since I float along with the ice, I am at rest with respect to that motion, and the ball, though moving diagonally as before, appears to me to go directly across the stream.* If the ball was rolled not directly across but obliquely, making any angle with the bank, if I stood upon the floating ice, and was therefore at rest with respect to one of the motions of the ball, I should see the other motion in the same manner as though I had stood on the shore and the ice had been at rest. But if when a ball is rolling towards me, and I strike it in a direction exactly opposite to its course, but do not stop it, can the blow be said to produce the same change of motion as though the body had been at rest when the blow was applied? Ans. If I had been moving in the same manner as the ball before the blow, then stopping a part of the motion of the ball, would have given it a relative motion in the opposite direction; since having none of my own motion stopped, I should leave it behind. This is what takes place when a cannon ball is fired in a direction contrary to that in which the earth is revolving about the sun. The cannon moves onward and leaves the ball behind. (Art. 226.)

264. The laws respecting the composition and resolution of motion which are demonstrated in Chap. III, p. 44, admit of being satisfactorily confirmed by experiment. Let two small wheels M, N, (Fig. 134,) be attached to a wall or board. Let a thread be passed over them, having weights A and B, hooked upon loops at its extremities. From any part P of the thread, let a weight C be suspended, in such a manner as to be in equilibrio with A and B. The weight C, therefore, is the resultant of the forces A and B; and since its direction is that of gravity, it will be represented by a line drawn directly upward from P. From P, on the line PO, take P*c*, having as many inches as there are ounces in C; and from *c* draw *ca* parallel to NP and *cb* parallel to PM.

Fig. 134

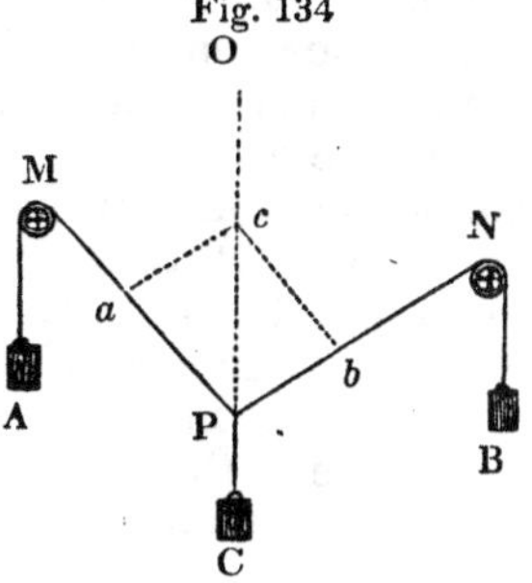

* In the same manner two persons sitting in a boat under sail, toss a ball from one to the other in the same manner as though they were at rest on land

If the sides P*a* and P*b* of the parallelogram thus formed be measured, it will be found that P*a* will consist of as many inches as there are ounces in A, and P*b* of as many inches as there are ounces in B; consequently, the lines P*c*, P*a*, and P*b*, have the same ratio to each other as the forces or weights C, A, and B. But P*c* is the diagonal, and P*a* and P*b* are the sides of a parallelogram. Hence the diagonal of a parallelogram represents a force equivalent to the two forces represented by the sides.*

This experiment is an illustration of the composition of *forces* rather than of the composition of *motions;* but a simple experiment will show that the same law holds good with respect to a body actually set in motion by two different forces. A ball is placed at one of the corners of a smooth table. To the same corner are attached two springs, respectively in the line of the two sides of the table, and capable of giving a simultaneous impulse to the ball. The springs moreover are so proportioned to each other, that one will drive the ball across one side of the table, in the same time that the other will drive the ball across the other side. Now on letting go both springs at once, the ball will pass, in the same time, across the diagonal of the table to the opposite corner.†

265. We daily observe examples strikingly illustrative of these laws. In crossing a river, the boatman heads up the stream, and so combines the direction of the boat with that of the current, as to move directly across in a line, which is the diagonal between the two. Rowing, swimming, and flying, are severally instances of motion in the diagonal between two forces. In feats of horsemanship, when the rider leaps up from his saddle, we are surprised not to see the horse pass from under him; but he retains the motion he has in common with the horse, and does not in fact ascend perpendicularly, but obliquely, rising in one diagonal, and falling in another. In the common feats of jumping through a hoop, and alighting again on the saddle, an inexperienced rider would be likely to project his body forward in the same manner as he would do in leaping through the same hoop from the ground. In such a case, instead of alighting on the saddle, he would alight either before the horse or upon his head or neck. All that is requisite in order to execute this feat, is to leap directly upward from the saddle to a sufficient height to clear the lower part of the hoop with the feet. By the speed which the rider has in common with the horse, his body will, without any exertion on his part, pass through the hoop, and he will alight again on the saddle, on the other side, in his descent.‡

266. The *sailing of a ship* affords an instructive illustration of

* Lardner's Mech. p. 51 † Lib. Useful Knowledge, Mechanics, p. 5. ‡ Ib.

the principles of the composition and resolution of motion. When a ship sails in the same direction as the wind, she is said to be *scudding*, or sailing *before the wind*, and if she had but one sail, it would act with the greatest advantage, when perpendicular or nearly so to the wind. When a ship advances against the wind, and endeavors to proceed in the nearest direction possible to the point of compass from which the wind blows, she is said to be *close hauled*. A large ship will sail against the wind with her keel at an angle of six points with the direction of the wind, and sloops and smaller vessels may sail much nearer. When a ship is neither sailing before the wind, nor is close hauled, she is said to be *sailing large*. In this case, her sails are set in an oblique position, between the direction of the wind and that of the intended course; as represented in the various plans of vessels in Fig. 135, where the direction of the wind is represented by the arrow, and the position of the yards and sails, which are necessary for proceeding on the various points of compass, are shown by the transverse line on each plan. The relation of the wind to the course of the vessel, is determined by the number of points of the compass between the course she is steering, and the course she would be steering if close hauled.

Fig. 135.

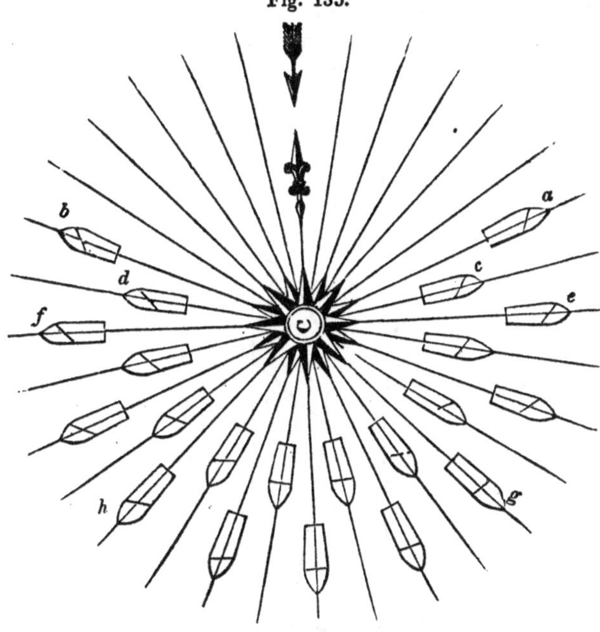

In Fig. 135, the ships *a* and *b* are close hauled, and the ships *c* and *d* (the former steering east by north, and the latter west by north) have the wind one point large. The ships *e* and *f*, one steering east and the other west, have the wind two points large. In this case, the wind is at right angles with the keel, and is said to be *upon the beam*. The ships *g* and *h*, steering southeast and southwest, have the wind six points large, or, as it is commonly termed, *upon the quarter*, and this is considered as a very favorable manner of sailing, because all the sails co-operate to increase the ship's velocity, whereas when the wind is directly aft, as in the vessel *m*, it is partly intercepted by the after sails, and prevented from striking with its full force on those which are forward.*

267. To one who has never studied the doctrine of the composition and resolution of forces, it is apt to appear mysterious that a ship is able to sail with a wind partly ahead, and still more that two ships are able to sail in exactly opposite directions by the same wind. It is proposed to explain these phenomena. Let AB (Fig. 136) represent the keel of a ship, and CD the sail; and let the wind come in from the side, in the direction of HD. Let DE represent the whole force of the wind, and resolve DE into two forces, viz. into EF perpendicular, and FD parallel to the sail DC. Then it is manifest that EF alone represents the effective force of the wind upon the sail. But EF is not wholly employed in urging the ship forward, since it is oblique to her course; therefore, again resolve EF into FG parallel with the course and GE at right angles with it. The latter force is lost by the lateral resistance of the water, while FG is employed in propelling the ship on her way.

Fig. 136.

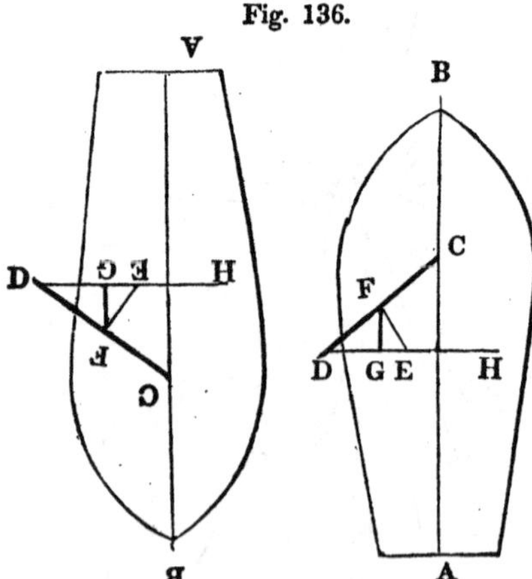

By inspecting Fig. 136, it will readily be seen that another ship may sail in the opposite direction by the same wind. When

* Bigelow's Elements of Technology, p 218.

the wind strikes the sail at right angles, or in the direction EF then only one resolution is necessary; for if EF represents the whole force of the wind, FG will represent the force that propels the ship forward, while GE will represent the part which is lost by the lateral resistance of the water.

Since, resolving the force of the wind after the foregoing manner, the effective part of the force, viz. FG, will not wholly disappear until the wind is directly ahead, it might seem possible to sail much nearer the wind than is found to be actually practicable. But, though, on account of the peculiar shape of vessels, the forward resistance is much less than the lateral, yet it is *something*, and therefore requires more or less of the force that acts parallel with the keel to overcome it.

268. The doctrine of the composition and resolution of motion, by reducing a great number of complicated motions to one, or by enabling us to estimate the precise influence of forces that act obliquely, has greatly simplified inquiries in Mechanics, and proportionally advanced the science. It is only by such means that the complex motions of the heavenly bodies, and the equally diverse forces that control them, could ever have been understood. The subject has also an extensive and important application in estimating the powers of machinery. Since, for example, any one side of a triangle or polygon, is always less than the sum of all the remaining sides, it follows, other things being equal, that a mechanical effect will always be more economically produced by a single force acting in the proper direction, than by a number of forces acting in different directions.

269. By Art. 55, it appears, that *a body acted upon at the same time by three forces, represented in quantity and direction by the three sides of a triangle taken in order, will remain at rest.* A kite at rest in the air, is commonly mentioned as an example of this, the three forces being, the direction of the wind, the weight of the kite, and the action of the string. Let AB, (Fig. 137,) be a kite, held by the string AS. Let DF represent the force of the wind blowing horizontally, and resolve it into two forces, viz.

Fig. 137.

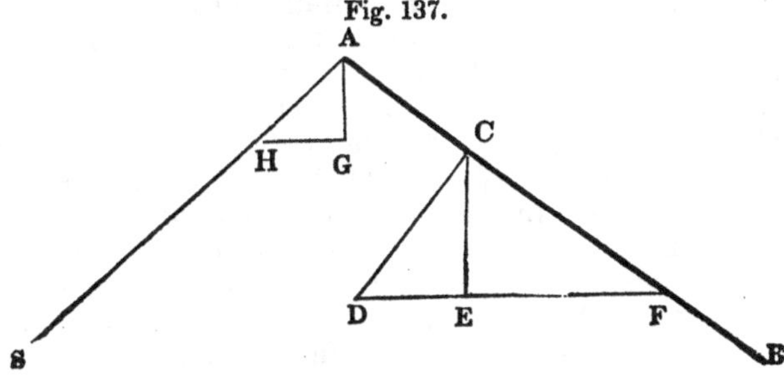

DC perpendicular, and CF parallel to the kite. Then DC will be the only effective part of the wind, since that part which acts parallel to the kite, can have no influence on its motions. Again, resolve CD into two forces, namely, CE perpendicular, and DE parallel to the horizon. Then CE will represent the upward force of the wind, and DE its force in a horizontal direction. Now when the string AS makes such an angle with the kite that its downward force AG, added to the weight of the kite, shall equal CE, and its horizontal force HG, equal DE, the kite will be at rest.

270. When two motions which are not in the same straight line, are combined, one of which is uniform, and the other accelerated, the moving body describes a *curve*. (Art. 48.) If two accelerating forces act, in which the rate of acceleration is the same, the motion is *rectilinear;* but if the rate of acceleration is different, the motion is still *curvilinear.* The nature of the curve will depend upon the proportion between the uniform and accelerating forces.

CHAPTER III.

OF THE CENTER OF GRAVITY.

271. The principles which have been discovered respecting the composition and resolution of forces, and respecting the center of gravity, have alike contributed greatly to simplify the doctrines of Mechanics. It is characteristic of a great and penetrating mind, to devise means of divesting intricate subjects of their complexity, and thus to bring easily within the grasp of the mind, subjects otherwise too much involved to be within its comprehension. By the rule of simple multiplication, we easily multiply any number by one thousand: indeed, it is nothing more than to annex three ciphers to the number itself; but how tedious would be this process, were the rule of multiplication undiscovered, and we were unacquainted with any other method of arriving at the result, except to add the given number to itself one thousand times. In like manner, by means of the rules for the composition of motion, we are enabled to reduce a thousand different motions to one; and by the doctrine of the center of gravity, we are taught how we may make a force, situated at one single point, equivalent to an infinite number of forces,

situated in as many different points. And, instead of pursuing the endless diversities of motions to which the different parts of a complicated system of bodies may be subject, we are taught how to follow merely the motions of a single individual point. By the earth's attraction, all the particles which compose the mass of a body, are solicited by equal forces in parallel directions downward. If these component particles were placed in mere juxtaposition, without any mechanical connection, the force impressed on any one of them, could in no wise affect the others, and the mass would, in such a case, be contemplated as an aggregation of small particles of matter, each urged by an independent force. Then, according to Art. 60, the *resultant* constitutes another force parallel to the others. But the bodies which are the subjects of investigation in mechanical science, are not found in this state. Solid bodies are coherent masses, the particles of which are firmly bound together, so that any force which affects one, being modified according to circumstances, will be transmitted through the whole body.*

272. As all bodies which are subjects of mechanical inquiry, on the surface of the earth, must be continually influenced by terrestrial gravity, it is desirable to obtain some easy and summary method of estimating the effects of this force. To consider it, as is unavoidable in the first instance, the combined action of an infinite number of equal and parallel forces, soliciting the elementary molecules downward, would be attended with manifest inconvenience. An infinite number of forces, and an infinite subdivision of the mass, would form parts of every mechanical problem. To overcome this difficulty, and to obtain all the ease and simplicity which can be desired in elementary investigations, it is only necessary to determine some force, whose single effect shall be equivalent to the combined effect of the gravitation of all the molecules of the body.† Such a force is obtained by supposing all the action to be concentrated in the center of gravity. We have already defined it thus: *the center of gravity of a body, is that point about which, if supported, all the parts of a body (acted upon only by the force of gravity) would balance each other in any position.*‡ (Art. 64.)

273. *To find the center of gravity by experiment*, several differ-

* Lardner's Elements of Mechanics, p. 107. † Ib. p. 108.

‡ Others define it to be "the point through which passes the resultant of all the particular forces exerted by the gravity of the several parts of the body, or system of bodies, in whatever position the body or system is placed." (Cambridge Mech. p. 45.)

"The resultant of any number of parallel forces continues of the same intensity, and passes through the same point, whatever be the direction of the forces; hence it is called the Center of Parallel Forces. When the body is moving forward in a straight line, under the action of forces other than gravity, it is called the *center of Inertia;* when the body is acted upon by gravity, it is called the center of gravity." (Renwick on the Steam Engine, p. 14.)

ent methods present themselves. We will first suppose the body to be in the shape of a piece of board of uniform thickness. Suspend it by one corner, and from the same corner let fall a plumb line, and mark its line of direction on the surface of the board. Suspend the board from any other point, and mark the line of direction of the plumb line as before, and the point where these lines intersect each other, must obviously be the center of gravity, since that center is in both of the lines. (Art. 65.)

But when the body is not of uniform thickness, but is any irregular solid, suspend the body by a thread, and let a small hole be bored through it, in the exact direction of the thread, so that if the thread were continued below the point where it is attached to the body, it would pass through this hole. The body being successively suspended by several different points in its surface, let as many small holes be bored through it in the same manner. If the body be then cut through, so as to discover the directions which the several holes have taken, they will all be found to cross each other at one point within the body. Or the same fact may be discovered thus: a thin wire which nearly fills the holes being passed through any one of them, will be found to intercept the passage of a similar wire through any other.*

A convenient method of finding the center of gravity of a body is *to balance it in different positions across a thin edge*, as the edge of a knife or a prism. The same thing may be effected, when the shape of the body will admit of it, by laying it on the edge of a table, and letting so much of it project over the edge, that the slightest disturbance will cause it to fall. The center of gravity is the point in which the several lines marked on the body, where the edge cuts it, intersect one another. From some or all of the foregoing trials, the center of gravity of bodies may be nearly ascertained; but in order to find it with absolute exactness, we are frequently obliged to resort to intricate mathematical processes.

274. By whatever method the center of gravity of a body has been ascertained, we shall find that when that is supported, the body will remain at rest in every position. Thus a globe will stand securely on a very small perpendicular support, since that support will necessarily be under the center of gravity; a lever, as the beam of a balance, poised on its center of gravity, will be at rest in every position it takes while turning round the fulcrum, and however irregular the body may be, it will, when balanced on its center of gravity, obstinately maintain its position.

When a body is suspended by an inflexible rod from a center of motion, and revolves around it, it will be at rest only *when*

* Lardner's Mech. p. 110

the center of gravity is either directly below, or directly above the center of motion. For it is only in these two cases, that the center of gravity will be in the line which is drawn through the center of motion perpendicular to the horizon. The stationary point above the center of motion is very *unstable*, since the slightest disturbing force throws the body out of the line of direction, when, by the force of gravity, it immediately descends to the lowest point it can reach, and vibrates about that point until it finally settles itself with the centre of gravity immediately under the point of suspension; and whenever it is thrown out of this position the same vibrations are renewed until it resumes it. When therefore the center of gravity is at the lowest point it is capable of reaching, the equilibrium is *stable*, since the body obstinately maintains that position. On this principle, gates which have their center of gravity raised as they are opened, shut spontaneously.

275. The stability of a body not only requires that the center of gravity should be low, but that *the line of direction* (or, the line which is drawn through the center of gravity perpendicular to the horizon) *should fall within the base.* (Art. 69.) The farther it falls from the extremity of the base, the more stable is the position. Hence the stability of a pyramid, when standing on its broad base, and its instability when inverted. For the same reason, all broad vessels, as steamboats, are difficult to upset, while vehicles with narrow bases are easily overturned. When a load is so situated as to raise the center of gravity, it increases the liability to upset, because it increases the facility with which the line of direction is thrown without the base. Thus carts loaded with hay, or bales of cotton, are very liable to be overturned. The same is true of stages carrying passengers or baggage on the top. On the other hand, a large ship well supplied with ballast, is capsized with great difficulty, since the center of gravity of all parts of the ship is so low, as to render it difficult to throw the line of direction without the base. Yet if the center of gravity is very low, a ship will rock excessively in a rough sea, since the upper parts near the deck, move over a greater space in proportion as their distance from the centre of gravity is greater.

276. There are many remarkable structures which *lean* or incline a little; but so long as the line of direction falls within the base, and the parts of the mass have sufficient tenacity among themselves to hold together, the structure will stand. The famous tower of Pisa was built intentionally inclining, to frighten and surprise: with a height of one hundred and thirty feet, it overhangs its base sixteen feet.* This circumstance greatly en-

* Some travellers, however, are of opinion that the inclination of the tower of Pisa,

hances the emotion of the spectator from its summit. Many ancient spires and other tall structures, are found to have lost something of their perpendicularity.

Rocking stones are rocks which are sometimes found so exactly poised upon their center of gravity, that a very small force is sufficient to put them in motion.* The rocking of a balloon when it begins to ascend, affords an illustration of the tendency of bodies to vibrate around the center of gravity.

277. The *motions of animals* are regulated in conformity with the doctrines of the center of gravity. A body is seen tottering in proportion as it has great altitude and a narrow base; but it is a peculiarity in man to be able to support his figure with great firmness on a very narrow base, and under constant changes of attitude. This faculty is acquired slowly, because of the difficulty. The great facility with which the young of quadrupeds walk, is ascribed in part to their broad supporting base. Many of our most common motions and attitudes, depend for their ease and gracefulness upon a proper adjustment of the center of gravity. The erect posture of a man carrying a load upon his head—leaning to one side when a heavy weight is carried in the opposite hand—leaning forward when a weight is on the back—or backward when the weight is in the arms; these are severally examples in point. When a man rises from his chair, he brings one foot back, and leans the body forward, in order to bring the center of gravity over the base; and without adjusting it in this manner, it is hardly possible to rise. A man standing with his heels close to a perpendicular wall, cannot bend forward sufficiently to pick up any object that lies on the ground near him without himself falling forward.

The art of rope or wire dancing, depends in a great degree upon a skilful adjustment of the centre of gravity. The rope dancer frequently carries in his hand a stick loaded with lead, which he so manages as to counterbalance the inclinations of his body which would throw the line of direction out of the base. Upon a similar principle, the equestrian balances himself on one foot on a galloping horse.

The *vegetable creation* is subject also to these general laws of nature. Trees, by the weight and height of their tops, would seem peculiarly liable to fall; but their roots afford a corresponding breadth of base, while their perpendicular trunks, and the symmetrical disposition of the branches, conspire to increase their stability.

278. The position of the center of gravity of any number of

is the effect of time. It is said that an ancient picture of the tower represents it as erect. Arnott's El. Phys. p. 121.

* Amer. Journal of Science, Vol. 7

separate bodies, is never altered by the *mutual action* of those bodies on each other. If, for example, two bodies, by mutual attraction, approach each other, the center of gravity remains at rest, until finally the bodies meet in this point. If, by their mutual action, they contribute to make each other revolve in orbits, it is around their common center of gravity. Thus the earth and moon revolve around a common center of gravity: the same is true of the sun and all the bodies that compose the solar system. Were the centrifugal force to be suspended, and the bodies abandoned to the mutual action of each other, they would all meet in their common center of gravity. (Art. 81.) This naturally results from the principle that the momenta on opposite sides of the center of gravity are equal, and that bodies by their mutual action produce equal momenta in each other.

279. The doctrines of the center of gravity suggest the readiest method of solving a great number of PRACTICAL PROBLEMS. Suppose three persons were carrying a stick of timber, (A by himself supporting one end, and B and C by a handspike lifting together toward the other end,) and it were required to determine at what distance from the end of the stick the handspike must be placed in order that the three persons might bear equally. A stick of timber being a body of regular shape and uniform density, has its center of gravity coincident with the center of magnitude. We may therefore proceed on the supposition that the entire weight is collected in the center. Now in order that B and C may together lift twice as much as A, they must be twice as near the center. But the distance of A from the center is one half the length of the stick; therefore the distance of the required point from the center is one fourth the length of the stick, and consequently it is one fourth the same length from the end of the stick.

The result thus obtained from theory, may be easily submitted to the test of experiment. For if we take the weight of the stick of timber with a pair of steelyards, and then, resting one end on some support, attach a cord at the distance of one fourth of the length of the stick from the other end, and thus connect the stick with the steelyards, we shall find the weight equal to two thirds of the whole.*

280. The method of finding *the areas of the surfaces and the solid contents of bodies*, particularly of such as are of unusual figure, may frequently be very much simplified, by applying the

* This experiment may be repeated with much precision in the following manner Take a Gunter's scale, well made, and ascertain its weight by a delicate balance. Let one end rest on a sharp edge, as the edge of a prism, and attaching a string at the distance of one fourth the length of the scale from the other end, connect it with one of the arms of the balance: the weight will be exactly two thirds of the whole.

principles of the center of gravity. This is sometimes called the *centrobaryc* method. It was discovered by Pappus, an ancient mathematician of Alexandria, but was more completely discussed and illustrated by Guldinus, Professor of Mathematics at Rome, about the year 1640.* This remarkable property of the center of gravity is expressed in the following propositions:

1. *If any line whatsoever revolves about a fixed point, the* SURFACE *which it generates is equal to the product of the given line into the circumference described by its center of gravity.*

2. *If a surface having any figure whatsoever, revolves about an axis, the* SOLID *generated, is equal to the product of that surface into the circumference described by its center of gravity.*

Fig. 138.

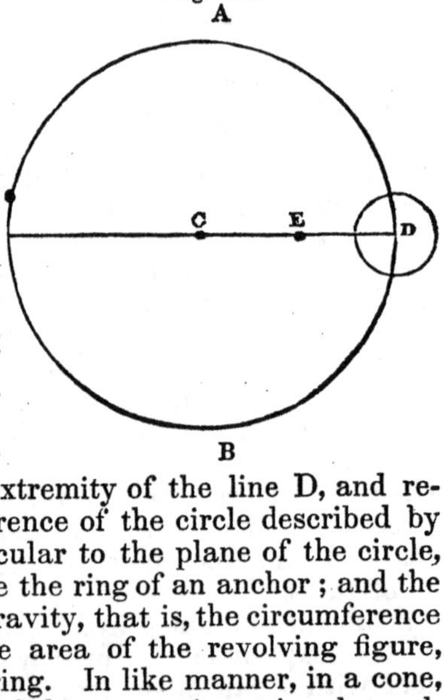

Thus, the straight line CD, (Fig. 138,) revolving about the center C, describes a circle whose surface is equal to CD into the circumference of the circle described by its center of gravity E. This is evident also from the consideration that, since E is the center of the line CD, the circumference described by it will be half the length of the circumference ADB; and the area of a circle is equal to the product of the radius into half the circumference.

Again, the small circle, having its center coincident with the extremity of the line D, and revolving round on the circumference of the circle described by CD, being everywhere perpendicular to the plane of the circle, would describe a solid figure like the ring of an anchor; and the line described by the center of gravity, that is, the circumference of the circle, multiplied into the area of the revolving figure, would give the solidity of the ring. In like manner, in a cone, the solidity is equal to the *area* of the generating triangle, multiplied into the circumference of the circle, formed by the revolution of the center of gravity of the triangle; and the surface is equal to the product of the *perimeter* of the same triangle, multiplied by the circumference described by the center of gravity of the same perimeter.

* Hence these properties of the center of gravity are sometimes called *Guldinus' Properties*

CHAPTER IV.

OF MACHINERY.

281. The organs employed in communicating motion, are tools, machines, and engines. *Tools* are the simplest instruments of art; these, when complicated in their structure, become *machines;* and machines, when they act with great power, take the name of *engines.* Among the ancients, machines were confined chiefly to the purposes of *architecture* and *war;* and they were moved almost exclusively by the *strength of animals.* Thus, in building one of the great Pyramids of Egypt, vast masses of stone were raised to a great height, amounting together to 10,400,000 tons. In this labor were employed 100,000 men for 20 years. The advantage which man has gained by pressing into his service the great powers of nature, instead of depending on his own feeble arm, is evinced by the fact, that by the aid of the steam engine, one man can now accomplish as much labor as 27,000 Egyptians, working at the rate at which they built the pyramids.* In war also, while the use of gunpowder was unknown, engines of great power were invented for throwing stones and javelins, and for demolishing fortifications. Such were the Catapulta, the Balista, and the Battering Ram, of the Romans. Yet it is remarkable, that during many ages, while such powerful auxiliaries were employed in architecture and in war, the ancients should have made so little use as they did of machinery in the ordinary processes of the arts. Nor did the philosophers of antiquity cultivate science with any reference to the improvement of the arts, an object which they considered below the dignity of true philosophy; but Lord Bacon was the first to show that "a principle in science is a rule in art."

282. The *Mechanical Powers,* being the principal instruments of art, will first require our attention. They have already been considered *theoretically;* we are now to consider them *practically.* It will be recollected that, when two forces act on one another by means of any machine, that which gives motion is called the *power;* that which receives motion, the *weight.* (Art. 97.) The weight includes not only the proper weight of the body or bodies moved, but also every kind of resistance opposed to the action of the power, whether it arises from the quantity of matter in the body moved, or from the inertia of the machine itself, or from the air, or from friction. It will be further recollected, that an equilibrium is produced between two forces, when their *momenta* are equal. (Art. 149.) Now, since momentum is compounded of quantity of matter and velocity, a given momentum

* Dupin.

may be produced, either by giving a great velocity to a small weight, or a small velocity to a great weight. The consideration of this subject will be resumed hereafter.

THE LEVER.

283. The principle of the lever has a most extensive application in the arts, and the forms under which it occurs are very various. We may contemplate it as having equal or unequal arms. The *balance* affords the most common example of a lever with equal arms. The necessity of arriving at the weight of bodies with the greatest degree of accuracy in pecuniary transactions, and more especially in delicate scientific researches, as those of chemical analysis, has induced men of science, and artists, to bestow great and united attention upon the construction of this instrument, until they have brought it to an astonishing degree of perfection.

Fig. 139.

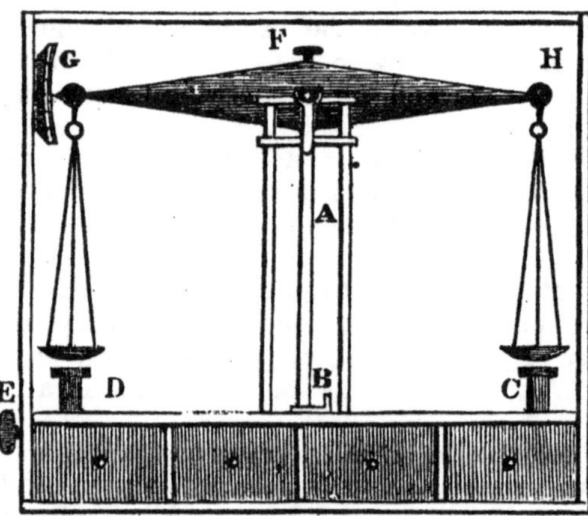

284. The principal parts of the balance are the beam GH, (Fig. 139,) the points of suspension G and H, and the fulcrum F. In order to construct a perfect balance, the most important particulars to be attended to, are the length of the arms, that is, of the beam; the situation of the center of gravity of the whole instrument, with respect to the fulcrum or center of motion; and the position of the points of suspension.

(1.) The sensibility of the balance is increased by increasing the *lengths of the arms;* but unless the arms, when long, are at the same time of considerable weight, they will not have the requisite strength, but will be liable to bend; and an increase of weight, adds to the amount of friction on the center of motion

It is not common, therefore, to make the arms of a very delicate balance more than nine inches in length; and, for the purpose of uniting lightness with strength, the beam is composed of two hollow cones placed base to base, as in Fig. 139.

(2.) The *center of gravity* of the instrument must be a little below the center of motion. For if the beam is balanced on its center of gravity, it will remain at rest in every position, (Art. 64,) whereas it must be at rest only when in a horizontal position. If the center of gravity is above the center of motion, the position is too unstable, (Art. 274,) and upon the least disturbance of the equilibrium, the beam will be liable to upset. Finally, if the center of gravity is too far below the center of motion, the equilibrium will be too stable. Hence, in very delicate balances, the center of motion is placed a little above the center of gravity.

(3.) The *points of suspension* must be in the same right line with the center of motion. For since when weights are added to the scales, the effect is the same as though they were concentrated in the points of suspension, (Art. 99,) were those points above the center of motion, the center of gravity would be liable to be shifted above the center of motion, when the beam would upset; and if the same points were below the center of motion, the center of gravity would be too low, and the equilibrium too stable.

In order to prevent friction as much as possible, the *fulcrum* is made of hardened steel, and shaped into a triangular prism, or knife edge, smoothly rounded, and turning on a plane of agate or steel, or some other very hard and polished substance.

285. It is only by a nice attention to all these particulars, that artists have been able to give to the balance so great a sensibility. Some have been made to turn with the 1000th part of a grain.* By loading the beam, the sensibility of the instrument is diminished; it is customary, therefore, to estimate its power, by finding what part of the weight with which it is loaded it takes to turn it. Thus, if when loaded with 7000 grains, it will turn with one grain, its power is $\frac{1}{7000}$. A balance constructed by Ramsden, a celebrated English artist, for the Royal Society, turned with the *ten millionth* part of the weight.† Delicate balances are usually covered with a glass case to prevent agitation from the air, and to secure them from injury. Fig. 139, represents an instrument of this kind made for the Royal Institution of Great Britain.‡

* Nicholson's Chemical Dict.—Kater in Lardner's Mechanics.
† Young's Lectures on Nat. Phil. I, 125.
‡ For a full account of the most accurate balances, see Kater on "Balances and Pendulums" in Lardner's Mechanics, p. 278.

286. The *bent lever balance* is represented in Fig. 140. The weight C acts as though it were concentrated in the point D, (Art. 104,) and the weight in the scale acts at K; hence an equilibrium will take place, when the article weighed has to C the same ratio as DB has to BK. Now every increase of weight added to the scales causes C to rise on the arc FG, and D to recede from B. Hence the different positions of C, according as different weights are added to the scale, may be easily determined, and the corresponding numbers marked on the scale FG.

Fig. 140

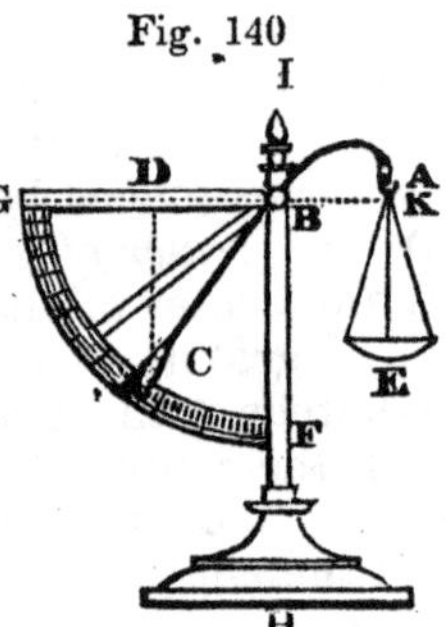

287. It is essential to an accurate balance, that the two arms should be precisely equal in length. The *false balance*, which is sometimes used with a design to defraud, has its arms unequal. The dealer turns such an instrument to his account, both in buying and selling. In buying he puts his weights on the longer side, for then it takes more than an equivalent to balance them; and, in selling, he puts his weights on the shorter side, because less than an equivalent will produce an equilibrium. The fraud may be detected by making the weights and the merchandise change places. The true weight may be determined from such a balance by the rule given on p. 94; or more conveniently, by putting the article whose weight is to be determined into one scale, and counterpoising it with sand, shot, or any convenient substance, in the other scale, and then removing the article, and finding the exact weight of the counterpoise. It is evident that the weight of the merchandise will be the same as that of the weights employed to balance its counterpoise.

288. The *steelyard* is a lever having unequal arms, in which the same body is made to indicate different weights, by placing it at different distances from the fulcrum. A pair of steelyards has usually two graduated sides for determining smaller or greater weights. It will be seen that on the greater side, the weight is placed nearer the fulcrum. Consequently, the weight indicated by the counterpoise, when at a given distance from the fulcrum, will be proportionally greater. This instrument is very convenient, because it requires but one weight. The pressure on the fulcrum, excepting that of the apparatus itself, is only that of the article weighed, whereas in the balance, the fulcrum sustains a double weight. But the balance is susceptible of more sensibility than the steelyard, because the subdivisions of its weights can be effected with a greater degree of precision than the subdivision of the arms of a steelyard.*

* Kater.

289. The *spring steelyard* is a very convenient instrument for weighing, in cases where the subdivisions of the weights are large. It depends on the elasticity of a spiral steel spring, to compress or extend which requires a force proportioned to the degree of compression or extension. The manner of applying it will be easily understood from the representation in Fig. 141. After continued use, especially when loaded with heavy weights, the elasticity of the spring is liable to be impaired, and the accuracy of the instrument diminished. When made, however, in the best manner, spring steelyards retain their accuracy for a long time.

Fig. 141

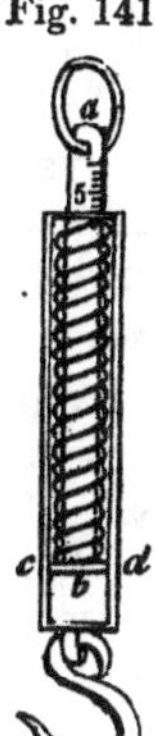

290. In Fig. 142, is represented a vertical section of a large *Weighing Machine*, such as is used for loads of hay, cotton, or other heavy merchandise.

AB, a section of the platform, resting loosely on a frame.

CN, DN, levers of the second kind, having their fulcrums at C, D, and resting on a bar at N.

W, W, pins which press upward against the platform when the levers are raised.

EF, a lever likewise of the second kind, having its fulcrum at E, and connected by a perpendicular arm, with the beam of a pair of steelyards at G.

Fig. 142.

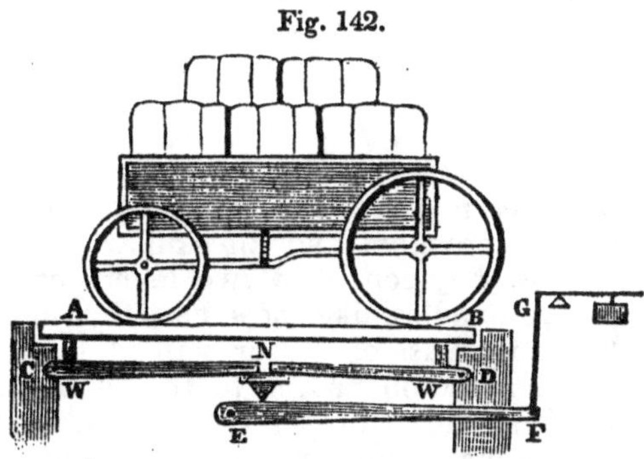

Four levers are usually employed, proceeding from the four corners of an immovable frame, or having their fulcrums firmly set in masonry. The levers all rest on the common support at N.

Suppose a load of merchandise is placed on the platform to be weighed. By the steelyards, we ascertain that the weight exerted at G is 100 pounds, which force is that exerted at F to raise the lever EF. Supposing, for convenience of computation, that the levers have their longer ten times the length of their shorter arms, then 100 pounds at F balances a force of 1000 pounds at N. This force is still further multiplied by the four levers so as to become 10,000 pounds, which is the weight of the load including that of the platform. If the platform rested on a single lever, this would of course sustain a weight of 10,000 pounds; but as the levers severally sustain the same part of the weight, each one bears only one fourth of the load, or 2,500 pounds.

291. When a weight is supported by a lever which rests on two props, the pressure upon both fulcrums is equal to the whole weight. This principle is sometimes applied in ascertaining the weight of a body too heavy for the steelyards. The body is suspended immovably near the center of a pole, and the steelyards are applied to each end of the pole separately, the other end meanwhile resting on its fulcrum. The two weights being added together, make the entire weight of the body. If the body is suspended exactly in the center of the pole, it will be sufficient to obtain the weight of one end and double it. The weight of the lever should, in both cases, be subtracted from the entire weight.

292. Since when a weight is sustained between two props, the *part sustained by each prop is inversely as the distance of the weight from it*, it follows that a load borne on a pole, between two bearers, is distributed in this ratio. As the effort of the bearers, and the direction of the weight, are always parallel, it makes no difference whether the pole is parallel to the horizon or inclined to it. Whether the bearers ascend or descend, or move on a level plane, the weight will be shared between them in the same constant ratio. (See Fig. 72.)

293. *Handspikes* and *crowbars* are familiar examples of levers of the first kind. A *hammer* affords an example of the *bent lever;* and shears, pliers, nutcrackers, and all similar instruments, are *double levers;* that is, they consist of two levers united. A pair of shears, with long handles, like those used by tinners, exhibit very strikingly the increase of power gained by bringing the weight or substance acted on nearer to the fulcrum. The jaws of animals exhibit a similar property. An oar, applied to a boat rowed by hand, a wheel-barrow, and a door shut by the hand applied to the edge remote from the hinges, severally furnish instances of levers of the second kind, where the weight is between the fulcrum and the power.

The *crane* is a lever of the second kind, which is much used when great weights are transported for a short distance, as heavy boxes of merchandise from a vessel to the wharf, or great masses of stone from the quarry to a car or boat. An example of the crane, on a small scale, is seen in the apparatus of a kitchen fire-place.

294. When one raises a *ladder* from the ground by one of the ower rounds, the ladder becomes a lever of the third kind, the power being applied between the weight and the prop. Since in all the mechanical powers, the power and weight have equal momenta, and since, in the third kind of lever, the weight has more velocity than the power, the power is as much greater than the weight as the velocity with which it moves is less. The difficulty experienced in raising a ladder from the ground by taking hold of the lowest round, or of shutting a door by applying the hand to the side next to the hinges, shows the mechanical disadvantage under which a lever of this kind acts. Yet it is very useful in cases when it is required to give great *velocity* to the body moved. *Sheep-shears* consist of two levers of this kind united. Here the whole force required is so small that to save it is of no consequence, while so soft and flexible a substance as wool, requires the shears to be moved with considerable velocity. A pair of tongs is composed in the same manner; and therefore it is only a small weight that we can lift with them, especially when the legs are long.

295. One of the most remarkable applications of the third kind of lever, is in the *bones of animals.* These are levers, the joints are the fulcrums, and the muscles are the powers. The muscles are endowed with a strong power of contraction, by which they are made to pull upon a tendon or cord, which is inserted in the bone near the fulcrum. Thus, the fore-arm moves on the joint near the elbow as a fulcrum, a little below which is inserted a tendon, connected with a muscle between the elbow and the shoulder which gives it motion. The arrangement may be well represented by attaching a small cord to one of the legs of a pair of tongs, near the joint. It will require a considerable force to lift the leg by pulling at the string, especially if the string be pulled in a direction nearly parallel with the leg, as it ought to be, since the tendon which lifts the fore-arm, acts in such a direction with respect to the arm. The muscles therefore act in moving the bones under a double mechanical disadvantage, their force being applied both obliquely and very near the fulcrum. The force which the muscles exert in raising a weight held in the palm of the hand, is enormous, as will be comprehended from the following illustration. Let AB (Fig.

Fig. 143.

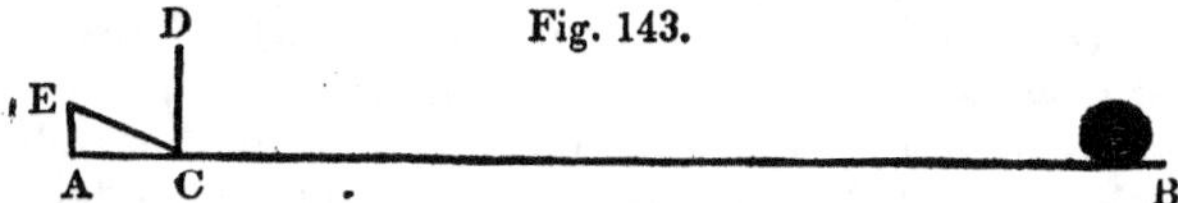

143) represent the fore-arm moving on the elbow joint at A, and having the tendon inserted at C, which we will suppose to be ten times nearer to A than B is to A. Consequently, a weight of 10

pounds at B, would require a force at C, acting directly upward, of 100 pounds. But the force of the tendon does not act *directly upward* in the direction of CD, but very obliquely, as in the direction of CE, of which the part EA only can contribute to support the weight. Suppose this part to equal $\frac{1}{10}$th of the whole force CE, and it follows that the muscular force exerted to raise a weight of ten pounds in the palm of the hand, would, were it to act without any mechanical disadvantage, be sufficient to raise a weight of 1000 pounds. Yet Dr. Young informs us, that a few years ago there was a person at Oxford, who could hold his arm extended for half a minute, with half a hundred weight hanging to his little finger.*

But by giving to the muscle the position it has, the greatest possible *compactness of structure* is obtained, while, by making it act so near the fulcrum, what is lost in force, is gained in *velocity;* and while the power acts through a small space, the hands are moved quickly through a great distance. In consequence of the dominion which man can gain over the stronger animals, and especially over the great powers of Nature, he has little occasion to exert great strength with his naked hands: the celerity of their movements is to him a far more important endowment.

WHEEL WORK.

296. When a lever is applied to raise a weight, or to overcome a resistance, the space through which it acts at one time is small, and the work must be accomplished by a succession of short and intermitting efforts. The common lever is, therefore, used only in cases where weights are required to be raised through small spaces. When a continuous motion is required, as in raising ore from the mine, or in weighing the anchor of a vessel, some contrivance must be adopted to remove the intermitting action of the lever and render it continual. The *wheel and axle*, in its various forms, fully answers this purpose. It may be considered as a revolving lever.

In numerous forms of the wheel and axle, the weight is applied by a rope coiled upon the axle; but the manner in which the power is applied is very various, and not always by means of a rope. The circumference of a wheel sometimes carries projecting pins as in Fig. 80, to which the hand is applied to turn the machine. An instance of this occurs in the wheel used in the steerage of a vessel. In the common *windlass* the power is applied by means of a *winch*, which corresponds to the radius of a wheel. (See Fig. 97.) The axis is sometimes placed in a vertical position, and turned by levers moved horizontally. The *capstan* of a ship (Fig. 144) is an example of this. Levers an

* Young's Lectures on Nat. Phil., I, 129.

swering to the radii of a wheel, are inserted in holes mortised in the axis, and turned by several men working together. In some cases, as in the *treadmill*, the wheel is turned by the weight of animals walking on the circumference, with a motion like that of ascending a steep hill.

Fig. 144.

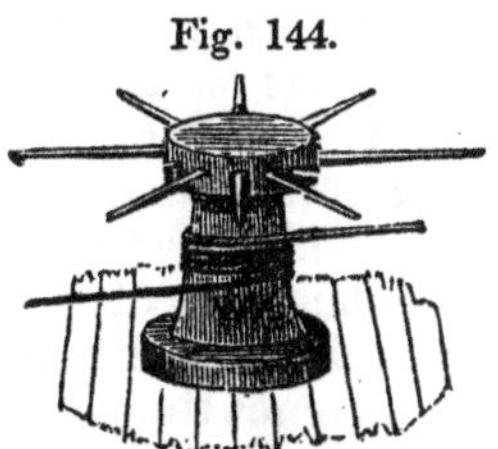

297. The power of the wheel and axle being expressed by the number of times the diameter of the axle is contained in that of the wheel, there are obviously two ways by which this power may be increased; either by increasing the diameter of the wheel, or by diminishing that of the axle. In cases where great power is required, each of these methods is attended with practical inconvenience and difficulty. If the diameter of the wheel is considerably enlarged, the machine will become unwieldy, and the power will work through an unmanageable space. If, on the other hand, the power of the machine is increased by reducing the thickness of the axis, the strength of the axis will become insufficient for the support of that weight, the magnitude of which had rendered the increase of the power of the machine necessary. To combine the requisite strength with moderate dimensions and great mechanical power, is therefore impracticable, in the ordinary form of the wheel and axle. This has, however, been accomplished by giving different thicknesses to different parts of the axle, and carrying a rope, which is coiled on the thinner part, through a wheel attached to the weight, and coiling it in the opposite direction on the thicker part, as in Fig. 145. To investigate the proportion of the power to the

Fig. 145.

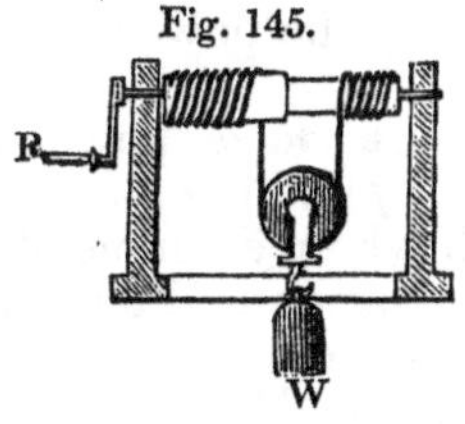

Fig. 146.

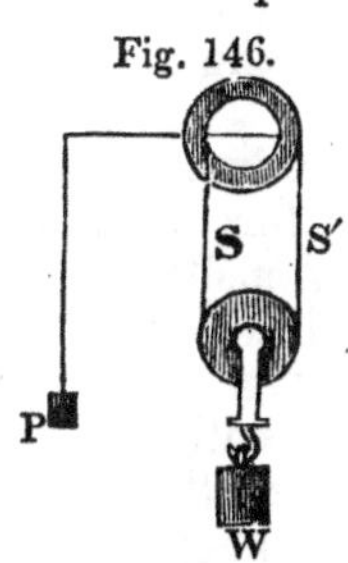

weight in this case, let Fig. 146 represent a section of the apparatus at right angles to the axis. The weight is equally suspended by the two parts of the rope S and S′, and therefore each part is stretched by a force equal to half the weight. The momentum of the force which stretches the rope S, is half the weight multiplied by the radius of the thinner part of the axis. This force being on the same side of the center with the power, co-operates with it in supporting the force which stretches S′, and which acts on the other side of the center. Now the momenta of P and S together, must be equal to the momentum of S′, (Art. 21,) and therefore if P be multiplied by the radius of the wheel, and added to half the weight multiplied by the radius of the thinner part of the axis, we shall obtain a sum equal to half the weight, multiplied by the radius of the thicker part of the axis. Hence the power multiplied by the radius of the wheel, is equal to half the weight multiplied by the difference of the radii of the thicker and thinner parts of the axis.*

298. A wheel and axle constructed in this manner, is equivalent to an ordinary one, in which the wheel has the same radius, and whose axis has a radius equal to half the difference of the radii of the thicker and thinner parts.† The power of the machine is expressed by the ratio which the radius of the wheel bears to half the difference of these radii; and therefore this power, when the diameter of the wheel is given, does not, as in the ordinary wheel and axle, depend on the smallness of the axle, but on the smallness of the *difference* of the thinner and thicker parts of it. The axle may, therefore, be constructed of such a thickness as to give it all the requisite strength, and yet the difference of the diameters of its different parts may be so small as to give it all the requisite power.‡

We see here strikingly exemplified the principle, that the weight sustained by a given power may be increased as its velocity is diminished. By inspecting Fig. 146, it will be seen that the string connected with the thinner part of the axle *unwinds*, while that connected with the thicker part *winds up*, by which means the ascent of the weight may be rendered slow in any degree, and a proportionally greater quantity of matter may be added to balance the constant momentum of the power.

299. It is sometimes desirable to make a *variable* power produce a constant force. This may be done by making its velocity increase as its intensity diminishes. We have an example of this in the reciprocal action between the main-spring and fusee of a watch. (Fig. 147.) The main-spring is coiled up in the box A, and is connected with the fusee B by a chain. When the watch is first wound up, the spring acts with its greatest intensity, but then as the wheel B turns, it uncoils with the least velocity; but on account of the varying diameters of the wheels of the

---

* Let W=weight. P=power. R=radius of the wheel. r=radius of the thicker part. r'=radius of the thinner part.

Then $P\times R+\frac{1}{2}W\times r'=\frac{1}{2}W\times r$,

$\therefore P\times R=\frac{1}{2}W\times r-\frac{1}{2}W\times r'=\frac{1}{2}W(r-r')$.

† $P\times R=\frac{1}{2}W(r-r')=W\times\frac{1}{2}(r-r')$.

‡ Lardner's El. Mech. p. 181.

Fig. 147.

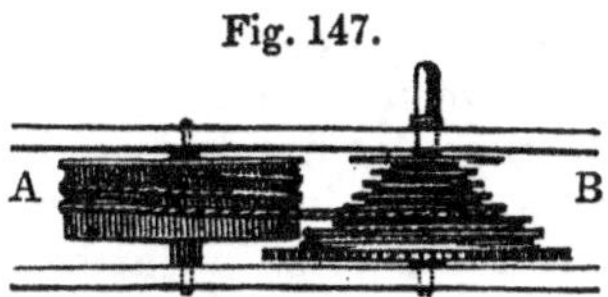

fusee, the velocity is continually increased as the intensity of the spring is diminished. In a similar manner a *varying* weight may be moved by a constant power.

Communication of Motion by Wheel-work.

300. Motion may be transmitted by means of wheel-work in several different methods, the principal of which are, the friction of the circumference of one wheel upon that of another—the friction of a band—and the action of teeth.

One wheel is sometimes made to turn another, by the mere *friction of the two circumferences.* If the surfaces of both were perfectly smooth, so that all friction was removed, it is obvious that either would slide over the surface of the other, without communicating motion to it. But, on the other hand, if there were any asperities, however small, upon their surfaces, they would become mutually inserted among each other, and neither the wheel nor axle could move without causing the asperities on its edge to encounter those which project from the surface of the other; and thus both wheel and axle would move at the same time. Hence if the surfaces of the wheel and axle are by any means made rough, and pressed together with sufficient force, the motion of either will turn the other, provided the load or resistance be not greater than the force necessary to break off these small projections which produce friction.

In some cases where great power is not required, motion is communicated in this way through a train of wheel-work, by rendering the surfaces of the wheel and axle rough, either by facing them with buff leather, or with wood cut across the grain. The communication of motion between wheels and axles by friction has the advantage of great smoothness and evenness, and of proceeding with little noise; but this method can be used only in cases where the resistance is not very considerable, and therefore it is seldom adopted in works on a large scale. Dr. Gregory mentions an instance of a saw-mill at Southampton, where the wheels act upon each other, by the contact of the end grain of the wood. The machinery makes very little noise and wears well, having been used not less than twenty years.*

301. Wheel work is extensively moved by the *friction of a band.*

* Gregory's Mech. II, 537.—Lardner's El. Mech 183

When a round cord is used, any degree of friction may be produced, by letting the cord run in a sharp groove at the edge of the wheel. When a strap or flat band is used, its friction may be increased by increasing its width. The surface at the circumference of a wheel which carries a flat band, should not be exactly cylindrical, but a little convex, in which case if the band inclines to slip off at either side, it returns again by the tightening of its inner edge, as may be seen in a turner's lathe. When wheels are connected in the shortest manner by a band, they move in the same direction: if the band be crossed, they will move in opposite directions.* (Fig. 148.) Wheels are sometimes turned by *chains* instead of straps or bands, and are then called *rag wheels.* The chains lay hold upon pins, or enter into notches, in the circumference of the wheels, so as to cause them to turn simultaneously. They are used where it is necessary that the velocities should be uniform, and where great resistance is to be overcome, as in locomotive steam engines, chain water wheels, &c.†

Fig. 148.

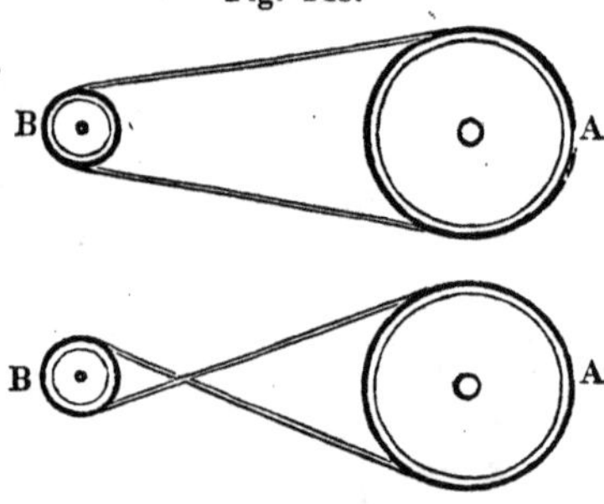

302. But the most common mode of moving wheel-work, is by means of *teeth* cut in the circumference of the wheels. The wheels of necessity turn in opposite directions. The connexion of one toothed wheel with another is called *gearing.* In the formation of teeth, very minute attention must be given to their figure, in order that motion may be communicated from one wheel to another, without rubbing or jarring. If the teeth are ill matched, as in Fig. 149, when the tooth A comes in contact

Fig. 149.

Fig. 150.

with B, it acts obliquely upon it, and as it moves, the corner of

* Bigelow's Technology, p. 229. † Ib. p. 230.

B slides upon the plane surface of A in such a manner as to produce much friction, and to grind away the side of A, and the end of B. As they approach the position CD, they sustain a jolt the moment their surfaces come into full contact; and after passing the position CD, the same scraping and grinding effect is produced in the opposite direction, until by the revolution of the wheels the teeth become disengaged. To avoid these evils, the surfaces of the teeth are frequently *curved* so as to roll on each other with as little friction, and with as uniform force and velocity as possible. (Fig. 150.) Much pains and skill have been bestowed on this subject by mathematicians, with the view of ascertaining the kinds of curves which fulfil these purposes best.*

Regulation of Velocity by Wheel-work.

303. Wheel-work serves the purpose, not only of forming a convenient communication of motion between the power and the weight, but also of *regulating its velocity.* Thus, when the connection is formed by means of a band, as in Fig. 148, the velocity of the wheel B, that carries the weight or sustains the pressure, may be altered at pleasure, by altering the ratio between the diameters of the two wheels. If the diameters are equal, the wheels will revolve with equal velocity; if A remains the same, while the diameter of B is diminished, the velocity of B will be increased in the same ratio; or if B remains the same, while the diameter of A is changed, the velocity of B will be changed in the same manner.† We see familiar examples of the application of this principle in the common spinning wheel, and the turner's lathe. In the spinning wheel a band passes round a large wheel and a small one called a spool, having the spindle for its axis; and in consequence of the great disparity in the size of the wheels, a great velocity is given to the spindle by a comparatively slow revolution of the wheel. In a turner's apparatus, machinery for spinning cotton, and the like, a large hollow cylinder or *drum,* is fixed horizontally, which is kept revolving by the moving power, and from which, motion is conveyed by bands to lathes, spindles, &c., to which any required velocity is given, by altering the diameter of the small wheel that is connected with them and turns them. Sometimes a change of velocity is effected by making the drum of a conical shape, and then the velocity imparted to the lathe or the spindle, will be greater or less, according as the band proceeds from the larger or smaller part of the drum.

---

* See Blake on the form of the teeth of cog wheels, Am. Journ. Sci., VIII, 86.

† This would be accurately true, in case the band did not slip or slide; but since it usually does slide more or less, the velocity of the driven wheel is commonly a little less in proportion, than that of the wheel which drives it.—Bigelow, El. Tech p. 229.

304. A more exact method of regulating the velocity of motion, is by means of *wheels and pinions.** An example of this kind is seen in Fig. 151, where A, B, C, are three wheels, and *a, b, c,* are the corresponding pinions. As the leaves of the pinions successively pass between the teeth of the wheel, the divisions of the two circumferences must correspond to each other, and the number of teeth in the wheel will be as much greater than in the pinion, as the circumference of the wheel is greater than that of the pinion. Therefore it follows, that the number of teeth in a wheel, and of leaves in the pinion that acts upon it, expresses the ratio of the circumference or radius of the wheel to that of the pinion. Hence, in an equilibrium, the power multiplied by the product of the numbers expressing the amount of teeth in all the wheels respectively, is equal to the weight multiplied by the product of the several numbers denoting the leaves in each of the pinions. (Art. 117.)

Fig. 151.

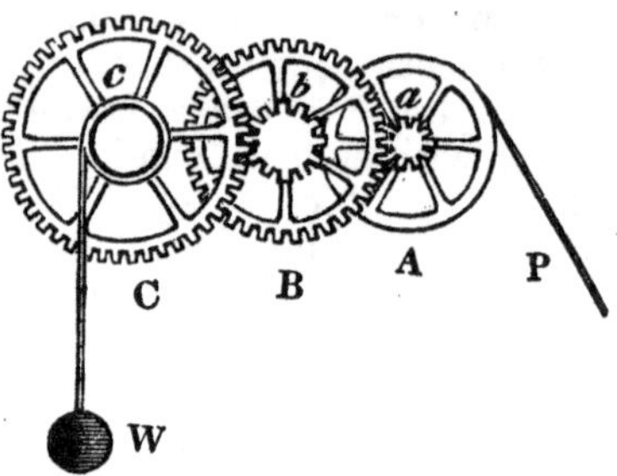

It is further evident that the velocity of the wheel and that of the pinion connected with its circumference, will be inversely as the *number of teeth* in each. Thus in Fig. 151, if the pinion *a* has 10 teeth, and the wheel B has 100, *a* will move ten times as fast as B. For the same reason *b* will move ten times as fast as C; so that, in this arrangement, the power moves with 100 times the velocity of the weight. By varying the ratio between the number of leaves in the pinion, and the number of teeth in the wheel with which it is connected, we may vary the velocity of any wheel at pleasure.

305. A familiar instance of this is afforded in the mechanism of a common clock. A pendulum by falling gains a quantity of motion sufficient to carry it, on the other side, to the same height as that from which it fell; and were it not for the resistance of the air and the impediments, a pendulum when once set in motion would continue to vibrate by its own inertia, (Art. 19,) and would thus afford, without the aid of any machinery, an exact measure of time. But, in order to continue its vibrations, some small force must be applied to it to compensate for the loss of motion from friction and resistance. This force is applied to the

---

* Pinions are smaller wheels acting on the circumferences of larger. The teeth of a pinion are called *leaves.* They are most commonly raised on the axis of one wheel, and form the communication between that wheel and the next in the series.

pendulums of clocks by the *weight*, and an analogous force is supplied to the *balance wheel* of watches and chronometers by *springs*. In Fig. 152, let AB be a wheel having 30 teeth, and let N, M, be a pendulum, connected with the wheel by the *pallets* I, K; and to the axis *a*, let a weight be hung. It is evident that this weight, were there nothing to arrest it, would descend by the force of gravity with accelerated velocity. It *endeavors* thus to descend, and hence exerts the required force on the pallets of the pendulum. For, every time the pendulum performs a double vibration, (returning to the same point from which it set out,) a tooth of the wheel escapes,* and the wheel runs down until the next tooth strikes upon the pallet, and thus gives it the impulse which is necessary to keep up the vibrations.

Fig. 152.

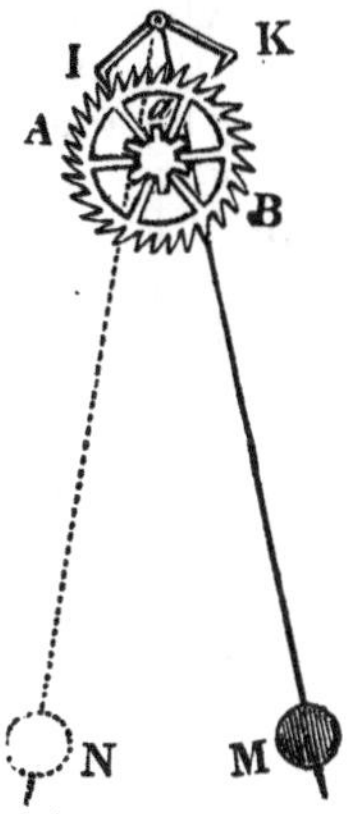

It would seem therefore that, for beating seconds, only a single wheel is necessary; nor would any more be absolutely indispensable; but in this case the weight would descend so fast, as soon to reach the floor, and the clock would require to be wound up again every few minutes. Hence a series of wheels are interposed between the pendulum and the weight, by which the descent of the latter is retarded upon the principle explained in Art. 304, and the descent of the weight is slower in proportion as the series is more extensive. In cheap clocks, as some of those made with wooden wheels, the series is short, or the number of wheels employed for retarding the descent of the weight is small, and such clocks require frequent winding up; but in clocks of finer workmanship, a greater number of wheels is interposed, and such clocks require to be wound up less frequently. Many go eight days, and some are made to go a whole year without winding.

Wheel Carriages.

306. When a loaded carriage is moving on a horizontal plane, free of obstacles, the resistance to be overcome does not consist of the weight of the load, directly, but of the friction occasioned by the weight. For, since the weight acts in a direction perpendicular to the plane, it cannot oppose the motion of the carriage in a direction parallel to the plane. Nor would increasing the weight to any extent, make any difference, were it not that we should thus increase the friction, which (as will be explained more fully hereafter) is proportioned to the weight.

* Hence this wheel is called the *scapement.*

When a carriage wheel is made to *slide* on the ground, (as when a wheel is *locked*,) the whole amount of the friction is encountered without bringing in to our aid any mechanical advantage; but when a wheel turns on its axle, the friction is transferred from the ground to the axle, and each spoke of the wheel successively becomes a lever, turning on the ground as a fulcrum, while the power or force of the team is exerted on the end next to the axle. By thus transferring the friction from the ground to the axle, each spoke, in its turn, is made to aid in overcoming that friction. Thus, in Fig. 153, let C be the axle, CP the line of draught, and R the point where the wheel touches the plane. The force applied in the direction CP, acts on CR at C, and turns it on its fulcrum at R. This is the force by which the wheel is made to advance. But the friction on the axle at C, reacts in the opposite direction, having a leverage equal only to the radius of the axle, while the power which overcomes this, has a leverage equal to the radius of the wheel. Hence, in the wheel, there is a mechanical advantage gained in overcoming the friction, in the ratio of the radius of the wheel to the radius of the axle. Moreover, the axle may be made of such materials, and lubricated with such substances, as to render the actual amount of friction much less than it would be were the wheel made to slide on the ground.

Fig. 153.

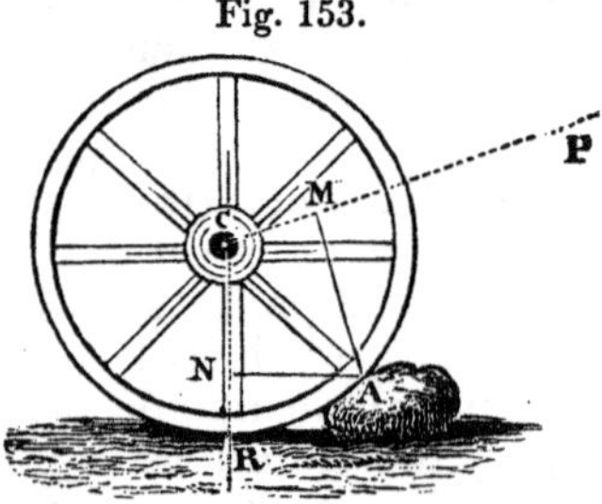

307. But wheels have another important advantage, namely, *in overcoming obstacles;* in which case they act on the principle of the bent lever.

Thus let A be an obstacle, as a stone for example. From A let fall the perpendiculars AN, AM, upon CR, CP, and conceive MAN to be a bent lever, turning on A as a fulcrum, the power being applied at M in the direction CP, and the weight resting on N, which supports the center of gravity. Now, the mechanical advantage gained, will be in the ratio of MA to NA. It will therefore be increased (and of course the force necessary to overcome the obstacle be diminished) as the point A is nearer to R; and the mechanical advantage will be lessened as the point A recedes from R. When the obstacle is so large as to make AM only equal AN, then no mechanical advantage is gained, but the whole weight of the load must be lifted by the former; and when AM becomes less than AN, the wheel involves a mechanical disadvantage, and the difficulty of carrying the wheel over the obstacle becomes very great. It is further obvious that *large wheels* have the mechanical advantage, both as regards overcoming the friction, and overcoming obstacles, in a higher degree than small wheels, since these afford a greater leverage than the others on account of the increased length of the spokes. But in practice very large wheels cannot be employed, since they would be either weak or too heavy, and the increased height of the axle would carry the center of gravity too high, and enhance the danger of upsetting. The difficulty of turning might also render unusually large wheels ineligible ; and the axle might be raised so high, as to make the horse draw obliquely downward and increase the pressure on the ground, whereas the line of draught ought to be so adjusted as to lighten that pressure, especially where the road is soft and yielding.

When a wheel sinks below the surface, the force is rendered strikingly inefficacious from several causes. The fulcrum on which each spoke successively turns gives way, and diminishes greatly the mechanical advantage otherwise gained by transferring the friction from the ground to the axle, as before explained. Likewise, the mud or sand into which the wheel has sunk, opposes in front of the wheel an obstacle like that represented at A in Fig. 153, while the fulcrum on which the bent lever turns in the effort to lift the wheel over the obstacle gives way as in the other case, and a great part of the mechanical advantage is lost. From these considerations, it is easy to understand the reason of the superior advantages of hard and smooth roads.

308. The *line of draught* should not be horizontal, but inclined upward toward the breast of the horse, in an angle not less than 15 degrees with the horizon. This brings the strain nearly at right angles with the collar, whereas a horizontal draught lifts the collar upward, by which the force is wasted and the animal is choked.* The angle of draught, however, should be less than the above when the road is very smooth. The general rule is, that the angle should be the same as the inclination of a hill, down which the carriage would roll spontaneously. Consequently, in smooth Macadamized roads, the line of draught should be at a small angle, and on railways nearly horizontal.†

309. The effect of suspending a carriage on *springs*, is to equalize the motion by causing every change to be more gradually communicated to it, and to obviate shocks. Springs are not only useful for the convenience of passengers, but they also diminish the labor of draught ; for whenever a wheel strikes a stone, it rises against the pressure of the spring, in many cases without materially disturbing the load, whereas without the spring, the load, or a part of it, must rise with every jolt of the wheel, and will resist the change of place with a degree of inertia

* Fuller on Wheel Carriages.

† Moseley's Mechanics applied to the Arts.

proportionate to the weight, and the suddenness of the percussion. Hence springs are highly useful in baggage wagons, and other vehicles used for heavy transportation.*

A pair of horses draw more advantageously abreast than when one is harnessed before the other. In the latter case, the forward horse, being attached to the ends of the shafts, draws in a line nearly horizontal; consequently he does not act with his whole force upon the load, and moreover expends a part of his force in a vertical pressure on the back of the other horse.

THE PULLEY.

310. Pulleys are divided into FIXED and MOVABLE. In the *fixed* pulley, as has been demonstrated in Art. 120, no mechanical advantage is gained, but its use consists in furnishing a convenient mode of *changing the direction* of the power. Thus, it is far more convenient to raise a bucket from a well by drawing downward, as is the case when the rope passes over a fixed pulley above the head, than by drawing upward leaning over the well. By means of the pulley, great facilities are afforded for managing the rigging of a ship. The sails at mast head can be easily raised while the hands stand upon the deck, whereas, without the aid of ropes and pulleys, the same force removed to mast head would operate under very great disadvantages. Similar facilities are afforded by this kind of apparatus for raising heavy weights, as boxes of merchandise, or

Fig. 154.

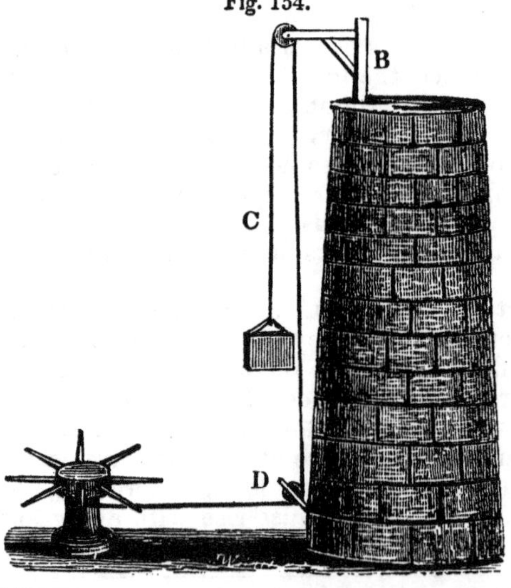

* Bigelow, El. Tech., p. 200.

heavy blocks of stone in building. Fig. 154, represents a convenient method of elevating large masses of stone in building a high tower, as a lighthouse or a monument. The *crane* at B enables the workmen, when the weight is raised, to swing it round to the point where it is to lie, or to a platform near to it. The lower end of the rope CD, is connected with a wheel and axle in the figure, but it is obvious that different methods of applying the power might be adopted, to suit the convenience of the workmen. For example, instead of the wheel and axle we might attach a horse or a yoke of oxen to the rope CD; or we might attach a long *sweep* to the top of the axis, and join a team of horses or cattle to the end of it, and raise the weight by driving them round, as in a common cider-mill.

Fire escapes sometimes consist merely of a pulley fixed near the window of the apartment, around which a rope may be easily placed, having a basket attached to the end. The man seats himself in the basket, grasping, at the same moment, the rope on the other side of the pulley, and thus he lets himself gradually down.

311. The *movable* pulley, by distributing the weight into separate parts, so that it is supported at several different points at once, is attended by a mechanical advantage, proportioned to the number of such points of support. Movable pulleys may be arranged according to several different systems, which increase the efficacy of a given power in different ratios.* It will be observed, however, that the ascent of the weight is in all cases retarded in proportion as the efficacy of a given power is increased. It must be further observed that in using any system of movable pulleys, the whole weight of the pulleys themselves, together with the resistance occasioned by the rigidity and friction of the rope, all act against the power, and so far lessen the weight which it is capable of raising. In the more complex systems of pulleys, it is estimated that at least two thirds of the power is expended on the machinery itself. On account therefore of the slowness of the motion which the weight receives, and the loss of power from the resistance of the ropes and blocks, such systems of pulleys are seldom employed. It is only in raising vast weights, such as large ships, or great masses of stone from a quarry, that they are ever used. For managing the rigging of a ship, the combination usually employed consists of not more than two or three movable pulleys. From its portable form, however, its cheapness, and the facility with which it can be applied, especially in changing or modifying the direction of motion, the pulley is one of the most convenient and useful of the mechanical powers.

---

* See Part I, Arts. 122—125.

THE INCLINED PLANE.

312. The inclined plane becomes a mechanical power in consequence of its supporting a part of the weight, and of course leaving only a part to be supported by the power. Thus the power has to encounter only a *portion* of the force of gravity at a time,—a portion which is greater or less, according as the plane is more or less elevated. When a plane is perfectly horizontal, it sustains the entire pressure of a body that rests on it; that is, the pressure on the plane is equal to the whole force of gravity acting on the body. As one end of the plane is elevated, this force is resolved into two, one of which is parallel and the other perpendicular to the plane. In proportion as the plane is more elevated, the part of the force which acts parallel with the plane is increased, until, when the plane becomes perpendicular to the horizon, it no longer sustains any portion of the weight, and the latter descends with the whole force of gravity.

313. The simplest example we have of the application of the inclined plane, is that of a plank raised at the hinder end of a cart for the purpose of rolling in heavy articles, as barrels or hogsheads. The force required to roll the body on the plank, setting aside friction, is as much less than that required to lift it perpendicularly, as the height of the plane above the ground is less than its length. Every one knows how much the facility of moving heavy loads is increased by such means, and how the force required to move them is diminished, by increasing the length of the plane while the height remains the same. Long inclined planes, constructed of plank, are frequently employed in building, especially where high walls are built of large masses of stone, the materials being trundled upon the plane on wheelbarrows, or transported on heavy rollers. It is even supposed, that in building the Pyramids of Egypt, the huge masses of stones were elevated on inclined planes. *Roads* also, except when they are perfectly level, afford examples of this mechanical power. When a horse is drawing a heavy load on a perfectly horizontal plane, what is it that occasions such an expenditure of force? It is not the weight of the load, except so far as that increases the friction; for gravity, acting in a direction perpendicular to the horizon, can oppose no resistance in the direction in which the load is moving. The answer is, that the force of the horse is expended chiefly in overcoming friction, and the resistance of the air. But when a horse is drawing a load up a hill, he has not only these impediments to encounter, but has also to overcome more or less of the force of gravity; that is, he *lifts* such a part of the load as bears to the whole load the same ratio, that the perpendicular height of the hill bears to its

length. If the rise is one foot in twenty, he lifts one twentieth of the load, and therefore encounters so much resistance in addition to those which he had to overcome on the horizontal plane. If the ascent were one foot in four, and the load were a ton the additional force required above what would be necessary on level ground, would be 560 pounds.

314. *Railways* afford another striking exemplification of the principles of the inclined plane. By means of them the irregular surface of a country, however hilly and uneven, is reduced to horizontal levels and inclined planes. These are sometimes inclined at so slight an angle, that the tendency of the cars down the plane, is only just sufficient to balance their friction, and they would remain at rest of themselves in any part of the plane, while a small force would move them either way. In other places the inclined planes are very steep for a short distance; and the cars ascending them are sometimes drawn up by means of a power (a steam engine, for example) stationed on the summit, and sometimes cars descending on one side are made to draw up others on the other side, the two being connected by a chain or rope which passes round a pulley on the summit.*

315. When railways first came into use, the power of one horse was considered only equal to a load of 10 tons; but it is now estimated that, upon a level railroad of the best construction with carriages of the most perfect finish, a horse-power is equivalent to a load of 22½ tons, although an average load is considered to be about 16 tons. The resistances to be overcome have thus been reduced to only $\frac{1}{280}$ of the weight, and may be safely taken at $\frac{1}{200}$, while upon the best common roads it is never less than $\frac{1}{40}$, and is, in most cases, as great as $\frac{1}{20}$. The advantage of a good railroad over a common turnpike, when horses are employed, is therefore in a tenfold proportion. But railroads derive their greatest value from the employment of steam as a moving power, as exhibited in the locomotive.† Horses tire at a moderate speed, but steam never tires, and is therefore peculiarly adapted to the transportation of passengers, where great expedition is required. By means of this force, a speed of 20 miles an hour is easily gained, and in some extreme cases, it has been pushed as high as 60 miles an hour. But such immense loads moving with such great velocities, acquire a momentum that is truly formidable, and involves inevitable danger.

316. In slow motions, canals have some advantages over railways. A horse easily draws on a canal a load of 30 tons; and

---

* Strickland's Reports. † Renwick's Practical Mechanics

it is said that the employment of several horses to a boat is advantageous, since the weight drawn increases in a higher ratio than the number of horses. When, therefore, heavy loads, with slow motions, are to be transported, canals have an advantage over railroads of two to one.* But canal boats cannot be made to move with much speed, without injury to the canal and great loss of force by the increased resistance of the water, this being augmented in proportion to the square of the velocity. In railways, on the other hand, the resistance arising from the friction of the wheels on the road, is diminished as the velocity is increased, while the resistance of the air, although increased by an increase of velocity like any other fluid, is so small as to occasion no serious impediment, being at a velocity of 14 miles per hour, only about 1 lb. on every square foot of the front of the leading carriage.

THE SCREW.

317. When a road, instead of ascending a hill directly, winds round it to the summit, so as to lengthen the inclined plane, and thus aid the moving force, the inclined plane becomes a screw. In the same manner a pair of stairs, winding around the sides of a cylindrical tower, either within or without, affords an instance of an inclined plane so modified as to become a screw. These examples show the strong analogy which subsists between these two mechanical powers; or rather, they show that the screw is a mere modification of the inclined plane.

318. The screw is generally employed when severe pressure is to be exerted through small spaces, and is therefore the agent in most presses. Being subject to great loss from friction, (upon which however its chief utility depends, as will be shown hereafter,) it usually exerts but a small power of itself, but derives its principal efficacy from the lever, or from wheel-work, with which it is very easily combined. Thus, in Fig. 155, were the power applied directly to the screw, the mechanical advantage gained would hardly more than compensate for the loss by friction; but by means of the lever, (which may be lengthened or shortened at pleasure,) the power is greatly increased. Also, by means of the *endless screw*, Fig. 98, combined with the wheel and axle, a very powerful force may be exerted; and as the mechanical power of the screw depends upon the relative magnitude of the circumference through which the power revolves, and the distance between the threads, (Art. 139,) it is evident that, to increase the efficacy of the machine, we must either increase the length of the lever by which the power acts, or di-

* Renwick's Practical Mechanics.

minish the distance between the threads. Although, in theory, there is no limit to the increase of the mechanical efficacy by these means, yet practical inconvenience arises from the great space over which a very long lever traverses. If, on the other hand, the power of the machine is increased by diminishing the distance between the threads, and of course their size, the thread will become too slender to bear a great resistance. The cases in which it is necessary to increase the power of the machine, being those in which the greatest resistances are to be over come, the object will evidently be defeated, if the means chosen to increase that power, deprives the machine of the strength which is necessary to sustain the force to which it is to be submitted.*

319. These inconveniences are remedied by *Hunter's Screw*, which, while it gives to the machine all the requisite strength and compactness, allows it to have an almost unlimited degree of mechanical efficacy. This contrivance, which is represented in Fig. 155, acts upon a principle similar to that of the wheel and axle, as represented in Fig. 145, where the efficacy of the power is increased by diminishing the velocity of the weight, which is accomplished by making the rope unwind on one side while it winds up, with somewhat greater speed, on the other side. In the case before us, the screw is likewise composed of a smaller and a larger thread, the former turning *upward* while the latter turns *downward* with a little greater velocity, and consequently the screw, on the whole, advances with the *difference* between the larger and the smaller threads; and since this difference may be small to any extent, so the efficacy of the power may be increased indefinitely. It will be seen, however, that the motion of such a screw is exceedingly slow. Thus in Fig. 155, A descends, while B, playing in a concave screw in A, ascends; but the distance between the threads of A being greater than the distance between those of B, the screw, on the whole, advances with the difference. Suppose that A has 20 threads in an inch, and B 21; then during one revolution, A will *descend* through the 20th, while B *ascends* through the 21st part of an inch. The compound screw, therefore, will advance through a space equal to the difference: that is, through a space equal to $\frac{1}{20}-\frac{1}{21}=\frac{1}{420}$th of an inch. This small space is therefore, in effect, the distance between two contiguous threads; and the power of the machine is, as usual, ex-

Fig. 155.

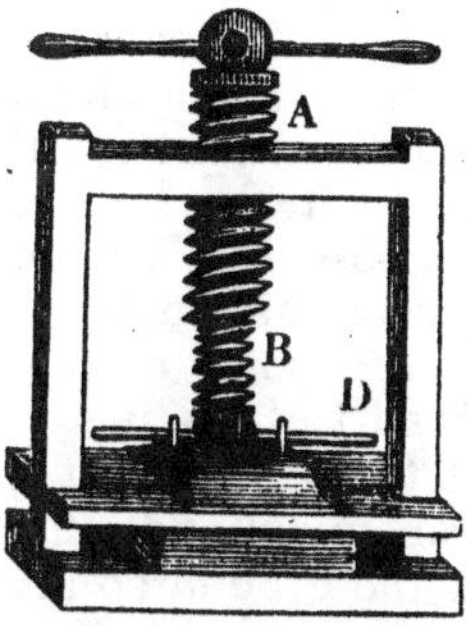

* Lardner's El. Mech., p. 220.

pressed by the number of times their distance is contained in the circumference described in one revolution of the power. For example, let the circumference of the circle be one foot; then $12 \div \frac{1}{420} = 5040$=the weight or resistance, the power being 1; or, in other words, the efficacy of the power is increased five thousand and forty times.

320. It is obvious, however, from principles already explained, that the power will in this case move over 5040 times as great a space as the weight. It is on this principle that the screw affords the means of measuring very minute spaces, and hence is derived the *Micrometer Screw.* The very slow motion which may be imparted to the end of a screw, while the power moves over a space vastly greater, renders it peculiarly adapted to this purpose. For example, suppose a screw to be so cut as to have 50 threads in an inch; then each revolution of the screw will advance its point through the 50th part of an inch, and if that point acted against a thread or wire, it would move it over a graduated space only that distance in a whole revolution of the screw. Now suppose the head of the screw to be a circle an inch in diameter, and of course something more than three inches in circumference. This circumference may easily be divided into a hundred equal parts, distinctly visible; and if a fixed index be applied to it, the hundredth part of a revolution of the screw may be observed, by noting the passage of one division of the head under the index. But the hundredth part of a revolution carries the point of the screw only through the ($\frac{1}{100}$ of $\frac{1}{50}$=) $\frac{1}{5000}$th part of an inch. Such an apparatus is frequently attached to the limbs of graduated instruments, for the purposes of astronomical and other observations; by which means, a portion of the graduated arc no greater than the 10th part of a second, can be estimated.

In like manner, any other small space may be measured by the aid of the Micrometer Screw. Thus, any aliquot part of a pound, or an ounce, in the steelyards, may be found by adapting the screw to the counterpoise so as to move it slowly over the space between two notches, and at the same time point out, by an index on its head, the exact portion of the space over which it passes.

THE WEDGE.

321. If instead of moving a load on an inclined plane, the plane itself is moved beneath the load, it then becomes a wedge. Thus, if a perpendicular beam have one end resting upon an inclined plane, (the beam being so secured as to be capable of moving only up and down,) and the plane be drawn under it, the beam will be elevated; and the power required to effect this

will be to that required to raise the beam when applied directly to it, *as the height of the plane to its length*:—or, considering the plane as a half wedge, the proportion will be, *as half the back of the wedge to its length.* (Art. 146.)

322. In the arts and manufactures, wedges are used where an enormous force is to be exerted through a very small space. Thus it is resorted to for splitting masses of timber or stone. Ships are raised in docks by wedges driven under their keels. The wedge is the principal agent in the oil-mill. The seeds from which the oil is to be extracted are introduced into hair bags, and placed between planes of hard wood. Wedges inserted between the bags are driven by allowing heavy beams to fall on them. The pressure thus excited is so intense, that the seeds in the bags are formed into a mass nearly as solid as wood. Instances have occurred in which the wedge has been used to restore a tottering edifice to its perpendicular position. All cutting and piercing instruments, such as knives, razors, scissors, chisels, nails, pins, needles, awls, &c. are wedges. The angle of the wedge in these cases, is more or less acute, according to the purpose to which it is applied. In determining this, two things are to be considered—the mechanical power, which is increased by diminishing the angle of the wedge, (Art. 146,) and the strength of the tool, which is always diminished by the same cause. There is, therefore, a practical limit to the increase of the power, and that degree of sharpness only is to be given to the tool, which is consistent with the strength requisite for the purpose to which it is to be applied. In tools intended for cutting wood the angle is generally about 30° ; for iron it is from 50° to 60° ; and for brass, from 80° to 90°. Tools which act by pressure may be made more acute than those which are driven by a blow ; and, in general, the softer and more yielding the substance to be divided is, and the less the power required to act upon it, the more acute the wedge may be constructed.*

323. In many cases, the utility of the wedge depends on that which is entirely omitted in the theory, viz. the *friction* which arises between its surface and the substance which it divides. This is the case when pins, bolts, or nails, are used for binding the parts of structures together ; in which case were it not for the friction, they would recoil from their places, and fail to produce the desired effect. Even when the wedge is used as a mechanical engine, the presence of friction is absolutely indispensable to its practical utility. The power generally acts by successive blows, and is therefore subject to constant intermission, and but for the friction, the wedge would recoil between the

* Lardner

intervals of the blows with as much force as it had been driven forward, and the object of the labor would be continually frustrated.

GENERAL REMARKS ON MACHINERY.

324. Archimedes is said to have boasted to King Hiero, that "if he would give him a place to fix his machine, (a που στῶ,) he would move the world." Yet there can be no machine by the aid of which Archimedes could move the world, in any other way, than by moving, himself, over as much more space than that over which he moved the earth, as his weight was less than that of the whole earth. If Archimedes had received the place he desired, and had also employed, what was equally desirable, a machine which operated free of all resistance, he must have moved with the velocity of a cannon ball, to have shifted the earth only the 27 millionth part of an inch in a million of years.*

325. Machines are divided into two classes, those intended simply to *sustain* a weight and those intended to *move* it. In machines of the first class, estimating the effect by the weight sustained, it is evident that the efficacy of the power is increased. By means of a lever, for example, a man may sustain a weight ten times as great as he could by his unaided strength. We may perceive, however, on closer examination, that he does not in fact bear the whole weight, but only one tenth part of it. Let it be a lever of the second kind, where the weight is ten times nearer the fulcrum than the end is to which the power is applied. Now the hand that supports this end performs the same office as the second fulcrum in a lever of the first kind; and since (Art. 102) the pressure on each fulcrum is inversely as its distance from the weight, therefore, in the present case, nine parts out of the ten are borne by the prop, and only one by the power. If (says Carnot) Archimedes had obtained his "fixed point," it would not, in reality, have been Archimedes, but the fixed point that would have sustained the earth.

In machines of the second class, the effect is not estimated simply by the weight moved, but we must take into the account the time occupied in moving it a certain distance, that is, the *velocity*. Hence, the effect of moving powers is estimated by the product of the weight moved multiplied by the velocity, or it is measured by the *momentum* produced. Moreover, in the former case, all resistance from friction, the rigidity of ropes, and so on, conspire with the power in sustaining the weight; but in machines of the second class, all such resistances oppose the action of the power, and require a greater power for a given weight

* Ed. Encyc., XII, 578.

than would be necessary, if the power were applied directly to the weight.*

326. Hence it will be inferred, that no *momentum*, or effective force, is gained by any of the mechanical powers, or by any machine. If a man, with his naked hands, can lift to a given height, as one foot, only 150 pounds in one second, it is impossible for him to perform any more labor than this by any mechanical contrivances.† On the contrary, when the structure of the machine is complicated, there is a loss of force, by employing the machine instead of the naked hands, proportioned to the resistance of the parts of the machine itself. It is to be remarked, however, that this doctrine proceeds on the supposition that the *useful effect* produced is estimated from the joint product of the *force, velocity*, and *time*. Thus, $F\times T\times V=\frac{1}{2}F\times 2T\times V=2F\times\frac{1}{2}T\times V=F\times\frac{1}{2}T\times 2V$, &c.‡ A convenient method of estimating different forces is to draw a heavy weight out of a well, by a rope passing horizontally over a fixed pulley, near the top of the well. Suppose that a man can draw up a rock weighing 100 pounds, through the space of 50 feet in one minute. He would, of course, be able to draw up ten such masses in ten minutes, weighing in all 1000 pounds. Now by passing the rope over five pulleys, (allowing nothing for the friction of the pulleys,) he might with the same force lift the whole 1000 pounds at once; but it would rise ten times as slowly as the 100 pounds did before, and consequently would be ten minutes in reaching the top. Therefore, in a given time, it appears that the man would raise the same weight through a given space, with or without the aid of machinery. In the former case, the 100 pounds might have been raised during the ten minutes through the space of 500 instead of 50 feet; but $100\times 500\times 10=1000\times 50\times 10$: so that the labor performed would have been the same in both cases. Let us suppose that P is a power amounting to an ounce, and that W is a weight amounting to 50 ounces, and that P elevates W by means of a machine. In virtue of the property already stated, it follows, that while P moves through 50 feet, W will be moved through 1 foot; but in moving P through 50 feet, *fifty distinct efforts* are made, by each of which, if applied directly, 1 ounce would be moved through 1 foot.§

327. *What then*, it may be asked, *are the advantages gained by Machinery?* The advantages still are very great, for the following reasons.

(1.) By the aid of machinery, *we can frequently apply our force to much better purpose*. Thus in lifting a weight out of a well,

* Venturoli's Mechanics, p. 164.

† Emerson's Mechanics, p. 150 Cavallo, Nat. Phil. I, 251.

‡ Gregory, I, 343.

§ Lardner's Mechanics, p. 162.

or in raising ore out of a mine, it is obvious with how much more effect a man can work at the arm of a windlass, than he could draw directly upon the rope by stooping over the well So in raising a rock from its bed by means of a handspike or crowbar, we can easily see how much more effectually we can bring our force to bear upon it than we could do with our naked hands.

(2.) By the aid of machinery, a man may be able to perform works to which his naked strength would be wholly incompetent. Thus, as in the preceding example, one might be able to lift a rock from its bed with a handspike, upon which he could make no impression with his naked hands: or, by means of pulleys, he might raise a box of merchandise from the hold of a ship, which he could not start at all with his unassisted force. In each of these cases, *if the weight could be divided into small parcels*, and if the force could be as advantageously applied without machinery as with it, the labor would be performed as easily in a given time in one way as in the other. But it might not be possible or at least convenient thus to divide it. Or if, instead of dividing it into a number of parcels, the *same number of men* could act directly upon the weight at once, the amount of labor which they would all exert in raising the weight without machinery, would be the same as that which the single man before supposed would exert with his machinery. But it might not be convenient to assemble so many hands at a time; or perhaps such a number could not work advantageously together. A farmer has many occasions for lifting or removing great weights when his laborers are not more in number than two or three in all. These must therefore perform the labor of 50 times as many men by being 50 times as long about it. Thus, in the example given on page 118, of a combination of the mechanical powers employed to haul a ship on the stocks, where a single man turning on a winch, with the force of 100 pounds exerts a force on the ship amounting to 161½ tons, the ship would move as much slower than the hand as 100 pounds is less than 161½ tons; and consequently a great length of time would be required for an individual to perform this labor, even supposing no resistance to be encountered from the machinery itself.

(3.) Machinery frequently enables a man to exert his *whole force* in circumstances where, without such aid, he could employ but a part of it. Thus, in winding silk or thread, to turn a single reel might not require one fiftieth part of the force which the laborer is capable of exerting. Suitable machinery would enable him to turn fifty spools at once.

(4.) But the most striking advantage of machinery, is not found in the facilities which it lends to the personal strength of man. It lies in this, that it affords the means of calling in to his assistance the superior powers of the horse and the ox, of water, of wind, and especially of steam. Here we find the excellence of mechanical contrivances fully exhibited; and nowhere else has the inventive genius of man displayed itself to so great advantage. But here, as in all other cases, the various combinations of mechanical powers *produce* no force: they only *apply* it. They form the communication between the moving power and the body moved, and while the power itself may be incapable of acting except in one direction, we are able by means of cranks, levers, and toothed wheels, to direct and modify that force to suit our convenience or necessities. Every one may see examples of this in the construction of the most common saw-mill or flour-mill, turned by water. In a mill for grinding wheat, the stones are required to move horizontally, while the action of the waterfall is perpendicular. We therefore receive the whole force on the circumference of a wheel, and transmit it through several intermediate wheels to the revolving stone, where the grinding is performed. So in a saw-mill, the water first communicates a *rotary* motion to the wheel, and this motion is converted by means of a crank into what is called a *reciprocating* motion, as that of the saw in its ascent and descent. By means of wheel-work the *velocity* of the moving body is increased or diminished at pleasure.

328. In short, machines enable us to form a convenient communication between the power and the weight; to give to the weight any required direction or velocity; to apply force to the best advantage; to vary the circumstances of velocity and time as the amount of our force may require; and to bring to our aid the great moving powers that exist in nature. Our next object, therefore, will be to see by what particular methods these several purposes are accomplished.

CHAPTER IV.

REGULATION OF MACHINERY, AND CONTRIVANCES FOR MODIFYING MOTION.

329. It is highly important to the successful operation of any machine, that its motion should be regular and uniform. Jolts and irregular movements, waste the power, wear upon the machine, and perform the work unevenly. The sources of irregu-

larity are various, but they are chiefly the three following, viz. variations in the power, variations in the weight or resistance, and changes of velocity in parts of the machine itself. Thus in the steam engine, the fire may burn with more or less intensity, and produce corresponding quantities of the moving power; the load to be carried (as that of a steamboat) may be much greater at one time than at another, and be subject to sudden changes; and the motion of the piston, which carries the machinery, ceases altogether at the highest and lowest points, and would move a machine by *hitches* or separate impulses, were there no contrivance connected with it for keeping up a uniform motion.

330. The kinds of apparatus employed to obviate these difficulties, and to secure uniform movements to machines, are, in general, called REGULATORS. Large machines or engines themselves, in consequence of their inertia, acquire and maintain to a considerable extent, uniformity of motion. A flour-mill carried by water, when it has acquired a certain rate of going, will not suddenly change that rate by any alteration in the force of the stream; and a ship sailing between the opposite forces, arising from the impulse of the wind and the resistance of the water, will move steadily along, notwithstanding the breeze that carries it may fluctuate continually. We can see this principle sometimes operating on a smaller scale. A grindstone turned by a winch moves steadily, although the force applied at one part of the revolution is much greater than at another. Large grindstones exhibit the advantage of this principle much more than small ones. But in many instances, this natural tendency toward uniform motion is not sufficient, and artificial contrivances are introduced expressly for this purpose. As examples of regulators we may especially notice three, the Pendulum, the Fly Wheel, and the Governor.

Fig. 156.

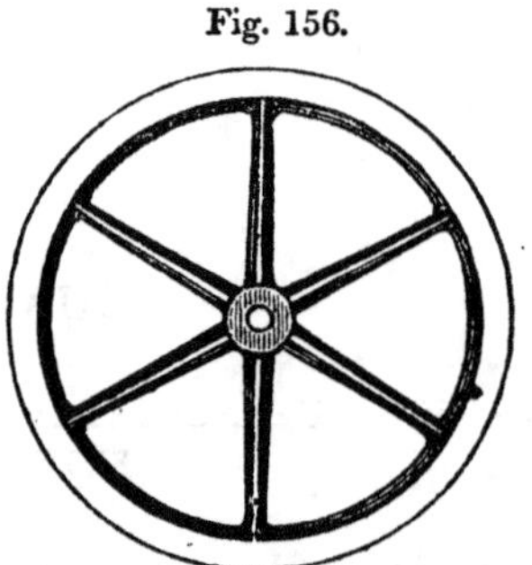

331. The *Fly Wheel* affords the most common and effectual method of equalizing motion, especially in heavy kinds of machinery. It consists of a heavy wheel, (Fig. 156,) affording as much weight as possible under as small a surface, in order that the inertia may be great, while the resistance from the air is small. It is therefore usually a heavy hoop of iron with thick bars of the same metal. The Fly is balanced on its axis, and so connected with the machinery as to turn rapidly around with it, and receiving a constant impulse from the moving power, it becomes a magazine or repository of motion. Con-

sequently, by its inertia, it is ready to supply any deficiency of power that may arise from the sudden diminution of the moving forçe, or to check any sudden impulse which may result from an accidental excess of that force. Suppose, for example, the handle of a pump to be connected with a water wheel, and to be carried by it. Here the power, namely the waterfall, is constant, while the weight is subject to continual alternations, amounting to a heavy load as the piston is ascending, but opposing scarcely any resistance while the piston is descending The motion, therefore, would vary between nothing and a highly accelerated velocity, and the machinery would be subject to constant strains and jolts. A Fly prevents these alternations, and renders the ascent and descent of the piston nearly uniform. In pile engines or stamping-mills a team of horses is sometimes employed to raise a heavy weight, which, when at a certain elevation, is suddenly disengaged, and falls with great force. As the disengagement is instantaneous, the horses would instantly tumble down were not their motion checked by some contrivance which should prevent the machinery from receiving any sudden increase of velocity. This purpose is completely answered by the Fly.*

332. Besides the use of the Fly Wheel in regulating the action of machinery, it is employed for the purpose of *accumulating* successive exertions of a power, so as to produce a much more forcible effect by their aggregation than could possibly be done by their separate actions. If a small force is repeatedly applied in giving rotation to a Fly Wheel, and is continued until the wheel has acquired a very considerable velocity, such a quantity of force will be at length accumulated in its circumference, as to overcome resistance and produce effects utterly disproportionate to the immediate action of the original force. Thus it would be very easy in a few seconds, by the mere action of a man's arm, to impart to the circumference of a Fly Wheel, a force which would give an impulse to a musket ball equal to that which it receives from a full charge of powder.†

The same principle explains the force with which a stone may be projected from a sling. The thong is swung several times around by the arm until a considerable portion of force is accumulated, and then it is projected with all the collected force. If a heavy leaden ball is attached to the end of a strong piece of cane or whalebone, it may easily be driven through a board: by taking the end of the rod remote from the ball in the hand, and striking the board a smart blow with the end bearing the ball, such a velocity may easily be given to the ball as will drive it through the board.‡

* Gregory's Mech., II, 14 † Library of Useful Knowledge. ‡ Ib.

333. The astonishing effects of a Fly Wheel, as an accumulator of force, have led some into the error of supposing that such an apparatus *increases the actual force of a machine.* So far from this, since a Fly cannot act without friction and resistance from the air, a portion of the actual moving force must unavoidably be lost by the use of this appendage. In cases, however, where a Fly is properly adjusted and applied, this loss of power is inconsiderable, compared with the advantageous distribution of what remains. As an accumulator of force, a Fly can never have more force than has been applied to put it in motion. In this respect it is analogous to an elastic spring. In bending a spring, a gradual expenditure of power is necessary. On the recoil, this power is exerted in a much shorter time than that consumed in its production, but its total amount is not altered. In this way the Fly Wheel is used. Thus, in mills for rolling metal, the water wheel or other moving power, is allowed for some time to act upon the Fly alone, no load being placed upon the machine. A force is thus gained which is sufficient to roll a large piece of metal, to which, without such means, the mill would be quite inadequate. In the same manner, a force may be gained by the arm of a man acting on a Fly for a few seconds, sufficient to impress an image on a piece of metal by an instantaneous stroke. The Fly is, therefore, the principal agent in coining presses. Its power is often transmitted to the working point by means of a screw. At the extremities of the cross arm AB, (Fig. 157,) which works the screw, two heavy balls of metal are placed. When the arm AB is whirled round, those masses of metal acquire a momentum, by which the screw, being driven forward, urges the die with immense force against the substance destined to receive the impression. Some engines used in coining have Flies with arms four feet long, bearing one hundred weight at each of their extremities. By turning such an arm at the rate of one entire circumference in a second, the die will be driven against the metal with the same force as that with which 7500 pounds weight would fall from the height of 16 feet; an enormous power, if the simplicity and compactness of the machine is considered. By the action of a Fly, working in this manner, is produced the open work of fenders, fire grates, and sometimes ornamental articles wrought in metal. The cutting tool, shaped according to the pattern to be executed, is attached to the end of a screw; and the metal being held in a proper position beneath it, the Fly is made to urge the tool downward with such force as to stamp out pieces of the required figure. When the pattern is compli-

Fig. 157.

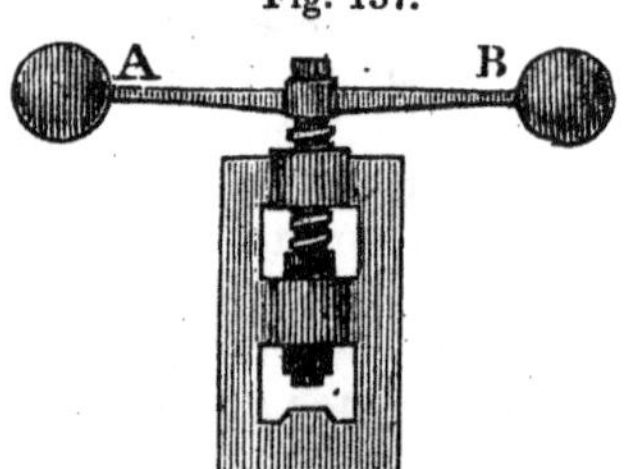

cated, and it is necessary to preserve with exactness the relative situation of its different parts, a number of punches are impelled together, so as to strike the entire piece of metal at the same instant, and in this manner the most elaborate work is executed by a single stroke of the hand.*

Fig. 158.

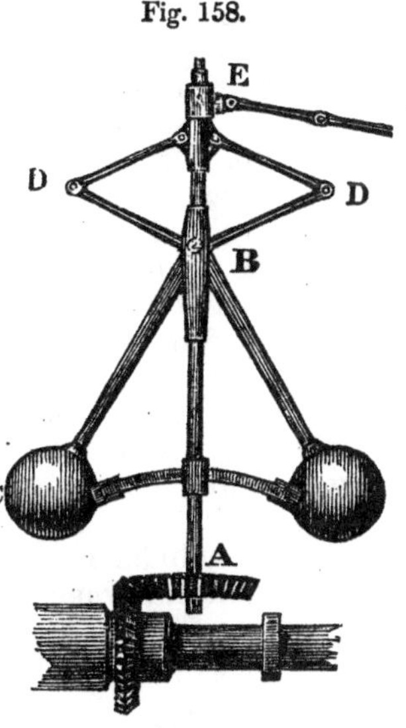

334. To maintain a uniform velocity with a varying resistance, one of the most beautiful contrivances ever used is the *Governor*, (Fig. 158,) an instrument used in mill work, but the application of which is most conspicuous in the steam engine, when that machine is applied to manufacturing purposes. The principle on which the efficacy of this instrument depends, is easily explained. Let AB be a vertical axis which is made to revolve by the wheel A, acted on by the other parts of the machinery, and so that it always revolves with a velocity proportional to that of the fly wheel. Two heavy balls C, C are attached to metal rods, which work on a pivot at B, so that they are capable of receding from the axis AB. As they recede from the axis, the joints D, D′ recede from one another, and the joint E is drawn down. This joint E is connected with the end of a lever or a system of levers, the action of which we shall presently explain.

Now by the revolution of the spindle or axis AB, the balls C, C acquire an obvious tendency to fly off from the axis, and this tendency is resisted by their weight; so that, when the instrument is revolving with a certain velocity, the balls remain suspended, and neither move to or from the axis. A greater velocity, by giving a greater centrifugal force, would cause the balls to fly further off, (Art. 238,) and a less velocity would cause them to fall toward the axis. This is strictly true only when the range of balls is small, compared with the length of the rods to which they are attached, which, however, is always the case in practice. If, therefore, the action of the levers with which the joint E is connected is directed upon the first mover in such a manner, that its energy is diminished when E is depressed, and increased when E is elevated, it is plain that the uniformity of velocity which is sought may be obtained. Let

* Lardner.

us suppose that the levers on which E works communicate with a valve which admits steam to the piston of a steam engine, to which this Governor is applied; and suppose that when E is raised, and the balls C, C rest in their seats, the valve is fully open, so as to allow the steam to flow in a full stream to the piston; but that, according as E is depressed, the levers gradually close the valve, so as to admit the steam in a constantly diminished quantity. Now suppose that the engine has been working twenty printing presses, and that the action of ten of them is suddenly suspended. The engine thus loses half its load, and would, if the same power of steam continued to be admitted, move with about twice its former velocity. But the moment an increased velocity is perceived in the machine, the balls C, C recede from the axis, draw down the joint E, partially close the valve, and check the supply of steam to the cylinder. The impelling power is thus diminished; and if it be diminished in exactly the same degree as the load, the machine will move with its former velocity; but if it should, at first, be more diminished, the velocity will be less than its former velocity, and the balls will again move toward the axis and open the valve, and will, at length, settle into that position in which the steam admitted to the cylinder is exactly proportioned to the load on the machine; and the proper velocity will thus be restored.*

CONTRIVANCES FOR MODIFYING MOTION.

335. In Chapter III, we have already explained the mode in which motion is communicated, and its velocity regulated by *wheel-work*. We proceed now to consider a few examples of the more special contrivances by which motion is modified to suit particular purposes, recommending it to the student of mechanics to make himself acquainted with other contrivances of the same nature, by the actual inspection of machinery as opportunity may offer.

336. The motion required for a particular purpose may be *rectilinear*, as that of a carriage or bucket drawn out of a well, or *rotary*, as in ordinary wheel-work, or *reciprocating*, as in a saw-mill or a pendulum.

The simplest mode of producing rectilinear motion is by means of a rope or chain, instances of which are familiar to every one. The simplest mode of *changing the direction* is by means of pulleys; but toothed wheels are also extensively employed for the same purpose.† The connection of one toothed wheel with another is called *gearing*. (Art. 294.) When both wheels with

* Library of Useful Knowledge. † Emerson's Mechanics, Prop. CX.

their teeth are in the direction of the same plane, it is called *spur gearing*, (Figs. 149, 150, and 151 ;) if the teeth, instead of being cut on the circumference in a direction parallel to the axis, are cut obliquely, so that if continued they would pass round the

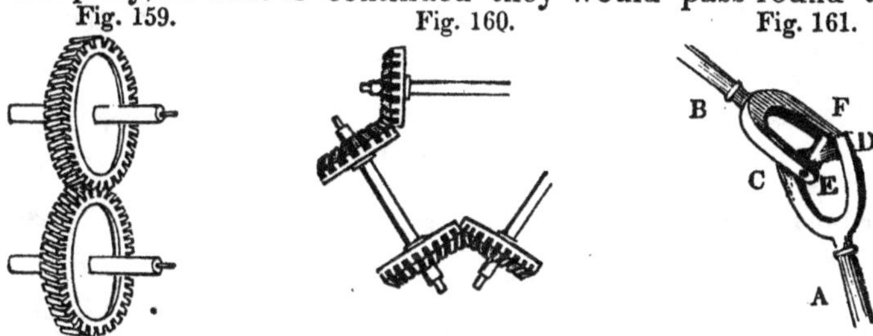

Fig. 159. Fig. 160. Fig. 161.

axis like a screw, it is called *spiral gearing*, (Fig. 159 ;) and when wheels are not situated in the same or parallel planes, but form an angle with each other, the wheels themselves are sometimes shaped like frustra of cones, having their teeth cut obliquely, and converging toward the point where the apex of the cone would be situated, and it is then called *bevel gearing*.* (Fig. 160.)

337. The *universal joint* consists of two shafts or arms, each terminating in a semicircle, and connected together by means of a cross upon which each semicircle is hinged, (Fig. 161.) When one shaft is turned, either to the right or left, the other shaft turns in the same direction.

The *ratchet wheel*, (Fig. 162.) is used to prevent motion in one direction while it permits it in the opposite. The teeth are cut with their faces inclining as in the figure, and a *catch* is so placed as to stop the wheel in one direction, while it slides over the teeth without obstruction in the opposite direction.

Fig. 162.

Fig. 163.

338. The *eccentric wheel*, (Fig. 163,) revolves about an axis which is more or less removed from the center, and, consequent-

* Bigelow's Tech., 232.

ly, the different portions of the circumference move with different degrees of velocity. Hence if this wheel is made to act upon a shaft or pinion, as in the figure, it will carry it with a corresponding movement. In orreries, such wheels are employed for indicating the variable velocities of the heavenly bodies, as they revolve about their centers of motion.

339. Reciprocating Motion is produced in various ways. The most common method is by means of the *crank*. In Fig. 164, a shaft AB is urged backward or forward, (either vertically or horizontally,) by means of the crank *ab*, moving on a wheel H, which may be turned by water or any other power acting at H. By considering the different positions of the crank during the revolution of the wheel, it will be readily seen that the shaft will move up and down like the saw in a saw-mill, or backward and forward, a use to which it is applied in polishing plane surfaces, as marble.

Fig. 164.

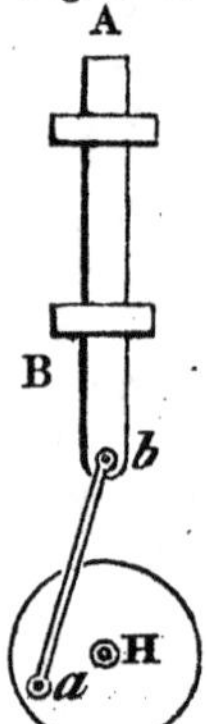

The motion produced by cranks is easy and gradual, being most rapid in the middle of the stroke, and gradually retarded toward the extremes; so that shocks and jolts in the moving machinery are diminished, or wholly prevented by their use.*

The steam-engine, as seen in steamboats, furnishes to the student of mechanics a valuable opportunity of observing various contrivances for producing, regulating, and modifying motion. Levers and wheels of various kinds and variously connected; fly wheels and cranks; circular and reciprocating motions; and numerous other particulars which appertain to the "elements of machinery," are there seen to the greatest advantage.

340. The *arch head*, (Fig. 165,) is a circular form given to the end of a lever, that moves on a gudgeon or pivot in a vertical plane, (like the working beam of a steam-engine,) by means of which the piston, or weight, whatever it be, is made to ascend and descend perpendicularly in a *straight* line, with a *uniform* motion, while the end of the lever itself works in the arc of a circle, and while the power, if fixed in the usual way at one point at the extremity of the lever, would lose a part of its efficacy, as its direction became oblique to that of the lever. In the figure, BD represents the arch head of a working beam, which turns on a gudgeon at C. At the top B is applied a flexible chain resting loosely upon the arch, and always maintaining a

Fig. 165.

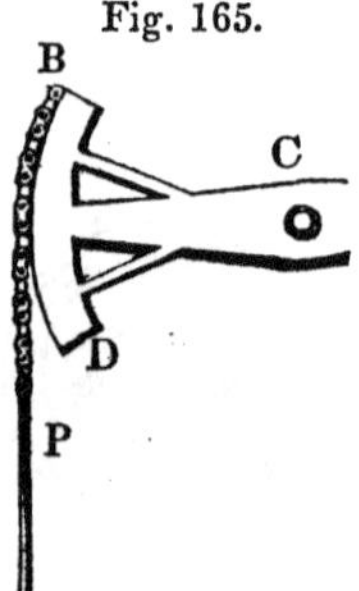

* Bigelow

vertical position in every situation of the beam; and, being a tangent to the circle at the point of contact, always acting at right angles to the lever, and consequently imparting to the weight a *uniform* motion. In cases where force is to be exerted only in one direction, as in working a pump, the power is applied while the piston is descending and drags up the other end of the lever. Here, in the descent of the working end, so small a resistance is encountered, that it is sufficient to give a slight preponderance to that end, and it will drag up the piston after the moving force is withdrawn which carried it down. But in cases where a continued force is required, both in the ascent and descent of the piston, (as in most kinds of manufacturing operations,) then the flexible chain, being incapable of forcing the weight upward, is inapplicable, but the object is attained by arming the arch head with *teeth* that act on *rack work* in the upper end of the piston rod.

Fig. 166.

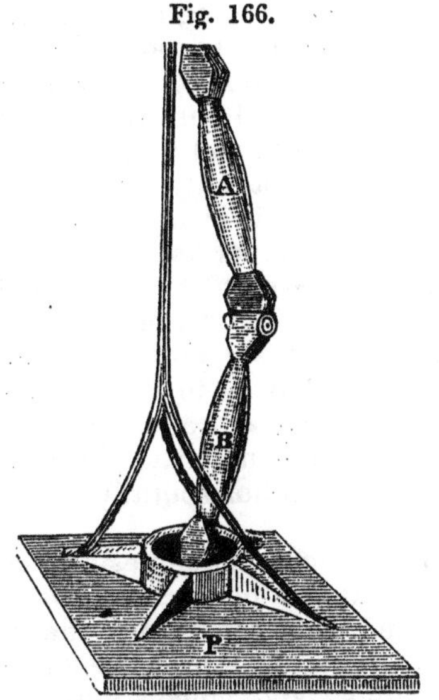

341. The *knee joint*, (Fig. 166,) is a contrivance by which a constantly increasing resistance is overcome by a force which acts nearly uniformly. In the figure, we have a representation of it as exemplified in printing presses. Here, when the *platen* (P) first descends upon the sheet, the resistance is very slight, but as the lever by which the press is worked is pulled still further, the resistance rapidly increases, until, without mechanical aid, it would require a laborious and exhausting effort, to render the contact sufficiently close to give a perfect impression. This difficulty is completely obviated by means of a combination of levers resembling the knee joint, where the efficacy of the power is continually increased as the levers approach nearer to the same straight line; so that without any additional effort on the part of the workman, the pressure is augmented a thousand fold. The action of this joint is exemplified in opening a pair of compasses, or a Gunter's scale which opens on a joint at the center. At first, the *thrust* exerted by the ends is slight; but as the two parts approach the direction of a straight line with each other, the thrust rapidly increases, until it becomes immensely great.*

* See an able article on this subject by Prof. Fisher, *American Journal of Science*, Vol. III, p. 311.

We shall conclude this chapter with a few practical rules for the construction and management of machinery, selected chiefly from Emerson's Mechanics.

342. If the given power is not able to overcome the given resistance when *directly* applied, then we must employ a longer time by making the power move over a greater space than the weight, in order to make the momentum of the power exceed that of the weight. This is effected by means of machinery We must consider that the power must have a momentum which is equal to the amount of all the resistances to be overcome, in order just to *sustain* or counterbalance the weight; and to put the machine *in motion*, a greater power must be allowed, more or less, according to the velocity with which the machine is required to move. The ratio between the power and the weight in a given machine, when in equilibrium, may be ascertained by observing the *comparative spaces* over which they move in a given time; the power being as much less than the weight, as its space is greater. A due proportion between the two must also be observed; for if the machine or engine is competent to overcome the resistance, and perform its work in a convenient time, it is sufficient for the end proposed; and to increase the power any further, must not only be a needless expense, but the engine would lose time in working. If a weight is to be moved but a little way, the lever is the most simple, easy, and ready instrument. If the weight be very great, the common screw is preferable. But if the weight is to be moved over a great space, the wheel and axle, the pulley, or a system of pulleys, or the endless screw, (Fig. 97,) is to be employed. When great wheels are wrought by men or cattle, their axes are most advantageously placed perpendicularly, as in Figure 154; if wrought by water, they are placed horizontally.

343. But most machines are combinations of some or all of the mechanical powers. Thus the lever is combined with the screw in a common press; the wheel and axle with pulleys in various ways, and with the endless screw; pulleys are combined with pulleys, and wheels with wheels. The wedge is the only one among the mechanical powers, that does not admit of combination with others. In wheels with teeth, the number of teeth that play together in two wheels ought to be *prime* to each other, that the same teeth may not meet at every revolution, but as seldom as possible. The strength of every part of a machine ought to be made proportional to the stress it is to bear; and no part must be stronger or heavier than is necessary, for all superfluous matter is nothing but a dead weight upon the machine, and serves for nothing but to clog its motion. The accomplished mechanic contrives all the parts to last equally well, so that when the machine fails, every part shall be worn out.

344. Every machine ought to be made of as *few parts*, and as simple as possible, to answer its purpose ; not only because the expense of making and repairing will be less, but it will be less liable to get out of order. Any useless motions also waste some portion of the power. Uniformity or steadiness of motion is carefully to be preserved. All these advantages are more easily attained in large than in small machines. All mechanical errors have a less ratio to the motion of the machine in great machines than in small ones, and these will therefore work with more uniformity and exactness, although, being proportionally weaker, they are less able to resist any violent shocks.*

CHAPTER V.

OF FRICTION.

345. The term Friction, in its usual acceptation, being generally understood, we have already employed it in the foregoing pages, but we proceed now to inquire more particularly respecting its nature, the laws of its action, and its effects upon machines.

In investigating the mathematical principles of Mechanics, we first proceed on the supposition that the forces in question act without any impediments; that the surfaces which move in contact are perfectly polished and suffer no friction; that axes and pivots are mathematical lines and points; that ropes are perfectly flexible; and, in short, that the power is transmitted through the machine to the working point without sustaining the least loss or diminution. Great simplicity is attained by first bringing the subject to this ideal standard of perfection, and afterward making suitable allowances for all those causes which operate in any given case to prevent the perfect action of a machine.

346. Surfaces meet with a certain degree of resistance in moving on each other, in consequence of *the mutual cohesion of the parts*, a principle which has the greater influence in any given case in proportion as the surfaces are smooth. But a much greater resistance arises from the asperities which the surfaces of all bodies have, though in very different degrees, according to their different degrees of smoothness. An extreme case is that of two brushes moving on each other, the hairs of which become interlaced, (especially when the brushes are

* Emerson's Mechanics, 4to. p. 175.

pressed together,) and oppose a great resistance. Even bodies apparently very smooth, as polished metals, exhibit under the microscope numerous inequalities. Under the solar microscope, the finest needle exhibits a surface as rough as the coarsest iron tools do when viewed by the naked eye. To these inequalities of surface, is principally ascribed the friction of bodies, when closely in contact; the prominent parts interlock with one another, or meet, and must be broken down before the surfaces can move. Hence, friction is diminished by processes which level these inequalities, either by polishing the surface, or by coating it with some lubricating substance which fills up the cavities.

347. Forces of this nature, which act by the resistance they occasion to motion, are called *passive* forces. They produce very different effects in machines when in a state of equilibrium, and in a state of motion. In the one case they assist the power; in the other case they oppose it. Thus, a weight placed on an inclined plane, will require a less power to *support* it in consequence of the friction of the plane; and a weight suspended by a rope passing over a pulley will require a less weight to *balance* it, on account of the friction of the axle. But the same passive forces operate in just the contrary way when a machine is to be put in motion; for then a power must be applied, which is sufficient not only to overcome the weight itself, but also the amount of all the resistances. For example, in order to draw a load up an inclined plane, we have to overcome not only the force of gravity by which the load endeavors to descend down the plane, but also the amount of the friction and all the other resistances which impede its motion, although the load would be kept from *descending*, that is, in a state of equilibrium, by a less force in consequence of these resistances. The principle is most strikingly observed in the wedge, where the difficulty of making the wedge *advance*, is greatly increased by friction, but the same cause operates to prevent it from *recoiling*.

348. Two philosophers of great eminence have severally performed an extensive series of experiments on friction, namely, M. Coulomb, member of the Academy of Sciences at Paris, and Professor Vince, of the University of Cambridge in England and upon their investigations is founded a great part of all that is known with precision respecting the laws of friction.

The forms under which this sort of resistance presents itself, are chiefly of two kinds, namely, that of bodies *sliding*, and of bodies *rolling* on each other. To the former of these let us first attend. Experiments on the friction of sliding bodies may be made, either by placing them on a table, and observing the weights which they respectively require to drag them along the table, or by placing them on an inclined plane, and observing at what angle the plane must be elevated in order that the body may *begin* to slide. In the former case, the table is prepared by attaching a vertical pulley to one edge over which a string is passed, one end being connected to the body in question, and the other end to a pan, like that of a balance, for containing weights. From this simple arrangement, a great variety of particulars may be ascertained respecting the friction of sliding surfaces. A body shaped like a brick, with a broader and a narrower side, may be tried on each of its sides separately, and thus it may be seen whether, in a given weight, the *extent of surface of contact* makes any difference; the body may be loaded with different weights, and hence may be learned the *influence of pressure* upon friction; the body may be tried as soon as it is laid on the table, and after remaining on it for a longer or shorter time, in order to learn whether this circumstance alters the friction; different kinds of bodies may be tried, and the *influence of different materials* ascertained; and finally, by dragging the body off the table with different degrees of velocity, the *relation of friction to velocity* may be investigated.

349. From experiments like the foregoing, endlessly varied, the following conclusions were established:

(1.) In a given body, *extent of surface* makes no difference in regard to friction; a brick laid on its edge meets with the same resistance from this cause as when laid on its side.

(2.) Friction is proportioned to the *pressure*. If the pressure of the brick is doubled or trebled by laying weights upon it, the amount of friction will be increased in the same ratio.

(3.) Friction is increased by bodies *remaining for some time in contact with each other*. In some cases it does not reach its maximum under four or five days. This principle therefore affects slow motions much more than such as are rapid. In the mutual contact of metals, the friction attains its maximum almost instantaneously. But when metal rubs against wood, or one piece of wood against another, the friction is always increased by resting. Two pieces of wood acquire the utmost friction in an hour or two; while iron running on oak will have its friction augmenting for five or six days. The application of a coat of tallow seems to protract the limit of friction. This limit is attained by the greased surfaces of iron and copper in four minutes; while pieces of wood, treated in the same way, will have their friction gradually augmented during nine or ten days.*

(4.) The friction is less between surfaces of *different* kinds of matter, than between those of the *same* kind. Copper slides on copper, or brass on brass, with greater difficulty than copper on

* Leslie, El. Nat. Phil. I, 225.

brass; and it is a general rule never to let two substances of the same hardness move upon each other. To this rule cast steel is said to form the only exception; in other cases, pivots revolve with less resistance on either harder or softer substances, than upon those of the same material with themselves.* When between the surfaces of wood newly planed, the friction would be equal to one half the pressure, and when between two metallic surfaces it would be equal to one fourth, between the wood and metal it would amount to only one fifth the pressure.†

(5.) Friction is much greater at the first moving of a load, than after it is brought freely into motion. In many instances, it is reduced, when a body has attained its final velocity, to less than one half of what it was at first.‡ With regard to different degrees of velocity over a given space, it is a *general* principle, that *the friction is the same for all velocities*; that a carriage, for example, in travelling from one place to another, would encounter the same resistance from friction, whether it performed the journey in one hour or in ten. The amount of friction, however, is augmented in very slow motions, and greatly diminished in those that are very swift. In this instance, the increase in the one case and the diminution in the other, appears to have some relation to the principle, that the friction of bodies is increased by their remaining in contact. From some observations of Professor Playfair, made at the slide of Alpnach, where large fir-trees are carried with great velocity down an inclined plane eight miles in length, it would appear that, in the case of very great velocities, friction is not, according to the common doctrine either proportioned to the pressure, or independent of the velocity; but that the ratio to the pressure is greatly diminished, and the actual resistance is far less than at common velocities. Thus, none but large trees could descend the plane at all; and when a tree broke into two pieces, the larger part would proceed while the smaller would stop; and the trees acquired in their descent a rapidity of motion, incompatible with the supposition that "friction acts as a uniformly retarding force," which has been considered as an established principle.§

The foregoing considerations are in favor of rapid travelling, whether on common roads or on railways, since the amount of the resistance is so much less than in slow movements; and accordingly it is said that the great speed given to stage coaches in England, amounting, in some instances, to 10 or 12 miles per hour, has not been attended with the degree of exhaustion to the teams that would have been anticipated.‖

---

* Allen's Mechanics, p. 137. † Leslie. ‡ Leslie, Nat. Phil. I, 218.

§ Playfair's Works, Vol. I, or Ed. Phil Jour VI, 345 Also see Nicholson's Oper Mech. II, 225.

‖ Nich. Oper. Mech.

350. The angle at which an inclined plane must be elevated, in order that a given body resting on it may be on the point of sliding, is called the *angle of friction*, or sometimes the *angle of repose*. Let ABC (Fig. 167) be the inclined plane, and let FD,

Fig. 167.

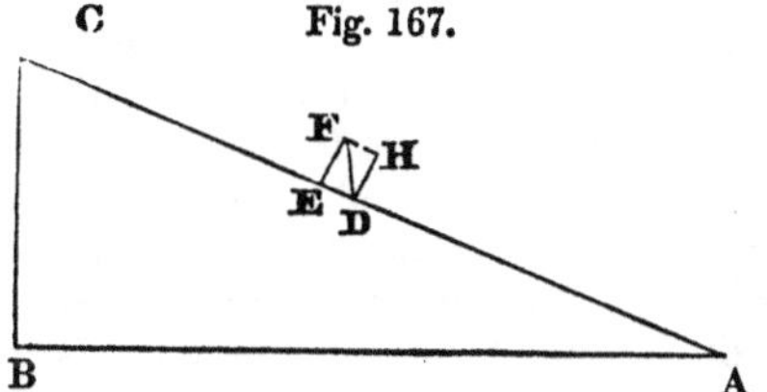

parallel to BC, represent the whole weight of the body. FD being resolved into FE perpendicular to the plane, and ED coincident with it, ED will be the force with which the body tends to slide down the plane, and, of course, that which is resisted by the friction, and hence it may be taken as the representative of the friction. But ED is to FE as BC to BA, or as the height of the plane to the base, or as the tangent of the angle of friction to radius. Consequently, putting F for the friction, P for the whole pressure on the plane, and α for the angle of friction, then,

$$F : P :: \tan. \alpha : \text{rad.} \therefore F = P \times \tan. \alpha ;$$

that is, the friction is always such a part of the entire pressure, as is denoted by the product of that pressure into the tangent of the angle of friction. This angle often determines the figure which natural objects spontaneously assume. Hence, sand hills are more sloping than eminences composed of ordinary mould, the movable particles arranging themselves in obedience to the foregoing law. The angle of friction of iron pressing on iron, is found to be 16 degrees. The angle given to the threads of a screw, is regulated on this principle.*

351. The laws of friction in *rolling* bodies were ascertained by Coulomb, by comparing the forces necessary to roll a cylinder upon a table under various circumstances; and by similar experiments, were found the modes in which friction takes place in bodies *revolving* on an axis. The comparative loss of power in these three cases is as follows:

Friction of the	sliding body	equal to $\frac{1}{4}$ the pressure, or	25 per ct
do.	revolving	do.	15
do.	rolling	do.	5

In the case of hollow cylinders revolving on an axis, the *leverage* of the wheel aids in overcoming friction, as has been already explained in Art. 306.

352. *Friction wheels*, a contrivance by which friction is dimin

* Leslie.

ished in the greatest degree possible, owe their efficacy in part to the operation of the same principle.* Here the axis of a wheel, instead of revolving in a hollow cylinder, or instead of rubbing against a fixed surface, rests, at each of its extremities, on the circumference of two wheels placed close by the side of each other, with their circumferences intersecting. The axis rests at the point of intersection, and as it revolves, the wheels revolve with it with the same velocity, and thus all friction between them and the axis is prevented; and what remains in the machine in consequence of the weight of the wheels themselves is *transferred* to the axles, and therefore (Art. 300) is diminished in the ratio of the diameter of one of the wheels to that of its axis. This combination may be repeated by several pairs of friction wheels. Eight wheels would contract the friction to the thousandth part.†

353. Other more common methods of diminishing friction are, by rendering the surfaces smooth, by using rollers, and by lubricating the parts in contact. The amount of friction in the several mechanical powers is very different. In the lever it is very small, especially when the turning edge is of hardened steel, and shaped like a knife or prism, and turns upon a hard and smooth basis. The wheel and axle acting upon the same principle as the lever, occasions but little friction. The stiffness of the cordage, however, and the friction of the gudgeons‡ of the axis, have an effect in most cases equal to about 8 or 10 per cent. of the entire resistance. The pulley is attended with great loss from this source. It is rarely less than 20 per cent. and often exceeds 60. The inclined plane involves but little friction when bodies simply roll on it; but when heavy bodies rest on axes, as in wheel carriages, the resistance from friction takes place in the same manner as upon plane surfaces. The transportation on inclined planes, as railways, is usually by means of wheels, since the resistance to sliding movements is too great to permit the use of them. The screw is attended with a great amount of friction. Those with sharp threads have more than those with square threads, and the endless screw has most of all.§ In both the screw and the wedge, the friction evidently exceeds the resistance; otherwise they would not retain their position.

354. Friction is not, therefore, in all cases to be considered as unfavorable to the operation of machinery. It is, in many instances, a highly useful force. Many structures, as those of brick and stone, owe no small part of their stability to the roughness of the materials of which they are composed; without this re-

* A fine example of the application of friction wheels is seen in Atwood's Machine.
† Leslie, El. Nat. Phil. I, 233.
‡ Pivots when large take the name of *gudgeons*. § Emerson.

sistance, the screw and the wedge would lose their efficacy, and wheels could not advance, nor could animals walk on the ground; and nails would lose their power of binding separate parts together. The art of polishing surfaces depends on the same cause, and the edges of most cutting instruments are saws, the teeth of which are more or less fine, and act on a similar principle. Even in certain rotary motions, friction, by a rope or otherwise, becomes a moving force, and urges a body in particular directions contrary to the force of gravity.

The resistance which moving bodies sustain from the air, and from water, will be considered hereafter.

CHAPTER VI.

OF PROJECTILES AND GUNNERY.

355. Early in the 17th century, Galileo, a celebrated Italian philosopher, established the mathematical theory of projectiles, as given in the former part of this work. Since missiles thrown by instruments of warfare are among the number of projectiles, these principles were supposed to constitute the foundation of the art of gunnery; and several of the kings of Europe held out munificent encouragement to experiments in this department of science.* Galileo had indeed intimated that the *resistance of the air*, (which is not taken into the account in the mathematical theory of projectiles,) might occasion some deviation in bodies from the curve of a parabola; but this fluid appears so light in comparison with the heavy metals of which balls are made, being 10,000 times lighter than lead, that for a long time afterward it was totally neglected.

356. About the year 1740, Mr. Robins, an eminent English mathematician, instituted a series of experiments on gunnery, which showed that the parabolic theory of projectiles is so modified by the resistance of the air, as to be wholly inapplicable to practice. Experiments on the same subject have since been performed by Count Rumford, and by Dr. Hutton of the Royal Military Academy at Woolwich, which have confirmed the results obtained by Mr. Robins. By the labors of these several gentlemen, the parabolic theory of projectiles has received its proper modifications.

* Robison's Mech. Phil. I, 167.

357. It is ascertained, in general, that projectiles moving *slowly* describe curves which are nearly parabolas; while such as move *swiftly*, deviate very far from this curve. Indeed, the curve which a body moving with great velocity in the air describes, is so complicated, that the utmost resources of the calculus have hardly been able to investigate its nature;* and it is a remarkable fact, that we understand the motions of the heavenly bodies better than we do those of a cannon ball, and can trace the path of a planet better than we can that of an arrow. The parabolic figure described in the case of projectiles moving slowly, may be observed in tracing the path of a small stone thrown into the air, and more especially in the curves described by jets of water, spouting upward, as in fountains. But when the jet of water is more rapid, and spouts at a high angle, as 45 degrees for example, we can plainly see that the curve deviates greatly from a parabola. The remote branch of the curve is seen to be much less sloping than the rising branch; and even in very great jets, which are to be seen in some great water-works, the falling branch is almost perpendicular at its remote extremity; and the highest point of the curve is far from being in the middle between the spout and the place where the water falls. This unequal division of the curve by its highest point may also be observed in the flight of an arrow or a bomb-shell.†

358. The following facts also show the discordance between the parabolic theory of gunnery and experience. A cannon ball, fired in such a direction and with such a velocity, that its random or horizontal range ought to be 24 miles, comes to the ground short of one mile. The times of rising and falling, if that theory held good, ought to be equal; but the time of rising is greater than that of falling at great elevations, and at small elevations, less than that of falling. According to the theory, the greatest random is at an angle of elevation of 45 degrees, (Art. 191,) but in practice it is found to be much below this. The greatest random of an arrow is when the elevation is about 36 or 38 degrees. Indeed, the angle for the greatest horizontal range, may be at all degrees from 45° to 30°; the slowest motions and the largest shot being almost at 45°, but gradually more and more below that degree as the shot is smaller and the velocity is greater; till at length, with the most rapid motions and the smallest shot, the angle is little above 30 degrees.‡ The following experiments were made in France by Borda, with a 24 pounder, with the same charge of powder in each experiment.

* Hutton's Tracts, No. 37.

† Robison's Mech. Phil. I, 185.

‡ Hutton, Tract 37.

Elevation.	Range.
15°	1950
30	2235
45	2108
60	1700
75	950

Whence it appears that at the elevation of 15 and 75, the randoms, instead of being the same, (being equally distant from 45,) were as the numbers 1950 and 950.*

359. All this discordance between theory and practice is owing to the resistance of the air, which, when the projectile moves with great velocity, becomes enormous. Nor will it be difficult, on a little reflection, to comprehend the reason why this resistance should be so great. The force with which a projectile strikes the air at rest, is the same as that with which the air moving with equal velocity would strike the body at rest. This, in the case of a cannon ball, would greatly exceed the most violent hurricane. Again, as a ball moves through the air, it displaces, that is, gives motion to, great quantities of air; yet whatever motion it imparts to other bodies is extinguished in itself. The loss of motion therefore increases very fast with the velocity. It is said to be in general as the square of the velocity: so that a body moving through the air with ten times the velocity of another body, would encounter one hundred times as much resistance. In very swift motions, the resistance was ascertained by Robins to be even much greater than in the ratio of the square of the velocity.

360. The researches of Mr. Robins were made chiefly by the aid of an instrument of his own invention, called the *Ballistic Pendulum.* It consists of little more than a large block of wood, like a log, suspended after the manner of a pendulum. Now if a bullet be fired into the block, as the bullet will be stopped, and as it imparts to the block whatever motion it loses, consequently the momentum of the block, after the stroke, is precisely that of the ball before the stroke. Hence the weight of the block and that of the ball being known, and the velocity imparted to the block being readily determined by observation, it is easy to find the velocity of the ball; for the weight of the ball is to the weight of the block, as the velocity of the block is to the velocity of the ball.†

---

* Robison's Mech. Phil. I, 183.

† That is, putting W, w, for the weight of the block and ball respectively, and V, v, for their velocities, then $W \times V = w \times v$, and $w : W :: V : v$. In other experiments on this subject, the gun itself has sometimes been suspended like a pendulum, and the velocity of the ball estimated by the distance to which the gun recoiled, since action and reaction are equal, and in opposite directions.

Example.—On firing an 8 pound cannon shot into a ballistic pendulum that weighed 500 pounds, the pendulum was made to move over a space of 16 feet per second: What was the velocity of the ball? Ans. 1000 feet per second.

361. This simple apparatus is sufficient for ascertaining a great number of particulars relative to the art of gunnery. If the ball is fired nearly in contact with the block, we find with what velocity it leaves the gun; if at different distances from the block, we find how much the velocity is retarded by passing through the air, for those distances respectively. If at a given distance we vary the charge of powder, we find the respective changes which the velocity undergoes, and hence learn the ratio that ought to be observed between the powder and the ball, in order to produce the maximum effect. The effects resulting from variations in the length, shape, and bore of the gun, are also ascertained with equal facility.

362. The following are some of the practical results ascertained by the experiments of Mr. Robins, Count Rumford, and Dr. Hutton. A musket ball, discharged with a common charge of powder, issues from the muzzle of the piece with a velocity between 1600 and 1700 feet in a second.* The utmost velocity that can be given to a cannon ball is a little more than 2000 feet per second, and this it has only at the moment of leaving the gun. In order to increase the velocity from 1600 to 2000 it requires half as much more powder, which involves a hazardous strain upon the gun, and the velocity will be reduced to 1300 before the ball has proceeded 500 yards.†

363. Mr. Robins, in the course of his experiments on the resistance of the air, made a curious observation. In very moderate velocities, the retardations were nearly as their squares. As the velocities were increased, the resistances increased at a somewhat greater rate, but with a certain observable regularity, till the velocity exceeded 1100 feet per second. But when the velocity is increased from 1100 to 1200, the increase of resistance is prodigious. After this, the resistance goes on increasing nearly with its former regularity. Mr. Robins accounts for this singular fact in the following manner. As the ball rushes through the air, the air falls in behind it, being pressed in by the weight of the surrounding air. But the ball may move so rapidly that the air cannot instantaneously fill up the place left by the ball. In this case the ball is retarded, not only by the resistance of the

* Space fallen through to acquire the velocity of 1600 feet per second $=\frac{(1600)}{64\dagger}$ Art. 34, p. 38,) =40,000 *nearly*=7.6 miles

† Robison's Mech. Phil. I, 201.

air which it displaces, but also by its having the pressure of the atmosphere on one side and not on the other. It is found by calculation, that the velocity with which air rushes into a vacuum, is about 1100 or 1200* feet per second; consequently, it is when the ball is moving at this rate, that it loses the entire pressure of the atmosphere behind it.† It is remarkable (says Dr. Robison) that this is also nearly the velocity of sound, and there is an observation of this kind, which seems to have some connection with the mechanical fact observed by Mr. Robins. If a person stand in such a direction from a cannon, when it is discharged, that the ball may pass him at no great distance, he will hear the noise made by the ball rushing through the air at the time of its flight, and as the ball approaches him, the noise should become more audible. But he will hear the noise loudest at the very first, immediately following the report of the gun; and after about two seconds, he may observe the sound change all at once, and not only become more faint, but even change its kind, after which the sound increases as the ball comes nearer. It seems highly probable that this abrupt alteration in the sound takes place just at the time when the resistance undergoes such a change; and that it is owing to the difference in the nature of the undulations, when there is a void behind the ball, and when there is not.‡

364. From the foregoing considerations it is inferred that great charges of powder are absolutely useless in the service of artillery, especially when the distance of the object is considerable, and that a velocity exceeding 1100 should not be aimed at. The maximum service charge is $\frac{2}{3}$ the weight of the ball. In close naval engagements great velocities are injurious, for the ball may then pass through both sides of the vessel without lodging, and the number of splinters produced by a ball in rapid motion, is much less than is caused by one moving more slowly. By reducing the charge we may also reduce the size and strength of the gun; and hence guns are made of smaller dimensions now than formerly, in order to do the same execution. The velocity with which a charge of powder expands itself at first, is estimated by Hutton as high as 5000 feet per second.§ As it expands, this velocity is of course constantly diminishing, but will exceed that of the ball while the latter is passing through the barrel of the gun, and will act as a constantly accelerating force. Long guns therefore give to balls a greater velocity than short ones; but the gain secured in this way after a moderate length is so small,‖

* The velocity with which air *begins* to rush into a void, has been estimated at 1338 feet. (See Cambridge Mechanics, p. 377.)

† Robins, Tract on Gunnery, p. 181.

‡ Robison's Mech. Phil. I, 200.

§ Hutton, Tract 37.

‖ The random, according to Dr. Hutton, increases only as the fifth root of the length, which is so small an increase as to amount to only about a seventh part more range for a double length of gun. (Hutton, Tract 37.)

(there being also some disadvantages peculiar to long guns,) that cannon have of late years been much shortened. In the naval service, *carronades* have been introduced. These are a short kind of gun, with small bore, requiring for a charge of powder, only one twelfth the weight of the ball. Their weight and thickness are proportionally reduced, yet in close action they produce effects superior to those of long guns.*

365. It has been found that no difference is caused in the velocity or range, by varying the weight of the gun, nor by the use of wads, nor by different degrees of ramming, nor by firing the charge of powder in different parts of it; but that a very great difference in the velocity arises from a small degree in the *windage.*† Indeed, with the usual established windage only, viz. about $\frac{1}{20}$ of the calibre, no less than between $\frac{1}{4}$ and $\frac{1}{3}$ of the powder escapes and is lost, and as the balls are often smaller than the regulated size, it frequently happens that half the powder is lost by unnecessary windage.‡ To this cause also, namely, too great windage, Dr. Hutton ascribes a great part of the sideways deviation of a ball; since, if in passing through the barrel of the gun, it is knocked from side to side, it will finally take the last direction which it happened to have at the muzzle of the gun.§ Another cause of this deviation from the line of direction, arises from a want of perfect sphericity in the ball, by which means the two sides do not meet with equal resistance. Rifles owe their superiority over common guns, chiefly to their obviating this deviation. They have a spiral groove cut in their bore, making about a turn and a half in the whole length of the barrel. The ball, which is made to fit close to avoid too great windage, has a corresponding motion impressed on it, which it retains after it leaves the gun, continuing to revolve around the line of direction. Whatever inequalities, therefore, may exist in the ball, their effects are neutralized, by their being first on one side and then on the other of this line.

366. When a ball is projected from a piece of ordnance at a small angle of elevation, and falls upon water, or on a plane of hard earth, its flight will not cease, but it will rise again and describe a second curve similar to the first, but less; and it will continue to rebound, until the whole of its projectile velocity is destroyed. This species of firing is called *Ricochet.* It is applied with great advantage from sea-coast batteries upon shipping, and in the attack of fortresses. The pieces are fired with small charges of powder and elevated only from 3 to 6 degrees. The

* Renwick, Heads of Lectures, in Literary and Scientific Repos. IV, 281.

† By windage is meant, the difference between the diameter of the ball and that of the bore of the gun.

‡ Hutton, Tract 3, 255.

§ Hutton, Tract 38, Prob. 4.

word signifies *duck and drake*, or rebounding; because the ball or shot thus discharged, goes bounding and rolling along, killing or destroying every thing in its way, like the bounding of a flat stone along the surface of water when thrown almost horizontally.*

CHAPTER VII.

APPLICATIONS OF THE PENDULUM.

367. The pendulum has three most important and interesting uses, viz. as affording a *measure of time*, the means of determining the *figure of the earth*, and a *standard of weights and measures.*

368. Time is any portion of indefinite duration, from which it may be considered as separated, as a given number of yards of cloth are measured off from a piece of unknown length. For the measure of time we may employ any instruments or marks that divide it into equal portions. The period occupied by a given quantity of sand in running through a funnel, as in the hour-glass—the successive intervals through which the surface of a column of water descends while discharging itself by an aperture in the bottom or side of the vessel—the beats of the hand, as in music, or even the pulsations of the wrist, are so many different modes of measuring time. But as the intervals measured by any of these modes are not perfectly uniform, and as they are incapable of that minute subdivision which is sometimes required, none of them is adapted to the purposes of the astronomer, who seeks by this method to estimate the motions of the heavenly bodies, or of the mariner, who depends on his time-piece for the means of ascertaining his longitude at sea.

369. The adaptation of the pendulum to the measuring of time, was first noticed by Galileo on observing the vibrations of a lamp suspended from the ceiling of a church. The theoretical doctrines of the pendulum revealed the fact, that all the vibrations of a pendulum of given length, are performed in nearly equal times, and, when made to move in the arc of a cycloid, (Art. 180,) in times that are exactly equal. Hence the pendulum becomes a measure of time. Galileo, indeed, used this instrument by itself, counting the number of its vibrations; but Huygens, an astronomer of Holland, first connected it with clock-

* Hutton's Math and Phil. Dict., II, 374.

work, by which its motions are indefinitely continued, more precisely regulated, and the number of its vibrations exactly registered. Pendulums are usually made of such a length as to beat seconds, or to indicate, at each vibration, the $\frac{1}{86400}$ part of a mean solar day; but as the period occupied by the passage of a star from the meridian round to the same meridian again (called a sidereal day) is less than that occupied by the sun, so the pendulum regulated to solar time, may, by being shortened a little, be made to indicate the corresponding aliquot parts of a sidereal day. Since, moreover, the times of vibration are as the square roots of the lengths, (Art. 181,) a pendulum may be made to beat half-seconds by being made only one fourth the length of the seconds pendulum, or to beat once in two seconds by making it four times as long.

370. As the vibrations of the pendulum in circular arcs are not exactly equal to each other, it has been attempted to make it vibrate in the arc of a cycloid, (Art. 179;) but the practical difficulties involved in making the pendulum rod adapt itself to the cheeks of a cycloid are so great, that this method has been abandoned in practice, and is at present regarded only as a theoretical curiosity. Where the arcs of vibration are very small, they will not differ sensibly from portions of a cycloid, and the times of vibration may without sensible error be considered as equal. In cases where extreme accuracy is required, the exact allowance, mathematically determined, may be applied to vibrations performed in circular arcs to reduce them to the corresponding vibrations in a cycloid; and a still further allowance may be made for the resistance of the air, so as to make the vibrations correspond to such as they would be if performed in a vacuum.

371. The cycloid is the curve of *swiftest descent;* that is, a body will descend from a given height in less time on this than on any other curve, or even than on an inclined plane. This proposition, in general, is demonstrated by the aid of the calculus; but the cycloid may be readily compared with the inclined plane by experiment, as in the annexed figure, where two balls let off

Fig. 168.

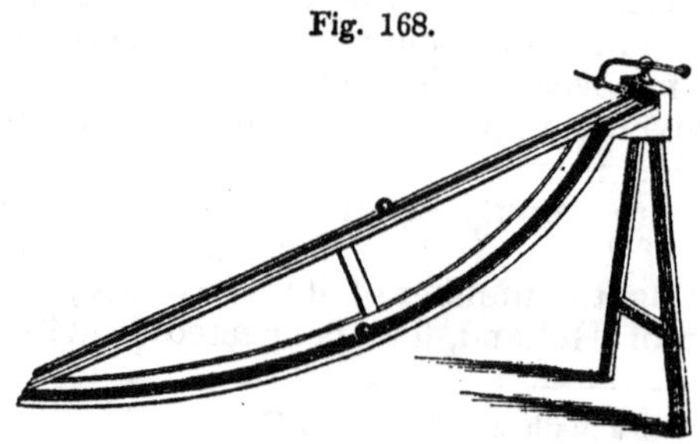

at the same instant by raising a joint at the top of the figure, roll down the plane and the curve respectively. The ball reaches the bottom soonest by way of the curve. The velocity acquired at first by a direction more nearly perpendicular, more than compensates for the greater length of the course.

Moreover, if both the balls be let off at any different heights on the cycloid, they will reach the lowest points at the same instant, since it is a property of this curve, that the accelerating force at any point in it is proportional to the distance of that point from the lowest point; consequently, the upper ball being urged by a force as much greater as its distance is greater, reaches the bottom at the same instant with the lower ball. On the same principle the pendulum performs all its vibrations in a cycloid in equal times.

372. The pendulum is liable to some sources of inaccuracy, which it has cost much labor and skill to obviate. The rod and bulb are both subject to expand by heat and contract by cold. In the former case the center of oscillation (Art. 170) is carried too far from the center of motion, that is, the pendulum becomes too long, and the clock goes too slow; in the latter case, the pendulum being shortened, the clock goes too fast. With a pendulum having an iron rod, a difference of temperature of 25 degrees would make a difference of six seconds in 24 hours in the rate of the clock.*

Pendulums so constructed as to remedy these defects arising from change of temperature, are called *Compensation Pendulums.* The general principle of these instruments is as follows: we connect with the pendulum rod, some substances which are made to expand or contract in a direction opposite to that in which the rod itself contracts or expands, and thus maintains the center of oscillation, at a uniform distance from the center of suspension. The length of the theoretical or simple pendulum, it may be remarked, depends wholly on the distance between these two points, and not upon the length of the pendulum rod. If then we make the pendulum rod of one kind of metal, as steel, and connect with the bulb another kind of metal, as brass, which is expanded more by the same amount of heat, a less length of the latter expanding upward, will compensate for a greater length of the former expanding downward, and the center of oscillation will be kept at the same constant distance from the center of suspension. Now it is found by experiment that the lengths of these substances, which are equivalent to each other, are as 1 : .6091, or as 100 : 61 nearly. Therefore, if we connect 61 inches of brass with 100 inches of steel, the former so arranged as to expand upward while the latter expands downward, we shall have

* Kater.

a compensation pendulum. The first instrument constructed on this principle is called the *Gridiron Pendulum*, a name which it derives from the fancied resemblance which the parallel bars of steel and brass have to the gridiron. In the following figure, the five bars marked *s* are of steel, and the four marked *b* are of brass. It will be seen by the figure that the rods of steel can elongate themselves only downward, while those of brass can expand only upward; and being combined in the ratio of 100 to 61, they will exactly compensate each other. A clock furnished with the gridiron pendulum is found capable of keeping very accurate time. Indeed a clock of this kind, constructed by Harrison, the inventor of the gridiron pendulum, gained the great premium offered by the British Board of Longitude, for a time-keeper which would keep time for a given period to a certain degree of accuracy.

Fig. 169.

Another form of the compensation pendulum consists of a jar of mercury, suspended from the rod in the place of the bulb. By means of a kind of stirrup the jar is so connected with the rod as to expand upward while the rod expands downward. The mercury being a very expansible metal, the length of the mercurial column required for the compensation of a steel rod, of 42 inches, (including the stirrup,) is only 6.31 inches.* This compact and simple form of the compensation pendulum, makes this variety peculiarly eligible.

373. In stationary time-pieces, the pendulum affords the most eligible, as it is the most accurate, measure of time. But where the instrument is required to be portable, as in watches and chronometers, a spiral spring called the main-spring is the moving power, instead of a weight, and a balance-wheel takes the place of the pendulum. The hair-spring attached to this maintains its vibrations, coiling up in one direction and uncoiling in the opposite, while the main-spring supplies the power in the same manner as the weight of a clock.

374. The use of the pendulum *in investigating the figure of the earth*, results from its power of measuring the intensity of the force of gravity in any given place. (Art. 182.) Now as this force varies inversely as the square of the distance from the

* Kater in Lardner's Mechanics, p. 331. Baily, Astron. Trans., 1824, p. 381.

center of the earth, when the force of gravity is ascertained at a great number of places remote from each other, a comparison of these observations indicates the respective distances of these places from the center of the earth, and of course, when the observations are sufficiently multiplied, they indicate, collectively, the figure of the earth.

If, therefore, one were to start from the equator with a pendulum which there vibrated seconds, and should proceed with it to the north pole, and there count the number of vibrations it would make in an hour, he might thus ascertain the respective forces of gravity at those two points, and hence learn the ratio between the equatorial and polar diameters. For, according to Art. 183, the number of vibrations performed by a pendulum in any given time, are as the square roots of the forces of gravity. Although it might be convenient, were it practicable, to derive the ratio between the equatorial and polar diameters of the earth directly from observations made at these two points, yet as that cannot be done, we can only obtain the law of curvature, and hence derive the figure of the earth, by observations made in many different latitudes. Such observations have been made with the greatest accuracy by Kater, Sabine, and others; and, although the result shows that the figure of the earth is that of an oblate spheroid, yet the difference between the two diameters, or the *ellipticity* as it is technically called, is somewhat greater when determined by the pendulum, than when estimated by measuring the length of the respective degrees of the meridian from the equator to the pole. By a comparison of a great number of pendulum experiments, Baily, an English astronomer, makes the ellipticity of the earth $\frac{1}{285}$ of the equatorial diameter, while that derived from a similar comparison of measures of the meridian, is $\frac{1}{298}$. Hence, the pendulum would lead us to the conclusion that the diameter of the earth when taken through the equator is about 34 miles greater than when taken from pole to pole, while actual measurements would make the same excess about 26 miles.*

Measurements of arcs of the meridian, made for the express purpose of determining the true figure of the earth, have been executed with an astonishing degree of accuracy, in various countries, and in different latitudes, from the equatorial regions to points within the Arctic circle. By these means, the nature of the curve that encompasses the earth would be ascertained, and the figure of the earth determined. The combined result of all these measurements is as already stated.

375. A third important application of the pendulum is, *as a standard of linear measures.*

* Herschel's Astronomy.

In order to insure confidence in business transactions, it is very essential that the weights and measures employed, should be and remain of a certain known amount or length,—a condition which cannot be attained otherwise than by adapting them to a fixed and invariable standard. To fix on such a standard is a matter of more difficulty than would at first be imagined. All things on the face of the earth are more or less subject to change, not only the works of man, but even those of nature. The elements decompose or wear down, in time, the hardest rocks. In ancient times, the standards of weights and measures were derived from the parts of the human body; as, a *hand's breadth*, a *cubit*, a *digit*, &c. The English linear measures were, for a long time, referred to the length of a yard-stick taken from the length of the arm of Henry the VIIth, and preserved in the tower of London.* Of the defectiveness of such a standard, it is sufficient to mention the impossibility of verifying it after the death of the king. The idea of standards involves two conditions that are indispensable,—the constancy of the thing itself, and the power of verification, should any change in it even be suspected. These conditions are secured by connecting the standard with the immutable laws of nature. In the year 1790, the French government undertook to effect a complete change in the weights and measures in use throughout the world, and to derive an unalterable standard from the dimensions of the earth itself. For this purpose, they undertook the determination of the exact length of a quadrant of the meridian, extending from the equator to the North Pole. A certain aliquot part (the ten millionth) of this they denominated a *metre*, which was to be the unit of all *linear* measures. The square of this furnished a measure of *surfaces* and the cube a measure of *solids*. Measures of *weight* were derived from the weight of a certain volume of water.

376. We may exemplify the precautions necessary in order to make the pendulum an accurate standard of measures, by reviewing the conditions under which it is adopted as a standard in the state of New York.

The standard is the pendulum vibrating seconds, in a cycloidal arc, and in a vacuum, in Columbia College in the city of New York. The vibrations are required to be in a *cycloidal* arc, because those performed in circular arcs are not absolutely uniform. But the vibrations performed in circular arcs may be *reduced* to the corresponding cycloidal arcs upon mathematical principles. As the resistance of the air might be unequal at different times, such an allowance must always be made for this, as will neutralize its effect. A particular spot is designated, because local

* Adams on Weights and Measures.

causes have some influence upon the vibrations of the pendulum.

The unit of linear measures is the *yard*, which is of such magnitude as to bear to the pendulum the proportion of 1 to 1.086158. Some have proposed to make the seconds pendulum itself the unit of linear measures, and to make all other measures of length parts and multiples of this; but it is thought advisable to retain the common measures to which the habits of society are adapted, determining their exact amount by referring them to this invariable standard. The usual subdivisions of the yard into feet, inches, and so on, remain as they are. The standard *temperature* of the standard yard-stick, is that of melting ice. It is necessary to attend to this circumstance, because all bodies are expanded by heat and contracted by cold, so that the yard-stick is of a uniform length only at a given temperature.

The unit of measures of *weight* is the avoirdupois pound, of such magnitude that a cubic foot of pure water, at its maximum density,* shall weigh 1000 ounces, or 62½ pounds.

The unit of dry measures of *capacity* is the gallon, a vessel of such magnitude as to hold exactly 10 lbs. of pure water, at its maximum density. The bushel therefore holds 80 lbs. The unit of liquid measures is also a gallon, containing eight pounds of distilled *water*, at its maximum density.†

The government of the United States has adopted, for the different standards, the following bases.

Standard of Length. The yard of 3 feet or 36 inches, from the scale of Troughton, which is a brass scale of 82 inches, made by a celebrated English artist for the survey of our coast. Hence the yard adopted as the standard, is identical with the British imperial standard.

Standard of Weight. The troy pound, containing 5762.38 grains, used as a standard at the mint of the United States.

Standard of Dry Measure. The British Winchester bushel of 2150.4 cubic inches, equal to 77.6274 pounds of distilled water, at the maximum density.

Standard of Liquid Measure. The English wine gallon of 231 cubic inches, equal to 8.339 pounds advoirdupois of distilled water, at the maximum density.‡

* Water expands as its temperature rises above or falls below a certain point; about 40° of Fahrenheit. Hence the necessity of employing it as a standard at its *maximum density*.

† See an able view of the applications of the Pendulum in Renwick's Mechanics

‡ North Amer. Rev. for 1837, p. 290.

PART III.—HYDROSTATICS.

377. The principles of Mechanics demonstrated and explained in the foregoing pages, are *universal* in their application, extending alike to all bodies, whether solid or fluid. But in addition to those properties which fluids have in common with solids, and which bring them under the general laws of Mechanics, they have also properties peculiar to themselves, and which give rise to a distinct class of mechanical principles, not applicable to solid bodies. These are embraced under the heads of Hydrostatics and Pneumatics, the former division comprising the doctrine of liquids, and the latter that of aëriform bodies or gases.*

In Mechanics, after having ascertained a few fundamental principles by experiment and observation, the superstructure is raised chiefly by mathematical reasoning, and thus the great body of truths in that science are established; but in Hydrostatics, and the other subjects of Natural Philosophy which follow, we are much more dependent on *experiment*, which frequently affords us more satisfactory evidence, than we can obtain by the application of abstract mechanical principles.

378. *A fluid is a body whose particles move easily among themselves, and yield to the least force impressed; and which, when that force is removed, recovers its previous state.*†

In accordance with the example of Sir Isaac Newton, which has been followed by most writers on Hydrostatics, we have in troduced the foregoing definition of a fluid. Although the best perhaps that the subject admits of, yet it is not very discriminating, since such substances as quicksand, or the powder of magnesia, have their particles as movable as those of tar or syrup, while yet the former are solid and the latter fluid bodies. The fact is, that no definition can be given of a fluid which would

* In some treatises these subjects are distributed under the heads of Hydrostatics and Hydrodynamics, the former comprehending the mechanical properties of fluids *at rest*, and the latter those of fluids *in motion*. In other works, what relates to liquids or non-elastic fluids, is divided into *Hydrostatics* and *Hydraulics*, (the latter denoting the mechanical powers and agencies of running water and of machines carried by water,) and *Pneumatics*. The most scientific division is that adopted in the Edinburgh Encyclopœdia, where the term *Hydrodynamics* (from Ὑδωρ and Δυναμις) is used to denote in general the mechanical powers and agencies of fluids; and this head is divided into the two, Hydrostatics and Hydraulics. Pneumatics is treated of in a separate article. But we prefer to follow the example of those who arrange these subjects under the two heads specified in the text.

† Vince.

convey to one unacquainted with these bodies any adequate idea of the distinction between them; and it may be doubtful whether any better definition can be given of a fluid, than that *it is a generic term, comprehending liquids and gases*, while liquids are defined to be *bodies in the form of water*, and gases *bodies in the form of air*. The property included in the definition of Professor Vince, which we have copied, namely, that a fluid when disturbed by any force impressed *recovers itself*, is as characteristic of this class of bodies as the mobility of their parts.

Since water, wind, and steam, are the only fluids that are usually employed as mechanical agents, the doctrines of Hydrostatics and Pneumatics have regard chiefly to them; but the principles established respecting these, are applicable also to all analogous bodies.

379. It has been usual to denominate liquids and gases respectively *non-elastic* and *elastic* fluids, on the supposition that water and other liquids are nearly or quite incompressible. An experiment performed by the Florentine academicians, as long ago as 1650, seemed to prove that water is wholly incompressible. They filled a hollow ball of gold with water, and subjected it to a strong pressure.* The water, not yielding to the compression, oozed through the pores of the gold. Considering the great density and compactness of this metal, the experiment was for a long time held as proving decisively that water is wholly incompressible. Although this experiment shows that water is compressed with great difficulty, yet later experiments have proved, that it is still capable of compression. The most decisive evidence of this point has been recently afforded by the experiments of Mr. Perkins. It had been previously ascertained that by a pressure equivalent to that of the atmosphere, or about fifteen pounds to the square inch, water is compressed about one part in twenty-two thousand. Mr. Perkins, by methods to be described hereafter, applied successive degrees of pressure up to that of two thousand atmospheres, or 30,000 pounds to the square inch, and found the contraction of volume to be nearly *one twelfth* of the whole.

With these preliminary remarks, we may now enter upon the immediate consideration of the principal subject before us.

---

* The method of applying the pressure is said to have been by means of a screw working through a water-tight joint.—*Partington.* As the screw was forced into the water, the latter must either be compressed or make its escape from the ball. (Art. 2.) An easier mode of applying the pressure would have been, to put the ball into a vise and flatten it, by which its capacity would have been diminished.

CHAPTER I.

OF LIQUIDS OR NON-ELASTIC FLUIDS AT REST OR IN EQUILIBRIUM.

380. HYDROSTATICS *is that branch of Natural Philosophy which treats of the mechanical properties and agencies of* LIQUIDS.

381. *Fluids at rest press equally in all directions.*

A point in a mass of fluid, taken at any depth, exerts and sustains the same pressure in all directions, upward, downward, or laterally. This is the most remarkable property of fluids, and is what particularly distinguishes them from solids, which press only downward, or in the direction of gravity. This property naturally results from the freedom of motion that subsists between the particles of fluids; for if, when a fluid is at rest, the pressure on any given portion were not equal in all directions, that portion would move in the direction in which the resistance was least. But by the supposition it does not move: therefore it is kept at rest by equal and contrary forces acting on all sides. But the most satisfactory evidence of this truth is obtained from experiments. On opening an orifice in the side of a vessel of water, and estimating the force with which the water issues, it is found to be equal to the weight of the incumbent fluid; and the upward pressure of water at a certain depth is found to sustain the heaviest bodies when exposed to its action alone, the column above the bodies, and of course the downward pressure, being removed.

382. *A given pressure or blow impressed on any portion of a mass of water confined in a vessel, is distributed equally through all parts of the mass.*

A given pressure, as that made by a plug forced inward upon a square inch of the surface of a fluid confined in a vessel, is suddenly communicated to every square inch of the vessel's surface, however large, and to every inch of the surface of any body immersed in it. Thus if I attempt to force a cork into a vessel full of water, the pressure will be felt not merely by the portion of the water directly in the range of the cork, but by all parts of the mass alike; and the liability of the body to break, supposing it to be of uniform strength throughout, will be as great in one place as in another, and it will break at the point where it happens to be the weakest, however that point may be situated relatively to the place where the cork is applied; and the effect will be the same whether the stopper be inserted at the top, the bot tom, or the side of the vessel.

383. It is this principle which operates with such astonishing effect in the *Hydrostatic Press*, by means of which a single man can exert a force which is adequate to crush the hardest substances, or to cut in two the largest bars of iron.* Its construction is as follows.† Fig. 170, represents a press made of the strongest timbers, the foundation of which is commonly laid in solid masonry. AB is a small cylinder in which moves the piston of a forcing pump, and CD is a large cylinder in which also moves a piston, having the upper end of its rod pressing against a movable plank E, between which and the large beam above is placed the substance to be subjected to pressure, as for example a pile of new-bound books. By the action of the pump handle, water is raised into the small cylinder, and, on depressing the piston, it is forced through a valve at B into the larger cylinder and raises the piston D, which expends its whole force on the bodies confined at E. Now, since whatever force is applied to any one portion of the fluid, extends alike to every part, therefore the force which is exerted by the pump upon the smaller column, is transmitted unimpaired to every inch of the larger column, and therefore tends to raise the movable plank E with a force as much greater, in the aggregate, than that impressed upon the surface of the smaller, as this surface is smaller than that of the larger column; or (which is the same thing) as the number of square inches in the end of the piston B is less than that of the piston D. The power of such a machine is enormously great; for, supposing the hand to be applied at the end of the handle, with a force of only ten pounds, and that this handle or lever is so constructed as to multiply that force but five times, the force with which the smaller piston will descend will be equal to 50 lbs.; and let us suppose that the head of the large piston contains the smaller 50 times, then the force exerted to raise the press-board will equal 2500 lbs. A man can indeed easily exert many times the force supposed, and can therefore exert a force upon the substance under pressure, equal to many tons.

Fig. 170.

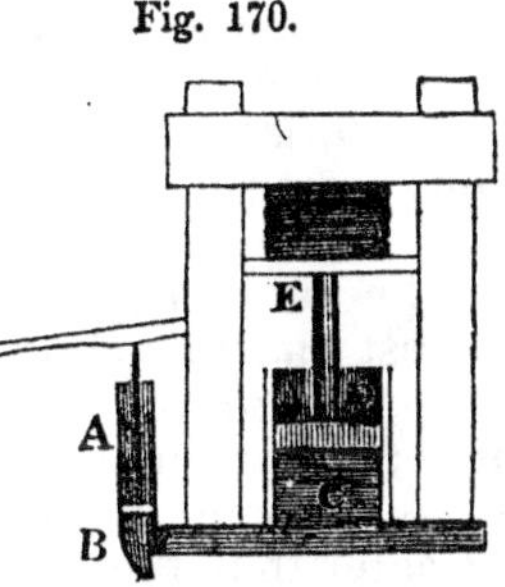

The hydrostatic press involves far less loss from friction than any other species of press, and it is said that the naked force of man is more effective when applied in this way than in any

* Partington's Manual of Nat. Phil.

† For a more complete description of the "Bramah Press," see Webster's *Principles of Hydrostatics*, p. 151.

other. By the mere weight of a man's body, when leaning on the extremity of the lever, a pressure may be produced of upwards of 2000 tons. It is the simplest and most easily applicable of all contrivances for increasing human power; and the only limit to the force which may be called into action by it, is the want of materials of sufficient strength to enable us to apply the enormous pressure which it generates.*

384. The rationale of the principle of the hydrostatic press will be best understood by recurring to the following principles, —that opposite forces are in equilibrium when their *momenta* are equal; that a small power may be made to balance a great weight, by making it move in a given time, over a space as much greater than the larger does, as its weight is smaller; and that it may be made to overcome that resistance or weight and give motion to it, if its velocity is greater than that of the latter in a still higher ratio. Now to apply these principles to the case before us, it is evident that any quantity of water forced out of the smaller into the larger cylinder, must rise in the latter as much slower as the area of the horizontal section is larger. If, for example, the capacity of the larger cylinder were ten times that of the smaller, then a quantity of water one inch in height transferred from the smaller to the greater cylinder, would occupy only the height of one tenth of an inch, consequently, the depression of the small piston one inch would raise the large one only the tenth of an inch. This case therefore resolves itself into that general principle, according to which a vast force is exerted through a short distance, by moving a small force through a distance much greater.

This press is used for the extraction of oils, either vegetable or animal, for pressing paper or books, and for packing cotton and other substances. Hay intended as food for cattle on shipboard, is reduced by this press almost to the state of a solid, and enormous quantities are thus brought into an inconceivably small compass. There would seem to be no force known to us which may not be made to yield to this power. It requires but one of its least efforts to tear up a tree by its roots, or to break a large beam asunder.

385. *The surface of a fluid at rest is horizontal.*

The evidence of the truth of this proposition is threefold.† *First*, this result is a natural consequence of the mobility of fluids, since, if any portion is raised above the rest, having nothing to support it, and being acted on by gravity, it must de-

---

* Mosely, *Mech. applied to the Arts*, Art. 197.

† It might appear superfluous to offer so much proof of a point so plain; but the several modes of reasoning will serve to instruct the young learner in the peculiar properties of fluids.

scend in the same manner as a body placed on a perfectly smooth inclined plane. *Secondly*, whenever a body is free to move, its *center of gravity* will descend as low as possible. (Art. 70.) When, therefore, any portion of a fluid is raised above the general level, the center of gravity of the mass is raised, and drawn out of the line which passed perpendicularly through it and the point of suspension, (Fig. 43,) and it must return to that line before the fluid can be at rest. Thus, let ABCD, (Fig. 171,) be a body of water, contained in a cylindrical vessel; let I be the center of the surface, and IGH the perpendicular line passing through the center of gravity G and the base. Through I, suppose a plane, movable on a hinge at E, to pass, and to be depressed into the situation FIE, by which means the water will be depressed on the side AF and raised on the side EB, while the center of gravity will be removed to the point G′. Now let the plane be removed, and the center of gravity being free to move, it will vibrate around I as a point of suspension, until it finally recovers its situation at G, and the surface of the fluid will return to its original level. *Thirdly*, experience shows that the proposition is true, since fluids, when free to move, always settle themselves with their surfaces parallel to the horizon.* It must be understood, however, that the surface of large bodies of water is not, strictly speaking, a horizontal level, but is a portion of the convex surface of the earth; for since the center of gravity of every portion of the fluid will descend as low as possible, the whole will dispose itself around the center of attraction, so as to form a portion of the earth's surface. For small distances the curvature is so slight that it may be neglected, not amounting to one second of a degree for 100 feet;† and for the distance of a mile, the deviation from a straight line, drawn in the direction of a tangent, is not more than 8 inches. The amount of the depressions for different distances, is estimated by the following formula, $D=\frac{2L^2}{3}$, where L represents the number of miles, and D the depression in feet. Thus the depression of two miles is $\frac{2\times4}{3}=2\frac{2}{3}$ feet. Let FE (Fig. 172,) or

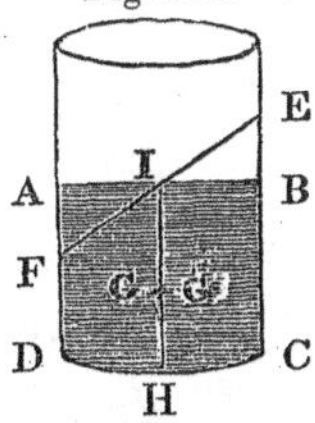

Fig. 171.

* The only exceptions to this law, are those arising from the *attraction of cohesion* among the particles of liquids, and the attraction exerted by small tubes, called *capillary attraction*. In consequence of cohesion, small portions of a liquid form themselves into drops, and large portions present a convex surface, as is strikingly exhibited in a wine-glass filled with quicksilver. By capillary attraction, the surfaces of liquids are made concave, a phenomenon to which we shall attend more particularly hereafter. But both these causes operate on a scale comparatively very small.

† $\frac{360\times60\times60\times100}{25000\times5280}=\frac{1296}{1320}=\frac{9''}{10}$.

its equal GB, be the depression for the distance BE. For moderate distances this arc may be considered as equal to its chord. Hence GB(D) : EB(L) :.: EB : AB; or $D=\frac{L^2}{AB}$, or D being expressed in feet and L and AB in miles, $D=\frac{L^2\times(5280)^2}{AB\times 5280}=\frac{L^2\times 5280}{7912}=\frac{2L^2}{3}$, nearly.

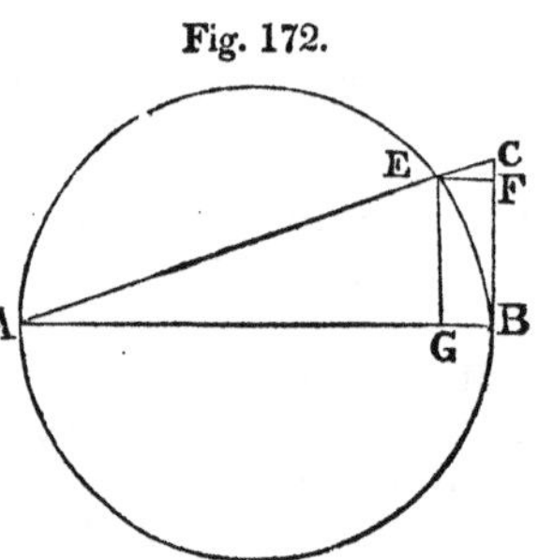

Fig. 172.

1 mile - - $\frac{2\times 1^2}{3}=\frac{2}{3}$ feet. 3 miles - - $\frac{2\times 3^2}{3}=6$ feet.

2 " - - $\frac{2\times 2^2}{3}=2\frac{2}{3}$ " 100 " - - $\frac{2\times 100^2}{3}=6666\frac{2}{3}$ do.

386. A practical application of this principle is made in the art of *levelling*. A level is sometimes made by merely cutting a groove or channel in a flat piece of board and filling it with water. When the board is brought into such a situation that the water in the groove remains stationary, the position is horizontal. But the *spirit level* is the instrument more commonly employed for this purpose. This consists of a small bent cylindrical tube of glass, from two to six inches long, filled with spirits of wine or ether, except a small space which is occupied by a movable bubble of air. When such a tube is placed horizontally, the bubble of air will remain stationary in the center of the tube, at a fixed mark; but whenever the tube is inclined, in the least degree, the bubble will ascend toward the elevated end. Spirit levels are much used for adjusting astronomical, surveying, and other delicate instruments. Figure 174 represents a levelling staff used in surveying and grading lands. The liquid in the two arms of the tube A and B being precisely on a level, any two objects, P and Q, may be brought accurately to the same horizontal level by *sighting* P with the eye at A, and then sighting Q with the eye at B.

Fig. 173.

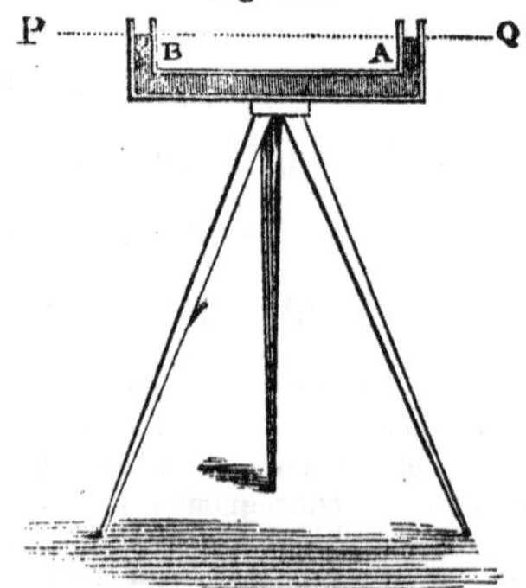

Fig. 174.

387. *The pressure upon any particle of a fluid of uniform density, is proportioned to its depth below the surface.*

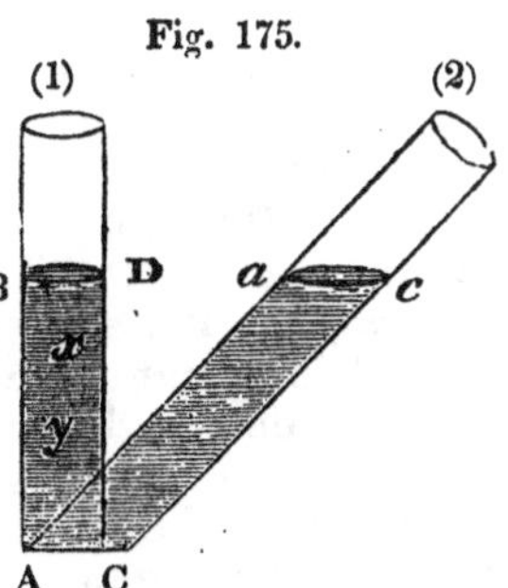

Case 1. Let the column of fluid ABCD, (Fig. 175,) be perpendicular to the horizon. Take any points, x and y, at different depths, and conceive the column to be divided into a number of equal spaces by horizontal planes. Then, since the density of the fluid is uniform throughout, the pressure upon x and y, respectively, must be in proportion to the number of equal spaces above them, and consequently in proportion to their depths.

Case 2. Let the column be of the same perpendicular height as before, but inclined as is Fig. 175, (2); then its quantity, and of course its weight, is *increased* in the same ratio as its length exceeds its height; but since the column is partly supported by the plane, like any other heavy body, the force of gravity acting upon it is *diminished* on this account in the same ratio as its length exceeds its height; therefore as much as the pressure on the base would be augmented by the increased length of the column, just so much it is lessened by the action of the inclined plane; and the pressure on any part of C*c* will be, as before, proportioned to its perpendicular depth; and the pressure of the inclined column AC*ac* will be the same as that of the perpendicular column ABCD.

There are various experiments also by which this proposition is fully established.*

388. According to Art. 381, the lateral is equal to the downward pressure; and consequently, on this principle, may easily be estimated the amount of pressure on the sides of any column of water, or on the banks of rivers, canals, &c. At the depth of 8 feet, the pressure on a square foot is equal to the weight of a column of water whose base is 1 foot and depth 8 feet, and consequently its solid contents 8 cubic feet; and since 1 cubic foot of water weighs 1000 ounces, or $62\frac{1}{2}$ lbs., therefore the weight of the column$=8\times62\frac{1}{2}=500$ lbs. Hence the pressure on a square foot, at different depths, will be as in the following table:

Depth in feet.	Pressure on a square foot.	Depth in feet.	Pressure on a square foot.
8	500 lbs.	56	3500 lbs.
16	1000	64	4000
24	1500	72	4500
32	2000	80	5000
40	2500	88	5500
48	3000	96	6000

1 mile or 5280 feet, - - - 330,000 lbs.
5 " - - - - - - - - 1,650,000

* Experimental illustrations are, in this part of the work, supposed to be given by the instructor.

Hence it appears that at the moderate depth of 64 feet, the pressure of a column of water on the bottom or sides of the containing pipe, becomes 4000 lbs. to the square foot; and the pressure on the bottom of the sea, where it is one mile in depth, is 330,000 lbs. to the square foot, and where it is 5 miles deep, that pressure is no less than 1,650,000 lbs.* From these considerations, we may readily apprehend the cause of the great difficulty experienced in confining a high column of water; and hence also may be inferred the immense pressure that is exerted on the bottom of the sea. It is said that the Greenland whale sometimes descends to the depth of a mile, but always comes up exhausted, and blowing out blood, showing that the pressure had so acted upon the vessels as to cause them to discharge a portion of their contents into the lungs.†

389. Indications of this vast pressure in deep waters, are manifested by several interesting facts. It has long been known to mariners, that if a common square bottle be let down into the sea, its sides are crushed inward before it has reached the depth of ten fathoms. If a stronger bottle, (a common junk bottle, for example,) be filled with water, corked close, and let down to a certain depth, either the cork will be forced inward, or if that be secured in its place, the salt water will make its way into the bottle in spite of it, either by compressing the cork or by forcing in water through it. It was by sinking an apparatus to the depth of 500 fathoms, that Mr. Perkins first proved the compressibility of water, as mentioned in Art. 379. The apparatus consisted of a hollow brass cylinder, resembling a small cannon,‡ and furnished with a stopper so contrived as to indicate, when the apparatus was drawn up, how far it had been driven in while at the lowest depth. The same experiments were afterwards repeated on shore, a pressure being applied to the plug, by means of the hydrostatic press, equivalent to 2000 atmospheres.

The increase of pressure in proportion to the depth of the fluid, renders it necessary to make the sides of pipes or masonry, in which fluids are to be contained, stronger the deeper they go. The same remark applies to dams, flood-gates, and banks.

At the depth of one mile the compression of water is $\frac{1}{1342}$ of its bulk, and its specific gravity is increased in the same ratio; so that bodies which sink near the surface of the sea, may float at a certain depth before they reach the bottom. On the other hand, a porous body, that is light enough to float near the surface, will have so much water forced into its pores, when it is sunk to a great depth, as never to rise. This is the case with ships that

* Allowance must also be made for the saltness of the sea, salt water being heavier than fresh.

† Ed. Phil. Jour., Jan. 1832.

‡ A cannon itself was afterwards employed in these experiments.

are wrecked in deep water; the parts of the wreck do not rise to the surface, as they do in shallow water.*

390. *The pressure of a fluid against any surface, in a direction perpendicular to it, varies as the area of the surface multiplied into the depth of its center of gravity below the surface of the fluid.*

When a portion, as a square foot, of the lateral surface of a column of water, is taken, all parts of it are not equally distant from the surface of the fluid; and, in this case, the *average* depth, or (which is the same thing) the depth of the center of gravity, is to be understood.

Let *m, n,* (Fig. 176,) be a given portion of the vessel ABCD, filled with water or any liquid, and let us conceive this portion to be occupied by any number of particles *m, o, p, n,* &c.; then the pressure produced by all these particles will be (Art. 83) $m \times mu + o \times ox + p \times py + n \times nz$, &c.: but by a property of the center of gravity,† the sum of the products is equal to the sum of the particles, that is, the area of the surface, multiplied into the distance of the center of gravity from the surface of the fluid.‡

Fig. 176.

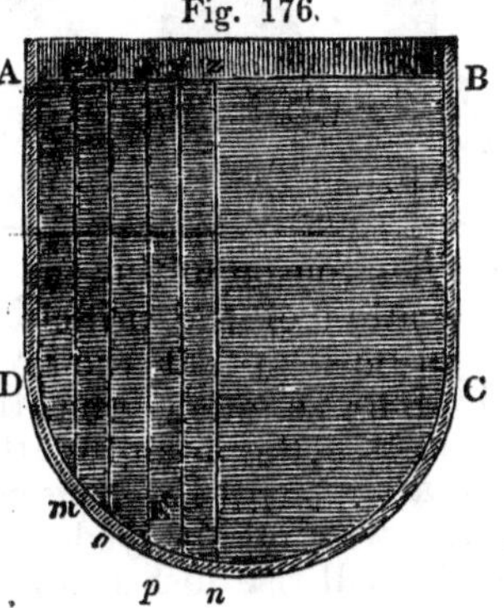

Hence, the pressure on the side of a cubical vessel, filled with fluid, is one half the pressure against the bottom; and the whole pressure against the sides and bottom, is equal to three times the weight of the fluid in the vessel.

Fig. 177

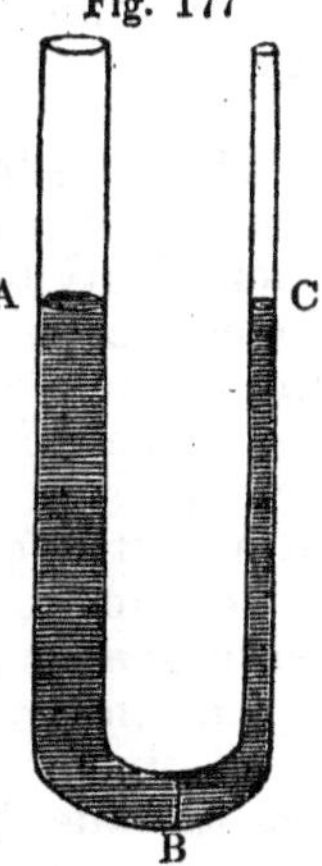

391. *Fluids rise to the same level in the opposite arms of a recurved tube.*

Let ABC, (Fig. 177,) be a recurved tube: if water be poured into one arm of the tube, it will rise to the same height in the other arm. For, by Art. 386, the pressure upon the lowest part at B, in opposite directions, is proportioned to its depth below the surface of the fluid. Therefore, these depths must be equal, that is, the height of the two columns must be equal, in order that the fluid at B may be at rest; and unless this part is at rest, the other parts of the column cannot be at rest. Moreover, since the equilibrium depends on nothing else than the *heights of* the respective columns, therefore, the opposite columns may dif-

* Arnott.

† See the last step in the demonstration of Art. 83, from which $GK \times (p+p+p'') = p \times px + p \times p'x' + p'' \times p''x''$.

‡ Ed. Encyc., 'Hydrodynamics'

fer to any degree in quantity, shape, or inclination to the horizon. Thus, if vessels and tubes very diverse in shape and capacity, as in Fig. 178, be connected with a reservoir, and water be poured into any one of them, it will rise to the same level in them all.

Fig. 178.

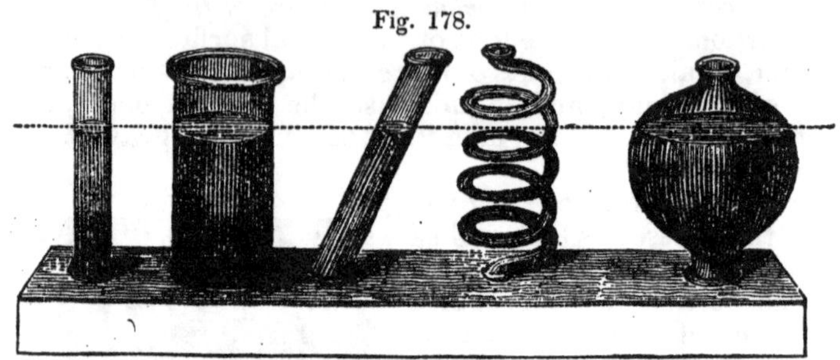

The reason of this fact will be further understood from the application of the principle of *equal momenta*, (Art. 149;) for it will be seen that the velocity of the columns, when in motion, will be as much greater in the smaller than in the larger columns, as the quantity of matter is less; and hence the opposite momenta will be constantly equal.

Hence, water conveyed in aqueducts or running in natural channels, will rise just as high as its source. Between the place where the water of an aqueduct is delivered and the spring, the ground may rise into hills and descend into valleys, and the pipes which convey the water may follow all the undulations of the country, and the water will run freely, provided no pipe is laid higher than the level of the spring. Waters running in natural channels in the earth are governed by the same law.

392. The aqueducts constructed by the ancient Romans, were among the most costly monuments of their arts. Several of them were from thirty to one hundred miles in length, and consisted of vast covered canals built of stone. They were carried over valleys and level tracts of country upon arcades, which were sometimes of stupendous height and solidity.* From the fact that the ancients built aqueducts with so much labor, raising them to a great height in crossing valleys, instead of availing themselves of the principle under consideration, some have supposed that they were unacquainted with this principle. It appears nevertheless that they were acquainted with it, and even understood the use of pipes in conveying water; but probably the expense of pipes, and the difficulty of making them strong enough to resist the pressure when laid at a considerable depth below the source, prevented their general use.

* Bigelow, El. Tech., 303.

393. *The pressure upon the horizontal base of any vessel containing a fluid, is equal to the weight of a column of the fluid, found by multiplying the area of the base into the perpendicular height of the column, whatever be the shape of the vessel.*

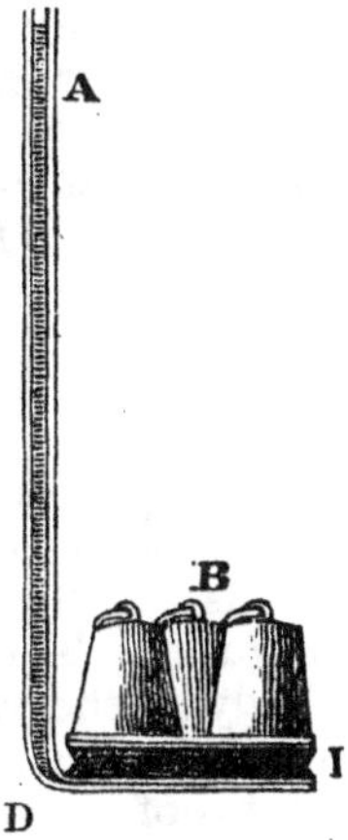

This follows from Art. 390, since here, the distance of the center of gravity of the base from the surface of the fluid, is the same as the perpendicular height of the column. With a given base and height, therefore, the pressure is the same whether the vessel is larger or smaller above, whether its figure is regular or irregular, whether it rises to the given height in a broad open funnel, or is carried up in a slender tube. Hence, *any quantity of water, however small, may be made to balance any quantity, however great.* This is called the *hydrostatic paradox.* The experiment is usually performed by means of a water-bellows, as represented in Fig. 179. When the pipe AD is filled with water, the pressure upon the surface of the bellows, and consequently the force with which it raises the weights laid on it, will be equal to the weight of a cylinder of water, whose base is the surface of the bellows, and height that of the column AD. Therefore, by making the tube small, and the bellows large, the power of a given quantity of water, however small, may be increased indefinitely. The pressure of the column of water in this case corresponds to the force applied by the piston in the hydrostatic press, (Art. 381,) and the explanation according to the principle of equal momenta, is the same in both cases.*

Fig. 179

394. The principle of the Hydrostatic Paradox, is sometimes exemplified by pouring liquids into casks, through long tubes inserted in the bung holes. As soon as the cask is full, and the water rises in the pipe to a certain height, the cask bursts with violence. The same cause is supposed sometimes to produce great effects in nature, such as splitting rocks, heaving up mountains, and other effects resembling earthquakes. For, suppose that in the interior of a mountain there were an empty space ten yards square, and only an inch deep, in which the water had lodged so as to fill it entirely; and suppose that a crevice in the earth should extend from this spot two hundred feet above, which should also become filled with water by rain or otherwise: the

* The bellows rises through so small a space, that its motion is hardly perceptible, but it may be rendered very striking by connecting with the bellows (as is done in the lecture room at Yale College) a lever, and several multiplying wheels, which give a rapid motion to a pointer.

force exerted would be adequate to shake the mountain and perhaps rend it asunder.*

395. Although the weight of a given quantity of water would not be altered by varying the shape of the vessel, yet the pressure which it exerts on the bottom of the vessel will be greater in proportion as the altitude of the mass is greater, and of course greater in a narrow vessel than in a wide one. If it be asked why the weight is not increased as the downward pressure is increased, the answer is, that the pressure in that direction is exactly counterbalanced by an equal pressure in the opposite direction.

OF SPECIFIC GRAVITY.

396. *The Specific Gravity of a body, is its weight compared with the weight of another body of the same bulk, taken as a standard.*

Water is the standard for all solids and liquids, and common air for gases. Therefore the specific gravity of a solid or a liquid body, is the ratio of its weight to the weight of an equal volume of water; and the specific gravity of an aëriform body, is the ratio of its weight to the weight of an equal volume of air. But a ratio is expressed by a vulgar fraction, whose numerator is the antecedent, and whose denominator is the consequent. If, therefore, the weight of a body is made the numerator, and the weight of an equal volume of water the denominator, the value of the fraction, that is, the quotient, will express the specific gravity of the body. Hence, the weight of a body being given, and being made the numerator, every process for finding the specific gravity, consists in finding for the denominator the weight of an equal bulk of water or air. The principles upon which the methods of doing this depend, are now to be explained.

397. *A body immersed in a fluid, loses as much weight, as is equal to the weight of an equal volume of the fluid.*

Fig. 180.

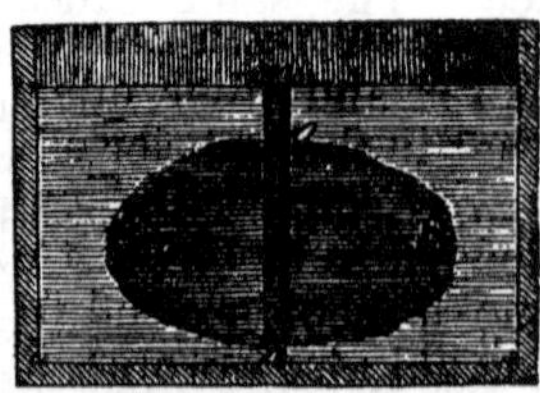

Let EF, (Fig. 180,) be a solid body immersed in a vessel of water or any fluid, and suppose it divided into an indefinite number of perpendicular columns reaching to the surface of the fluid, as *mon*. Now the upward pressure at *n* is as its depth, and the downward pressure at *o* as its depth; therefore the upward pressure exceeds the downward, by the weight of a column of water equal to *no*. The same is true of all the columns, however numerous they may be, that

* Library of Useful Knowledge, 'Hydrostatics.'

can be drawn parallel to *no;* but these columns, taken collectively, make up a body of water equal in bulk to the solid. Hence the solid is pressed upward more than downward, by the weight of a quantity of water of the same magnitude, and consequently loses so much of its weight. Hence the specific gravity of any solid body that will sink in water, is found by the following

RULE.—*Divide the weight of the body by its loss of weight in water.*

398. When the body whose specific gravity is required is lighter than water, as a cork, for example, the object is still to find the weight of an equal bulk of water, since that will constitute the denominator, or divisor, as before. To ascertain this, suspend any heavy body, as a mass of lead or glass, in water, and find its weight. Attach to it the lighter body. Now the cork will not only lose its own weight, but will diminish the weight of the heavy body; and the weight of an equal bulk of water will be indicated by the whole of what the cork loses, namely, its own weight added to the loss occasioned to the other body. Whence we have the following

RULE.—To find the specific gravity of a body lighter than water, *Divide its weight by the sum of its weight added to the loss of weight which it occasions in a heavy body previously balanced in water.*

399. A solid which is soluble in water, as a lump of salt, is protected from solution by covering it with oil or a thin coat of beeswax; and solids that are very porous and would absorb water, and thus increase their specific gravities, as certain kinds of wood, are first covered with varnish.* The specific gravity of solid substances, which are too minutely divided to be weighed in water, separately, as grains of sand or shot, may be found by weighing them in a small bucket previously balanced in water

400. The specific gravity of *liquids* may be ascertained by several different methods.

RULE I.—*Weigh equal volumes of the liquid and of water, and divide the former result by the latter.*

RULE II.—*Ascertain the loss of weight of any solid body first in the liquid and then in water, and divide the former result by the latter.*

Both these rules obviously depend upon the same principles as those explained in Art. 396, the weight of the liquid being immediately compared with that of an equal bulk of water; but there is another method, founded on the following proposition.

---

* Cavallo, I, 212.

401. *Two columns of fluids of different specific gravities, pressing freely on each other at their bases, balance one another when their heights are inversely as their specific gravities.*

Let AB, (Fig. 181,) be a recurved tube, and let the height of the column of the fluid B be as much greater than that of A, as the fluid B is lighter than the fluid A; the two columns will then be in equilibrio.

Fig. 181.

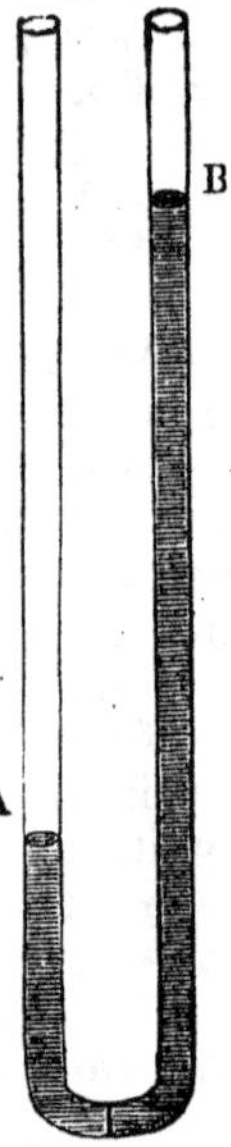

If the tube be of uniform bore throughout, then the proposition is manifestly true, because the quantities of matter pressing on each other in opposite directions will be equal and will have equal momenta; but from the peculiar nature of fluids, (Art. 391,) the opposite pressures will be the same, when the heights of the columns are the same, whatever may be the shape or capacity of the tube. If we introduce mercury into one arm of the tube and water into the other, the graduated scale will indicate that the water stands 13½ times as high as the mercury. Therefore the specific gravity of mercury is 13½. Proof spirit will stand at .923; sweet oil at .915; and their specific gravities are the same, water being 1.

402. These comprehend the most common methods of determining the specific gravities of bodies; but when great accuracy is required, the process of finding the specific gravity becomes one of much delicacy. The balance must be very sensible; the body, when suspended in water, must hang by a fine thread or hair; and the water must be of a standard temperature, since it alters its density by a change of temperature, so that the quantity of water which we wish to find, (namely, a quantity equal in magnitude to the given body,) will weigh more when the water is cold and less when it is warm.*

As expeditious methods of finding the specific gravity are sought for in commerce and the arts, various *instruments* have been invented for this purpose, which are called in general *hydrometers.* But before we can understand the principles on which these instruments depend, we must attend to the theory of floating bodies, so far as it is contained in the following propositions.

403. *If a body floats on a fluid, it displaces as much of the fluid as is equal to its own weight.*

---

* The standard temperature is not universally agreed on. The temperature of 60° is usually understood, at which a cubic foot of water weighs 62.353 lbs. or a little less than 1000 ounces. A better standard is that of 40°, at which temperature the weight of a cubic foot is still nearer to 62½ lbs. or 1000 ounces.

For the body P+Q* is supported by the pressure of the fluid, upward against the part immersed. But before the body was immersed, the same pressure supported a quantity of the fluid which occupied the same space and was therefore of the same magnitude with Q; and since the same pressure must sustain the same weight, when there is an equilibrium, the weight of the body must be equal to the weight of a quantity of the fluid equal in bulk to Q.

This proposition is also easily established by experiment; for if into a vessel full of water a floating body, as a piece of wood, be introduced, the quantity of water displaced will be found to be exactly equal in weight to the body. Or if the vessel full of water be accurately balanced in a scale, and then removed and the piece of wood introduced, on restoring the vessel to the scale it will still remain in equilibrium, the wood exactly compensating for the water it displaced.

404. *If a body floats on a fluid, the part immersed bears the same ratio to the whole body, as the specific gravity of the body bears to that of the fluid.*

Let S be the specific gravity of the fluid, and s that of the solid; then the absolute weight of the body (P+Q)* will equal $(P+Q)\times s$,† and the weight of a quantity of the fluid equal to Q is $Q\times S$; and by Art. 403, these weights are equal to each other, that is, $(P+Q)\times s=Q\times S \therefore Q : P+Q :: s : S$.

405. *A floating body will be at rest, only when its center of gravity is in the same vertical line with the center of gravity of the part of the fluid displaced.*

As the space from which the fluid has been displaced, exactly copies the figure of the part of the body immersed, the center of gravity of the displaced fluid is the same with that belonging to the figure of the segment. Let this center of gravity be at C, (Fig. 182,) while that of the entire body is at G. Now the fluid,

Fig. 182.

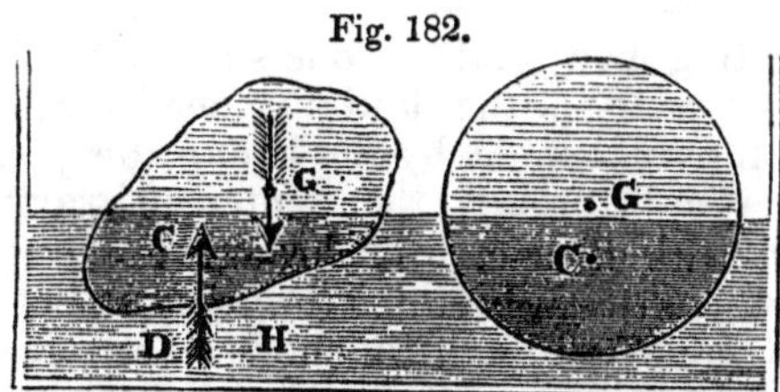

previous to its removal, was sustained by an upward force equal

* Q represents the part immersed; P the other part.

† s denotes the specific gravity of the body, water being 1; and $s=\frac{W}{P+Q}$, (Art. 396,) $\therefore W=(P+Q)\times s$.

to its own weight, acting in the vertical line DC; and the same upward force now acts upon the floating body at the point C. But the body, being free to move, is carried downward by a force acting in the direction of the vertical line GH. Were these two forces exactly opposite and equal, they would keep the body at rest; but this is the case only when the points C and G are in the same vertical line: in every other position of these points the two parallel forces tend to turn the body round. (Art. 70.)

When the floating body is a regular solid, as a sphere, the space occupied by the immersed segment being always the same, its center of gravity is immovable. If the body is of uniform density, so that the center of gravity coincides with the center of magnitude, it will remain at rest in every position; for the center of gravity of the whole body and of the segment, will always be in the same vertical diameter. But if the sphere is of unequal density, then it will turn around the center of gravity of the segment as around a fixed point, and be at rest only when its center of gravity is either directly above or directly below that of the segment.

Persons have sometimes attempted *to walk on the water*, by attaching to the feet inflated bladders or air bags of India rubber. The body can be kept from reversing its position, only by maintaining its center of gravity directly over that of the fluid displaced,—a point of such delicacy, as to require great skill and dexterity. The feat is rendered easier by carrying a staff having a blown bladder at the end of it, by which three points of support are commanded, and consequently a greater breadth of base is secured. Most persons, however, in attempting this feat, would shortly find their heads downward, and their feet out of water.

Life preservers, consisting of air bags attached to the waist, act on the foregoing principles, and, in numerous instances, have been instrumental in saving the lives of shipwrecked mariners.

405′. When a floating body rolls on one side, so as slightly to disturb the center of gravity of the fluid displaced, (or that which occupied the space now occupied by the body,) the point of intersection of the vertical through the center of gravity of the fluid displaced, with the vertical through the center of gravity of the body when at rest, is called the *Metacenter*. Let Fig. 183 represent a vessel turned a little on one side; and let G be the center of gravity of the vessel, AB the surface of the water, and C the center of gravity of the fluid displaced. When the vessel floated at rest, GC was vertical; but when it rolled slightly on one side, GC becomes inclined to the vertical in a small angle, and the center of gravity of the fluid displaced, shifts its position tc C′. Let the vertical through C′ meet GC in M, then M is the

Fig. 183.

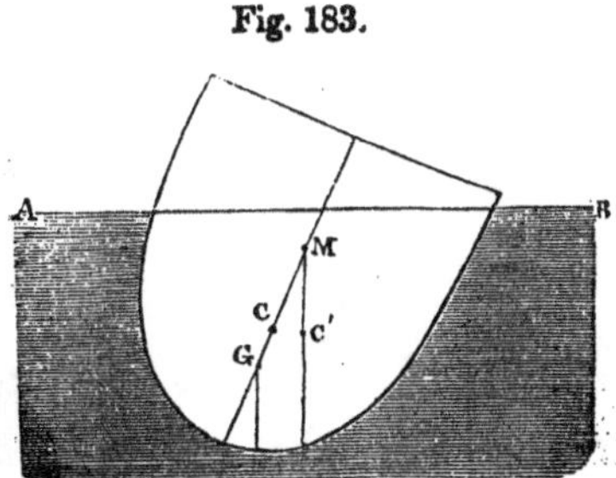

Metacenter. Now the weight of the vessel acts downward through G, while the upward pressure of the water acts upward in the direction of C′M, and both forces tend to *right* the vessel, or to bring it back to its original position. Hence, when the Metacenter is above the center of gravity of the vessel, the equilibrium is stable, and the vessel readily and forcibly recovers its fixed position. But suppose M and G to change places, then these two forces will conspire to turn the vessel over still more, and the equilibrium is unstable. If the Metacenter coincides with the center of gravity, the equilibrium is stable in any position of the vessel. Balloons in rising from the ground forcibly right themselves in accordance with the foregoing principle.

Hence the necessity of having not only the heavier parts of a ship's cargo stowed at the bottom of the vessel, but also of having the vessel ballasted, or the keel heavily laden, is apparent. For the masts and rigging may raise the center of gravity of the vessel above the center of gravity of the water displaced, in which case the vessel will be very liable to upset. The danger from the rolling of large vessels arises from the liability of parts of the cargo to shift, in which case the equilibrium may cease to be stable; and the danger of standing up in a small boat is quite apparent, for the elevation of the body will certainly raise the center of gravity of the floating mass above the center of gravity of the water displaced.*

406. The Hydrometer, an instrument used in commerce for determining at once the specific gravity of liquors, depends on the principle enunciated in Art. 404.

The most common hydrometer is represented in Fig. 184. It consists of a hollow ball, with a long slender stem, and since it will sink until it has displaced a quantity of the fluid equal in weight to itself, it will of course sink to a greater depth the lighter the fluid. From the depths to which it sinks, therefore, as indicated by the graduated stem, the corresponding specific gravities are estimated according to the formula in Art. 404, and

* Webster's "*Principles of Hydrostatics,*" p. 59.

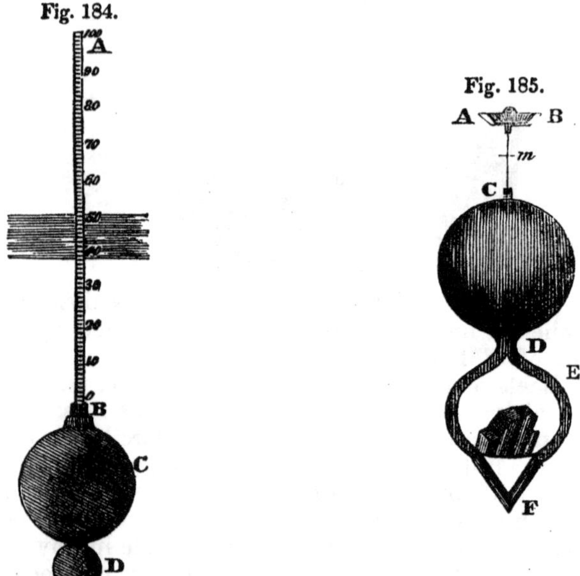

arranged in a table. *Nicholson's Hydrometer*, (Fig. 185,) is the most useful of this class of instruments, since it may be applied to finding the specific gravities of solid bodies. In addition to the hollow ball of the common hydrometer, it is furnished with a dish AB for receiving weights, and a stirrup EF for holding the substance under trial. The instrument is so adjusted that when 1000 grains are placed in the upper dish, the whole will sink in distilled water at the temperature of 60° Fahr. to the point *m* in the middle of the stem. If, therefore, in the case of different *fluids*, we add weights until *m* stands at the level of the fluids, the quantities displaced will be the same in all, and the weight of each quantity will be that of the instrument W together with the weight *w* added to the dish, which will constitute the *numerator* of the fraction expressing the specific gravity, (Art. 396,) while W+1000 expresses the *denominator*, being the weight of an equal bulk of water; and since the weight of the instrument is known, the only thing to be determined in any case is the weight to be added, in order to sink the point *m* to the level of the fluid. Then $\frac{W+w}{W+1000}$=the specific gravity.

To determine the specific gravity of a *solid* body not exceed ing 1000 grains in weight, we first place the instrument in water, put the body into the dish, and add weights to sink the point *m* to the level of the water. Then, since the weight of the body itself and the weight added, together equal 1000 grains, the weight of the body=1000—the weight added. Now we transfer the body to the stirrup EF, and ascertain what weight must be added to the dish to sink it to the same level as before. This weight wil. show the loss of weight of the body in water; and the weight in air, divided by the loss of weight in water, gives the specific gravity. As the cylindrical stem of the instrument is only one fortieth of an inch in diameter, the instrument will rise or fall nearly one inch by the subtraction or addition of one tenth of a grain. It will therefore indicate a change of weight less than one twentieth of a grain, and give the specific gravity correct to five places of decimals.*

407. An accurate knowledge of the specific gravities of bodies is of great use for many purposes of science and the arts, and they have therefore been determined with the greatest possible precision. The heaviest of all known substances is platina, whose specific gravity, in its state of greatest condensation, is 22,† water being 1; and the lightest of all ponderable bodies is hydrogen gas, whose specific gravity is .073, common air being 1. Also, air is 818 times lighter than water. Hence, by calculation, it will be found that platina is about 247,000 times as heavy as hydrogen, and a wide range is allowed to the various bodies which lie between these extremes. The metals, as a class, are the heaviest bodies; next to these come the metallic ores; then the precious gems; and finally, minerals in general, animal, liquid, and vegetable substances, in order, according to the following table.

Metals, (pure,) not including the bases of the alkalies and earths, from - - - - - - - - - 5 to 22

Gold,	19.25	Steel,	7.84
Quicksilver,	13.58	Iron,	7.78
Lead,	11.35	Tin,	7.29
Silver,	10.47	Zinc,	7.00
Copper,	8.90		

Metallic Ores, lighter than the pure metals, but usually above - - - - - - - - - 4.00‡

Precious Gems, as the ruby, sapphire, and diamond, 3—4

Minerals, comprehending most stony bodies, - - 2—3

Liquids, from ether highly rectified to sulphuric acid highly concentrated, - - - - - - - $\frac{3}{4}$—2§

Acids in general, heavier than water.

Oils do. lighter; but the oils of cloves and cinnamon are heavier than water; the greater part lie between .9 and 1. - - - - - - - - .9—1

* Ed. Encyc. X, 781.

† Ib. 794.

‡ In a few instances, metallic ores, when largely combined with foreign ingredients, fall below 3.

§ Ed. Encyc.

Milk, - - - - - - - - - - - - -	1.032
Alcohol, (perfectly pure,) - - - - - - - -	.797
Do. of commerce, - - - - - - - - -	.835
Proof Spirit, - - - - - - - - - - -	.923
Wines; the specific gravity of the lighter wines, as Champagne and Burgundy, is a little less, and of the heavier wines, as Malaga, a little greater than that of water.	
Woods, cork being the lightest, and lignum vitæ the heaviest, - - - - - - - - - -	$\frac{1}{4}$ to $1\frac{1}{3}$*

408. If we balance, in a pair of scales, a tumbler filled with water, to a certain mark near the top, and then, turning out all the water except a small quantity, introduce any solid body, (as a tumbler a little less than the first,) so as to raise the water on the sides to the same mark as before, the equilibrium will be restored. Here, the space occupied by the solid immersed, is the same with that before occupied by the water. On the same principle, a ship is floated in a dock with a very small quantity of water, and still rides as freely as on the ocean. By the ascent of the water on the sides, the upward pressure on the bottom is increased, on the same principle as in the Hydrostatic Paradox, (Art. 393.) Though, in this case, we cannot say that a quantity of water is *displaced* equal in weight to the solid, (since the whole of the water originally in the dock may not have been nearly sufficient to fill the space occupied by the ship,) yet the effect is the same, in regard to the pressure on the water below the ship, and of course on the upward pressure, (Art. 381,) as though the space occupied by the ship below the level of the fluid on its sides, were filled with water. On this principle, the weight of a loaded boat in the lock of a canal is easily estimated.

Boats are sometimes made of iron instead of wood, their thickness being so much less, that the entire weight of the boat is not greater than when made of wood.

409. The human body, when the lungs are filled with air, is lighter than water, and but for the difficulty of keeping the lungs constantly inflated, it would naturally float.† With a moderate degree of skill, therefore, swimming becomes a very easy process, especially in salt water. When, however, a man plunges, as divers sometimes do, to a great depth, the air in the lungs becomes compressed, and the body does not rise except by muscular effort. The bodies of drowned persons rise and float after a few days, in consequence of the inflation occasioned by putrefaction.

---

* The lightest kinds of wood, are poplar, pear, and sassafras, all of which are below .5; the heaviest are ebony, cedar, mahogany, oak-heart, box, and pomegranate, which are heavier than water. Wood increases in density by age. Knots have been known to reach the specific gravity of 1.76.

† See Dr. Franklin's remarks on this subject, in his Works.

Quadrupeds swim much more easily than man, because the motions of the limbs necessary to sustain themselves, nearly coincide with their natural motions in walking, while the body maintains nearly its usual posture.

410. *If a body is held beneath the surface of a fluid, the force with which it will ascend, if it is lighter than the fluid, or with which it will descend, if it is heavier, is equal to the difference between its own weight and the weight of an equal quantity of the fluid.*

Let s be the specific gravity of the solid and S that of the fluid, and M the weight of an equal bulk of water. Now the body held beneath the water obviously descends with its own weight$=M \times s$,* while it is pressed upward with the weight of the quantity of fluid displaced$=M \times S$; consequently, the force with which it ascends must be $(M \times S)-(M \times s)$, and the force with which it descends is $(M \times s)-M \times S$, which are the differences between the weight of the body and the weight of the fluid displaced.†

411. On the foregoing principle is founded the construction of a machine called the Camel, for raising sunken vessels, or for lifting ships over sand banks. A similar effect is exhibited in rivers, where the ice is formed upon the stones at their bottom. Ice is specifically lighter than water, and therefore when it accumulates to a certain degree round the stones, the upward pressure upon the stones exceeds their pressure downward, and they are brought to the surface, having been sometimes torn up with great force. Huge masses of stones appear in many cases to have been floated by the ice adhering to them, and carried to a great distance from the place of their formation.‡

412. Rocks and stones being only a little more than twice as heavy as water, of course nearly half their weight is sustained while they are immersed in water; and hence the increased weight which is felt when a large stone is lifted from the bed of a river, as soon as it reaches the surface. Large masses of rocks are transported with far greater facility by torrents, on account of their diminished weight. On the same principle, the limbs feel very heavy after lying for some time in a bath. Life-boats have a large quantity of cork mixed in their structure; or of air-tight vessels of thin copper or tin plate, so that, even when the boats are filled with water, a considerable part still floats above the surface.§

When light bodies are attached to heavier for the purpose of making them float, some caution is necessary, since, on account of the tendency of the center of gravity to seek the lowest point,

* For $s=\frac{W}{M} \therefore W=M \times s$. † Ed. Encyc. X, 722. ‡ Ib. § Arnott

there will be danger of upsetting. Thus, persons endeavoring to swim by attaching to their persons blown bladders, sometimes have their heads turned downward; and a man undertaking to walk on the water in a pair of cork boots, would shortly disappear, and nothing would be seen except the boots sticking out of the water.* (Art. 405.)

413. The *magnitude* of bodies may often be most conveniently and accurately estimated from the doctrine of specific gravities. Suppose we wish to ascertain the exact number of solid inches contained in a stone of rude and irregular shape, we should find great difficulty in applying to it any linear measurement; but if we ascertain its loss of weight in water, then we have the weight of an equal bulk of water, and since 1000 ounces contain 1728 cubic inches, we may easily find how many cubic inches correspond to the weight of water of equal magnitude with the body in question.

We may estimate the quantity of ice in an island of ice, such as occur in the northern seas, by applying the principle demonstrated in Art. 404.

414. As examples of the manner in which questions are solved by the principles of Hydrostatics, we subjoin a few problems of the more difficult class, with their solutions, to which we shall add a variety of questions as an exercise for the learner.

PROBLEMS.

1. What is the weight of a chain of pure gold, which raises the water 1 inch in height, in a cubical vessel whose side is 3 inches? and suppose a chain of the same weight were adulterated with $14\frac{1}{2}$ oz. of silver, how much higher would it raise the water in the vessel?

The contents of the chain are $3\times3\times1=9$ cubic inches. Since a cubic foot or 1728 inches of distilled water weigh 1000 oz.† avoirdupois, and since 192 oz. avoirdupois=175 troy, in which gold and silver are weighed; therefore, 1728 inches of water weigh 911.458 oz. troy; or 1 cubic inch weighs .5274 oz. The specific gravity of gold being 19.25, and that of silver being 10.47, therefore 1 *inch of gold weighs* 10.15 *oz.* and 1 *inch of silver weighs* 5.52 *oz.* Hence the chain weighs 91.35 oz. Again,

10.15 oz. : 1 in. : : 14.5 oz. : 1.428=cub. inch. of gold in 14.5 oz.
5.52 : 1 : : 14.5 : 2.626= do. silver do.

$\therefore$ $9-1.428+2.626=10.198$ = column of water raised by the

* Arnott.

† By act of Parliament, 1825, 19 cubic inches of distilled water, temp. 50° Fahr., weigh 10 oz. troy. These numbers may be used instead of those in the text

chain; and $\frac{10.198}{9}=1.133$=whole height, and $1.133-1=.133$. Ans.

2. The specific gravity of lead being 11.35; of cork, .24; of fir, .45: How much cork must be added to 60 lbs. of lead, that the united mass may weigh as much as an equal bulk of fir?

Let x=weight of the cork in pounds.

Then $60+x$=weight of united mass=weight of an equal bulk of fir. $\therefore \frac{60+x}{.45}$=do. of an equal volume of water.

And $\frac{60}{11.35}+\frac{x}{.24}$=the same, and therefore$=\frac{60+x}{.45}$.

Hence, $x=65.8527$ lbs. Ans.

3. A cone, whose height is 6 inches, is immersed in water, with its vertex downward. Its specific gravity being $\frac{1}{8}$th that of the water, what part of the axis will be immersed?

Let x=part of the axis immersed:

Then $x^3 : 6^3 :: 1 : 8 \therefore x=\left(\frac{216}{8}\right)^{\frac{1}{3}}=\frac{6}{2}$=*one half the axis.* Ans.

4. A sailor had a 10 gallon cask half full of brandy, and not being allowed to keep liquors on board, he threw it overboard for the purpose of concealment, attaching to it a mass of lead just sufficient to keep it under water: Required *the weight of the lead,* the wood of the cask containing 216 cubic inches, the specific gravity of the brandy being .886, that of sea water 1.030, that of the lead 11.35, and that of the wood .8?

The whole quantity of water displaced$=231\times10+216=2526$ inches; and $231\times5=1155$=quantity of brandy.

Then, 1728 : 1030 :: 2526 : 1505.66=wt. of water.
1728 : 886 :: 1155 : 592.2 =wt. of brandy.
1728 : 800 :: 216 : 100. =wt. of wood.

Then $1505.66-(592.2+100=692.2)=813.46$=weight of the water above that of the brandy and cask, which is the buoyant power to be counteracted by the lead.

Now, how much lead will it take to weigh 813.46 in water?

Let x=wt. of lead required; then $\frac{x}{11.02*}$=weight of an equal bulk of water. Therefore $x-\frac{x}{11.02}=813.46$, and $x=894.64$ oz. or *nearly* 56 lbs. Ans.

QUESTIONS ON HYDROSTATICS.

1. In a hydrostatic press, (Fig. 170,) the height of the small column AB on which the power acts is 2 feet above the bottom

* The weight of lead above sea water.

of the large piston CD; the diameter of the cylinder AB is 1 inch, and of the cylinder CD 1 foot. By means of a screw turned by a lever, a man can exert a force on A equal to 500 pounds: What amount of pressure can he apply with the aid of this press, combining his own strength with the pressure of the column of water AB?* *Ans.* 72098.17 *lbs.*

2. A junk bottle, whose lateral surface contained 50 square inches, was let down into the sea 3000 feet: What pressure would the sides of the bottle sustain, no allowance being made for the increased specific gravity of the sea water?

Ans. 65104.166 *lbs.*

3. A Greenland whale sometimes has a surface of 3600 square feet: What pressure would he bear at the depth of 800 fathoms?

Ans. 1080,000,000 *lbs., or more than* 482,142 *tons.*

4. A mill-dam, running perpendicularly across a river, slopes at an angle of 25 degrees with the horizon. The average depth of the stream is 12 feet, and its breadth 500 yards: Required the amount of pressure on the dam?

Ans. 15971906 *lbs., or* 7130+ *tons.*

5. A mineral weighs 960 grains in air, and 739 grains in water: What is its specific gravity? *Ans.* 4.344.

6. What are the respective weights of two equal cubical masses of gold and cork, each measuring 2 feet on its linear edge?

Ans. The gold weighs 9625 *lbs.*=4.297 *tons; the cork weighs* 120 *lbs.*

7. On one of the peaks of the Alps, is a single mass of granite rock of a nearly globular shape, which is estimated by measure to contain 5949 cubic feet. The whole mass is so nicely balanced on its center of gravity, that a single man may give it a rocking motion. By trial made upon a small fragment, its specific gravity was found to be 2.6: What is its weight?

Ans. 431.568 *tons.*

8. An island of ice rises 30 feet out of water, and its upper surface is a circular plane, containing ¾ths of a square acre. On the supposition that the mass is cylindrical, required its weight, and depth below the water, the specific gravity of sea water being 1.0263, and that of ice .92?

Weight 242900 *tons; depth* 259.64 *feet.*

9. Wishing to ascertain the exact number of cubic inches in a very irregular fragment of stone, I ascertained its loss of weight in water to be 5.346 ounces: Required its dimensions?

Ans. 9.238 *cubic inches*

10. Hiero, king of Syracuse, ordered his jeweller to make him a crown of gold containing 63 ounces. The artist attempted a fraud by substituting a certain portion of silver; which being

* It is obvious that the elevation, and consequently the pressure, of this column would be continually diminishing. The question respects only the commencement of the process.

suspected, the king appointed Archimedes to examine it. Archimedes, putting it into water, found it raised the fluid 8.2245 inches; and having found that the inch of gold weighs 10.36 ounces, and that of silver 5.85 ounces, he discovered what part of the king's gold had been purloined: It is required to repeat the process? *Ans.* 28.8 *ounces.*

CHAPTER II.

OF LIQUIDS OR NON-ELASTIC FLUIDS IN MOTION.

415. THAT branch of Natural Philosophy which treats of fluids *in motion*, is usually denominated *Hydraulics*, from ὕδωρ, *water*, and αὐλὸς, *a torrent.* It embraces, therefore, the phenomena exhibited by water issuing from orifices in reservoirs—projected obliquely or perpendicularly—flowing in pipes, canals and rivers, —oscillating in waves—or opposing a resistance to the progress of solid bodies.*

In this part of the doctrine of fluids, the deductions of *theory alone* are of little value, and are in fact so discordant with experience, that little reliance can be placed on them, except as modified by experiment. When thus modified, the truths learned respecting the laws that govern the motions of fluids, have a high degree of practical utility.

The causes of this discordance between theory and experiment are such as the following: the different states in which the same fluid is found in consequence of changes of temperature, and different degrees of purity—the attraction existing between its particles and preventing that perfect fluidity which is contemplated by the definition, and upon which our reasonings are founded—the friction against the sides of the vessel—the resistance of the air—the size of the vessel in proportion to the aperture—the shape of the aperture itself—the different directions in which the various parts or *filaments* (as they are called) of the fluid run toward the aperture—and the vortices, or irregular motions which are communicated to the fluid by a variety of causes.†

416. The manner in which vessels of water discharge themselves through small orifices in the bottom or sides of the vessels, has been carefully observed by introducing into a column of water contained in a glass vessel, small solid particles, which render the currents of the fluid visible. From such observations it

* Ed. Encyc., Art. *Hydrodynamics.*

† Cavallo, I, 267.

appears, that the particles of fluid descend in vertical lines, until they arrive within three or four inches of the orifice, when they gradually turn into a direction more or less oblique, and run directly toward the orifice. When the surface of the descending column comes very near to the orifice, a funnel-shaped hollow or cavity makes its appearance, and the various particles which rush toward the orifice, converge to a point without the orifice, *at the distance of the semi-diameter of the orifice itself*, where is the point of greatest contraction, called the *vena contracta.*

From the great number of propositions which contain the doctrine of hydraulics, we shall select those which appear to be most valuable on account of their practical utility, adding such remarks as will serve to show the modifications to which they are subject in practice.

417. *If a fluid runs through any tube, pipe, or canal, and keeps it constantly full, its velocity, in any part of its course, will be* INVERSELY AS THE AREA OF THE SECTION *at that part.*

Let A and a be the areas of two cross sections of a tube of unequal bore, and let V and v denote the velocities of the fluid as it flows through A and a respectively. Now the quantity of fluid which passes through any section, must depend upon the area and velocity conjointly; and since the tube is, by the supposition, kept constantly full, the same quantity runs through every section of the tube in any given time. Hence $A \times V = a \times v$ $\therefore A : a :: v : V$; that is, the velocity is inversely as the area of the section.

It follows from the foregoing proposition, that the velocity of a stream increases as either the breadth or the depth decreases. In a tube, the parts near the axis move with greater velocity than near the sides; and in every canal or river, the velocity of the stream is greater in the central parts, or channel, than at the sides, and greater at the surface than at the bottom. The *mean velocity*, therefore, is to be inferred from the combination of at least three distinct measurements. For example, the velocity of a stream was found to be

On the surface of the channel, - -	5	miles per hour.
At the bottom, - - - - - - - -	3	" "
At the sides, - - - - - - - - - -	$3\frac{1}{2}$	" "

Therefore, the mean velocity $= \frac{5+3+3.5}{3} = \frac{11.5}{3} = 3.83$ miles per hour.

It is important to avoid all unnecessary expansions as well as contractions, in pipes or canals, since there is always a useless expenditure of force in restoring the velocity which is lost in the wider parts.*

* Young's Lectures on Nat. Phil., I, 283.

418. To find the *quantity* of water which flows in a river, we may first ascertain the area of a section by taking the depth in various parts of the section, and dividing the sum of all the depths by the number of measurements, which gives us the mean depth. This multiplied into the breadth of the stream, gives the area of the section, which multiplied into the average velocity, (Art. 417,) shows the quantity required.

Example.—Wishing to know the quantity of water that discharged itself per minute through a certain strait, I chose a place below the strait where the water flowed at a nearly uniform rate, where, by noting the time occupied by light substances in floating through various parts of the stream for a given distance, I found the average velocity to be 96 feet per minute. The mean depth of a section was $8\frac{1}{2}$ feet, and the breadth 560 feet. Hence,

$$8\tfrac{1}{2}\times560\times96=456960 \text{ cubic feet per minute.}$$

419. The phenomena of RIVERS have sometimes been explained on the supposition that rivers are bodies falling freely down inclined planes. But the conclusions deduced from this doctrine are so at variance with experience, as to be of no value.* Were every part of the bed of a river uniform, like a tube, the channel or portion which moves with the greatest velocity, would be in the center of the surface; but inequalities in the sides and bottom usually throw it out of the center, and incline it to one side or the other. The increased velocity of a stream during a freshet, while the stream is confined within its banks, exhibits something of the acceleration which belongs to bodies falling freely down an inclined plane. It presents the case of a river flowing upon the top of another river, and consequently meeting with much less resistance than when it runs upon the rough unequal surface of the earth itself. The augmented force of a stream in a freshet, arises from the simultaneous increase of the quantity of water and the velocity. In consequence of the friction of the banks and beds of rivers, and the numerous obstacles they meet with in their winding course, their progress is very slow, whereas, were it not for these impediments, it would become immensely great, and its effects would be exceedingly disastrous. A very slight declivity is sufficient for giving the running motion to water. Three inches per mile, in a smooth, straight channel, gives a velocity of about three miles per hour. The Ganges, which gathers the waters of the Himalaya Mountains, the loftiest in the world, at the distance of eighteen hundred miles from its mouth, is only eight hundred feet above the level of the sea,—that is, about twice the height of St. Paul's church in London; and to fall these eight hundred feet, in its

* See Robison's Mech. Phil., or Encyc. Brit., Art. *Theory of Rivers.*

long course, the water requires more than a month. The great river Magdalena, in South America, running for a thousand miles between two ridges of the Andes, falls only five hundred feet in all that distance.* The Croton aqueduct, that waters the city of New York, falls one foot per mile. The motion of rivers soon becomes uniform, even in very great inclinations, the sum of the resistances forming an equilibrium with the acceleration produced by the descent on an inclined plane. The smallest inclination capable of giving motion to water is a descent of one foot in a million; or about $\frac{1}{15}$th of an inch per mile. A fall of 3 feet per mile, makes a mountain torrent.†

420. *The velocity with which a fluid issues from a small orifice in the bottom or side of a vessel, kept constantly full, varies as the* SQUARE ROOT OF THE DEPTH *below the surface of the fluid.*

The *pressure* at different depths varies as the depth, (Art. 385;) also the *momentum* is as the force impressed, and likewise therefore as the depth. Hence, let Q, Q′, denote the quantities of fluid discharged at each orifice respectively; V, V′, the corresponding velocities, and AB, AC, the depths. Then, since the momentum is as $Q \times V$, $Q \times V : Q' \times V' :: AB : AC$.

But, through a given orifice, the quantity of fluid discharged evidently varies as the velocity, or $Q \propto V$; hence,

$$V^2 : V'^2 :: AB : AC, \text{ or}$$

$$V : V' :: \sqrt{AB} : \sqrt{AC}.$$

It appears, therefore, that the fluid issues with the velocity which a body would acquire by falling freely from the surface of the fluid to the orifice, for that is also as the square root of the space. (Art. 29.) And since a body when projected upward with a certain velocity, will rise to the same height as that from which it must have fallen to acquire that velocity, consequently, if a fluid issue from the side of a vessel through a spout bent upward, it would, were it not for the resistance of the air, and friction at the orifice, rise to the level of the fluid in the reservoir.

It follows from the foregoing proposition, that an orifice sixteen inches from the surface will discharge twice as much fluid in a given time as one at the depth of four inches; and this is conformable to the law that the pressure varies as the depth; for since twice the quantity flows with twice the velocity, of course the pressure or momentum is four times as great at the depth of sixteen as at that of four inches. In order, therefore, to draw off from a cistern four times as much water as before, we must place the gate sixteen times as deep. A gate opened in a reservoir at the depth of sixty four inches, will discharge only four times as much as at the depth of four inches.

421. In the construction of water works it is customary to con-

* Arnott's El. Phys., I, 260. † Rennie, *Reports of Brit. Assoc.*, 1834.

duct the stream, or such part of it as is required, into a cubical cistern, and to let it issue from the side of this, near to the bottom, and thus fall upon the main wheel. Instead of admitting the water to the wheel in this manner, it has sometimes been supposed that an advantage might be gained by letting the water fall down a height equal to that of the top of the cistern, perpendicularly upon the wheel, on the supposition that we might thus avail ourselves of the force acquired by the water in falling. But according to the preceding proposition, the force would be the same whether the water issued from the cistern and thus applied itself to the wheel, or whether it fell upon the wheel from a height equal to that of the surface of the water in the reservoir above the orifice. This is true in *theory*, but in *practice* it would be found more advantageous to take the water out of the cistern, to avoid loss from the resistance of the air.

422. *If a cylindrical or prismatic vessel, of which the horizontal section is everywhere the same, is filled with fluid, and empties itself by an orifice, the velocity with which the surface descends, and also the velocity with which the water issues, is uniformly retarded.*

The velocity with which the surface descends is proportional to that with which the fluid issues from the orifice, and therefore is as the square root of the depth. (Art. 420.) But the velocities of bodies projected perpendicularly upward are in the same ratio to their spaces, (Art. 30,) and therefore a body projected perpendicularly upward is in the same relative circumstances as the descending surface of the fluid; and as the projected body is uniformly retarded, the same is true of the descending surface.

On this principle is constructed the *Clepsydra*, or water-clock. Since the descent of the surface is uniformly retarded, the spaces which it describes in equal times, reckoning from the bottom, are as the odd numbers, 1, 3, 5, 7, &c.; and if a cylindrical vessel of water be furnished with an orifice at the bottom which will exactly discharge the whole column in twelve hours, and the sides of the vessel be divided into spaces corresponding to the foregoing numbers, the successive heights of the column become measures of time.*

423. *If we accurately mark the time in which a cylindrical or prismatic vessel, whose horizontal section is everywhere the same, discharges itself to the level of a given orifice, and then draw off for the same time, keeping the vessel constantly full, we shall obtain double the quantity of fluid in the latter case as in the former.*

When the vessel is kept constantly full, the velocity at the orifice (and of course the quantity discharged) continues uni-

* An excellent description and representation of various forms of the Clepsydra may be found in the Encyclopædia Metropolitana, Art. *Hydrodynamics*

formly the same as at first; and since the circumstances of this case are exactly analogous to those of a body projected perpendicularly upward; and since if a body thus projected were to continue to ascend with the first velocity, it would pass over a space twice as great in the same time as when uniformly retarded; therefore, the truth of the proposition is manifest.

424. *A fluid spouting from the side of a vessel, describes the curve of a parabola.*

The fluid is precisely in the same circumstances as a projectile acted on by the force of projection (viz. the pressure of the incumbent fluid) and by the force of gravity. Therefore, according to Art. 48, it describes the curve of a parabola. As in the case of other projectiles, the proposition holds good, whatever may be the angle of elevation of the jet.

425. *If a semicircle be described on the perpendicular side of a vessel, as a diameter, and a fluid spout horizontally from any point in the diameter, its random will equal twice the length of the ordinate to that point.*

Fig. 186.

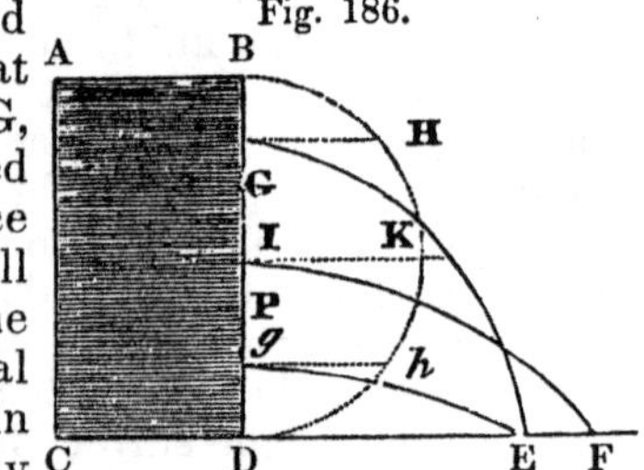

The velocity with which the fluid issues from the vessel, being that which is due to the height BG, (Fig. 186,) is such as if preserved would carry the jet through a space equal to 2BG, in the time of the fall through BG; but after leaving the orifice, it arrives at the horizontal plane in the same time as that in which a body would fall freely through GD. And since, in falling bodies, the times are as the square roots of the spaces, (Art. 29,) therefore, $\sqrt{BG} : \sqrt{GD} :: tBG : tGD$,* that is, the time employed in reaching the plane.† But in the time of describing BG, the jet would be carried uniformly and horizontally over a space equal to 2BG; therefore, to find how far it will proceed in a horizontal direction, while it is descending through GD, we have

$$\sqrt{BG} : \sqrt{GD} :: 2BG : DE=\frac{2BG\times\sqrt{GD}}{\sqrt{BG}}=2\sqrt{(BG\times GD)}=2GH,$$

that is, double the ordinate to the point G.

The greatest random will be when the fluid spouts from the center, for then the ordinate (and of course twice the ordinate) is

* *t* signifies the time of falling down the spaces BG and GD.

† For the jet will reach the plane in the curve, acted on by the two forces of projection and gravity, in the same time in which it would descend through the perpendicular height, urged by gravity alone.

greatest; and the randoms will be equal at equal distances above and below the center, for at these points the ordinates are equal.

426. When a fluid is contained in any vessel, it presses equally on opposite points of the vessel, and is thus maintained at rest. Now if we remove the pressure from one of the opposite points while it remains on the other, the force exerted on the latter will tend to move the vessel in that direction. Thus if we suspend a bottle of water, like a pendulum, and open a small jet on one side, the pressure on this side being taken off, but still remaining on the other, the bottle will swing toward the side opposite the orifice, and remain suspended at a certain height above its former position.

Barker's Mill acts on the foregoing principle.* Its construction is as follows: AB (Fig. 187) is a hollow cylinder movable about a vertical axis MN. PP′ is another cylinder placed at right angles to the former, and communicating internally with it. Near its extremities, which are closed, two apertures are made in the sides of this horizontal cylinder, opening in opposite directions. That at P is supposed to front the reader, and that at P′ to lie on the opposite side of the tube. Water is supplied by the spout O, keeping AB constantly nearly full. As the water flows out at P and P′, the unbalanced pressure on the sides of the cylinder opposite to those openings, acts on the respective arms PB, P′B, and sets the horizontal cylinder revolving, carrying along with it the perpendicular shaft MN, and any machinery connected with it. The power resulting from the pressure of a column of water is here applied to the best advantage, for since the arms of the horizontal shaft BP and BP′ may be lengthened at pleasure, while the power is still applied at the extremity of each, the circumstances are the same as when the power is applied to the end of a lever or the circumference of a wheel, and the power gains a similar advantage. Moreover, the *centrifugal* force acquired by the revolving fluid, being greatest at the extremities of the shaft, acts under a similar advantage and conspires with the simple pressure. This machine is said by writers on mechanics to be the most effective

Fig. 187.

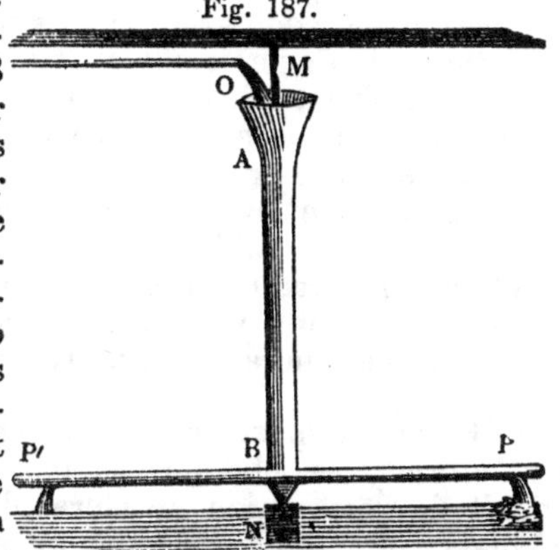

* Machines acting on this principle, are erroneously said to go by *reaction*.

known for applying the power of a given quantity, and a given fall of water, to the working of machinery.*

427. The term FRICTION is applied to the obstruction occasioned to the passage of fluids in the same manner as it is to solids; and it exists to such an extent as to become an object of considerable inconvenience in practice. It can be obviated only by making the conveying pipe of much larger dimensions than would otherwise be necessary, so as to allow the free passage of a sufficient quantity of fluid through the center of the pipe, while a ring or hollow cylinder of water is considered to be at rest all around it. Other circumstances beside friction likewise tend to diminish the quantity of fluid which would otherwise pass through pipes,—such as the existence of sharp or right-angled turns in them, permitting eddies or currents to be formed, or not providing for the eddies or currents that form naturally, by suiting the shape of the pipe to them. It follows, therefore, that whenever a bend or turn is necessary in a water-pipe, it should be made in as gradual a curve or sweep as possible; that the pipe should not only be sufficiently capacious to afford the necessary supply but should be of a uniform bore throughout, and free from all projections or irregularities against which water can strike, and form eddies or reverberations, since these will impede the progress of the fluid as effectually as the most solid obstacles.

428. Fluids, it must be recollected, are subject to the same law of gravity as solid bodies, and a mass of fluid descending vertically has its motion accelerated in the same manner as a solid mass; and the momentum generated is the product of its quantity of matter and velocity. If a column of water move through either a vertical or an inclined pipe, it acquires a velocity, which from the friction of the pipe will soon become uniform, and the momentum generated will be measured by the mass multiplied into this uniform velocity. Now force is also necessary for the destruction of motion, and the shorter the time through which it acts the greater is the effect produced. Thus a small hammer with a hard face is much more effective in driving a nail, than a mallet of twenty times its weight, and moved with the same velocity. For in consequence of the hardness of the face, the motion is destroyed instantly and is instantly received by the nail. By this means the momentum is confined to the nail, whereas were the motion communicated gradually, it would be diffused more or less over the body into which the nail is driven. (See Arts. 231, 250.) The sudden destruction of motion in a fluid mass is attended with effects precisely analogous. When the motion of a large mass of water is suddenly

* Mosely's *Mechanics applied to the Arts*, p. 231.

stopped, the surface which stops it sustains a very great force. The operation of this principle is seen when the gates of a lock are instantly closed, and when the stop-cock of an aqueduct which discharges a rapid jet of water, is suddenly shut. In the latter case, the violence is sometimes such as to burst the pipe nearest the opening. A powerful engine for raising water, called the *Hydraulic Ram*, acts on this principle.*

429. An unexpected facility is gained in the discharge of a fluid from the bottom or side of a vessel, by applying a *pipe to the orifice*. On account of the friction known to occur in the passage of a fluid through a tube, it might be supposed that a simple orifice made in the vessel might be more favorable to the discharge of the fluid than an opening prolonged by a tube; but it has been found by experiment, that a vessel of tin, with a smooth hole formed in its bottom, did not discharge water as rapidly as another containing the same weight of water, and an orifice of the same dimensions, to which a short pipe was applied. By varying the length of the pipe, it is found that when its length is twice its diameter, it produces the most rapid discharge, delivering in this case 82 quarts of water in 100 seconds, while the simple hole delivered but 62 quarts in the same time. If, however, the pipe projects into the vessel, the quantity discharged is diminished instead of being increased by the pipe.†

When water is conveyed through a straight cylindrical pipe of any length, the discharge of water may be increased by only altering the *shape of the terminations* of that pipe, viz. by making the end of the pipe which is close to the reservoir, or the entrance to it, of the conical shape of the *vena contracta*, (Art. 416,) and by making the other extremity of the pipe, where the water issues, of a trumpet-shape.‡ By this means, the quantity of water which is discharged in a given time, is more than doubled.§

WATER WHEELS.

430. Three kinds of water wheels are employed under different circumstances, namely, the *overshot*, the *undershot*, and the *breast wheel*.

The *overshot* wheel is used when the supply of water is scanty, since this construction admits of a more economical use of the

* See Webster's *Principles of Hydrostatics*, p. 145.

† Ed. Encyc., Art. *Hydrodynamics*.—Millington's Epitome of Nat. Phil., 181.

‡ Experience shows that the divergency of this termination must not exceed a certain degree, for in that case it will prove rather disadvantageous than useful. It appears that when the divergency is greater than an angle of 16 degrees, the effect ceases entirely; and that the greatest quantity of fluid is discharged, when the divergency is equal to an angle of about 3 degrees.—*Cavallo.*

§ Cavallo, I, 276. See Ewbank's Hydraulics.—Rennie's Report to the British Association, 1834.

water than either of the others. Fig. 188, represents a section of an overshot wheel at right angles to the axis. Its diameter is usually nearly equal to the whole fall of the water. It is placed under the head of water in such a way as to receive its whole force into buckets, connected with the rim of the wheel. These buckets are made of such a shape as to retain as much of the water as possible until they reach the lowest point of the wheel, but none at all after passing that point. By this means the weight of the water in the buckets is made to exert as much weight as possible on one side of the wheel, thus causing it to descend, while they oppose little resistance to the ascent of the opposite side of the wheel. Let us trace the effect of a single bucket in its revolution. Were it to receive the water directly on the top at H, the only effect would be to cause an increased pressure on the axis of the wheel, while it would not contribute to turn the wheel; and, indeed, within a certain distance from H towards *a*, the weight of the water increases the resistance from friction on the axle more than its force tends to turn the wheel. It is evident, therefore, that the water must begin to fall on the wheel so far toward the side, that its leverage, measured from O on the line OF, may enable it to overcome the friction and all other impediments. As the wheels revolve, the weight of the water acts with a constantly increasing leverage, until at F it acts with its greatest force. From this point, the force declines from two causes, namely, the loss of water from the buckets as their position is gradually reversed, and the diminution of the leverage or effective distance from O on the line OF, until, before it reaches the lowest point L, it may again slightly act as an impediment by increasing, from its weight, the friction on the axle more than it contributes to turning the wheel.

Fig. 188.

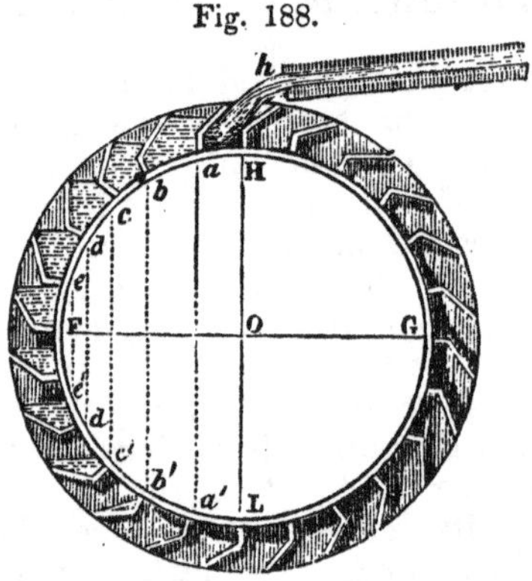

431. There is a certain velocity with which an overshot wheel should move in order to produce the greatest effect. If, on the one hand, the wheel is loaded so heavily that the weight of water is insufficient to move it, then of course the effect is nothing; and if, on the other hand, the velocity of the wheel were to equal that with which water would fall freely, then its pressure on the buckets would become nothing and its moving power nothing. The best velocity that can be given to an overshot wheel is found to be three feet per second.

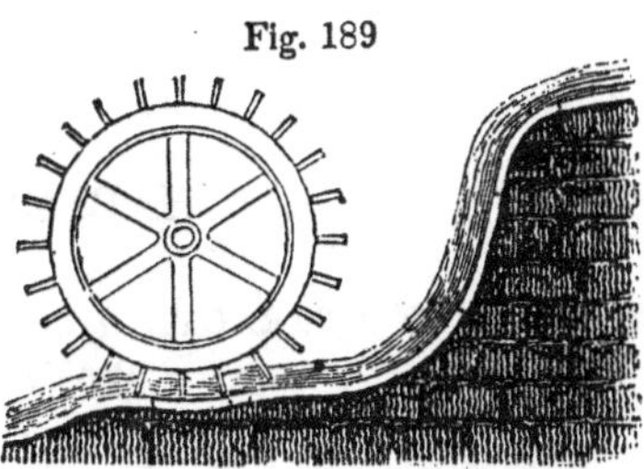
Fig. 189

432. The *undershot wheel* (Fig. 189) is carried, not by the weight of the water simply, as is the case in the overshot wheel,) but by the momentum or force of running water. Instead of close buckets for holding water, it is furnished with *float boards*, which receive the impulse of the stream. When these are placed, as in the figure, with their planes at right angles to the rim of the wheel, the latter may turn either way; and this, therefore, is the form of wheels employed in tide mills. When the wheel is required to turn only in one direction, an advantage is gained by placing the float boards so as to present an acute angle toward the current, by which means the water acts partly by its weight, as in the overshot wheel. The undershot wheel is adapted to situations where the supply of water is always abundant. It acts, moreover, with the greatest effect, when its velocity is half that of the stream.

Fig. 190.

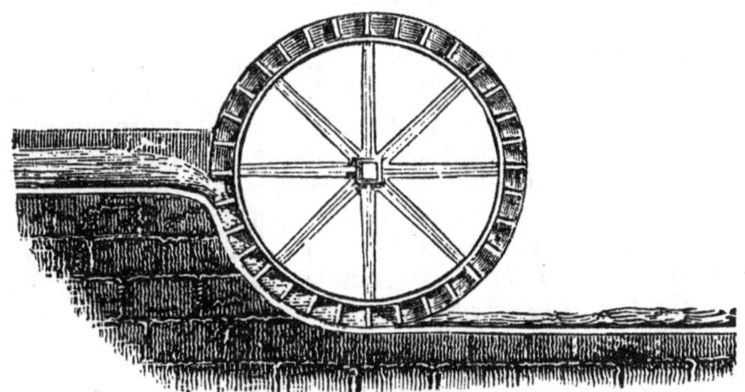

433. The *breast wheel* (Fig. 190) combines the advantages of both the others, and is therefore adapted to situations where the supply of water is generally sufficient, but not always abundant. The planes of the float boards are at right angles to the rim of the wheel, and are brought so near to the mill course that the float boards hold water like buckets.

According to Smeaton, the effect of overshot wheels, under the same circumstances of quantity and fall, is, at a medium, double that of an undershot wheel.

CHAPTER III

OF CAPILLARY ATTRACTION, RESISTANCE OF FLUIDS, AND WAVES

434. THE definition of a fluid, (Art. 376,) proceeds on the supposition that fluids are destitute of cohesion, and that their particles move among themselves without the slightest impediment. All *liquids*, however, have in fact more or less cohesion or mutual attraction among their particles. This is apparent in their forming drops, and in the viscidity of certain liquids, as oil and tar, which on account of this property are sometimes denominated *semi-fluids*. It is owing to this property that water so readily forms itself into drops, and that its surface when viewed in a small cup or wine glass, appears convex. Both of these properties are still more observable in quicksilver, which when poured on a table, forms numerous globules of a perfectly spherical figure; and the convex figure of the surface, as seen in a wine glass, is very striking. When we dip a glass tube into water, it comes out covered with drops of the fluid, which are held by the attraction of the glass for water; but the tube when dipped into quicksilver comes out dry, because the cohesion between the particles of quicksilver for one another is greater, than the mutual attraction that exists between the metal and the glass. Hence, a solid body, when immersed in a fluid, is sometimes wet by it and sometimes not, according as the attraction between the solid and the fluid is greater or less than that which exists between the particles of the fluid for one another.

435. If a disk or thin plate of almost any solid substance, as of glass or metal, be suspended from the arm of a balance and counterpoised, upon bringing it into contact with the surface of a fluid capable of wetting it, it will adhere with a considerable force, the amount of which will be indicated by the weights required to be added to the opposite scale, in order to detach it. This experiment shows that a lamina of water is held to a contiguous lamina by a strong force, and hence when a cause operates upon the surface of a fluid to draw up the lamina, a column of fluid may rise along with it, in consequence of the mutual cohesion of the successive laminæ. We see the same principle strikingly exemplified in viscid fluids, as tar, where, on drawing up a small portion of the surface, a column of the fluid follows it. The foregoing fact will lead us to an understanding of th causes of capillary attraction.

436. CAPILLARY ATTRACTION *is the attraction which causes the ascent of fluids in small tubes.*

The tubes must be less than one tenth of an inch in diameter, and tubes whose bores are no larger than a hair, (*capillus*,) present the phenomenon the most strikingly. But though the rise of water above its natural level, is most manifest in small tubes, it appears, in a degree, in all vessels whatsoever, by a ring of water formed around the sides with a concavity upward.*

The following are the leading *facts* respecting capillary attraction.

(1.) *When small tubes, open at both ends, are immersed perpendicularly in any liquid, the liquid rises in them to a height which is inversely as the diameter of the bore.* Though tubes of glass are usually employed in experiments on this subject, yet tubes made of any other material exhibit the same property. Nor does the thickness of the solid part of the tube, or its quantity of matter, make the least difference, the effect depending solely on the attraction of the surface, and consequently extending only to a very small distance.

(2.) *Different fluids are raised to unequal heights by the same tube.* Thus, according to Gay Lussac, a tube which will raise water 23 inches will raise alcohol only 9 inches.

(3.) A tube $\frac{1}{100}$ of an inch in diameter raises water 5.3 inches, and since the height is reciprocally as the diameter, *the product of the diameter into the height is a constant quantity,* namely, the .053th part of an inch square.†

(4.) *Fluids rise in a similar manner between plates of glass, metal, &c.,* placed perpendicularly in the fluids, and near to one another. If the plates are parallel, the height to which a fluid will rise, is *inversely as the distance between the plates;* and the whole ascent is just *half that which takes place in a tube* of the same diameter. If the plates be placed edge to edge, so as to form an angle, and they be immersed in water, with the line of their intersection vertical, the water will ascend between them in a *curve* having its vertex at the angle of intersection. This curve is found to have the properties of the *hyperbola.*

(5.) If a capillary glass tube be immersed in mercury, *the mercury, instead of rising, sinks to a lower level within than without, and its surface is convex instead of concave.*

(6.) *Tubes which are wider at bottom than at top, elevate fluids to the same height as though the bore were throughout only equal to that of the smaller part.* As this experiment does not succeed *in vacuo,* when the wider end is immersed, the column in this case is supported by the pressure of the atmosphere.‡

437. Such are the leading facts ascertained respecting capil-

* Playfair, *Outlines* I, 177 † Ib. ‡ Biot

lary attraction. Various explanations of them have been attempted, but that of La Place is most generally received. According to this high authority, the action of the sides of the tube draws up the film of fluid nearest to it, and that film draws along with it the film immediately below it, and so each film drags along with it the next below, until the weight of the volume of fluid raised exactly balances all the forces which act upon it. The fact that the elevation of the water between the parallel plates, is exactly *half* that in a tube of the same diameter, clearly indicates that the force resides in the surrounding body; and the additional fact that the thickness or quantity of that body makes no difference, proves that the force resides in the surface, and that the action extends only to a very small distance. The concave surface exhibited by water and all fluids capable of wetting the tube, (where, of course, the attraction between the fluid and the tube, is greater than between the particles of the fluid among themselves,) still further indicates a force acting in the direction of the surface of the tube, while the convex surface and depression of mercury, are such effects as might be anticipated from the cohesion of its parts, which greatly exceeds its attraction for glass and other substances.

438. Various *phenomena* in nature and art are explained upon the principles of capillary attraction. Capillary action is not confined to tubes, but is exerted among all substances which are perforated by pores, or subdivided by fissures or interstices. On this power depend chiefly the functions of the excretory vascular system in plants and animals, and hence also the ascent of humidity through the shivered fragments of rocks, unglazed pottery, gravel, earth, and sand. Thus if the pores of the human skin (which are known to be exceedingly small) are estimated at the $\frac{1}{3000}$th part of an inch in diameter, they will support the fluids that circulate through them to the height of 120 inches, or ten feet, or higher than is required for the animal system.* The ascent of the sap in trees has usually been ascribed to capillary attraction, their circulating vessels being a congeries of small tubes; but some physiologists maintain that this action is dependent, not on the mechanical structure, but upon something which they denominate the *living principle* of vegetables.

439. According to Professor Leslie, if a soil of gravel contains pores 100th part of an inch in diameter, water will ascend in it by capillary action more than four inches; and supposing the particles of coarse sand to have interstices of the 500th part of an inch, the water would rise through a bed of sixteen inches; and if the pores were diminished to the 10,000th part of an inch, water would rise twenty-five and a half feet. Hence, in agri-

* Leslie, Elem. Nat. Phil., Vol. I.

culture, are derived the advantages of deep and perfect tillage; since, the more effectually a soil is pulverized, the better fitted it is to raise and retain water near the surface.

440. Several familiar examples of capillary attraction may be added. A piece of sponge, or a lump of sugar, touching water by its lowest corner, soon becomes moistened throughout. The wick of a lamp lifts the oil to supply the flame, to the height of several inches. A capillary glass tube, bent in the form of a syphon, and having its shorter end inserted in a vessel of water, will fill itself and deliver over the water in drops. A lock of thread or of candle-wick, inserted in a vessel of water in a similar manner, with one end hanging over the vessel, will exhibit the same result. An immense weight or mass may be raised through a small space, by first stretching a dry rope between it and a support, and then wetting the rope.*

RESISTANCE OF FLUIDS.

441. The resistance to a body moving in a fluid, arises from the inertia, from the cohesion, and from the friction of the fluid, admitting the particles to be in contact. The influence of this last cause, granting it to exist, is probably very small; and the second is in most fluids inconsiderable, when compared with the inertia. The resistance, therefore, which we shall here consider, is that which arises from the inertia of the fluid.†

442. *The resistance which a plane surface meets with while it moves in a fluid, in a direction perpendicular to its plane, is proportioned to the square of its velocity.*

Whatever motion or momentum is imparted to the fluid, exactly the same amount is extinguished in the moving body, constituting the resistance (R.) But the momentum is proportioned to the quantity of matter and velocity conjointly; or $M \propto Q \times V$. Again, in the present case, the quantity of fluid displaced must evidently be proportioned to the velocity of the moving body; that is, $Q \propto V$. Therefore, M or $R \propto V^2$.

This proposition is found to hold good in practice, where the velocity is very small, as the motions of boats or vessels in water; but when the velocity becomes very great, as that of a cannon ball, the resistance increases in a much higher ratio than as the square of the velocity. (See Art. 359.) Since action and reaction are equal, it makes no difference, in the foregoing proposition, whether we consider the plane in motion and the fluid at rest, or the fluid in motion and striking against the plane at rest.

---

* Arnott's El. Phys., I, 19—Robison's Mech. Phil. † Vince's Hydrostatics.

443. On account of the rapidity with which the resistance increases as the velocity is augmented, when a vessel or a steamboat is moving in water, it is only a comparatively moderate velocity that can possibly be given to it. A vessel driven by a wind which moves 60 miles an hour, is not carried forward faster than at the rate of 12 or 14 miles per hour. Steamboats are sometimes urged forward at the rate of 16 miles an hour; but to gain the additional speed over and above 12 miles, requires a great expenditure of force. If a steam-engine of 20 horse power give a motion of 4 miles an hour, it would require one of 180 horse power to increase the speed to 12 miles an hour. But, it must be observed, that the resistance decreases as fast when the velocity is diminished, as it increases when the velocity is augmented; and consequently, that canals may have the advantage over railways, when heavy articles are to be transported by very slow motions, although railways, encountering only the resistance of the air instead of water, have greatly the advantage when the motion is swift.* A cannon ball, on the other hand, meets with so much resistance on striking the water as to rebound.

Reckoning the resistance to increase only as the square of the velocity, it follows that twice the speed encounters four times the resistance; four times the speed, sixteen times; and ten times the speed one hundred times the resistance. Hence a body descending freely through the air by gravity, for a great distance, does not continue to be accelerated throughout the whole distance, but is finally brought, by the resistance of the air, to a uniform motion.

Notwithstanding the difficulties attending the mathematical theory of hydraulics, so much has already been done by the assistance of practical investigations, that we may in general, by comparing the results of former experiments with our calculations, predict the effect of any proposed arrangement, without an error of more than one fifth, or perhaps one tenth, of the whole,—a degree of accuracy fully sufficient for practice.†

FORMATION OF WAVES.

444. When the surface of water is pressed upon unequally, in parts contiguous to one another, the columns most pressed are shortened, and sink beneath the natural level of the surface, while those that are least pressed are lengthened, and rise above that level. As soon as the former columns have sunk to a certain depth, and the latter have risen to a certain height, their motions are reversed, and continue so, until the columns that were at first most depressed have become most elevated, and those that were most elevated have become most depressed.‡

* See Leslie, Nat. Phil., I, 443.

† Dr. Young's Lectures on Nat. Phil., I, 277.

‡ Playfair's Outlines, I, 203.

445. *The alternate elevations and depressions of the surface of a body of water, produced by a force acting unequally on the surface, are called waves.*

The water in the formation of waves has a vibratory or reciprocating motion, both in a horizontal and in a vertical direction, by which it passes from the columns that are shortened to those that are lengthened, and returns again in the opposite direction. *Progressive* motion is not necessary to undulation.*

446. Sir Isaac Newton first observed the analogy between the motions of waves and the vibrations of a column of water in a recurved tube, and upon this analogy he founded his theory of waves. Let AFGB, (Fig. 191,) be a bent tube, of equal bore throughout, having its sides parallel to each other and perpendicular to the horizon. Suppose it to be filled with water or any fluid to the height MM′. By any pressure applied at M′, let the column be depressed to N′ and raised to E in the opposite arm The pressure being removed, the longer column EF will preponderate and seek to regain its original level, but the ascending column will not stop at M′, but on account of its inertia, will ascend to E′, that is, to the same height as that from which it descended on the other side. It will now descend again, and these reciprocal motions will continue until destroyed by the natural impediments to motion. On account of these, each successive vibration is shorter than the preceding, but all of them, like those of a pendulum, are performed in equal times; for the moving force is obviously proportioned to the column EM, that is, to the space through which the whole column vibrates; and when the forces are as the spaces, the times are equal.

Fig. 191.

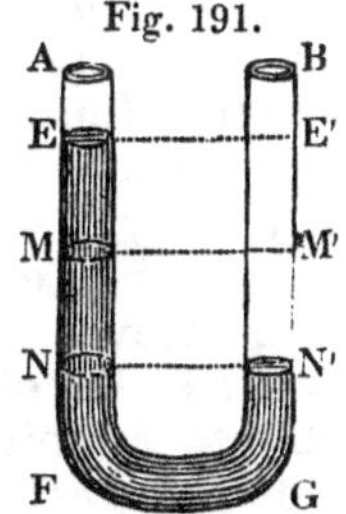

447. Now when the surface of water is smooth and at rest, if any force (as the action of the wind or the fall of a stone) depress that surface in any particular place, the contiguous water will necessarily rise all around that place. The water which has thus been elevated, descends soon after in consequence of its gravity; and by the time it has reached the original level, it will have acquired velocity sufficient to carry it lower than that level; therefore, it now acts as another original moving force, in consequence of which, the water will be raised on both sides of it. And for the same reason, the descent of those elevated parts will produce other elevations contiguous to them, and so on. Thus the alternate rising and falling of the water in ridges, will expand all around the original place of motion; but as they recede from

* Playfair.

that place, so the ridges, as well as the adjoining hollows, grow smaller and smaller until they vanish. This diminution of size is produced by three causes, namely, by the want of perfect freedom of motion among the particles of water, by the resistance of the air, and by the remoter ridges being larger in diameter than those which are nearer.

448. From a variety of experiments and observations, it appears that the utmost force of the wind cannot penetrate a great way into the water; and that even in violent storms the water of the sea is slightly agitated at the depth of twenty feet below the usual level, and probably not moved at all at the depth of thirty feet.* Therefore, the actual displacing of the water by the wind cannot be supposed to reach nearly so low; and hence it would seem that the greatest waves could not be so very high as they are often represented by navigators. But it must be observed that in storms waves increase to an enormous size from the *accumulation of waves upon waves;* for, as the wind is continually blowing, its action will raise a wave upon another wave, and a third wave upon a second, in the same manner as it raises a wave upon the flat surface of the water. In fact, at sea, a variety of waves of different sizes are frequently seen one upon the other, especially while the wind is actually blowing. When it blows fresh, the tops of the waves, being lighter and thinner than the other parts, are impelled forward, broken, and turned into a white foam, particles of which, called *spray*, are carried to a great distance.†

449. While the depth of the water is sufficient to allow the oscillation to proceed undisturbed, the waves have no progressive motion, and are kept, each in its place, by the action of the waves that surround it. But if, by a rock rising near to the surface, or by the shelving of the shore, the oscillation is prevented, or much retarded, the waves in the deep water are not balanced by those in the shallower, and therefore acquire a progressive motion in this last direction, and form *breakers*. Hence it is that waves always break against the shore, whatever be the direction of the wind. Breakers formed over a great extent of shore, are distinguished by the name of *surf*. The surf is greatest in those parts of the earth where the wind blows always nearly in the same direction.‡

---

* Boyle's Works, Vol. III, in Cavallo, I, 260. † Cavallo. ‡ Playfair

PART IV.—PNEUMATICS.

CHAPTER I.

OF THE MECHANICAL PROPERTIES OF AIR.

450. PNEUMATICS *is that branch of Mechanics, which treats of the equilibrium and motion of* ELASTIC *fluids.*

Those laws of equilibrium which are founded on the peculiar nature of fluids arising from the mobility of their particles, are equally applicable to Hydrostatics and Pneumatics. But certain additional properties result from the *elasticity* of vapors and gases, which may be conveniently considered under the latter head.

Vapors are elastic fluids, which are produced from liquid or solid bodies by the agency of heat, and which readily become liquid or solid again on the application of cold. Thus steam is raised from boiling water, and is again easily condensed by cold into the liquid state. *Gases* are permanently elastic fluids. They are never met with in nature either in the liquid or solid state, and it is only by means of extraordinary degrees of cold or pressure, that they can be made to give up their elasticity and become liquids. Atmospheric air is a body of this class; and since air and steam are, with slight exceptions, the only elastic fluids employed as mechanical agents, it is to these, chiefly, that our attention will be devoted.

451. The effects of HEAT upon all bodies, are usually treated of in Chemistry; but a few of those effects which are strictly mechanical, especially such as are produced on aëriform fluids, may be advantageously considered in this place.

The most general mechanical effect of heat is *expansion.* Heat expands all bodies, whether solid, liquid, or aëriform. *Aëriform bodies are expanded equally by equal additions of heat.** The increase of volume is continued without limit, as the heat is

---

* This and various other propositions in Pneumatics, are proved by experiment. It is supposed that most of the instructors who use this work, will have the means of illustrating or proving the truth of these propositions, by the aid of appropriate apparatus. But even when this is not the case, we conceive that very little benefit can accrue to the learner from the bare description of experiments.

augmented. The elasticity of a confined portion of air, as that contained in a close bottle or flask, for example, is uniformly increased by equal additions of heat. This is true of steam, however, only when the vessel is free from water; for, if steam is heated in contact with water, in a close vessel, where new portions of steam are continually added, without any enlargement of volume, its density and elasticity are *rapidly* increased, in a geometrical ratio, and its mechanical force shortly becomes so great as to burst almost any vessel that can be employed to contain it.

452. The properties of air may be exhibited under the form of a few simple propositions.

(1.) *Air is material.*

The two essential properties of matter are extension and impenetrability. (Art. 2.) That air has extension, needs no proof. That it is impenetrable, or has the property of excluding all other matter from the space which it occupies, is proved by experiment. Thus if we depress in water a tall jar, or a tumbler, we shall find that the water rises only through a certain *part* of the vessel, to whatever depth we immerse it; and if to a hollow cylinder, made smooth and closed at the bottom, we fit closely a stopper or solid cylinder, called a piston, moving freely in it, on applying the piston, no force will enable us to bring it into contact with the bottom of the cylinder, unless we permit the air within it to escape. Two other properties exhibited by air, likewise indicate that it is material: these are *inertia* and *weight.* The inertia of air is manifested by the resistance it opposes to bodies moving in it; as, for example, an open umbrella moved through the air, in a direction parallel with the staff; and the weight of the air is shown by the fact that a vessel, as a bottle, from which the air has been withdrawn, (by methods to be described hereafter,) weighs less than before. A vessel of the capacity of a wine quart, weighs about eighteen grains less after the air is exhausted, than before. One hundred cubic inches of air weigh thirty and a half grains.

(2.) *Air is a fluid.*

This property is manifested not only by the great mobility of its parts, but also by the distinguishing properties of fluids, (Art. 379,) viz. that any portion of air at rest, presses and is pressed equally in all directions; and that a pressure or blow applied to any part, is propagated through the whole mass, and affects every part alike. (Art. 382.)

(3.) *Air is an* ELASTIC *fluid.* (Art. 84.)

Thus, when an inflated bladder is compressed, it immediately restores itself to its former situation. Indeed, since air when compressed restores itself, or tends to restore itself, with the same force as that with which it is compressed, it is a perfectly elastic body.* (Art. 84.)

452′. *The volume of a given weight of air is inversely as its compressing force.*†

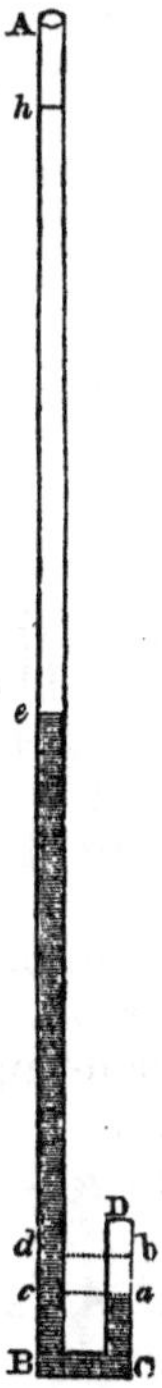

Let ABCD be a glass tube open at A and closed at D. Let quicksilver be poured into the tube; it will *tend* to rise to the same height in both arms of the tube, (Art. 391,) but the air in CD, by its elasticity, will resist its ascent, so that, when at rest, it will stand much higher in AB than in AC. Let it stand at *e* when the air in CD is compressed into half its original bulk above the quicksilver at *a*. Then the column of quicksilver C*a* will just balance the equal column B*c*, and the column *ce* will measure the elastic force of the compressed air. Add more quicksilver, and the column C*a* will rise. Let it rise to *b*, so that the air shall be compressed into D*b*, one fourth its original volume. It will be found, on measuring, that *dh* which measures the elastic force of the air in D*b* is just twice the height of *ce*. Consequently, a double pressure is required to reduce a given quantity of air to half its volume; and in the same manner it may be shown that three times the pressure reduces the volume to one third; and, universally, that the volume is inversely as the compressing force.

Fig. 192.

Since the compressing force is in each case a measure of the elastic force of the compressed air, it follows, that *the elastic force of a given weight of air, is as the compressing force,* and also *inversely as the volume.* And since the density of a given quantity of matter is inversely as its volume, hence, *the elastic force of a given quantity of air is as its density.*

453. Before we proceed further, it is necessary for the learner to be made acquainted with the apparatus by which the mechanical properties of air are illustrated.

THE AIR-PUMP.

The Air-Pump is an instrument used for the purpose of exhausting the air from any given space. Though of several different forms, yet the most common construction is that represented in Fig. 193. The chief parts are the *plate* A, the *barrels*

* The phrase perfect elasticity is used here in its technical sense, and does not preclude the idea that the elastic force of air is susceptible of increase and diminution.

† This is commonly called the *Law of Mariotte.*

Fig. 193

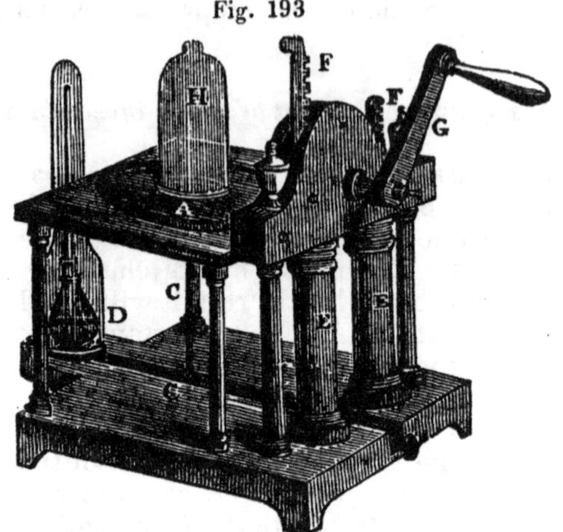

E, E, and the *pipe* or *canal* CC, leading from the plate to the barrels. The glass vessels which are set upon the plate, are called in general *receivers*. A *gauge* is sometimes employed (as represented by D in the figure) to indicate the degree of exhaustion; but the nature of this appendage will be better understood hereafter. Such is the construction of the air-pump in general; but the importance of this apparatus entitles it to a more minute description. In order, then, fully to understand the principle of the air-pump, and other kinds of apparatus designed for producing a vacuum, we must learn the construction of *valves*, and of the *cylinder* and *piston*.

454. *A* VALVE *is a contrivance which permits a fluid to pass in one direction, but prevents its passing in the opposite direction.* The clapper seen on the under side of a pair of bellows, is a familiar example of a valve. The valve employed in the air-pump, usually consists merely of a strip of oiled silk, tied over a small orifice. The air by pressing *outward* from the orifice raises the silk, opens the valve, and makes its escape; while by pressing *inward* upon the orifice, it keeps the strip of silk close to the orifice, and is therefore prevented from passing in that direction. The piston and cylinder are exemplified in a common syringe. It consists of a hollow cylinder, or barrel, to which is fitted a short solid cylinder called the piston, which is moved up and down the barrel by means of a projecting handle called the piston rod, and is fitted so closely to the barrel as to be air tight. Suppose now that the cylinder is in a perpendicular position, closed below, but open above, and that the piston rests on the bottom. On drawing up the piston, the air above it is lifted out, and the space below it is a vacuum. If a small orifice be made in the bottom

of the barrel, then as the piston is drawn upward, the air will flow in and no vacuum will be formed; and as the piston is depressed again, the air is forced back. But by attaching a valve to the orifice, we may admit or exclude the external air at pleasure. If the strip of silk be tied on the *outside*, then, on drawing up the piston, the air will not follow, but the piston will go up heavily, since it lifts up the entire weight of the column of air that rests upon it, (there being nothing below it to act as a counterpoise,) and if the hand be withdrawn from the piston rod, the piston will descend spontaneously. Again, if the valve be placed on the *inside*, then the external air will follow the piston as it rises, and no vacuum will be formed. If now the piston be depressed, the air cannot be expelled, (since the valve closes on the orifice in that direction,) and the piston cannot be forced down to the bottom of the barrel, unless a valve is placed in the piston itself, opening outward; in this case the air of the barrel may be expelled by depressing the piston.

455. We have been thus minute in the description of the construction of valves, and of the cylinder and piston, because when these things are clearly understood, the learner will easily comprehend the principle of the air-pump, of the common house pump, of the steam-engine, and of every other species of pneumatic apparatus. Let us now return to the *air-pump*.

Fig. 194.

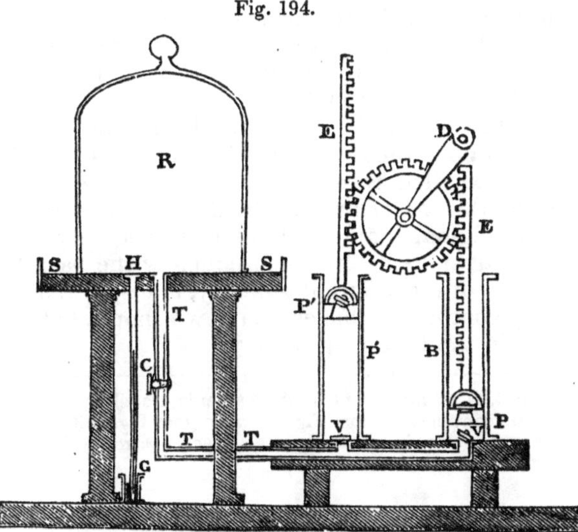

In the barrels, two pistons play up and down, each of which is furnished with a valve opening upward into the open space, through which the piston rods move. Another valve is placed at the bottom of each barrel, opening into the barrel. The piston rods are indented bars, to which a toothed wheel (concealed in Fig. 193, but seen in Fig. 194) is adapted, which, being turned backward and forward by means of the winch G, (Fig. 193,) alternately raises and depresses the two pistons, as is represented in the preceding figure. Suppose now the receiver to be placed on the plate of the pump, one of the pistons being at the top, and the other at the bottom of the barrel. We turn the winch, the piston rises, and the air of the receiver opens the valve at the bottom of the barrel, and diffuses itself equally through the barrel and the receiver. We turn the winch in the opposite direction, the piston descends, compresses the air in the barrel before it, which, as it cannot go back into the receiver, opens the valve in the piston itself, and escapes into the vacant space in which the arm of the piston moves. This process is repeated every time the piston rises and falls; and it is the same in both barrels, the two being employed to accelerate the process of exhaustion, and to facilitate the working of the pump, since the pressure of the atmosphere on the descending, counteracts the effect of the same pressure on the ascending piston.*

456. *The exhaustion proceeds at a rate, which increases in a geometrical ratio.*

Suppose, for example, that the capacity of one of the barrels is just one ninth part of that of the receiver, including that of the pipe which leads from the receiver to the barrel. When the piston is first raised from the bottom to the top, the air which previously occupied the receiver, expands so as to diffuse itself equally through the receiver and barrel. The barrel, therefore, will contain a tenth part of the whole of the enclosed air, and nine tenths will remain in the receiver. On depressing the piston, this tenth part is expelled through the piston valve. On elevating the piston, the air remaining in the receiver (which is nine tenths of the original quantity) diffuses itself equally through the receiver and barrel, as before; consequently the barrel contains $\frac{1}{10}$ of $\frac{9}{10}=\frac{9}{100}$ of the original quantity, and $\frac{81}{100}$ remain in the receiver. By continuing this estimate, we should obtain the results expressed in the following table.†

---

* The letters in Fig. 194, are inserted to aid the learner in describing the air-pump, which can be done more conveniently from Fig. 194 than from Fig. 193.

† The estimate is made for a single barrel: in the double-barreled air-pump, the rate of exhaustion will be just doubled.

Number of strokes.	Part of the air expelled at each stroke.	Part remaining in the receiver.	Whole quantity expelled.
1	$\frac{1}{10}$	$\frac{9}{10}$	$\frac{1}{10}$
2	$\frac{9}{100}$	$\frac{81}{100}$	$\frac{19}{100}$
3	$\frac{81}{1000}$	$\frac{729}{1000}$	$\frac{271}{1000}$
4	$\frac{729}{10,000}$	$\frac{6561}{10,000}$	$\frac{3439}{10,000}$
5	$\frac{6561}{100,000}$	$\frac{59049}{100,000}$	$\frac{40951}{100,000}$
6	$\frac{59049}{1,000,000}$	$\frac{531441}{1,000,000}$	$\frac{468559}{1,000,000}$
7	$\frac{531441}{10,000,000}$	$\frac{4782969}{10,000,000}$	$\frac{5217031}{10,000,000}$

The numbers in the second column denote the rate of exhaustion, and it is evident that they compose a geometrical series, the constant ratio being $\frac{9}{10}$. Also the quantities remaining in the receiver after each stroke, compose a similar series, the ratio being the same. After seven strokes, the quantity remaining in the receiver is less than one half the original quantity. If we had taken a smaller receiver, the rate of exhaustion would have been much more rapid. Thus, if the receiver had only the capacity of the barrel, the series would have been $\frac{1}{2}$, $\frac{1}{4}$, $\frac{1}{8}$, $\frac{1}{16}$, $\frac{1}{32}$, $\frac{1}{64}$, $\frac{1}{128}$, $\frac{1}{256}$, $\frac{1}{512}$, $\frac{1}{1024}$; so that, with ten strokes of the piston, the air of the receiver would have been rarefied more than one thousand times.

As this series never terminates, it is evident that a complete exhaustion can never be effected by the air-pump. Indeed, in practice, the vacuum is far less perfect than the theory would make it by the repetition of the blows of the piston; for when the air in the receiver becomes very much rarefied, it has not elasticity sufficient to raise the valve at the bottom of the barrel: or even if that difficulty is obviated by a different construction of the valve, still the difficulty of making the joints and valves perfectly air-tight, is such as to impair the perfection of the void. In the most improved air-pumps, the valves are made of small pieces of metal, which are opened and closed by the action of the piston itself. Also to prevent the corrosion of brass, arising from the action of the oil employed to lubricate the parts, in the place of this material, glass is now used for the barrels and the plate of the pump, and the piston is made of steel.

457. By means of this instrument, we may obtain very striking illustrations of the mechanical properties of air.

(1.) The *pressure* of the air acts with great force on all bodies at the surface of the earth, amounting, as we shall show hereafter, to nearly 15 pounds upon every square inch, or more than 2000 pounds upon a square foot. Upon so large a surface, therefore, as that of the human body, the pressure amounts to no less than 13 or 14 tons; but being so uniformly distributed within and without, and on all sides, it is, when the air is at rest, scarcely perceptible.* In consequence of this pressure, the air insinuates itself into all fluids, and fills the pores of all solids except the most dense, as gold or platina. The pressure of the air diminishes the tendency of fluids to pass into the state of vapor, and of course raises their boiling point. Warm water, at a temperature much below the boiling point, will be set a boiling under the receiver of an air-pump, or in a vacuum formed in any other way. Indeed, if it were not for atmospheric pressure, water would require only the moderate heat of 72 instead of 212 degrees of heat to make it boil; and the more volatile fluids, as alcohol and ether, would hardly be found in nature, in the liquid state.

(2.) The *elasticity* of the air is such, that the smallest portion of it may be expanded beyond any known limits, by removing the external pressure. By this means, a bubble may be made to fill a very large space. On the other hand, air has been condensed by pressure, until its density has been greater than that of water, still retaining the elastic, invisible state.† In consequence of its elasticity, air is set in motion by the least disturbance of its equilibrium, whether by condensation or rarefaction, thus giving rise to the phenomena of winds.

(3.) Air is essential to the support of *combustion*, and to the *respiration* of animals; and finally, it is the principal medium of *sound*. It may be further shown, that the weight of bodies is diminished by the buoyancy of air, (acting on the same principle as water, Art. 397,) and that light bodies are sustained in it, in consequence of its greater specific gravity, while, in a vacuum, bodies of various densities, as a guinea and a feather, fall toward the earth with equal velocities.

These are the leading truths which are established and illustrated by means of the air-pump, which the learner will better comprehend by witnessing the actual experiments, than by any description of them that could be offered.

458. The condensation of air is usually effected by means of the *Condensing Syringe*. This instrument is a cylinder and piston, the cylinder having a valve opening outward, while the piston is without a valve. The principle of its operation will be readily understood from the figure. Near the top of the cylinder, at E, is a small hole in the side, which is immediately below the piston, when this is drawn up to the top of the cylinder. On forcing down the piston, the air is driven before it, and expelled through the valve at the bottom. By connecting a bottle or other close vessel with the bottom, the air expelled

* Fishes are sometimes caught at the depth of 2600 or 2700 feet, where the pressure of the water is equal to 80 atmospheres, or more than 82 tons to the square foot; yet these fishes are not injured by such an immense weight, or sensibly impeded in their motions. (Camb. Mech. p. 352.)

† Gregory, I, 481

Fig. 195.

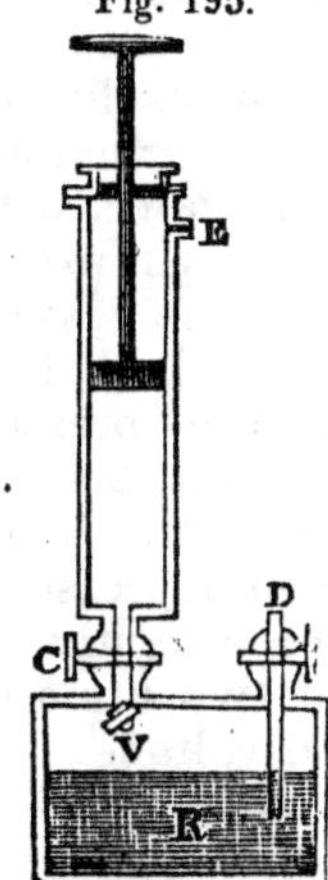

may be driven into that, its return being prevented by the same valve. The piston being drawn up again above the opening in the cylinder, another similar portion of air may be forced into the condensing bottle ; and thus the process may be continued indefinitely. Sometimes a valve, opening downward, is placed in the piston itself, and then the orifice at E is omitted.

The *Condensing Fountain* is a bottle, usually of copper, partly filled with water, upon the surface of which the air is condensed by means of the condensing syringe. The fluid being thus brought under a strong pressure, it tends to issue with great force whenever a pipe, that is inserted in the bottle, and extends below the surface of the water, is opened. The manner of its operation may be clearly understood from Fig. 195, where D represents the spout, having a long pipe descending into the water R, above which the air is condensed. The celebrated spouting springs of Iceland, called the Geysers, in which water accompanied by large masses of rock, is thrown to the height of 200 feet, arise from pneumatic pressure acting upon the surface of the water in the interior of the earth, the aëriform substance, whatever it may be, being produced by means of volcanic action.

The *Air-Gun* is an instrument in which condensed air is substituted as the moving force instead of gunpowder. By means of a condensing syringe, air is strongly condensed in a metallic ball furnished with a valve at the mouth, where it is screwed on the gun below the lock. As the lock is sprung, it falls upon a plug, and forces it upon the valve, which instantly opens, and the air rushes into the barrel of the gun, and by its sudden expansion, propels a ball much in the same manner as gunpowder would do in its place.

459. The *Diving Bell* is an apparatus employed for exploring the depths of the sea. It was formerly made in the shape of a bell, but is now more commonly made square at the top and bottom, the bottom being a little larger than the top, and the sides slightly diverging from above. The material is sometimes cast iron, the whole machine being cast in one piece, and made very thick, so that there is no danger either from leakage or fracture Sometimes the diving bell is made of planks of two thicknesses, with sheet lead between them. In the top of the machine are placed several strong glass lenses for the admission of light, such as are used in the decks of vessels to illuminate the apartments below.*

* Ed. Phil. Jour. V, 8. Amer. Jour. Science, xxii, 325.

The diving bell depends for its efficacy on that quality of air which is common to all material substances, *impenetrability;* that is, the exclusion of all other bodies from the space it occupies. The principle may be illustrated by depressing a tumbler or jar in water, with the mouth downward: it will be seen that the water will ascend so far as to occupy only a part of the capacity of the vessel, the upper part being occupied by air. As the diving bell descends in the water, the air inclosed in it is subject to its pressure, (which increases with the depth,) and by virtue of its elasticity, it will become condensed in proportion to this pressure. Thus at the depth of about thirty-four feet, the hydrostatic pressure will be equal to that of the atmosphere, and consequently, the air being under a pressure equivalent to that of two atmospheres, it will be condensed into one half its original volume. As the depth is increased, the space occupied by the air in the bell will be proportionally diminished. Seats are furnished for the workmen, and shelves for tools, and various other conveniences. Although at the depth of thirty-four feet, the water would occupy one half the capacity of the vessel, and more or less at different depths, yet by means of a forcing pump or condensing syringe, communicating between the atmosphere above and the machine, through a pipe, air may be thrown in so as to exclude the water entirely. By the same means fresh air may be conveyed to the workmen, the portion of air rendered impure by respiration being at the same time suffered to escape by opening a stop-cock in the top of the machine.*

Fig. 196.

460. Before we proceed to the consideration of the atmosphere, it is necessary for the learner to become acquainted with another important instrument, the Barometer, by means of which, as well as by means of the air-pump, our knowledge of the atmosphere has been greatly enlarged.

THE BAROMETER.

Let us take a glass tube, about three feet in length, closed at one end and open at the other. We fill the tube with quicksilver, and invert it in a vessel of the same fluid. The column of quicksilver falls to a certain height, about twenty-nine or thirty inches, where, after vibrating a few times, it remains at rest. The space in the tube above the quicksilver being void of air or any other substance, it is of course a vacuum, and is usually denominated the *Torricellian* vacuum,

* Lardner's Pneumatics.

from Torricelli, an Italian philosopher, who first discovered this method of producing a vacuum. Various precautions are necessary, in order to preserve this space free from air or any aëriform substance: when these precautions are taken, this vacuum is one of the most complete that we can command.

The column of quicksilver is sustained by the pressure of the atmosphere on the open mouth of the tube, which is immersed in the same fluid;* and it must have the same weight with a column of the atmosphere of the same base, otherwise it would not be in equilibrium with it. We hence arrive at an accurate knowledge of the actual weight and pressure of the air, since it is equal to the weight of a column of quicksilver of the same base, thirty inches in length. The weight of such a cylinder of quicksilver is easily ascertained. Since a cubic inch of water weighs 252.525 grains, and quicksilver is 13.57 times heavier than water, therefore a cubic inch of quicksilver weighs 3426.76 grains; and 30 inches weigh 102802.8 grains. But 7004 grains troy make one pound avoirdupois. Therefore, $\frac{102802.8}{7004}=14.7$ lbs. It results, that the pressure of the air on every square inch of surface is, as stated in Art. 457, about 15 lbs., or more than 2000 lbs. upon a square foot. Since different fluids balance each other in opposite columns pressing base to base, when their heights are inversely as their specific gravities, (Art. 401,) a column of water in the place of the mercury would stand at the height of about 34 feet. For quicksilver being 13.57 times heavier than water, the latter column must be 13.57 times higher than the other; that is, $30\times13.57=$ 407.1 inches=33.92 feet.

By observing from day to day the height of the column of quicksilver prepared as above, we shall find that it varies through a space of two or three inches, showing that the atmosphere does not always exert the same pressure, but that a given column of air is sometimes lighter and sometimes heavier. This instrument, therefore, enables us to ascertain the relative weight of the air at any given time, and hence its name, *barometer*.† For the purpose of indicating these variations with minuteness and precision, a graduated scale is attached to the barometer, divided into inches and tenths of an inch, and usually extending from twenty-seven to thirty-one inches,—a space which is more than sufficient to comprehend all the natural variations in the weight of the atmosphere.

* As young learners sometimes find a difficulty in conceiving clearly how the pressure of the air acts in this case, we subjoin a remark or two. It must be recollected, that any impulse or pressure exerted on the surface of the fluid in the vessel, extends alike to all parts of it, (Art. 382;) and since fluids act upward as well as downward, it is plain that the pressure acts in sustaining the column of mercury in the same manner as though it were applied directly to the mouth of the tube.

† From βαρος *weight* and μετρον *measure*.

461. As these changes of weight are sometimes very minute, a contrivance called a *vernier* is attached to the scale, by means of which the tenth of a tenth, that is, the hundredth part of an inch, may be estimated. The vernier consists of a small plate movable up and down by a screw upon the graduated part of the barometer, and is divided as follows. Let AB (Fig. 197,) represent the upper part of a barometer, the level of the mercury being at C namely, at 30.3 inches, and nearly another tenth. The vernier being brought (by a screw which is usually attached to it) to coincide with the surface of the mercury, we look along down the scale, until we find that the coincidence is at the 8th division of the vernier. Now as the vernier gains $\frac{1}{10}$ of $\frac{1}{10}=\frac{1}{100}$ of an inch at each division upward, it of course gains $\frac{8}{100}$ in eight divisions. The fractional quantity, therefore, is .08 of an inch, and the height of the mercury is 30.38. If the divisions of the vernier were such, that each gained $\frac{1}{60}$ (when 60 on the vernier would equal 61 on the limb) on a limb divided into degrees, we could at once take off minutes; and were the limb graduated to minutes, we could in a similar way read off seconds.

Fig. 197.

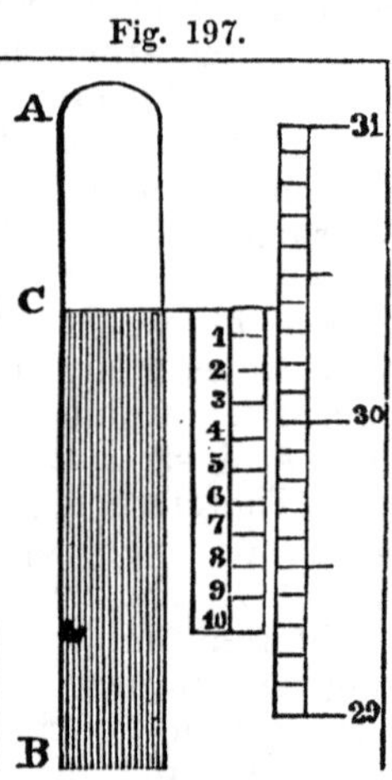

462. When the barometer is to remain *stationary* in a single place, the cistern containing the mercury is made of the form of a wide basin. In a vessel of small diameter, the rise of the mercury in the cistern, as it descended in a tube, would, by its reaction, tend to raise the mercury in the tube, for which effect a correction would be necessary. But in a wide cistern, the difference of level occasioned by the rise and fall of the mercury in the tube is so small, that it may be safely neglected.

But it is often desirable to have the barometer so constructed, that it may be conveniently carried from place to place without danger of derangement. Portable barometers are constructed in several different ways. In one, the mercury of the cistern is inclosed in a leathern bag, to the bottom of which is affixed a screw. On turning the screw, the mercury is forced up the tube until it completely fills it, and remains fixed. Over the mercury of the cistern an ivory float is placed, which is brought by means of the screw to a mark on the stem, which shows when the mercury is at the level whence the divisions on the scale were commenced.

The portable barometer, furnished with the above, or some similar contrivance for keeping the mercury steady, is sometimes made of the form of a walking cane, and thus becomes very convenient for taking the altitude of mountains.*

463. Since the variations of the barometer correspond to the variations in the weight of the air at the same place, and since these variations are connected with changes of weather, this instrument thus becomes a *weather glass*, and enables us in certain cases, to foresee changes of weather. The most important indications of the barometer are, that *its rise denotes fair*, and *its fall denotes foul weather*, whatever may be its absolute height. Also, a *sudden and extraordinary descent* of the mercury attends, and frequently precedes, a *violent wind*. The immediate cause of the descent of the barometer, is undoubtedly a *rarefaction* of the air at that place; but the cause of this rarefaction itself, it may be difficult to account for. The consideration of this point will be resumed hereafter.

The mean pressure of the atmosphere, as indicated by the barometer, is nearly the same, at the level of the sea, in all parts of the earth, corresponding very nearly to 30 inches of mercury. This fact has been verified by numberless observations, made with the barometer in both hemispheres, from the equatorial to the polar regions. The following results for several places, in different latitudes, corrected for temperature, elevation above the level of the sea, and the influence of the earth's rotation on its axis, are nearly uniform.

	Latitude.	Bar. Pressure.
Calcutta,	22° 35′	29.776
London,	51 31	29.827
Edinburgh,	55 56	29.835
Melville Island,	74 30	29.884

But, though the mean pressure of the atmosphere is nearly the same, at the level of the sea, over the whole globe, the extent of the variations to which it is liable, is exceedingly different in different parallels of latitude. In the equatorial regions, the range of the barometer is much more limited than within the polar circles; and in the frigid zones, it is more limited than in the temperate. Within the tropics, the fluctuations of the barometer do not much exceed ¼ of an inch.† At New York the variation does not much exceed 1½ inches, while in Great Britain it is as great as 3 inches.‡ The most extensive variations take place between the latitudes of 30° and 60°, being the zone in which

* To render the indications of the barometer entirely worthy of confidence, a number of practical precautions are requisite in the mode of constructing and filling, a good account of which may be found in *Renwick's Mechanics*, p. 382. Many of the cheaper kinds exposed for sale are very inaccurate. Daniell's barometer is one of the best.

† Daniell's Meteorology, I, 108.

‡ Renwick, p. 384.

the annual changes of temperature and humidity possess the widest range.*

The barometer also undergoes certain variations corresponding to the different hours of the day, called its *horary* variations. To ascertain the nature of these at any given place, a long series of observations must be made, from which the maximum and minimum height may be deduced. Mr. Redfield states the mean range of the diurnal oscillation between 10 A. M. and 6 P. M., to be at New York, .039 inches.†

464. Shortly after the invention of the barometer, it was observed that the mercury descends, when the instrument is carried to a more elevated situation. The descent is found to be about $\frac{1}{10}$ of an inch for 87 feet. From this observation, we may deduce nearly the specific gravity of air compared with mercury or water; for $\frac{1}{10}$ of an inch of mercury has, it appears, the same weight as 87 feet, or 1044 inches, of air. Consequently, 1 inch of mercury weighs as much as 10440 inches of air; that is, mercury is 10440 times, and water is $\left(\frac{10440}{13.57}=\right)$ 769 times, heavier than air.

465. The learner is now prepared to understand the principles on which are constructed the several *gauges* used in connection with the air-pump, *to indicate the degree of exhaustion.*

The gauge represented at D, Fig. 193, consists of a glass tube filled with mercury, and inverted in a small jar of the same fluid, and covered over with a receiver. This apparatus is placed upon the smaller plate of the pump, which is connected with the larger plate, by a horizontal pipe. Consequently, when the air in the receiver H is rarefied by working the pump, the air in the small receiver D, being rarefied in the same degree, will at length have its elasticity so much diminished, as to be unable to sustain even the short column of mercury in the tube. The mercury, therefore, will descend in the tube, and will approach toward the level of the fluid in the jar, and will come nearer to it in proportion as the exhaustion is more perfect.

The gauge exhibited in Fig. 194, G, (which is connected immediately with the receiver,) acts on a different principle. It consists of a tube, about 30 inches long, open at both ends, the lower end dipping into a small vessel of quicksilver, and the upper end opening into the receiver. On turning the pump, the pressure is diminished on the upper surface of the mercury in the tube, and the external pressure of the atmosphere forces up the fluid to a height corresponding to the degree of exhaustion. A scale, graduated into inches and tenths, is attached to the tube.

* Ed. Encyc. Art. 'Physical Geography'

† Amer. Jour. XXVIII, 158.

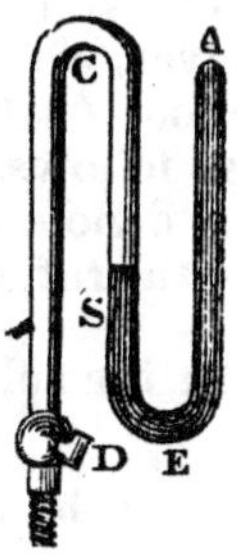

The *syphon gauge*, represented in Fig. 198, is screwed upon the small plate of the pump, instead of the apparatus exhibited at D, Fig. 193. Previous to exhaustion, the quicksilver is sustained in the arm A of the tube by the atmospheric pressure. When this pressure is diminished to a certain extent, the column of quicksilver descends, and in a perfect exhaustion would attain the same level in both arms of the tube. Consequently, the nearer it approaches to that level, the better is the exhaustion.

Fig. 198

466. The elasticity of air may be increased either by compressing it, or by heating it in a confined state; and its elasticity may be diminished either by lessening the pressure, or by cooling it. The elasticity of springs is known to be frequently impaired by continued action. This is not the case with air. Air has been left for several years very much compressed in suitable vessels, in which there was nothing that could have a chemical action upon it; and afterward, on removing the unusual pressure, and restoring the same temperature, the air has been found to recover its original bulk, which shows that the continuance of the pressure had not diminished the elasticity of it in the least perceptible degree.*

CHAPTER II.

OF THE ATMOSPHERE AND ITS PHENOMENA.

467. THE knowledge now acquired of the properties of elastic fluids, will qualify the learner to enter advantageously upon the study of the entire body of the air, which constitutes the atmosphere. Let us therefore now proceed to consider its *weight*,—its *extent* and *density*,—and its relations to *heat* and *moisture*, giving rise to the various phenomena of Meteorology.

468. The WEIGHT of the entire atmosphere may be easily estimated by means of the barometer; for, taking the medium height of the mercury at thirty inches, the weight of the atmosphere is equal to that of a sea of quicksilver covering the whole earth to the depth of two and a half feet. This would add five feet to the diameter of the globe, and the contents of the whole mass of quicksilver, in cubic feet, would be equal to the difference be-

* Cavallo, II, p. 225.

tween the solid contents of the globe, and those of a sphere of a diameter five feet greater. Having the number of cubic feet of quicksilver, we have only to multiply that number by the weight of one foot ($=13.57\times62\frac{1}{2}=848.125$ lbs.) The calculation proceeds as follows.

Let R denote the radius of the earth; r the height of the mercury; π the ratio of the circumference of a circle to its diameter, or 3.14159; the solidity of the globe$=\frac{4\pi R^3}{3}$;

Do. of the sphere, including the mercury$=\frac{4\pi(R+r)^3}{3}$;

Do. of the mercury$=\frac{4\pi(R+r)^3}{3}-\frac{4\pi R^3}{3}=4\pi(R^2r+Rr^2+\frac{r^3}{3})$.

But since r denotes but a very small fraction of R, the two last terms have so small a value, that they may be thrown out without materially affecting the result, and the contents of the mass of quicksilver will be $4\pi R^2r$. Substituting for these several quantities their numerical values, we have $4(3956\times5280)^2\times 3.14159\times2.5$=number of cubic feet in the mass of mercury; which being multiplied by $848\frac{1}{8}$, gives 11,624914,885408,838323 pounds, or more than eleven trillions of pounds, or five thousand billions of tons.*

Were the atmosphere of equal density throughout, it would be easy to determine its height, since opposite columns of different fluids are in equilibrium, when their heights are inversely as their specific gravities. (Art. 401.) Therefore, as the specific gravity of air is to that of quicksilver, so is the height of the column of quicksilver to the corresponding height of the column of air that balances it. That is, 1 : 10440 : : 2.5 : 26100 feet = 5 miles nearly.

But the atmosphere is very far from being throughout of uniform density. Several causes conspire to produce this result. 1. The different quantities of superincumbent air at different altitudes; 2. The decreasing attraction of the earth in proportion as the square of the distance from its center increases; 3. The influence of heat and cold; 4. The admixture of vapors and other fluids; 5. The attraction of the moon and other celestial bodies.† That the lower strata of the atmosphere are far more dense than the upper, will be obvious from this consideration, that the portions which rest on the surface of the earth, sustain the weight of the whole body of the atmosphere, which, as appears above, is immensely great. But the density of air is as the compressing force. (Art. 452′.) As we ascend from the earth,

* A less accurate method of finding the weight of the atmosphere, is to multiply the number of square inches on the surface of the globe by fifteen pounds.

† Cavallo, I, 227.

the weight sustained is constantly diminished, and the density lessened, according to the following law.

469. *The densities of the air decrease in a geometrical, as the distances from the earth increase in an arithmetical ratio.*

For, let us suppose that the strata of air are taken so thin, that the density of each may be considered as uniform throughout. Let the density of the inferior stratum be A, that of the second B, of the third C, and so on. Moreover, let a be the weight of the whole column of the atmosphere including A; b, the weight of the column when A is taken away; c, its weight when A and B are subtracted, and so on. Then the weight of the first stratum is $a-b$, that of the second, $b-c$, &c. Now the densities of two bodies of the same volume are as their weights. Therefore, $A : B :: a-b : b-c$. But since the densities are as the pressures, (Art. 452′) and the pressures are the weights of the incumbent volumes, therefore, $A : B :: b : c$. Hence, $a-b : b-c :: b : c$, $\therefore$ $ac-bc=b^2-bc$, $\therefore ac=b^2$, $\therefore a : b :: b : c$; that is, the weights, and consequently the densities of the successive strata, form a geometrical series. If, therefore, at a certain distance from the earth, the air be twice as rare as at the surface of the earth, at twice that distance it will be four times as rare, at three times that distance eight times as rare, &c.

470. By observations on the barometer at different altitudes, aided by calculation, it is ascertained, that at the height of seven miles above the earth, the air is only one fourth as dense as it is at the surface.* Hence if we take an arithmetical series, increasing by seven, to denote different heights, and a geometrical series whose constant multiplier is one fourth, to denote the corresponding densities, we may easily ascertain the density of the air at any proposed elevation.

Arithmetical series,	7	14	21	28	35	42	49
Geometrical series,	$\frac{1}{4}$	$\frac{1}{16}$	$\frac{1}{64}$	$\frac{1}{256}$	$\frac{1}{1024}$	$\frac{1}{4096}$	$\frac{1}{16384}$

From this table it appears, that at the height of twenty-one miles, the air is sixty-four times as rare as at the surface of the earth; at the height of forty-nine miles, sixteen thousand three hundred and eighty-four times as rare; and if we pursue the calculation, we shall find that its rarity at the moderate distance of only one hundred miles, is one thousand millions of times greater than at the earth,† and of course would oppose no sensible resistance to bodies revolving in it. De Luc ascended in a balloon to such a height that his barometer fell to twelve inches. Supposing the barometer at the surface to have stood, at that time, at thirty inches, it follows that he must have left three fifths of the whole atmosphere below him; for six inches being one fifth of

* Cotes, Hyd. Lect. p. 103.

† Rees's Encyc., Art. "Atmosphere"

thirty, twelve inches must be two fifths, and consequently three fifths of the whole must be below. His elevation was upward of twenty thousand feet.*

If there were an opening into the interior of the earth, which would permit the air to descend, its density would increase in the same manner as it diminishes in the opposite direction. At the depth of about thirty-four miles, it would be as dense as water: at the depth of forty-eight miles, it would be as dense as quick silver; and at the depth of about fifty miles, as dense as gold.

471. The foregoing law, however, does not afford *exact* data for estimating the density of the air at any given elevation, since the density is affected by the several other circumstances mentioned in Art. 468, which are not here taken into the account. Since the force of attraction diminishes as the square of the distance from the center of the earth increases, this diminution will occasion a corresponding decrease of density. However, as the force of attraction will be very nearly the same at such elevations as the highest mountains, as at the general level of the earth, (Art. 8,) no allowance is made on this account for barometrical measurements, except in cases when extreme accuracy is required. Changes of temperature produce a much greater effect, since heat expands, and cold contracts the air; and therefore, in estimating altitudes, the state of the thermometer is always to be taken into the account, in connection with the height of the barometer. Heat and cold also affect the height of the mercury in the barometer, independently of the pressure of the atmosphere without, and therefore it becomes necessary to reduce the observations to a fixed standard of temperature.

Owing to these different causes of irregularity in the density of the air at different elevations, it becomes a problem of much nicety and difficulty to obtain accurate measurements of heights, by means of the barometer; but the importance of the subject has led men of science to bestow very great attention upon it. We have room only to indicate the *general principles* on which such measurements depend, leaving the details to treatises of greater extent.†

472. With regard to the *actual height of the atmosphere* above the earth, it is a point not easily determined. Efforts have been made to ascertain its height by means of the twilight; but the student is not prepared to judge of the accuracy of this method, without a knowledge of Optics and Astronomy. The considera-

* Lardner's Pneumatics, Sec. 144.

† The necessary rules for barometrical measurements may be found in Robison's Mechanical Philosophy, Vol. III; Cavallo's El Nat. Phil. Vol. II; Gregory's Mechanics, Vol. I; Renwick's Mechanics, p. 386; and in most of the Encyclopedias under the article *Barometer*.

tion of it, therefore, belongs to a subsequent part of our course of instruction. We merely remark here, that no great reliance is placed upon this method by those who are most competent to judge of it.

If the decreasing densities of the air as we ascend from the earth were accurately expressed by a geometrical series, (Art. 470,) it is obvious that such an atmosphere would be unlimited, since such a series would never end. But several considerations render it probable, that the atmosphere is bounded by definite limits. Such are the following: (1.) The heavenly bodies move in void spaces; otherwise they would meet with resistance which would retard their motions, and the periods of their revolutions would not be unalterable, as is found to be the case. (2.) The expansion of air is owing to a mutual repulsion between its particles. This force is diminished as the particles are removed further asunder, by the enlargement of its volume; and we may conceive the repulsive force to be so much diminished at a certain distance from the earth, as to be counterbalanced by gravity, which being inversely as the square of the distance from the *center of the earth*, is nearly the same at all distances within a few miles of the earth's surface. (Art. 8.)* (3.) The condensation produced by extreme cold, such as is known to exist in the upper regions of the atmosphere, will oppose the expansion of the air, and counteract its enlargement of volume beyond a certain limit.

473. As we ascend from the earth, the temperature of the air constantly diminishes until we arrive at a region of frost, the lower limit of which is called the *term of perpetual congelation;* by which is meant a certain height above any place on the earth where, at a given time, water begins to freeze. The heights of the term of congelation for every parallel of latitude from the equator to the north pole, have been computed, partly from observation, and partly from the known mean temperature of each parallel, and the decrement of heat as we ascend in the atmosphere; and the result is expressed in the following table:—

Latitude. °	Mean height of the term of congelation in feet.	Difference for every 5 deg. of latitude.
0	15577	
5	15455	122
10	15067	388
15	14498	569

* This argument takes it for granted that the air consists of indivisible atoms; for were the air infinitely divisible, there would be no such *increase of distance between the particles*, and consequently diminution of repellent force, as is here supposed. But the existence of such atoms has been rendered extremely probable, and the conclusions deduced from the supposition of such atoms, are found to accord well with experience. (See Wollaston on the Finite Extent of the Atmosphere.—Phil. Trans for 1822.)

41

Latitude °	Mean height of the term of congelation in feet.	Difference for every 5 deg. of latitude.
20	13719	779
25	13030	689
30	11592	1438
35	10664	928
40	9016	1648
45	7658	1358
50	6260	1398
55	4912	1348
60	3684	1228
65	2516	1168
70	1557	959
75	748	809
80	120	628

From this table it appears, that the height of the region of perpetual frost at the equator is almost three miles; at the parallel of 35°, about two miles; and at the latitude of 54°, about one mile; while at the latitude of 80°, this region approaches very near to the earth, and at the pole it probably comes nearly or quite down to the earth. It is further to be remarked, that the different heights decrease very slowly as we recede from the equator, until we reach the limits of the torrid zone, when they decrease much more rapidly, the maximum being at the parallel of 40°. The average difference for every 5 degrees of latitude from 30° to 60°, is 1318, while from the equator to 30°, the average is only 664, and from 60° to 80°, it is only 891. Important meteorological phenomena depend on this fact.

474. *What is the cause of the cold that prevails in the upper regions of the atmosphere?*

It is found by experiment that radiant heat, like that of the sun, passes through a transparent medium without obstruction, and consequently does not heat that medium.* Were the air perfectly transparent, the heat of the sun would scarcely affect it at all; but the vapors, clouds, and other substances that diminish the transparency of the atmosphere, intercept a certain portion of the sun's rays. In general, however, the manner in which the air receives the heat of the sun is this: the sun's rays first communicate their heat to the surface of the earth; the stratum of air next to the earth imbibes a portion of this heat and rises, while colder currents descend or flow in laterally which in turn become heated and rise. Hence from the ground when heated by the sun, a current of air is constantly ascending. On the other hand, in the absence of the sun, the ground loses its heat by radiation, and becomes colder than the air im-

* Black's Lectures on Chemistry, Vol. I.

mediately above it. The air therefore now imparts a portion of its heat to the ground, is condensed, and remains in contact with the ground unless removed, as is commonly the case, by winds. The atmosphere, therefore, is, for the most part, heated and cooled *indirectly* by coming in contact with the surface of the earth.

475. The changes of temperature induced on the earth's surface by the sun's heat, are not sufficient to rarefy the air to any great extent. A part, moreover, of the heat received from the earth in the day time, is restored to it again at night; hence the rarefied portions of air do not ascend far above the earth until they find their equilibrium.

As a portion of air rarefied by heat at the earth's surface ascends, the diminishing pressure which it sustains as it rises, has a tendency to enlarge its volume. But, on the other hand, an enlargement of volume increases its capacity for heat, and lowers its temperature, which tends to condense it. At a moderate elevation above the earth, these causes operate to keep the air at rest, and thus the heat of the earth is incapable of raising the temperature of the air, except within a short distance, beyond which the region of frost prevails, and the cold continues to increase, until it probably reaches, at a comparatively moderate distance from the earth, an extreme intensity.*

RELATIONS OF THE ATMOSPHERE TO HEAT AND MOISTURE.

476. Air is set in motion by every cause which disturbs its equilibrium. It is more sensible than the most delicate balance, and moves with the slightest inequalities of pressure.

Air is put in motion by *the least change of temperature*. Heat rarefies it, and, as intimated in Art. 475, renders it specifically lighter than the neighboring portions, and it ascends, while colder and denser portions flow in to restore the equilibrium. On the other hand, if air be condensed by cold, it descends, or flows off, until it meets with air of the same density, where it rests. These effects naturally result from the perfect fluidity and elasticity of this substance.

477. An illustration of this principle is seen in the manner in which air circulates in the shaft or pit of a deep mine. Such a circulation is kept up briskly, even amounting sometimes to a strong wind, when two shafts or pits of unequal heights are made to communicate with each other by means of a horizontal gallery called a drift. The earth remains nearly at the same temperature

* According tc Fourier, the temperature of the planetary spaces is —58° Fahr.

Fig. 199.

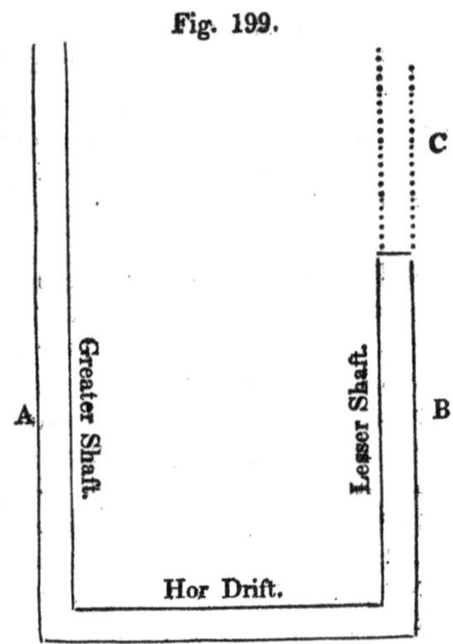

summer and winter, while the external air is hotter in summer and colder in winter, than that within the mine. Now were the air within the earth and without of the same density, then the air of the two shafts and of the drift would remain in equilibrio, the longer shaft A, being counterbalanced by the shorter shaft B, extended so as to embrace C, a portion of the external air, to the same height as the column A. But suppose it summer; then the air in A, becoming condensed by the influence of the colder earth, is rendered specifically heavier, and overpowers the columns B and C, the latter consisting of air more rarefied than that within the earth. Hence the air will flow down the longer, and out of the shorter shaft; and by bringing all parts of the mine into the circulation, the whole interior will be ventilated. Again, suppose it winter; then the air in the longer shaft being warmer and more rarefied than the compound column BC, the latter preponderates, and the air flows in the opposite direction; namely, down the shorter and out at the longer shaft. In spring and autumn, when the temperature of the atmosphere and the mine are nearly equal, the miners complain much of the suffocating state of the air.*

478. The contemplation of the motions of the atmosphere on a large scale, as they exist in nature, leads to the subject of winds; but we may see the same principle exemplified in *chimneys* and *fire-places.* The motion of air in chimneys may be understood by considering the chimney as one arm of a bent tube, while the external column, rising to the same height, is the other arm. Then the tendency to ascend will equal the difference in the densities of the columns of air in the opposite arms of the tube. When the air of the chimney is rarefied by heat from the fire-place, the cold air from below makes its passage upward into the partial void, and thus supplies air to the fire to support its combustion, and carries up along with it the smoke and vapors which proceed from the fire. The smoke, it will be remarked, is carried up, mechanically, by the ascending current of hot air; for smoke is itself heavier than air, and sinks or descends

* Robison's Mechanical Phil., III, 763.

when not thus supported.* The *draught* of the chimney, or tl strength and velocity of the ascending current, is influenced l several circumstances. (1.) Long chimneys have a strong draught than short ones, because they present a longer colun of rarefied air; but they may be so long as to cool the air t much before it has reached the top, in which case the smo falls by its greater specific gravity. In the case of large manufactories, where a great number of fires communicate with the same chimney, tall chimneys are used because they are easily kept hot throughout, and thus a very strong draught is maintained. They also serve the important purpose of conveying the noxious fumes to a great elevation in the atmosphere. Long horizontal pipes frequently have a bad draught, because they cool the smoke before it reaches the chimney. (2.) A narrow throat, opening into a large pipe or funnel, makes a strong draught, because the velocity of the ascending current is thus increased, it being in different parts of the chimney inversely as the area of the section. (Art. 417.) The throat of the chimney, however, must be wide enough to admit freely all the mixed products of the ascending current, including the rarefied air, smoke, watery vapor, and so on; and, consequently, a wider throat is required for green wood than for dry, and least of all for anthracite coal, where the amount of volatile substances expelled from the fuel is comparatively small. Small funnels, being more easily rarefied than large ones, are, in general, to be preferred. When reduced, however, beyond a certain limit, they encounter too much resistance from *friction*, especially when the surface is rough. Indeed, friction impedes the circulation of air over any surface, more than it does water, since the latter fills up the inequalities and deposits a film in contact with the surface which serves to lubricate it. Anthracite coal, on account of the small volume of gases produced in its combustion, admits of a smaller funnel than most other kinds of fuel. A large funnel is sometimes rendered unfavorable for carrying smoke, on account of a descending current of cold air from without, which meets it. This is especially the case when the top of the funnel is very large. Hence a chimney which is smoky from this cause is frequently cured by inserting in the top a smaller funnel of clay. (3.) A fire-place with a low front or breast, has a strong draught, because, in this case, no air can enter the chimney, except such as has felt the influence of the fire, and is thus fitted to keep the chimney warm; whereas, if the throat of the fire-place is high, much of the air that flows into it is cold and cools the chimney, and of course diminishes the degree of rarefaction in it. More-

* This fact is illustrated by an experiment suggested by Dr. Franklin, viz. by blowing the smoke of a tobacco pipe through water in a tumbler. The smoke, being cooled by this process, rests upon the surface of the water.

over, when the throat is near the fire, it becomes more intensely heated, and thus the degree of rarefaction of the current of air that passes through it is augmented and its velocity increased. In the structure of fire-places and stoves, it is an important principle, that *as little air as possible should get into the flue of the chimney, except what passes through the fire*; for if air which has not felt the influence of the fire, makes its way into the chimney, it cools the chimney, and diminishes the draught, which, other things being equal, is always proportioned to the difference of temperature between the air within the chimney and without. It is another important principle, in regard to the economy of fuel, that *no more air should traverse the fire than what is necessary to support the combustion.** All the air that passes through the fire, over and above what undergoes decomposition, cools it, and carries a portion of the heat up chimney. It is obvious that the air of an apartment must be denser than at the top of the chimney, otherwise the current will flow downward, as is sometimes the case when the room is very close, and the throat of the fire-place so large as to require a great quantity of air to fill the rarefied space, in which case, the air of the room is speedily exhausted. Hence the advantage, in close apartments, of small fire-places, or stoves which require but a small supply of air.†

478′. A stove constructed a few years since by the author of this work, having been found to answer a good purpose, and having come into extensive use, it is thought not improper to devote a few words to the explanation of its principles. It is particularly designed for anthracite coal.

The stove consists of two parts, a *furnace*, F, and a *radiator*, C. Figure 200 represents a front view of the apparatus, and Figure 201, a section. The radiator is employed to absorb and distribute the heat, instead of the long pipe that is frequently used for that purpose. It consists of two concentric cylinders of sheet iron, having a narrow space, *aa*, not exceeding an inch in diameter, between them. The inner cylinder (usually called the *air cylinder*) terminates below in a narrow pipe, like the neck of a bottle, which closes upon an opening in the bottom of the outer cylinder. Two vertical partitions pass down between the two cylinders, one of which is just behind the pipe that connects the furnace to the radiator, and the other is opposite to it, on the

* "It is a fact which ought never to be forgotten, that of the air which forces its way into a closed fire-place, that part only which comes into actual contact with the burning fuel and is decomposed by it in the process of combustion, contributes any thing to the heat generated; and that all the rest of the air that finds its way into and through a fire-place, is a thief that steals heat, and flies away with it up the chimney."—(Rumford, III, 172.)

† See Dr. Franklin's Remarks on the Causes and Cure of Smoky Chimneys, *Works*, Vol. II, p. 256. Also, Count Rumford's Essays, III, 465.

Fig. 200.

Fig. 201.

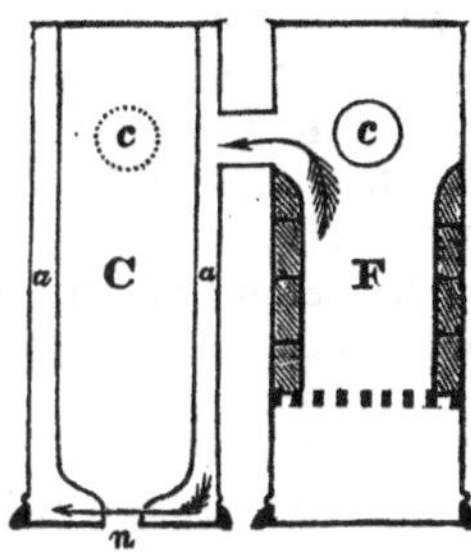

other side of the air cylinder. These partitions compel the heated current to pass down in front, and, flowing around the neck at *n*, to ascend in the rear, and go out by a pipe near the top, into the chimney. The arrows show the course of the current.

The objects of the inventor were threefold,—*economy*, *purity of air*, and *elegance of form*. The first end was attained by absorbing and diffusing the heat by means of the greatest extent of metallic surface that could be exposed under a given volume. In a large open pipe, there is a great loss of effect, since a great portion of the heated current that enters it from the furnace, flows through the central parts of the pipe without coming into contact with the absorbing surface. This evil is remedied by making the current flow into the narrow space between two concentric cylinders, (*aa*, Fig. 201,) by which means it is kept closely in contact with the surfaces of both cylinders. It is found, by calculation, that a flue of sufficient dimensions may be obtained to afford a free exit to the smoke and heated air, when the space between the cylinders is only an inch in diameter. Thus, suppose the outer cylinder is 12 inches in diameter, and the inner or air cylinder, 10 inches; then the value of the flue, compared with a pipe, will be as follows—

Section of outer cylinder,	$12^2 \times .7854$
Ditto of inner ditto,	$10^2 \times .7854$

Difference, $(144-100).7854=34.5576$ square inches=circular ring between the two cylinders. Or, since the space between the cylinders is bisected by vertical partitions, half this, or $17.2788=$ the effective value of the flue=a pipe 4.7 in diameter.

Hence it appears, that so small a space as one inch between two cylinders of twelve and ten inches, will afford a flue as great as a pipe of about $4\frac{3}{4}$ inches in diameter, and sufficient therefore for a furnace of corresponding dimensions. We thus find it in our power to transmit the heated current from a furnace through a space so narrow as to be brought very effectually into contact with the absorbing surfaces; consequently, in passing a comparatively small distance, under such circumstances, the heat is as fully absorbed and transmitted to the room, as in traversing a great length of large open pipe. Hence, there is a great saving of material. But, secondly, another incidental advantage of great importance is obtained by this construction, namely, great purity of air. The inner cylinder is no sooner heated than the air within is rarefied, and cold air flows in from below (at *n*) as into a chimney. This imbibes the heat of the cylinder itself, keeping that from becoming so hot as to contaminate the air of the room, while the air that circulates through it, flows out at the top into the room, in a current of warm air of most agreeable temperature. By this constant circulation of the air of the room through the radiator, a mild and uniform temperature is maintained throughout the apartment. If the apparatus is properly managed, no part of it ever becomes so hot as to burn the particles of vegetable or animal matter, more or less of which are usually floating around a stove, and which when burnt communicate an unwholesome effluvium to the atmosphere of the room, as is experienced in many close stoves which become highly heated. Thirdly, by substituting for a long pipe travers ing an apartment, (which usually presents a very unseemly ap pearance,) a symmetrical figure tastefully ornamented, the unpleasant aspect frequently accompanying close stoves is avoided.

479. But a much more extensive operation of the principles by which the atmosphere is put in motion, is exhibited to us by nature, in the phenomena of WINDS. Rarefaction by heat and condensation by cold are the chief causes of winds. Their distinct existence and modes of operation, can frequently be discovered; and, in cases where we can discover neither, we are authorized to infer the presence of such a cause, since it is so constantly connected with the same effects in very numerous examples that daily pass before our eyes, while we are unacquainted with any other adequate causes of the same phenomena. The motion of the air, however, producing a wind, may be merely *relative*, arising from the motion of the spectator. Thus a steamboat, moving at the rate of sixteen miles an hour in a perfect calm, would appear to one on board to be facing a wind, moving at the same rate in the opposite direction; or if, in the diurnal revolution of the earth on its axis, any point of the earth's surface should move faster than the portion of the atmosphere above it, a relative wind in the opposite direction would be the result. (Art. 11.) The *direction* of the wind may be modified by various causes, the actual direction being the *resultant* of two or more currents which meet from different directions, or of several different forces.*

* See some able remarks on this subject by Mr W. C. Redfield, Amer. Jour. xx. 17.

480. *Land and sea breezes* afford a striking exemplification of the principle in question. These winds prevail in most maritime countries, but more especially in the islands of the torrid zone, blowing off from the land at night, and toward the land in the day time. If we place a hot stone in a room, (says Dr. Robison,) and hold near to it a candle just extinguished, we shall see the smoke move toward the stone, and then ascend up from it. Now, suppose an island receiving the first rays of the sun in a perfectly calm morning; the ground will become warm, and will rarefy the contiguous air. If the island be mountainous, this effect will be more remarkable; because the inclined sides of the hills will receive the heat more directly. The midland air will therefore be most warmed; the heated air will rise, and that in the middle will rise fastest; and thus a current of air upward will begin, which must be supplied by air coming in on all sides, to be heated and to rise in its turn; and thus the morning sea breeze is produced, and continues all day. This current will frequently be reversed during the night, by the air cooling and gliding down the sides of the hills, and we shall then have the land breeze. Professor Mitchell has rendered it probable that this current is performed in a constant *gyration;* so that the air which flows in upon the land by day, rises, flows out above, and returns again in the same current; and that the process is similar by night, only the current is reversed.*

481. The *trade winds* afford an example of the operation of the same causes on a still greater scale. These winds prevail in the torrid zone and a little beyond it, extending to nearly 30° on both sides of the equator. When not affected by local causes, they blow constantly at the same place, in one and the same direction throughout the year. Their general direction is from northeast to southwest on the north side of the equator, and from southeast to northwest on the south side of the equator. They owe their origin to the combined agency of two causes, namely, the movement of the air on either side of the equator, northward or southward toward the place of greatest rarefaction, and the westerly tendency arising from the effect of the earth's diurnal rotation on its axis,† since they do not instantaneously acquire the greater velocity which the equatorial regions have in consequence of the earth's revolution on its axis.‡ The duration of the trade winds is variously modified in different parts of the world, but always in such a manner, that they blow toward the point of greatest rarefaction, and receive a relative motion from the effect of the earth's diurnal rotation.

* American Journal of Science, Vol. xix.

† See p. 61, *Problem* 3.

‡ For a more extended description and inquiry into the causes of the trade winds, see Daniell's Meteorology, p. 455, and American Journal of Science, Vol. xix.

482. The foregoing atmospheric phenomena arise chiefly from the relations of air to *Heat*; we are next to trace a few of the leading phenomena, which result from the relations of air to *Moisture*.

By the action of the sun's heat upon the surface of the earth, whether land or water, immense quantities of vapor are raised into the atmosphere, supplying materials for all the water that is deposited again in the various forms of dew, fog, rain, snow, and hail. Our limits will not allow us to enter largely into Meteorology, under which head the various phenomena of the atmosphere are included, but we shall be able barely to glance at the subject.

483. A view of the constitution of the atmosphere first proposed by Mr. Dalton, a distinguished English meteorologist, now generally prevails. It maintains, that the different aëriform substances of which the atmosphere consists, namely, oxygen, nitrogen, carbonic acid, and watery vapor, do not combine with one another, but co-exist as so many independent bodies; that the watery vapor in the atmosphere, is raised into it and maintained there, not by any force of attraction existing between that and the other elements, but simply by the agency of heat; and that it is precipitated, or returns to the state of water, merely by the reduction of temperature, or the application of cold. The quantity of vapor which can exist in the atmosphere at any given time, depends upon its elasticity, and that depends upon its temperature. The elasticity of vapor increases very rapidly as we heat it. At 32°, the quantity of vapor that can exist in the atmosphere is only $\frac{1}{150}$th of its volume; while at 93°, the extreme heat of summer, the quantity rises to $\frac{1}{20}$.* The increase of quantity, as the temperature is raised, is very slow in the lower degrees of the scale, but very rapid in the higher; so that when the air, in a hot summer day, rises from 80° to 90°, the amount of water is increased vastly more than it would be by rising in the winter through an equal number of degrees, as from 40° to 50°. Hence, hot air contains, in fact, a far greater amount of water than cold air, but being in the elastic invisible state, it is not obvious to the senses, but such air appears very dry. On the other hand, when air which is very hot, cools a few degrees, as from 90° to 80°, the amount of water precipitated is much greater than is occasioned by a similar reduction of temperature in the lower degrees of the scale, as from 40° to 30°. The temperature at which the vapor of the atmosphere is condensed at any given time is called the *dew point*. Thus, if we place a tumbler of cold water on the table, in a warm day, when the thermometer is at 80°, with a thermometer inserted in the tumbler,

* Thomson, on Heat and Electricity, p. 251.

and find that when the mercury has sunk to 74° moisture just begins to form on the outside of the tumbler, then we say, the dew point, for that time, is 74°.

484. *Dew is formed when the air comes in contact with a surface in a certain degree colder than itself.** This is the simplest deposition of moisture from the atmosphere. Thus dew is formed copiously on a cup of cold water during summer, particularly before a thunder shower; because then the air is hot, and saturated with moisture, a portion of which it deposits as soon as it is cooled. It is ascertained by actual observation that on those nights when copious dews occur, the ground becomes twelve or fourteen degrees colder than the air a few feet above it.† Consequently, whenever the air, by circulating over the surface of the ground, comes in contact with this colder surface, it deposits a portion of moisture upon it. The quantity actually deposited will of course be greater as the difference of temperatures between the air and the ground is greater, and as the air contains more moisture.

Dew is found to be deposited on different substances unequally, —more on vegetables than on dry sand; very little on bright metallic surfaces; and none at all on large bodies of water, as the ocean. In all cases, however, these surfaces are observed to maintain a corresponding difference in the temperature they acquire, some growing much colder than others equally exposed, while the surface of the ocean remains at the same temperature as the air incumbent upon it. The air therefore is not cooled by circulating upon it, and no dew is deposited.‡

485. *Fogs are produced by watery vapor coming in contact with air colder than itself.*

The vapor may be such as is just rising from the ground, or such as before existed in a body of common air that meets and mixes with the colder air. Thus, in a cold morning, smoke proceeds from various moist substances, as from the breath of animals, from a hole in the ice of a river, from wells, and from many other sources. In each case, the vapor meets with cold air, is condensed, and deposited in the form of fog. A striking example of fogs is seen over rivers, particularly in a summer morning, marking out their courses for a great distance. Here, since the temperature of the water changes but little during the night, while the neighboring land, and of course the air over the land, has become cold, the vapor which rises from the river during the

---

* It will be remarked that dew is deposited only at the *surfaces* of bodies, and not, like fog and rain, in the atmosphere itself.

† Wells, on Dew.

‡ Ibid.

night, and meets with cold air, is condensed into fog. The fogs formed over shoals and sand banks, as the banks of Newfoundland, are deposited from the warm and humid air of the ocean, which is cooled by mixing with the cold air over the banks. Fogs are phenomena of cold climates, and are not so common in hot countries; the vapor in the latter situations having too great a degree of elasticity to permit it to condense into a fog near the surface of the earth.

486. *Clouds are dependent on the same principles as fogs, consisting of vapor condensed by the cold of the upper regions.* They are formed over water or moist places, by vapor rising so high, as to reach a degree of cold sufficient to condense it; or they result from the mixture of warmer with colder air, proceeding always from the warmer portion.

487. *Rain is produced by the sudden cooling of air, containing large quantities of watery vapor.*

Suppose two bodies of air, a hotter and a colder portion, both near the dew point, (Art. 483,) to meet; the compound would assume a temperature which was the mean between the two; but the elasticity which the colder portion of air would gain, would not equal that which the warmer portion would lose, by the loss of the same amount of heat. (Art. 483.) Hence the elasticity of the mixture would be less than the average elasticities of the separate portions, and consequently water would be deposited. If the separate portions of air are not near the point of condensation, still the elasticity of the vapor in the mixture may be so much less than that of the constituents, as to render it unable to hold in the invisible state all the water they contained; and in this case, more or less water would be deposited.

488. This view of the general cause of rain, (which is commonly called Hutton's theory of rain, from Dr. Hutton of Edinburgh, who first proposed it,) is capable of being confirmed by a extensive induction of facts, by which it would appear, that *variable* winds, favorable to the mixture of air of different temperatures, are accompanied by rain, while *constant* winds are accompanied by dry weather.

489. *Hail is produced by the mixture of exceedingly cold air, with a body of hot and humid air.** The cold wind is supposed to be derived from an elevation considerably above the term of perpetual congelation, and to be suddenly transferred to a body of hot and humid air, from which it precipitates the hail. Or it

* See some remarks on Hail Storms, by the compiler of this work, in the Am. Jour. Science, Vol. XVIII.

may be supposed to result from a hot wind blowing from the torrid regions into the limits of perpetual frost, and thus having its watery vapor suddenly congealed. But probably the most fre quent mode by which violent hail storms are actually produced, is by the sudden transportation of a body of very hot and humid air to a great height in the atmosphere, as by whirlwinds. All that the theory requires, in order that hail should be precipitated, is, that *very hot* and *very cold bodies of air* should be mixed in any way whatsoever. Accordingly, hail is found to be most frequent and violent in those regions where hot and cold bodies of air are most easily mixed. Such mixtures are rarely formed in the torrid zone, since there the portion of *cold air* would be wanting; and a similar difficulty exists in the frigid zone, for there the *hot air* is wanting; but in the temperate climates, the heated air of the south, and the intensely cold winds of the north, may be much more easily brought together; and accordingly, in the temperate zones it is that hail storms chiefly occur. Even in these climates, they are most frequently found in places where such mixtures are most easily formed, as in the south of France, lying, as it does, between the Pyrenees and the Alps, which are covered with perpetual snows, while the intervening country is subject to become highly heated by the summer's sun, or is even visited, especially at a certain elevation, by occasional blasts of the hot winds that cross the Mediterranean.

490. Mr. Redfield has investigated, with great success, the phenomena of violent storms, especially of *Atlantic hurricanes*, and has shown that they are generally, if not always, great whirlwinds. They usually take their rise in the equatorial region eastward of the West India Islands; they spin like a top, (or more like those little whirls which we sometimes observe to take up leaves and other light substances,) advancing slowly to the northwest, until they approach the coast of the United States near the latitude of 30°, and then veer to the northeast, running nearly parallel to the American coast, and finally spend themselves in the northern Atlantic. It is a remarkable fact, that their rotary motion is always in one direction, namely, from right to left, or against the sun. This motion is also far more violent, especially in the central parts of the storm, than the progressive motion. The rotary motion may amount to 200 miles or more per hour, while the forward motion of the storm is not more than 20 or 30 miles. The able and ingenious papers of Mr. Redfield on this subject may be found in the different volumes of the American Journal of Science.

Mr. Espy does not admit the rotary motion of storms supposed by Mr. Redfield, but maintains that *the winds blow from all directions toward the center of the storm;* that the air ascends in the center in a continual column; that being cooled as it rises, its vapor condenses, forming cloud; that the latent heat given out on condensation expands the ascending column, and causes it to rise still higher, condensing more vapor; and finally, that when the original body of air is hot and largely charged with vapor, and rises rapidly to the higher and colder regions of the atmosphere, the effects are proportionally violent, giving rise to tornadoes, thunder storms, and water-spouts.*

CHAPTER III.

OF THE MECHANICAL AGENCIES OF AIR AND STEAM.

491. IN consequence of our power of forming a vacuum, either by the exhaustion of air or by the condensation of steam, and of directing the force with which these elastic substances rush into a void or press toward it, air and steam become important agents or prime movers, in various kinds of machinery. Many of the most useful machines involve in their construction the principles of both hydraulics and pneumatics, and therefore we have reserved an account of such machines to the present section.

Fig. 202.

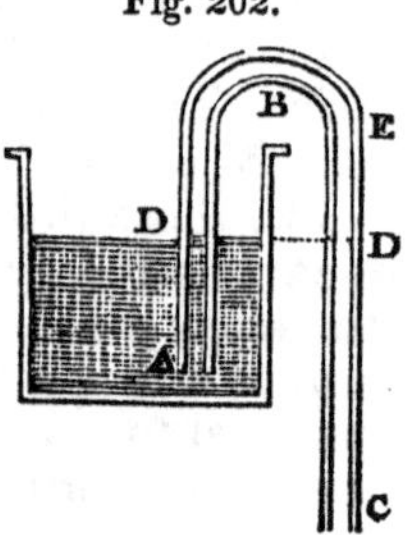

492. THE SYPHON.—If a tube having two arms, a longer and a shorter, be filled with water, or any other liquid, and the mouth of the shorter arm be immersed in water, the fluid will run out through the longer arm until the whole contents of the vessel are discharged. Such a tube is called a *syphon*. It may be filled with the fluid, either by suction or by pouring water into it, keeping the two orifices closed until the shorter arm is immersed. Or, when the syphon is large, each orifice is plugged, and water is poured in through an opening in the top of the bend. The opening being closed, the shorter leg is placed in the cistern, the plugs removed, and the fluid is discharged through the longer leg. The *principle* of the syphon is as follows. The atmosphere presses equally on the mouths of both arms of the tube; but this pressure on each orifice is diminished by the weight of the column of water in the leg nearest to it; consequently, more of the atmospheric pressure is overcome by the longer than by the shorter column, and

* Espy on the Philosophy of Storms.

therefore the *effective pressure*, (or what remains,) is less at the mouth of the longer than at that of the shorter column, and the fluid runs in that direction in which the resistance is least; and the constant pressure of the atmosphere on the surface of the fluid, causes the fluid to continue running until the surface has descended to the level of the mouth of the syphon. All this will be obvious by inspecting the figure.*

Were the shorter column thirty-four feet in height, it would counterbalance the entire pressure of the atmosphere on the surface of the fluid, and consequently, there would be no force remaining to drive the water forward through the tube. The syphon, therefore, can never raise water to a greater height than thirty-four feet, nor quicksilver higher than about thirty inches. It is obvious, also, that the place of delivery, that is, the mouth of the longer arm, must be at a lower level than the surface of the water in the reservoir; so that this instrument cannot be used for elevating, but only for decanting fluids, or transferring them from one vessel to another. Its chief use is by grocers, in transferring liquors from cask to cask. It is, however, sometimes employed to convey water from a well situated on rising ground, to a lower situation, or to carry it over a hill to a lower level on the other side.

493. *Intermitting Springs*, or springs which flow freely for a time, and then cease for a certain interval, when they flow again,

Fig. 203.

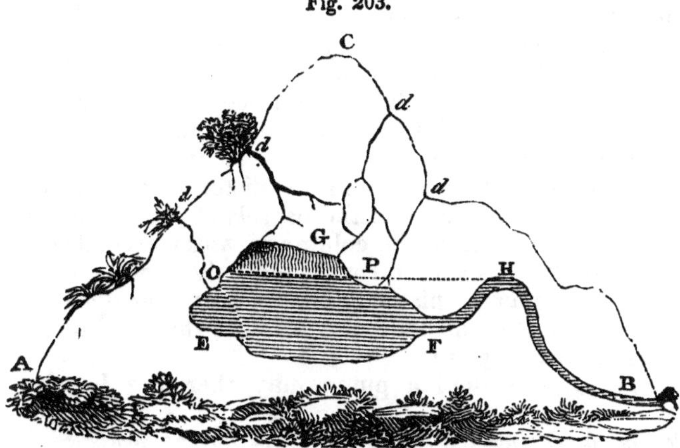

are explained on the principle of the syphon. The annexed cut represents a reservoir or hollow in the interior of a hill, having

* We prefer to describe such instruments *in general terms*, but the student will find it convenient to recite the explanation from the figure, and letters are annexed to the figures for that purpose.

a syphon-shaped outlet. It is obvious, upon hydrostatic principles, that no water will be discharged until the fluid has reached a level in the reservoir equal to the top of the bend in the outlet. Then it will begin to run out, and will continue to run, until the water has descended to the level of the outlet; after which, no more water will be discharged until enough has collected to reach the higher level, as before.*

494. THE COMMON SUCTION PUMP.—This pump consists of two hollow cylinders, placed one under the other, and communicating by a valve which opens upward. The lower cylinder (which has its lower orifice under water) is called the *suction tube.* In the upper cylinder, a piston moves up and down from the bottom to a spout in the side near the top. This cylinder we call the *exhausting tube.* Suppose, at the commencement of the operation, the piston is at the bottom of the exhausting tube, in close contact with the valve. On raising it, the air in the suction tube having nothing to resist its upward pressure, lifts the valve and expands, so as to fill up the void space, which would otherwise be left in the lower part of the exhausting tube. By this means, the air in the suction tube is rarefied, and no longer being a counterpoise to the pressure of the atmosphere on the surface of the well, the latter preponderates and forces the water up the tube, until enough has been raised exactly to counterbalance the excess of the elasticity of the external air above that of the tube. As the piston descends, the air below it is prevented from returning into the suction pipe by the valve which closes on its mouth, but escapes through a valve in the piston itself opening upward in the same manner as in the barrels of the air-pump. The piston being raised again, the column of water ascends still higher, until it makes its way through the valve into the exhausting pipe. Then as the piston descends, the water opens its valve, and gets above the piston, and is lifted to the level of the spout, where it is discharged.†

Fig. 204.

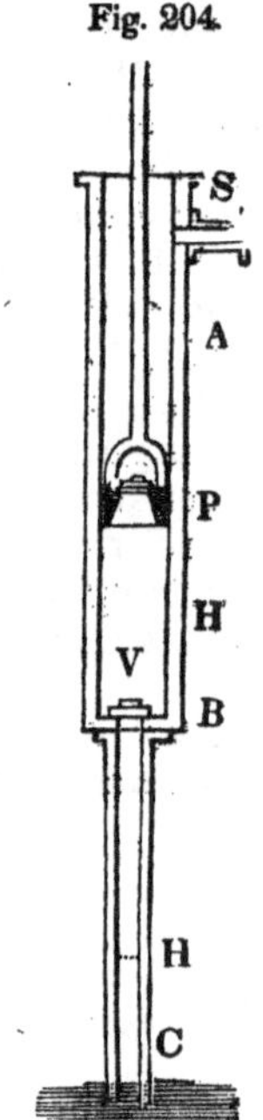

The principle of the suction pump may therefore be thus enunciated:

The water is raised into the exhausting pipe by the pressure of the atmosphere, and thence lifted to the level of the spout by means of the piston.

---

* Cavallo, I, 387.—Library of Useful Knowledge.

† The student is requested to describe from the figure. It is recommended to him, however, to form as distinct an idea as possible of the principle of a machine from the general description, before he resorts to the figure.

Since a column of water thirty-four feet in height, in the suction tube, would counterbalance the entire pressure of the atmosphere on the surface of the well, no force would remain to urge the column any higher, and therefore the valve at the top of the suction tube, must be less than thirty-four feet above the well.

495. Let us now consider the force which is required in each stage of the process, to elevate the piston, exclusive of the weight of the piston, rods, and the effects of the friction. Let the piston be at V, and the level of the water in the suction pipe at H. Let the number of feet in CH be called h. The elastic force of the air in BH will then be such as to exert a pressure on every square inch, equal to the weight of a column of water, whose base is a square inch, and whose height, expressed in feet, is $34-h$. In its ascent, therefore, each square inch of the section of the piston is pressed upward by this force. It is, on the other hand, pressed downward by the whole force of the atmosphere, which is equal to the weight of a column of water of the same base, and thirty-four feet high. The effective force then which resists the ascent of the piston, for every square inch, is the weight of a column of water, whose base is a square inch, and whose height is the difference between thirty-four feet and $34-h$ feet; that is, the effective force is h feet. Thus it appears, that it requires a force to lift the piston exactly equal to the weight of a column of water, whose base is equal to the section of the piston, and whose height is that of the water in the suction pipe, above the level of the water in the well. It follows, therefore, that as the water rises in the suction pipe, the force required to lift the piston is proportionally increased.

Let us next consider the force required to lift the piston in the second part of the process; viz. when the water raised has passed through the piston-valve.

Let the piston be at V, and the level of the water at H′; the downward pressure sustained by the piston, in this case, is evidently the weight of the incumbent water BH′, together with the weight of the atmosphere. Let h be the number of feet in the height BH′, and $34+h$ will express the number of feet in a column of water, whose base is equal to a section of the piston, and whose weight is equal to the whole downward pressure sustained by the piston.

On the other hand, the upward pressure is produced by the weight of the atmosphere pressing on the water in the reservoir, and transmitted through the column CB, to the lower surface of the piston. But as this pressure has to support the column BC, we must subtract from it the weight of this column, in order to obtain the effective upward pressure on the piston. From a column of water thirty-four feet in height, and with a base equal to the section of the piston, subtract as many feet as there are in BC, and we shall obtain a column whose weight is equal to the upward pressure.

The downward pressure equals $34+h$
The upward do. do. $34-\mathrm{BC}$

Remainder $h+\mathrm{BC}$

But $h+\mathrm{BC}=\mathrm{H'B}+\mathrm{BC}=\mathrm{H'C}$.

Thus it appears, that the force necessary to lift the piston, i the weight of a column of water, whose height is that of the column above the level of the water in the well, and whose base is equal to the section of the piston. This force, therefore, from the commencement of the process, continually increases, until the level of the water rises to the discharging spout, and thenceforward remains uniform.*

496. From the foregoing remarks, it is evident that the same force is expended in raising water by means of the pressure of the atmosphere, as when the force is applied directly. We lift upon the atmosphere, instead of lifting directly upon the column of water. This method of raising water from a well, is frequently more convenient than by a simple bucket, but the expenditure of force is the same in both cases.

To compute the actual force necessary to work a pump, (exclusive of the pump rods,) let the height of a discharging spout S, above the level of the water in the well, be expressed in feet, and let the number which expresses it be h. Let the diameter of the piston, expressed in parts of a foot, be d; then the section of the piston expressed in parts of a square foot, will be $d^2\times.7854$. If this product be multiplied by the number of feet h in the height, we shall obtain the number of cubic feet of water which it is necessary to lift at each stroke, since this number$=$ $d\times.7854\times h$. Now each cubic foot of water weighs about $62\frac{1}{2}$ pounds; hence $d^2\times.7854\times h\times 62\frac{1}{2}=$number of pounds required at each stroke to lift the piston.

The column of water discharged at each stroke, is equal to a column of water, whose base is the section of the piston, and whose altitude is the length of the stroke. The quantity may therefore be found, in cubic feet, by multiplying $d^2\times.7854$ by the number of feet in the length of the stroke. The weight of the water discharged may be ascertained in pounds avoirdupois, by multiplying this product by $62\frac{1}{2}$.

497. The Forcing Pump.—A cylinder ABC, (Fig. 205,) is placed with its lower end C in the reservoir. It has a fixed valve at V, opening upward, and a solid piston without a valve, play-

* Library of Useful Knowledge, Art. 'Pneumatics.'

ing air-tight in the upper barrel AB. It is connected with another barrel DE by a valve V′ opening upward and outward. The tube DE is carried to whatever height it may be necessary to elevate the water. Let us suppose that the solid piston P is in contact with the valve V, and that the water in the lower barrel is at the same level C with the water in the reservoir. Upon raising the piston, the air in BC will be rarefied, and the water will ascend in BC exactly as in the suction pump. Upon again depressing the piston, the air in PV will be depressed, and it will force open the valve V′, and escape through it. The process, therefore, until water is raised through V into the upper barrel, is precisely the same as for the suction pump, the valve V′ taking the place of the piston-valve in that machine. Now, let us suppose that water has been elevated through V, and that the space PV is filled with it. Upon depressing the piston, this water, not being permitted to return through V, is forced through V′, and ascends in the tube DE. By continuing the process, water will accumulate in the tube DE, until it acquires the necessary elevation, and is discharged. Or, to enunciate the principle of this machine in general terms:

Fig. 205.

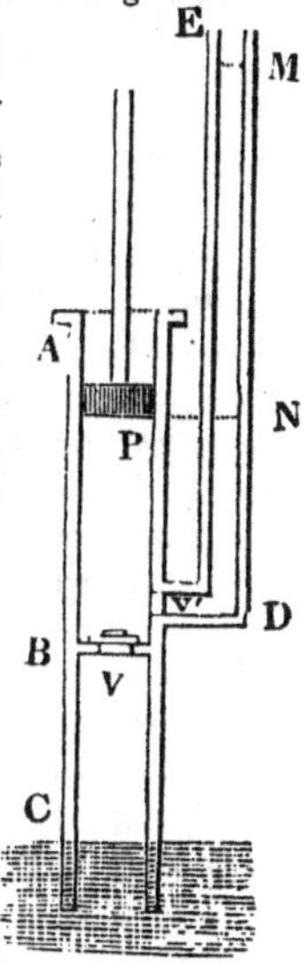

In the forcing pump, the piston has no valve, but the water being elevated into the exhausting tube, as in the suction pump, it is then forced, by the descent of the piston, into the ascending pipe through a valve placed in the side and at the bottom of the exhausting tube.

498. The force requisite to elevate the piston in this pump until the water reaches it, is computed in exactly the same manner as for the suction pump, and exclusive of the weight of the piston and its rods, and the effects of friction, it is equal to the weight of a column of water whose base is the section of the piston, and whose height is the distance of the level of the water in the barrel AC, above the level in the reservoir. It is evident also from what has been said on the suction pump, that the valve V should be less than thirty-four feet above the level of the water in the reservoir. If P express in pounds *av.* the weight of the piston and its rods, d be the diameter of a section of the piston expressed in parts of a foot, and h be the number of feet in AC, the force in pounds necessary to lift the piston will be $h \times d^2 \times .7854 \times 62.5 + P$.

Let us now examine the force necessary to depress the piston. Let the level of the water in ED be M. The atmospheric pressure on M will be balanced by the same pressure on the piston, by the power of transmitting pressure peculiar to fluids. This force may therefore be neglected; also the part PV′ will balance the part ND of the ascending column. Hence it appears, that the pressure exerted by the water in PV on the lower surface of the piston, is equal to the weight of a column of water whose base is equal to the section of the piston, and whose height is MN. This, therefore, is the force to be overcome in the descent of the piston, and the weight P of the piston and its rods assists in overcoming it. Let h' be the number of feet in MN, and the mechanical force necessary to be applied to depress the piston will be expressed in pounds by $h' \times d^2 \times .7854 \times 62.5 - P$.

From these observations, it appears that the weight of the piston and its rods assists the *forcing power* of the machine, but opposes its *suction power*. These effects, therefore, on the whole, neutralize one another.*

499. The entire force used in raising the water, will be found by adding the force necessary to elevate the piston to that which is necessary to depress it. As in this case the weight of the piston and rods increases the one as much as it diminishes the other, the entire force will be the weight of a column of water whose base is the section of the piston, and whose height is PC +MN, that is, the height of the level of the water in the ascending pipe above the level of the water in the reservoir; and expressed in pounds, this is $(h+h') \times .7854 \times d^2 \times 62.5$.

It appears, therefore, that, other circumstances being the same, the power of the forcing pump has the advantage over that of the suction pump, by the weight of the piston and its rods.

500. In forcing pumps, since the power is applied by separate impulses, the water would issue in jets were not some contrivance adopted to equalize its flow from the tube. This purpose is effected by means of an air-vessel, in which a portion of condensed air is made the medium of communication. The force imparted by successive blows of the piston is first received by this confined body of air, and this, by its elasticity, reacts on the surface of the water in the air-vessel, and forces it out by the conducting pipe or hose.

An example of this is afforded in the *Fire Engine*. The fire engine consists of two forcing pumps, which throw the water into an air-vessel, from which it is thrown out of the conducting hose by the elastic pressure of condensed air. Thus, (Fig. 206,) AB, AB are two forcing pumps, whose pistons P, P, are wrought by a beam whose fulcrum is at F; V, V, are valves which open upward from a suction tube T which communicates with a reservoir; t, t, are force pipes, which communicate by valves V′, V′, opening into an air vessel M. A tube L is inserted

* Library of Useful Knowledge.

in the top of this vessel, terminating in a leathern tube or hose, through which the water is forced by the pressure of air confined in M, which, in consequence of its elasticity, acts nearly uniformly on the surface of the water, and forces it through the hose in a continual stream.

Fig. 206.

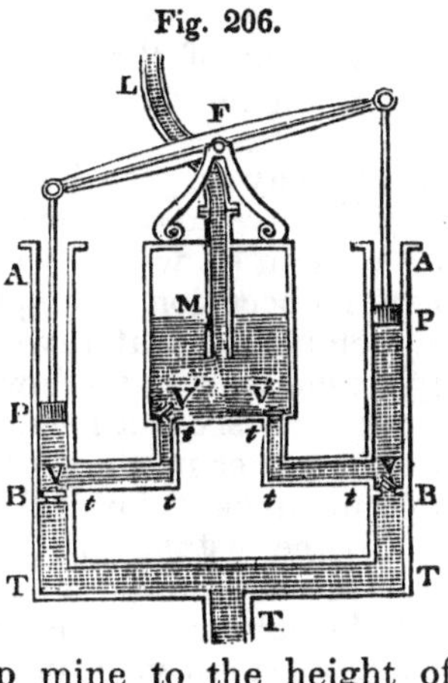

501. The Hungarian Machine.—This celebrated machine is employed in draining a mine at Chemnitz, in Hungary. We introduce a description of it here, on account of its affording a good illustration of several hydraulic and pneumatic principles. Here the object is, to raise water from a deep mine to the height of ninety-six feet, where it can be poured off by a horizontal channel. Now it is easy in such a case, to take a stream of water near the top of the pit, (and a very small stream will answer the purpose,) and to convey it into a pipe which shall descend into the mine, and afford, by hydrostatic pressure, any degree of force required to raise the water of the mine, not indeed to the top of the pit, for that is hardly ever necessary, but to such a height that it may be poured off by a horizontal drain. In the mine of Hungary, the water, which is to supply the required pressure, is taken at the height of two hundred and sixty feet above the surface of the water in the pit. From the cistern A, where the head of water collects, it enters the perpendicular pipe B, which descends nearly to the bottom of the air-vessel C. Flowing into this, it condenses the air before it, which, by its elasticity, receives and exerts the whole force created by the pressure of the column of water B. This force is transmitted through the air-pipe D, to the surface of the water contained in the well E, which is sunk into the water of the mine, admitting it freely by means of a valve in the bottom opening upward. This

Fig. 207.

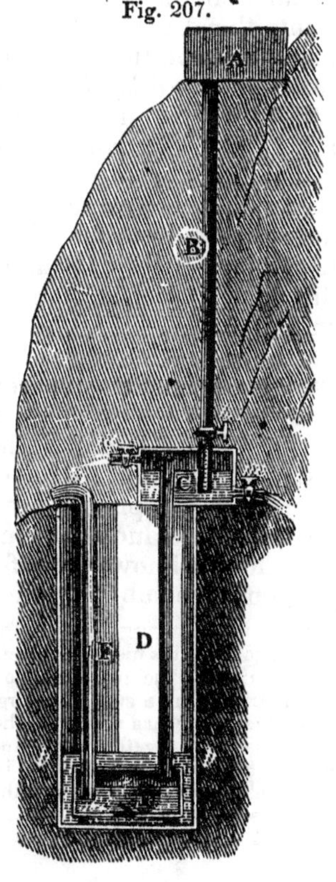

well and the air-vessel C, are made strong and air-tight. From near the bottom of the well proceeds a perpendicular tube F, reaching to the height of the drain.

502. We may now easily understand the operation of the machine. We have to raise water ninety-six feet, and we can command a column of water two hundred and sixty feet high; but we have no occasion to employ the whole of this force, and so long a column of water would require a pipe of very great strength, especially in the lower parts of it. A column one hundred and thirty-six feet long, is found by calculation competent to raise the water in the pit to the required height of ninety-six feet, and to make it flow off with considerable velocity into the drain. At the distance of one hundred and thirty-six feet from the reservoir, we interpose an air-vessel C, and receive the entire force of the column B upon the air of this vessel, which is compressed into a small space in the upper part of the vessel, and has its elasticity proportionally augmented. (Art. 452'.) This force, by means of the pipe D, is transmitted to the surface of the water in the well, and forces the water up the pipe F, which delivers it into the drain. The *principle* of the Hungarian machine, therefore, may be thus enunciated:

*Water is raised by the pressure of a column of water, longer than the column required to be raised, and at a higher level; the pressure being transmitted from one column to the other, through the medium of condensed air.**

When the water contained in the cistern E is raised and delivered through the pipe F, the pressure upon E is relieved by opening the stop-cock *n*, and closing *l*. Water is again let in by opening the stop-cock *l*, and closing *n*, and thus the process is continued.

503. Steam-Engine.—It belongs to Chemistry to investigate the properties of steam, and to Natural Philosophy to apply it as a mechanical agent. The steam-engine is the fruit of the highest efforts of both these sciences, and the most valuable present ever made by philosophy to the arts.† As it is impossible clearly to understand the principles and construction of this engine, without a knowledge of the properties of steam, on which they depend, we subjoin an account of a few of its leading prop-

* A remarkable fact is mentioned in connection with the Hungarian Machine, which shows very strikingly the increase of capacity for heat and consequent production of cold, which arises from a sudden enlargement of volume. When the efflux of the water from the pipe F has ceased, if the cock of the air-vessel C be opened, the water and air rush out together with prodigious violence, and the drops of water are changed into hail or lumps of ice. It is a sight usually shown to strangers, who are desired to hold their hats to receive the blasts of air; the ice comes out with such violence as frequently to pierce the hat like a bullet.—*Gregory's Mechanics, II,* 221.

† Dr. Black.

erties, referring to chemical authors* for a more detailed view of the subject.

(1.)† The great and peculiar property of steam, on which its mechanical agencies depend, is *its power of creating at one moment a high degree of elastic force, and losing it instantaneously the next moment.* This force, acting on the bottom of the piston which moves in the main cylinder, raises it, and fills the space below it with steam. The steam is suddenly condensed, and hence no obstacle is opposed to the descent of the piston, but it is readily forced down again by steam acting from above. This alternate motion of the piston, the rod of which is connected with the working beam, is all that is required in order to communicate motion to all parts of the engine.

(2.) *The elastic force of steam depends on its temperature and density conjointly; and the temperature necessary to its production depends upon the pressure incumbent upon the water during its formation.*

The reason why water boils at the temperature of 212° is, that at that temperature the vapor acquires just elasticity sufficient to overcome the atmospheric pressure. Hence, steam produced at the temperature of boiling water, has a force equal to the pressure of the atmosphere. When formed at a lower temperature, its elasticity diminishes in a geometrical ratio, and increases in the same ratio when it is formed at a higher temperature. Water boils, or is converted into vapor, at a temperature less than 212°, on high mountains, (Art. 457,) or under the receiver of an air-pump, or in other situations where the pressure of the atmosphere is diminished; and in a *vacuum* the boiling point of water is as low as 72°.

(3.) *Heat rapidly augments the elasticity of steam by increasing its density.* If we introduce a few grains of water into a flask, and place it over the fire, the water will soon be converted into steam, which will expel the air of the vessel and fill its whole capacity. If we now close the orifice of the flask and continue the heat, the steam will increase in elastic force in the same manner as air would do under similar circumstances, which is at a comparatively moderate rate, so that it might be heated *red hot* without exerting any very violent force. If, however, the vessel is partly filled with water, and the heat is continued as before, then the elastic force is rapidly augmented, and becomes at length so great as to burst almost any vessel that can be provided; for every new portion of vapor that is raised from the surface of the water, adds to the density of that which was before in the vessel, and proportionally increases its elasticity. In the experiments of Mr. Perkins, a confined portion of steam, not

* See, especially, Silliman's Chemistry, Vol. I.

† See Review of Renwick on the Steam-Engine, America Journal of Science Vol xx, 326.

in contact with water, was heated to the temperature of 1400°, and still its pressure did not exceed that of five atmospheres; but, by injecting more water, although the temperature was lessened, the elastic force was gradually increased to one hundred atmospheres.*

The elastic force of steam at 250° is twice that at 212°, so that the addition of only 38° of heat doubles the power; 25° more trebles it, and 18° more quadruples the force. Thus steam at 293°, has four times the elastic force of steam at 212°; and steam at 510° would be 50 times stronger, and would exert a force of 750 lbs. to the square inch, or 108,000 lbs.† to the square foot. Hence, the elastic force increases much faster than the temperature, and decreases very fast as we descend the scale below the boiling point. Vapor is formed at every temperature, even from ice and snow; but in this case it differs from that formed at the higher temperatures in possessing but very little elasticity, as will be apparent from the following table:

Table of the elastic force of Vapor from 32° *to* 80°.

Temperature.	Force of vapor in inches of mercury.	Temperature.	Force of vapor in inches of mercury.
32°	0.2000	57°	0.4657
33	0.2066	58	0.4832
34	0.2134	59	0.5012
35	0.2204	60	0.5200
36	0.2277	61	0.5377
37	0.2352	62	0.5560
38	0.2429	63	0.5749
39	0.2509	64	0.5944
40	0.2600	65	0.6146
41	0.2686	66	0.6355
42	0.2775	67	0.6571
43	0.2866	68	0.6794
44	0.2961	69	0.7025
45	0.3059	70	0.7260
46	0.3160	71	0.7507
47	0.3264	72	0.7762
48	0.3372	73	0.8026
49	0.3483	74	0.8299
50	0.3600	75	0.8581
51	0.3735	76	0.8873
52	0.3875	77	0.9175
53	0.4020	78	0.9487
54	0.4171	79	0.9809
55	0.4327	80	1.0120
56	0.4489		

* Renwick on the Steam-Engine, p. 95.

† Webster's Principles of Hydrostatics, p. 186

(4.) *The space into which a given quantity of water is expanded in becoming steam, depends upon the temperature, and of course upon the degree of pressure, at which it is formed.* Water converted into steam at the temperature of 212°, expands nearly one thousand and seven hundred* times; but when so confined as to be heated to 419°, its volume is only thirty-seven times that of the water from which it is formed. According to Dr. Thomson,† at a temperature not much higher than 500°, steam would not much exceed double the bulk of the water from which it is generated. The expansive force of such steam would be truly formidable. It would, when it issued into the atmosphere, suddenly expand six hundred and fifty times. We do not know at what temperature water would become vapor without any increase of volume, but we can estimate that it would then support a column of mercury three thousand two hundred and forty-three feet (or more than half a mile) high, and would exert a pressure of nearly *twenty thousand pounds* on every square inch.

(5.) *The absolute quantity of heat is always the same in the same weight of steam, whatever may be its temperature.* When vapor is formed at a low temperature, nearly all the heat that enters it is in the latent state; but as we heat it to a higher degree, its proportion of sensible heat is constantly augmented, and that of latent heat diminished in the same ratio, so that the sum of the two is the same constant quantity.

These preliminary principles being well understood and kept clearly in mind, it will be easy for the learner to comprehend the principles involved in the steam-engine, and the dangers with which it is environed. The general interest felt in this subject renders it one peculiarly deserving of the attention of the student, and induces us to devote a considerable space to the consideration of it.

504. The steam-engine owes its present form and perfection, chiefly to the genius and labors of the late *James Watt, Esq.*, of England. His inquiries on the subject commenced in the year 1763. The engine in use previous to that time, was what is now called the Atmospheric Engine. It has already been remarked, (Art. 503,) that the chief object in the use of steam is to cause the alternate ascent and descent of a piston moving in a cylinder, since this motion may, by the aid of machinery, be so modified as to answer all the purposes required of the engine. In the atmospheric engine, at the commencement of the operation, the piston remained drawn up to the top of the cylinder, being kept there by the preponderance of the opposite arm of the lever or working beam, to which it was attached. Steam being ad-

* It will assist the memory to consider a cubic inch of water as forming a cubic foot of steam, as is nearly the fact.

† Outlines of the Sciences of Heat and Electricity, p. 228

mitted through a valve into the cylinder, expelled the air and occupied its place. Cold water now being admitted, the steam was suddenly condensed, a vacuum formed, and the atmospheric pressure on the upper side of the piston, having nothing to counterbalance it on the lower side, forced it down to the bottom of the cylinder. Steam being again admitted below the piston, supplied an upward force equivalent to the downward pressure of the atmosphere on the piston, and the preponderance of the opposite arm of the lever dragged up the piston as before.

It is impossible to understand the reason of the construction of the different parts of Watt's steam-engine, without a knowledge of the imperfections of the atmospheric engine,—imperfections for which he sought and found a complete remedy. We therefore subjoin a brief notice of the successive steps by which Mr. Watt was led to his great improvements.*

505. Previous to the discoveries of Watt, the energy of this power was well known; but the waste of fuel consumed in generating it was so enormous, as to render it, in most situations, a more expensive power than the other forces generally employed to move machinery. The reason of this waste will be easily understood. When the steam fills the cylinder, so as to balance the atmospheric pressure on the piston, the cylinder must have the same temperature as the steam itself. Now, on introducing the condensing jet, the steam mixed with this water, forms a mass of hot water in the bottom of the cylinder. This water, not being under the atmospheric pressure, boils at a very low temperature, (Art. 457,) and produces a vapor which resists the descent of the piston. The heat of the cylinder itself assists this process; so that in order to produce a tolerably perfect vacuum under the piston, it was found necessary to introduce so considerable a quantity of condensing water, as would reduce the temperature of the water in the cylinder lower than 100°, and which would consequently cool the cylinder itself to that temperature. Under these circumstances, the descent of the piston was found to suffer very little resistance from any vapor within the cylinder. But then on the subsequent ascent of the piston, an immense waste of steam ensued; for on being admitted under the piston, the cold cylinder and water of condensation immediately condensed the steam, and continued to do so, until the cylinder became heated again up to 212°, to which point the whole cylinder must be again heated, before the steam would acquire sufficient elasticity to raise the piston. Here then was an obvious and an extensive source of the waste of heat. At every descent of the piston, the cylinder must be cooled to below 100°; and at every ascent, it must be again heated to 212°. It there-

* Abridged from Lardner's Lectures on the Steam-Engine.

fore became a question, whether the force gained by the increased perfection of the vacuum was adequate to the waste of fuel in producing the vacuum; and it was found, on the whole, more profitable not to cool the cylinder to so low a temperature, and consequently to work with a very imperfect vacuum, and a diminished power. Watt, therefore, found the atmospheric engine in this dilemma, either much or little water of condensation must be used. If much were used, the vacuum would be perfect; but then the cylinder would be cooled, and would occasion an extensive waste of fuel in heating it. If little were used, a vapor would remain, which would resist the descent of the piston, and rob the atmosphere of part of its power. The great problem then pressed itself on his attention, *to condense the steam without cooling the cylinder.* To the solution of this problem, Watt now gave his whole mind. The idea occurred to him of providing a vessel separate from the cylinder, in which a constant vacuum might be kept up. If a communication could be opened between the cylinder and this vessel, the steam, by its expansive property, would rush from the cylinder to this vessel, where, being exposed to cold, it would be immediately condensed, the cylinder meanwhile being sustained at the temperature of 212°. This happy conception formed the first step of that brilliant career, which has immortalized the name of Watt, and spread his fame throughout the civilized world. He states that the moment the notion of "separate condensation" struck him, all the other details of his improved engine followed in rapid and immediate succession; so that in the course of a day, his invention was so complete, that he proceeded to submit it to experiment.

506. His first notion was, as has been stated, to provide a separate vessel called a *condenser*, having a pipe or tube communicating with the cylinder. This condenser he proposed to keep cold by immersing it in a cistern of cold water, and by providing a jet of cold water to play within it. When the communication with the cylinder is opened, the steam, rushing into the condenser, is immediately condensed by the jet and the cold surface. But here a difficulty presented itself, viz. how to dispose of the condensing water and condensed steam, which would collect in the bottom of the condenser. Besides this, a certain quantity of air would inevitably enter, mixed with the condensing water, which, accumulating, would collect in the cylinder, and resist the descent of the piston. To remedy this, he proposed *to form a communication between the bottom of the condenser and a pump, which he called the* AIR-PUMP; *so that the water, the air, and the other fluids, which might be collected in the condenser, would thus be drawn off, and this pump could be worked by the machine itself.*

507. Another inconvenience was still to be removed. On the descent of the piston, the air which entered the cylinder from above would lower its temperature; so that, upon the next ascent, some of the steam which entered it would be condensed, and hence would arise a source of waste. To remove this difficulty, Watt proposed to close the top of the cylinder altogether, by an air-tight and steam-tight cover, allowing the piston rod to play through a hole furnished with a stuffing box, and *to press down the piston by steam instead of the atmosphere.* Watt's grand improvements in the steam-engine consisted, therefore, of three separate steps. The first was the introduction of the *condenser;* the second, the contrivance of the *air-pump;* and the third, the employment of *steam* instead of atmospheric pressure, to force down the piston. This third step totally changed the character of the machine. It now became really a *steam-engine* in every sense; for the pressure above the piston was the elastic force of steam, and the vacuum below it was produced by the condensation of steam; so that steam was used both directly and indirectly as a moving power; whereas, in the atmospheric engine, the indirect force of steam only was used, being adopted merely as an easy method of producing a vacuum.

508. The last difficulty respecting the economy of heat that remained to be removed, arose from the liability of the external surface of the cylinder to become cool by the circulation of the cold air around it. To obviate this difficulty, Mr. Watt first proposed casing the cylinder in wood, as being a substance which conducts heat slowly. He subsequently, however, adopted a different method, and enclosed one cylinder within another, leaving a space between them, which he kept constantly supplied with steam. Thus the inner cylinder was kept constantly up to the temperature of the steam which surrounded it. The outer cylinder is called the *jacket.*

Watt computed that in the atmospheric engine, three times as much heat was wasted in heating the cylinder, and the other parts of the machine, as was spent in useful effect. And since, in the improvements proposed by him, nearly all the waste was removed, he contemplated, and afterward actually effected, a saving of *three fourths* of the fuel.

With these things distinctly in view, the learner will now be prepared to understand the construction of this noble engine, in its most improved state.

509. The difficulty of understanding the construction and principles of the steam-engine, (as is the case also with many other machines where the parts are numerous,) is greatly enhanced by the variety of accidental trappings or appendages that are employed about the machine, to perform subordinate offices. As

these render the comprehension of the leading principles difficult, when the explanation is attempted from the engine itself, so these inferior parts are often so multiplied in diagrams as greatly to obscure the representation. We shall begin our explanation with a diagram which presents the naked principles, divested of all unnecessary appendages.

510. The chief parts of the engine are the *boiler* A, the *cylinder* C, the *condenser* L, and the *air-pump* M. B is the *steam pipe*, branching into two arms, communicating respectively with the top and bottom of the cylinder; and K is the *eduction pipe*, formed of the two branches which proceed from the top and bottom of the cylinder, and communicate between the cylinder and the condenser. N is a cistern or well of cold water, in which the condenser is immersed. Each branch of pipe has its own valve, as F, G, P, Q, which may be opened or closed as the occasion requires.

Fig. 208.*

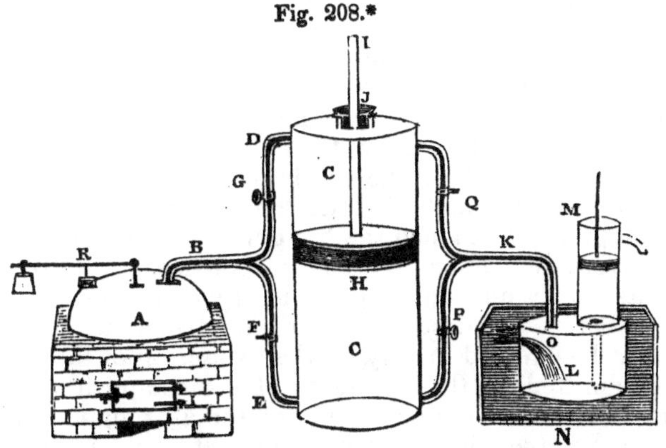

511. Suppose, first, that all the valves are open, while steam is issuing freely from the boiler. It is easy to see that the steam would circulate freely through all parts of the machine, expelling the air, which would escape through the valve in the piston of the air-pump, and thus the interior spaces would be all filled with steam. This process is called *blowing through;* it is heard when a steamboat is about setting off. Next the valves F and Q are closed, G and P remaining open. The steam now pressing on the cylinder, forces it down, and the instant when it begins to descend, the stop-cock O is opened, admitting cold water, which meets the steam as it rushes from the cylinder and effectually condenses it, leaving no force below the piston to oppose its descent.

* From Jones's Conversations on Chemistry, a work which contains a very luminous view of the elementary principles of the steam-engine.

Lastly, G and P being closed, F and Q are opened, the steam flows in below the piston and rushes from above it into the condenser, by which means the piston is forced up again with the same power as that with which it descended. Meanwhile the air-pump is playing, and removing the water and air from the condenser, and pouring the water into a reservoir, whence it is conveyed to the boiler to renew the same circuit.

512. The kind of valve chiefly employed in the steam-engine is that called the *puppet valve.* It resembles the stopper of a decanter, but is more obtuse. All these various appendages of the machine, are carried by the engine itself; the air-pump is worked by having its piston rod attached to one arm of the working beam, and the valves are opened at the instant required by means of levers, to which also motion is communicated from the same source.

513. Soon after the invention of these engines, Watt found that, in some instances, inconvenience arose from the too rapid motion of the steam piston at the end of its stroke, owing to its being moved with an *accelerated motion.** This was owing to the uniform action of the steam pressure upon it. For on first putting it in motion at the top of the cylinder, the motion was comparatively slow, but from the continuance of the same pressure, the velocity with which the piston descended was continually increasing, until it reached the bottom of the cylinder, when it acquired its greatest velocity. To prevent this, and to render the descent as nearly uniform as possible, it was proposed to cut off the steam before the descent was completed, so that the remainder might be effected merely by the expansion of the steam which was admitted to the cylinder. To accomplish this, he contrived by means of a pin on the rod of the air-pump, to close the upper steam valve when the steam piston had completed one third of its entire descent, and to keep it closed during the remainder of that descent, and until the piston again reached the top of the cylinder. By this arrangement the steam pressed the piston with its full force through one third of the descent, and thus put it into motion ; during the other two thirds of the way, the steam thus admitted acted merely by its expansive force, which became less in exactly the same proportion as the space, given to it by the descent of the piston, increased. Thus, during the last two thirds of the descent, the piston is urged by a gradually decreasing force, which in practice is found just sufficient to keep up in the piston a uniform velocity. Another advantage gained by this contrivance, independently of the uniformity of motion, was, that

* For since the steam *continues* to act upon the piston during its descent, its velocity would be constantly increased, like that of a ball in the barrel of a gun.

two thirds of the fuel was saved; for instead of consuming a cylinder full of steam at each descent of the piston, only one third of a cylinder full was necessary. Steam-engines constructed on this principle are said *to act expansively.*

514. From the foregoing account of the principles of the steam-engine, the learner will be able to give a full explanation of the construction and use of the various parts of this important and interesting machine, from the figure.*

Fig. 209.

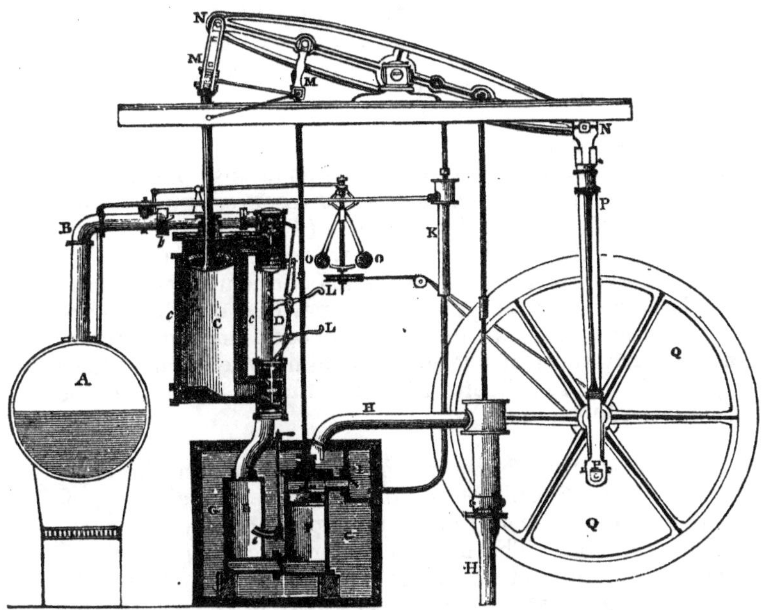

A. The Boiler.

B. The Steam Pipe, conveying the steam to the cylinder, having a steam-cock *b* to admit or exclude the steam at pleasure.

C. The Cylinder, surrounded with the *jacket, c c.*

D. The Eduction Pipe, communicating between the cylinder and the condenser.

E. The Condenser, with a valve *e*, called the *Injection-cock,* admitting a jet of cold water, which meets the steam the instant the latter enters the condenser.

F. The Air-Pump.

---

* One of the best descriptions of the steam-engine may be found in Millington's *Epitome of Natural Philosophy.*

G. G. COLD WATER CISTERN, for the Condenser, filled by
H. The COLD WATER PUMP.
I. The HOT WELL, containing water from the Condenser.
K. The HOT WATER PUMP, which returns the water of condensation to the boiler.
L. L. LEVERS, which open and shut the valves in the channel between the Induction Pipe, Cylinder, Eduction Pipe, and Condenser; which levers are raised or depressed by projections attached to the piston rod of the Air Pump.
M.M. Apparatus for PARALLEL MOTION, a beautiful contrivance, by which the piston rod is maintained constantly in a perpendicular position, while the end of the working beam which carries it moves in the arc of a circle.*
N. N. The WORKING BEAM.
O. O. The GOVERNOR. (Art. 334.)
P. The CRANK. (Art. 339.)
Q. Q. The FLY WHEEL. (Art. 331.)

The working beam is here represented as acting immediately upon the fly wheel, from which, as from a reservoir, motion may be distributed to all parts of the engine. (Art. 324.) It is obvious, however, that the same end of the working beam, instead of expending its force upon the fly wheel, may be connected directly with the piston rod of a pump for raising water, or with a horizontal shaft with wheels, as in the steamboat. In some steamboats, particularly those of a large size, the fly wheel is dispensed with, the inertia of the boat itself being sufficient to regulate the motion. (Art. 330.)

515. In steam-engines of the foregoing construction, the pressure introduced on one side of the piston derives its efficacy, either wholly or in part, from the vacuum produced by condensation on the other side. This always requires a condensing apparatus, and a constant and abundant supply of cold water. An engine of this kind must therefore necessarily have considerable dimensions and weight, and is inapplicable to uses in which a small and light machine only is admissible. If the condensing apparatus be dispensed with, the piston will always be resisted by a force equal to the atmospheric pressure, and the only part of the steam pressure which will be available as a moving power, is that part by which it exceeds the atmospheric pressure. Hence, in engines which do not work by condensation, steam of a much higher pressure than that of the atmosphere is indispensably

* In the engines constructed recently at New York, under the direction of Mr. R. L. Stevens, a substitute for the parallel motion has been introduced, that performs the task equally well, and is much less complex. On the head of the piston rod a bar is fixed, at right angles to it, and to the longitudinal section of the engine. The ends of this bar work in guides formed of two parallel and vertical bars of iron, by which the upper end of the piston rod is constrained to move in a straight line.—(*Renwick.*)

necessary; and such engines are therefore called *high pressure engines.* The steam, when it has produced its effect in raising or depressing the piston, escapes into the atmosphere, not being condensed and returned to the boiler as in low pressure or condensing engines. In these engines the whole of the condensing apparatus, viz. the cold water cistern, condenser, air-pump, &c. are dispensed with, and nothing is retained except the boiler, cylinder, piston, and valves. Consequently, such an engine is small, light, and cheap. It is portable also, and may be removed, if necessary, along with its load, and is therefore well adapted to locomotive purposes. Hence its use in small steamboats, and in locomotive carriages on railways.

516. Although the idea of propelling boats by the force of steam was entertained by different individuals, in different countries, long before it was carried into practice, yet the first successful effort at steam navigation was made by our countryman, *Robert Fulton,* in the year 1807. This year, the first steamboat, the *Clermont,* ascended the river Hudson to Albany. Fulton never contemplated a velocity for steamboats greater than eight or nine miles per hour; but the average speed now given to boats on the Hudson is no less than fifteen miles per hour, and sixteen miles is no unusual rate. Crossing the ocean by steam, as is now done with great speed and safety, is well placed among the highest enterprises of the present age. The first successful voyages across the Atlantic by steam, were performed by two steamships, the *Great Western* and the *Sirius,* in 1838.

The prevalence of westerly winds in the Atlantic, particularly at certain seasons of the year, rendered the passage from England to the United States uncertain, and frequently long and tedious; but the resistless force of steam has triumphed over this difficulty, and has made the passage between the two countries sure and speedy.

Steam is the most manageable of all the forces intrusted to man. It is his sole prerogative to develop and direct this power; and so pliant is it, that it is ever ready to perform for him the humblest or the mightiest of his works.

PART V.—ACOUSTICS.

517. ACOUSTICS* *is the science which treats of the nature and laws of* SOUND.

It has for its object to explain the origin or production of sounds—its propagation through different media—its reflexion from surfaces—and the philosophical principles of music.

CHAPTER I.

OF SOUND AND ITS MODES OF PRODUCTION.†

518. *Vibrations, in the sounding body, communicated to the organ of hearing, are the immediate cause of sound.*

Whatever may be the remoter cause of sound, vibrations must be considered as the *immediate* cause, since they always precede or accompany it, and since whatever affects the vibration of a body, produces a corresponding effect upon the qualities of the sounds which it emits, while those bodies whose sounds are similar, have something in common in their mode of vibration.

If we rub our moistened finger along the edge of a drinking glass, or draw a bow across the strings of a violin, we can in both cases procure sounds which remain undiminished in intensity, as long as the operation by which they are excited is continued. A similar fact takes place with respect to any other sonorous body, whose structure is not destroyed by the mode of excitation employed.

519. Though all bodies may, by some mode of excitation, be made to sound, there is a great difference among them in the *intensity* of the sounds which they produce during the operation, and in the *permanence* of these sounds after the excitation has ceased. Thus if we strike two bells, one of lead and the other of brass, the sound of the lead is feeble and momentary compared with that of the brass. Soft bodies, as wool, cotton, and down, cannot produce any sound; and those which are of the harder class, as rocks and stones, are not in general sonorous. Liquids, also, are incapable of producing much sound. The quality of producing sound belongs to *elastic* bodies, and those are the most sonorous which are most elastic. Glass, certain

* From ακουω, to hear.

† Edinburgh Encyclopedia, Art. *Acoustics*

metals, strings when strained, close, hard wood in thin layers, and the air itself, are at once among the most elastic and most sonorous bodies. Yet this quality is not an invariable attendant of elasticity. India rubber is extremely elastic, but not at all sonorous.

520. In comparing the properties of these substances, we shall find them distinguished from each other by the degree of *vibration* which they are capable of receiving, and by the *length of time* during which they can preserve a *vibratory motion;* those substances which are most capable of vibration being most sonorous, and those which can longest maintain a state of vibration, also persevering longest in emitting sound. Bodies, though of the same substance, differ in these respects according as their form varies; those forms which are most favorable to the production and continuance of a vibratory motion, being also most favorable to the production and permanence of sound. Thus a hollow globe of brass is far less sonorous than the hemispheres which are made by dividing it into two equal parts, since the structure of a globe is such that the parts mutually support each other, like a continued arch, while the form of the hemispheres, which approaches that of a bell, is peculiarly liable to a tremulous vibratory motion.* Indeed, when a body sounds powerfully, as a large bell, or the lowest string of a harpsichord, we can perceive that it actually vibrates; and even in cases where the vibration is imperceptible to the naked eye, we may detect it by the microscope, or by some other artifice. Thus, if we put some water into a glass tumbler or basin and make it sound, by applying the moistened finger as in Art. 518, the water will be agitated. If we hold the hand over the pipe of an organ, we feel a tremulous motion in the air passing through it. Such experiments may be extended to all solid bodies by placing upon them pieces of paper, or strewing them with fine sand.

521. Vibrations recurring at equal intervals, constitute *musical* sounds. All continued sounds, which remain in any degree uniform throughout their duration, are capable of being compared with each other in their degree of acuteness. *When sounds are equally acute, they are said to have the same pitch;* but when they differ in acuteness, that sound which is more shrill, is said to be acute, or to have a higher pitch; and that which is less shrill, is said to be *grave,* and to have a lower pitch, or a deeper tone. A difference in pitch forms the chief character by which *musical sounds* are distinguished from each other, and is the foundation of their use in music. In unmusical sounds, it generally holds a place subordinate to their other qualities.

---

* Millington's Natural Philosophy, p. 123.

Musical sounds have occupied the attention of philosophers more than any other class of sounds. The superior precision with which the ear can estimate any variation in pitch, renders these sounds more easily compared; and the vibrations of the sonorous bodies which produce them, are, on account of their superior simplicity of form, more easily investigated.*

522. Musical strings.—A musical string is of a uniform thickness, and is stretched between two points, by a force much greater than its weight. The stretching force is generally conceived as measured by the weight which would occasion an equal tension, on the supposition that the string is made fast at one end and passes over a pulley at the other, the latter being loaded with weights. In the usual mode of exciting a musical string, it vibrates on each side of its quiescent position, the extremities being the only points which remain at rest. The sound which the string gives in this mode of vibration is called its *fundamental* sound.

The pitch of the fundamental sound of musical strings, is found by experience to depend on three circumstances; the *length* of the string,—its *weight* or quantity of matter,—and its *tension*. The tone becomes more acute as we increase the tension, or diminish either the length or the weight. The operation of these several circumstances may be seen in a common violin. The pitch of any one of the strings is raised or lowered by turning the screw so as to increase or lessen its tension; or, the tension remaining the same, higher or lower notes are produced by the same string, by applying the fingers in such a manner as to shorten or lengthen the string which is vibrating; or, both the tension and the length of the string remaining the same, the pitch is altered by making the string larger or smaller, and thus increasing or diminishing its weight.

523. The time of a *double vibration*, is the time occupied by a string, in passing from a point to which it is stretched on one side to the opposite extreme, and returning to the same point again. It has been ascertained, that the time of a double vibration, expressed in parts of a second of time, will be found by the following operation.

Let L represent the length of the string in inches; w, the weight of an inch of the string; t, a weight equivalent to the force of tension; g, the rate of a falling body $=193$; and T the time of a double vibration expressed in seconds. Then

$$T=L\left(\frac{2w}{gt}\right)^{\frac{1}{2}}=L\left(\frac{2w}{193t}\right)^{\frac{1}{2}}.$$

As the distance of the string from its quiescent position does

* Ed. Encyc., Art. *Acoustics*.

not form an element of the algebraic expression, which is thus found for the time of a vibration, it follows that this time is independent of the distance. Hence, as in the pendulum,

The vibrations of a string, fixed at both ends, are performed in equal times, whether the length of the vibrations be greater or smaller.

Upon this uniformity in the times of vibration depends the *uniformity of tone;* for if we employ a string of unequal thickness, and consequently one whose vibrations are performed in different times, the sound is confused and variable, and any other mode by which we destroy the isochronism, produces a similar effect. The same law has been found to extend to all other cases of musical sounds; and, therefore, we may conclude, that *isochronism in the vibrations of sonorous bodies, is essential to their producing musical sounds.*

The number of vibrations performed by a string in a second of time, being inversely as the time of one vibration, it is expressed by the reciprocal of the formula denoting the time; so that if N represents the number of vibrations, we shall have the following expression: $N = \frac{\sqrt{(193t)}}{L\sqrt{(2w)}}$.

The frequency of vibration which this equation gives, is found to agree very exactly with the result of experiments performed with strings, whose vibrations are so slow as to admit of being numbered.

524. The *relation* between the number of vibrations, performed by *different strings,* may be expressed by a more simple formula; for *g* and the two numbers being constant quantities, they may, in this case, be rejected, and we get the following expression: $N \propto \frac{\sqrt{t}}{L\sqrt{w}}$. According, then, as we diminish the length of a string, and the weight of an inch of it, or increase its tension, we increase its frequency of vibration; but equal changes in these circumstances do not produce equal effects. Thus if in different strings their tension and the weight of an inch remain the same, *their frequency of vibration will be inversely as their lengths;* for then $N \propto \frac{1}{L}$. If we make the length one third, we triple the number of vibrations, and so for any other proportion. If the length and tension remain the same, $N \propto \frac{1}{\sqrt{w}}$, or, *the number of vibrations is inversely as the square roots of the weights:* consequently a string four times as heavy as another will vibrate half as fast. The bass strings in most instruments have fine wire twisted round them to increase their quantity of matter, otherwise greater length must be resorted to for the production of sim-

ilar tones. If the length and the weight of equal portions be the same, then N $\propto \sqrt{t}$, or *the frequency of vibration is as the square root of the tension.* Therefore, we must give the string of a violin four times the tension in order to make it vibrate twice as fast.

525. WIND INSTRUMENTS.—In wind instruments, a column of confined air itself is the vibrating body; and here the vibrations are longitudinal instead of lateral, as is the case with strings. That it is really the air which is the sounding body in a flute, organ pipe, or other wind instrument, appears from the fact, that the materials, thickness, or peculiarities of the pipe, are of no consequence. A pipe of paper and one of lead, glass, or wood, provided the dimensions are the same, produce, under similar circumstances, exactly the same tone as to *pitch*. If the *qualities* of the tones produced by different pipes differ, this is to be attributed to the friction of the air within them, setting in feeble vibrations their own proper materials.* The class of bodies vibrating *longitudinally*, is not only more diversified in its powers than the other classes of sounding bodies, but also more extensive in the range of substances which it comprehends. A uniform rod under any solid substance, or a column of air contained in a cylindrical tube, whose diameter is everywhere equal, may have its vibration limited at both extremities by an immovable obstacle; or both extremities may be at liberty; or one extremity may be confined and the other disengaged.

A column of air, or a rod of any substance, whether confined or free at both extremities, performs a double vibration in the same time that a minute impulse would occupy, when travelling in a medium of the substance through twice the length of the sonorous body; and a body fixed at one extremity only, will occupy double that time. Hence, the number of vibrations performed in a second of time by a given body, is the same, whether that body be fixed at both extremities, or free at both; and therefore its sound in these two cases should be the same. But if the body be fixed at one extremity and free at the other, its length must be reduced to one half, to make it give the same tone as in the two former cases. Thus, if we blow into a tube closed at one extremity, it will give the same tone as we procure by blowing into an open tube of double the length.

526. The different *pitch* of bodies vibrating longitudinally, and free at both extremities, depends on four circumstances, viz. their elasticity, the temporary rate at which their elasticity is increased by condensation, their length, and their specific gravity, the tone of a body being more acute, according as the elasticity, and the rate of its increase by condensation, are greater, or the

* Herschel.

length and specific gravity less. The *length* of the sonorous body is almost exclusively the only one of these circumstances which we have completely in our power; and with regard to ordinary wind instruments, and all musical instruments where common air is the vibrating body, the length is the circumstance of most importance, since the elasticity, rate of condensation, and specific gravity, are then nearly constant quantities. The change of specific gravity, however, to which the air is subject in consequence of changes of temperature, materially affects the pitch of wind instruments. The frequency of vibration of a column of air is found to be increased about $\frac{1}{33}$, by an elevation of 30° Fahrenheit. Thus, the tone of an organ has been found to be higher in summer than in winter; and flutes and other wind instruments become gradually more acute as the included air is heated by the breath.*

527. Bells.—If a bell be struck by a clapper on the inside, the bell is made to vibrate. The base of the bell is a circle; but it has been found that, by striking any part of the circle on the inside, that part flies out, so that the diameter which passes through this part of the base, will be longer than the other diameters. The base is changed by the blow into the figure of an ellipse, whose longer axis passes through the part against which the clapper is thrown. The elasticity of the bell restores the figure of the base, and again elongates the bell in a direction opposite to the former; and the two elliptical figures thus alternate with each other, growing smaller and smaller, like the vibrations of a pendulum when the moving force is withdrawn, until the sound dies away. We may be convinced by our senses, that the parts of the bell are in a vibratory motion while it sounds. If we lay the hand gently upon it, we shall feel this tremulous motion, and even be able to stop it; or if small pieces of paper be put upon the bell, its vibrations will set them in motion.†

We may conceive the bell to be formed of an infinitude of rings, placed one above another, from the base to the highest point. The rings situated nearer to the base, having a greater circumference, tend to perform their vibrations more slowly, while the rings nearer to the summit, whose circumferences are smaller, tend to produce vibrations oftener. These sounds will so coalesce as to produce a mixed sound, intermediate between those of the higher and lower rings.‡

528. The voice and the ear.—The *human voice* depends principally on the vibrations of the membranes of the *glottis*, excited by a current of air which they alternately interrupt and suf-

* Ed. Encyc., Art. *Acoustics*. † Haüy's Nat. Phil. I, 203. ‡ Ib. 305.

fer to pass; the sounds being also modified in their subsequent progress through the mouth.*

The parts of the *ear*, and the progress of sound to the sentient nerve, may be simply described as follows.

Fig. 210.

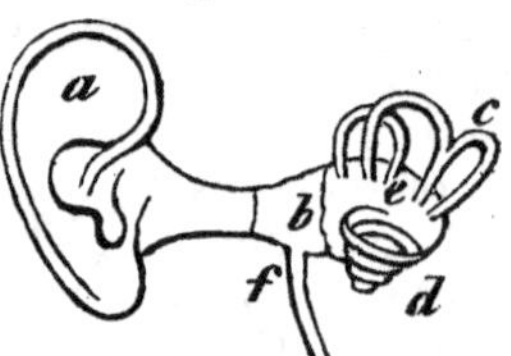

(1.) There is externally a wide-mouthed tube or ear trumpet, *a*, for catching and concentrating the pulses of sound. In many animals it is movable, so that they can direct it to the place from which the sound comes.

(2.) The sound concentrated at the bottom of the ear tube falls upon a membrane stretched like the top of an ordinary drum, over the tym panum or drum of the ear, *b*, and causes it to vibrate. That its motion may be free, the air contained within the drum has free communication with the external air by the open passage *f*, called the *Eustachian tube*, leading to the back part of the mouth. A degree of deafness ensues when this tube is obstructed by wax.

(3.) The vibrations of the tympanum are conveyed further inwards by a chain of four bones, (not here represented on account of their minuteness,) reaching from the center of the tympanum to the oval door or window of the *labyrinth*, *e*.

(4.) the labyrinth, or complex inner compartment of the ear, over which the nerve of hearing is spread as a lining, is full of water; and therefore, when the vibrations of the tympanum acting through the chain of bones (3) are communicated to this fluid, they are instantly felt over the whole cavity. (Art. 382.) The labyrinth consists of the *vestibule*, *e*, the three *semi-circular canals*, *c*, imbedded in the hard bone, and of a winding cavity, *d*, called the *cochlea*, like that of a snail shell, in which fibres, stretched across like harp-strings, constitute the *lyra*. The exact uses of these various parts are not yet perfectly known. The membrane of the tympanum may be pierced, and the chain of bones may be broken, without loss of hearing.†

CHAPTER II.

OF THE PROPAGATION OF SOUND.

529. Air is, in general, the medium of sound. A bell struck under the receiver of an air-pump, gives a feebler and feebler sound as the exhaustion proceeds, until, when the rarefaction is

* Young's Lectures, I, 401.

† Dr. Arnott, El. Phys., Vol. I, p. 507.

carried to a certain extent, it emits no sound at all.* Hence on the summit of high mountains, where the air is naturally rare, sound ought to be weaker than at the general level of the earth; and such is found to be the fact. Saussure relates that on the top of Mont Blanc, the firing of a pistol made a report no louder than that of a child's toy-gun. A fact mentioned by travellers in Alpine countries, is explained on this principle. They see distinctly a huntsman on a neighboring eminence, and observe the flashes of his gun, but can scarcely hear the report, even when comparatively near them.†

Yet meteoric bodies are said to give a distinct rumbling sound in passing through the air at the height of fifty miles, an altitude at which the air is rarefied to a degree exceeding the vacuum of the air-pump. Dr. Halley mentions an instance of a meteor, whose elevation was at least sixty-nine miles, exploding with a sound equal to "the report of a very great cannon, or broadside." Probably, however, these sounds do not emanate from the meteor itself, but from fragments projected from it, which fall through the air to the ground. If the "rumbling sound" above mentioned, proceeds from the body of the meteor, it is necessary to suppose that the air is condensed before it to a great extent. On the other hand, when the elasticity of the air is augmented, either by condensation or by heat, the force of sound is considerably increased. This effect has been experienced in the condensed air of diving bells.

530. *Air receives from sounding bodies vibrations, which it communicates to the organs of hearing.* Let us take, for example, a cord of a stringed instrument, and suppose it struck when played upon; immediately all the points of that string will deviate more or less from the position which they occupied when the string was at rest, according as they are more or less distant from the points where the string is fixed; and the string will go and return alternately on this side and on that side of its first situation, by a vibratory motion occasioned by its elasticity. The particles of air contiguous to the different points of the string, assume motions similar to those of the respective points, that is, they move backward and forward with them. Each particle communicates motion to that which is next to it, that to a third, and so on, until the particles of air are reached which are in contact with the tympanum, or drum of the ear. The air then acts upon that membrane, by communicating to it its own vibrations, which the drum transmits to the auditory nerve; and thence results the sensation of sound.‡

* Herschel in Encyc. Metrop. II, 747. † Partington, I. 263.

‡ Haüy, I, 203.—It is evident from the mechanical concussion attending loud noises, that sound consists in a motion of the air itself, communicated along it by virtue of its elasticity, as a tremor runs along a stretched rope.—*Herschel on Sound.*

531. In an open space, and in a serene atmosphere, sound is propagated from the sounding body in all directions. Sounds, even the most powerful, when thus transmitted freely through the air, diminish rapidly in force, as they depart from their sources, and within moderate distances wholly die away. What *law* this diminution follows, is not yet ascertained; and is indeed, in the present state of acoustics, incapable of determination. Some writers have supposed that sound follows the common law of emanations radiating from a center, (Art. 7,) and consequently, that its intensity at different distances from its source varies inversely as the square of the distance;* but we can estimate the force of sounds by the ear alone; an instrument of comparison whose decisions on this point vary with the bodily state of the observer, and whose scale expresses no definite relation but that of equality.

Though sound has in general, at its origin, a tendency to diffuse itself in all directions, it is sometimes more propagated in one direction than in others. A cannon seems much louder to those who stand immediately before it, than to those who are placed behind it. The same fact is illustrated by the speaking trumpet; the person toward whom the instrument is directed, hears distinctly the words spoken through it, while those who are situated a little to one side, hardly perceive any sound.

532. Sound is in a great measure intercepted by the intervention of any solid obstacle between the hearer and the sonorous body. Thus, if while a bell is sounding, houses intervene between us and the bell, we hear it sound but faintly, compared with what we hear after we have turned the corner of the building. From this fact sound would seem to be propagated in straight lines. If, however, we speak through a tube, the voice will be wholly confined by the tube, and will follow its windings however tortuous; hence we infer that sound is propagated not in right lines like radiant substances, as heat and light, but in *undulations*, after the manner of waves, such as follow when a stone is thrown into still water.

533. Though air is the most common medium of sound, yet it is not the only medium. Various other bodies, both solid and fluid, are excellent conductors of sound; and the fainter sound of the bell when buildings intervene, as in the case supposed, (Art. 532,) arises from the fact, that *sound passes with difficulty from one medium into another.*

If a log of wood is scratched with a pin at one extremity, a person who applies his ear to the other extremity will hear the

* Millington's Nat. Phil. p. 125, Epit.—Herschel on Sound, Encyc. Metropol. II, 773

sound distinctly; and when a long pole of wood is applied at one end to the teeth, the ticking of a watch may be heard at the other end, at a much greater distance than when there is no medium of communication but the air. The motion of a troop of cavalry is heard at a great distance by applying the ear close to the ground, and it is well known that dogs, by this method, first discover the approach of a stranger.

534. *The* VELOCITY *of sound is progressive.* Thus when a gun is fired at a distance from us, we perceive the flash some time before we hear the report. Thunder follows the lightning at a perceptible interval, although they are known to be cotemporaneous events. If a gun be fired at a certain known distance, and we observe the interval between the flash and the report, we may obtain the rate at which sound passes, that is, the velocity of the sound. Many years since, Dr. Derham made a number of accurate and diversified experiments on this subject, and fixed the velocity of sound at 1142 feet per second. The mean of a great number of experiments gives the average velocity of 1130 feet per second: but the velocity as determined by Derham, namely, 1142 feet per second, is that which has been generally admitted as the standard. Since, however, the transmission of sound depends on the elasticity of the medium, (Art. 519,) causes which affect the elasticity, likewise affect the velocity of sound. Thus the velocity is a little greater in warm than in cold air, and consequently is somewhat influenced by climate.* M. Goldingham, by a series of experiments made at Madras, found that the velocity of sound was affected even by the seasons of the year, increasing regularly from the coldest to the hottest months, and afterward regularly decreasing. Hence, for every degree of Fahrenheit's thermometer 1.14 feet is allowed for the velocity of sound per second. A similar gradation in the velocity of sound at different seasons of the year, was observed by Capt. Parry in his experiments on sound within the frigid zone.†

All sounds travel in the same air with the same velocity, otherwise there could be no such thing as harmony, an essential condition of which is that the sounds should reach our ears in a precise order, and at exact intervals.‡

535. *Sound moves with a uniform velocity;* that is, it passes over equal spaces in equal times. This important fact was first ascertained by Derham, who found that it held good whether the sound were strong or feeble; whether it proceeded from a hammer or a cannon; in short, that neither the strength nor the origin of the sound makes any difference. M. Biot caused several

* Partington, I, 263. † Phil. Trans. 1828, p. 97. ‡ Webster's El. Phys. p. 194.

airs to be played on a flute at the end of an iron pipe 3120 feet long, and the notes were distinctly heard by him at the other end, without the slightest derangement in the order or quality of the sounds.

The velocity of sound, however, when transmitted through the air, is slightly influenced by the strength and direction of the *wind*. Dr. Derham found that when the wind is blowing in the direction of the sound, its velocity must be added to the standard velocity of sound, and must be subtracted from it when opposed to it.* A *transverse* wind does not affect the velocity of sound in the slightest degree.

536. Several distinguished philosophers, both of France and Holland, have recently made experiments on the velocity of sound, under circumstances the most favorable to the attainment of accurate results. A difficulty experienced by the earlier experimenters, as Derham, arose from the want of a method of measuring a small fraction of a second, and yet this was necessary where a variation of one hundredth part of a second makes a difference of more than eleven feet in the result. The Dutch experimenters† employed a clock with a conical pendulum, capable of determining intervals to the hundredth of a second, by suddenly suspending the motion of the index without stopping the clock. In the French experiments a kind of watch was used, one of whose hands performed a revolution in a second, and by the sudden pressure of a small lever could be made to touch with its extremity the dial plate, at any instant, and leave there a dot, without interrupting its motion or rotation, to effect which, it carried with it a drop of printer's ink, in a peculiar and ingenious species of dotting pen. By the use of these instruments, it was found practicable to ascertain the interval between the sight of a flash, and the arrival of the report of a gun, with such precision as to destroy in the result all material error which might arise from this cause. Accordingly, their results afford a striking agreement,—the experiments of the French gave for the velocity of sound, per second, 1086.1 feet: those of the Dutch, 1089.42. both considering the air at the temperature of freezing water But it is found that the velocity is increased 1.14 feet for every degree of Fahrenheit; consequently, reducing the estimate to the temperature of 62½°, (which is the standard temperature of the British metrical system,) the velocity becomes 1124.19, as deter-

* If a stone be thrown into a small lake, the waves spread with equal rapidity in all directions, in circles whose center is the stone. If into a running river, still they form circles, but their center is carried down the stream; and, in point of fact, the waves arrive opposite to the point of the bank *above* the place where the stone fell, later than at a point at the same distance below it, in proportion to the rapidity of the stream.—*Herschel on Sound.*

† Moll, Vanbeck, &c.

mined by the Dutch experimenters, which is deemed the most accurate result hitherto obtained. It may, therefore, be stated in round numbers, that sound passes through air at the rate of nine thousand feet in eight seconds, or twelve miles and three fourths per minute, or seven hundred and sixty-five miles an hour, which is about three fourths of the diurnal velocity of the earth's equator. Hence, in latitude $42\frac{1}{2}°$, if a gun were fired at the moment a star passes the meridian of any station, the sound would reach any other station exactly west of it at the precise instant of the same star's arriving on its meridian; that is, it would keep pace with the velocity of the earth at that place, as it turns on its axis, in the diurnal revolution.*

From a knowledge of the velocity of sound, the *distance* of a sounding body may be estimated. Thus if the interval between seeing a flash of lightning, and hearing the thunder, be six seconds, the distance of the cloud is $6 \times 1130 = 6780$ feet, or $1\frac{3}{10}$ miles.

537. *The air is a better conductor of sound when humid than when dry.* Thus, a bell is heard better just before a rain; and this fact lends some countenance to an opinion of the ancients, that sound is heard better by night than by day. Humboldt was particularly struck with this fact, when he heard the noise of the great cataracts of Orinoco, which he describes as three times greater in the night than in the day.†

The *distance* to which sound may be heard, will of course vary with its force and various other circumstances which are incapable of being reduced to an exact law. Volcanoes, in South America, have sometimes been heard at the distance of three hundred miles; and naval engagements have been heard at the distance of two hundred miles. The unassisted human voice has been heard from Old to New Gibraltar, a distance of ten or twelve miles, the watch-word, *All's Well,* given at the former place being heard at the latter. Sounds are heard to a much greater distance over water than over land, and further on smooth than on rough surfaces.

538. *Liquids are good conductors of sound.* Indeed, sound is conveyed with far greater velocity in water than in air. Dr. Franklin, having plunged his head below water, caused a person to strike two stones together beneath the surface, and heard the sound distinctly at the distance of more than half a mile. By similar experiments, it has been ascertained, that, though water is a much better conductor of sound than air, yet the *sound is*

---

* Herschel on Sound, in Encyc. Metrop. II, p. 751.

† Humboldt, however, accounts for the greater audibility of sounds by night than by day, from the absence of those ascending and descending currents of air which, while the sun is shining, impair the uniformity of the medium, and thus diminish its conducting powers.

greatly enfeebled by passing out of one medium into the other The most accurate experiments on this subject, are those made in the year 1826 by M. Colladon, in the Lake of Geneva. He caused a bell to be rung under water, and found that although the sound of the blow was well heard in the air directly above the bell, yet the intensity of the sound diminished very rapidly as the observer removed from its immediate neighborhood, and at the distance of two or three hundred yards, it could no longer be heard at all.* To conduct the sound from the water to the ear, a tin pipe was employed, which was plunged into the water, and the ear brought close to the upper end. By this contrivance he was enabled to hear the strokes of the bell in water, at the distance of about nine miles. The velocity of sound under water, M. Colladon found to be four thousand seven hundred and eight feet, or nearly a mile, per second.†

539. *Solid substances convey sound with various degrees of facility, but in general much better than air, and as well or even better than fluids.* By placing the ear against a long dry brick wall, and causing a person at a considerable distance to strike it *once* with a hammer, the sound will be heard *twice*, because the wall will convey it with greater rapidity than the air, though each will bring it to the ear.‡ The rate at which *cast iron* conducts sound, was ascertained by M. Biot in the following manner. He availed himself of the laying of a series of iron pipes to convey water to Paris. The pipes were about eight feet in length, and were connected together with small leaden rings. A bell being suspended within the cavity, at one end of the train of pipes, on striking the clapper at the same instant against the side of the bell, and against the inside of the pipe, two distinct sounds were successively heard by an observer stationed at the other extremity. With a train of iron pipes two thousand five hundred and fifty feet, or nearly half a mile in length, the interval between the two sounds was found from a mean of two hundred trials, to be 1.79 seconds. But the transmission of sound through the internal column of air, would have taken 2.2 seconds; which shows that the sound occupied only .41 of a second in passing through the metal. From more direct trials, it was concluded that the exact interval of time, during which the sound performed its passage through the substance of the train of pipes, amounted to only the .26 of a sec-

* It is inferred from this experiment, that sound is reflected by the same laws as light; when the direction is perpendicular to the reflecting surface, (in this case, the surface of the water,) it passes without reflexion; but the quantity reflected increases, as the angle of reflexion is more oblique.

† Herschel.

‡ Millington, p. 125.

ond, showing that iron conducts sound about ten times as rapidly as air does.*

If a string be tied to a common fire-shovel, and the two ends of the string be wound around the fore fingers of each hand, and the fingers be placed in the ears, on striking the bottom of the shovel against an andiron or other solid body, very deep and heavy tones will be heard, and the vibrations of the metal will be clearly perceived.

540. Solids, as well as other bodies, owe their power of conducting sound to their elasticity. By elasticity in a solid, however, is not meant a power of undergoing *great* extensions and compressions, after the manner of air, or India-rubber, and returning readily to its former dimensions; but rather what is commonly called *hardness*, in contradistinction to toughness, a violent resistance to the displacement of its molecules *inter se* in all directions. Thus the hardest solids are, in general, the most elastic, as glass, steel, and the hard brittle alloys of copper and tin, and in proportion as they are elastic, they are adapted to the free propagation of sound through their substance. But an important condition in their constitution is, that their substance be *homogeneous*, and their structure *uniform*. By the want of homogeneity and uniformity in the conducting medium, the sonorous pulses are every instant changing their medium, and the general wave is broken up into a multitude of non-coincident waves, emanating from different origins, and crossing and interfering with each other in all directions. Thus, a glass vessel containing an effervescing liquor, cannot be made to ring, but gives a dead sound; but as the effervescence subsides, the tone becomes clearer, and when the liquid is perfectly tranquil, the glass rings as usual.†

The great power of solid bodies to conduct sound is exemplified in *earthquakes*, which are heard almost simultaneously in very distant parts of the earth. *Musical boxes* sound much louder when placed on a table or some solid support, than when the air affords the only conducting medium. It is easy to ascertain whether a kettle boils, by putting one end of a stick or poker on the lid, and the other end to the ear: the bubbling of the

* Herschel observes, that from this determination we may estimate the time it requires to transmit force, (whether by pulling, pushing, or by a blow,) to any distance, by means of iron bars or chains. For every eleven thousand and ninety feet of distance, (=the velocity of sound per second in iron,) the pull, push, or blow will reach its point of action one second after the moment of its first emanation from the first mover. In all moderate distances, then, the interval is utterly insensible. But, were the sun and the earth connected by an iron bar, no less than one thousand and seventy-four days, or nearly three years, must elapse before a force applied at the sun could reach the earth. The force actually exerted by their mutual gravity may be proved to require no appreciable time for its transmission. How wonderful is this connection!—*Herschel on Sound, Encyc. Metrop.* II, 773.

† Herschel.

water, when it boils, appears louder than the rattling of a carriage in the streets. A slight blow given to the poker, of which the end is held to the ear, produces a sound which is even painfully loud.*

A physician of Paris introduced into medical practice an instrument, depending on the power of solid bodies to conduct sound, called the *Stethoscope*,† the object of which is to render audible the action of the heart and the neighboring organs. It consists of a wooden cylinder, one end of which is applied firmly to the breast of the patient, while the other end is brought to the ear. By this means, the processes that are going on in the organs of respiration, and in the large blood-vessels about the heart, may be distinctly heard; and it is said that the stethoscope, when skilfully used, "becomes the means of ascertaining some diseases in the chest, almost as effectually as if there were convenient windows for visual inspection."‡

CHAPTER III.

OF THE REFLEXION OF SOUND.

541. SOUND is *reflected* by hard bodies, producing the well known phenomenon called an ECHO. If a straight line be drawn from the sounding body to the reflecting surface, representing the course of the sound before reflexion, and another straight line be drawn from the reflecting surface, in the direction of the sound after reflexion, these two lines will make equal angles with that surface; that is, when sound is reflected, *the angle of reflexion is equal to the angle of incidence.*

The surfaces of various bodies, solids as well as fluids, have been found capable of reflecting sounds, viz. the sides of hills, houses, rocks, banks of earth, the large trunks of trees, the surface of water, especially at the bottom of a well, and sometimes even the clouds.§ It is therefore evident that in an extensive plain, or at sea, where there is no elevated body capable of reflecting sounds, no echo can be heard. It is hence easy to see why the poets, who convert echo into an animated being, place her habitation near mountains, rocks, and woods.‖

542. An echo is heard when a person stands in a position to hear both the original and the reflected sound; and the interval

* Arnott's El. Phys. I, 497.

† στῆθος, *the chest;* σκοπέω, *to examine.*

‡ Dr. Arnot.t

§ Cavallo, II, 345.

‖ Haüy.

will be greater or less according to the distance of the reflecting surface from the sounding body and from the hearer, and hence the interval may be made a measure of the distance. If the sound of the voice returns to the speaker in two seconds, the distance of the reflecting surface is one thousand one hundred and thirty feet, and in that proportion for other intervals. Thus, the breadth of a river may be ascertained when there is an echoing rock on the further shore. A perpendicular mountain's side, or lofty cliffs, such as frequently skirt the sea coast, sometimes return an echo of the discharge of artillery, or of a clap of thunder, to the distance of many miles.* The number of syllables that can be pronounced in half the interval, will be repeated distinctly; but a greater number would be blended with the commencement of the echo.†

When a single obstacle reflects the sound, the echo is *simple;* when there are several obstacles disposed at suitable distances, the echo is *complex.* Echoes of the latter kind have been observed which repeated the original sound forty times.‡ Two parallel walls which mutually reverberate the sound, may produce a double or complex echo, with regard to an auditor placed in the intermediate space. The sound of artillery and of thunder, is frequently prolonged by reverberations in an uneven country.

543. The furniture of a room, especially the softer kind, such as curtains or carpets, impair the qualities of sound by presenting surfaces unfavorable to vibrations. A crowded audience has a similar effect and increases the difficulty of speaking. Halls for music or declamation, should be constructed with plain bare walls. Alcoves, recesses, and vaulted ceilings, produce reverberations which often greatly impair the distinctness of elocution. Indeed, the qualities of a room, in regard to sound, are modified by so many circumstances,§ that the science of acoustics is worthy of more attention from the architect than it has generally received. Plane and smooth surfaces reflect sound without dispersing it, convex surfaces disperse it, and concave surfaces collect it. The concentration of sound by concave surfaces, produces many curious effects both in nature and art. There are remarkable situations where the sound from a cascade is concentrated by the surface of a neighboring cave, so completely,

* Arnott. † Cavallo, II, 347. ‡ Haüy.

§ The famous Dr. Sanderson, formerly Professor of Mathematics in the University of Cambridge, who had been blind from the time he was a year old, possessed such acuteness of hearing, that he not only distinguished persons with whom he had ever once conversed, so long as to fix in his memory the sound of their voice, but he could also recognise places by observing the manner in which they modified sound. He could judge accurately of the size of a room, and of his distance from the wall; and if ever he had walked over a pavement in courts, or piazzas, and was conducted thither again, he could tell his exact situation, by the note which the place sounded.

that a person accidentally bringing his ear into the focus, is astounded by a deafening noise. Sound issuing from the center of a circle is, by reflexion, returned to the center again, producing a very powerful echo.* Such effects are observed in the central parts of a circular hall. An elliptical apartment conveys sound very perfectly from one focus to the other. A whisper uttered by a person in one focus of such a chamber will be audible to a person in the other focus, though not heard by persons between.

544. *Whispering Galleries*† are constructed on this principle. Domes, as that of St. Paul's Cathedral, in London, sometimes exhibit the same curious property.‡ Concave surfaces facing each other, as two alcoves in a garden, or covered recesses on opposite sides of a street, or bridge, will enable persons seated in their foci to converse by whispers, notwithstanding louder noises in the space between, and without themselves being overheard in that space.§ A notorious instance of a sound-collecting surface, was the *ear of Dionysius*, in the dungeons of Syracuse. The roof of the prison was so formed as to collect the words, and even whispers of the unhappy prisoners, and to direct them along a hidden conduit to the place where the tyrant sat listening. The wide-spread sail of a ship, rendered concave by a gentle breeze, is also a good collector of sound. Dr. Arnott relates an instance where the ringing of the bells at St. Salvador on the coast of Brazil, was heard on board a ship at the distance of one hundred miles from land.‖

545. The most frequent instances of the reflexion of sound, are from surfaces which may be considered as plane. In these, the sound issuing from any point seems, after reflexion, to proceed from a point equally distant, and similarly situated, on the other side of the reflecting surface; the phenomena differing a little according to the position of the speaker, with respect to the body which occasions the reflexion. If a person's voice strike any surface perpendicularly, it will be reflected back in the same line; and the time occupied between the utterance of the sound, and its arrival again at the speaker, will be equal to the time in which the sound travels through twice the distance between the speaker and the reflecting surface. The interval, therefore, be-

* If a spherical room could be constructed of perfectly solid materials, perfectly polished, and a sound were to issue from the voice of a person in the center, there would be an accumulation of echo at the center, which would probably be destructive of the organs of hearing.—*Latrobe in Ed. Encyc.*

† The *Hall of Secrets*, as it is called, in the Observatory at Paris, is a whispering gallery. This hall is of an octagonal form, with cloister arches, or arched by portions of a cylinder, which meet at angles, corresponding to those formed by the sides of the building. The speaker applies his mouth very near to the wall to one of the angles, and the person situated at the opposite angle hears his voice distinctly.

‡ Cavallo. § Arnott. ‖ El. Phys. I, 505.

tween setting out and returning, will be found by the following rule: Let x = the interval in seconds, and d = twice the distance from the sounding body to the reflecting surface; then $1 : 1130 :: x : d \therefore x = \frac{d}{1130}$. If, therefore, the distance is less than forty-eight feet, the interval of time between the speaker's hearing the direct and the reflected sounds, will be less than $\frac{1}{12}$ of a second, and the two sounds will seem to coalesce and form but one sound; but if the distance exceeds forty-eight feet, then the interval will be greater than $\frac{1}{12}$ of a second, and as this interval can be discerned by the ear, the two sounds will be separate, and will form an echo.*

546. The rolling of thunder has been attributed to echoes among the clouds; and that such is the case has been ascertained, by direct observation on the sound of a cannon. Under a perfectly clear sky, the explosion of guns is heard single and sharp; while when the sky is overcast, or when a large cloud comes overhead, the reports are accompanied by a continued roll, like thunder, and occasionally a double report arises from a single shot.†

The continued sound of distant thunder, which is sometimes prolonged for many seconds, is not always owing to reverberation, but frequently arises simply from the different distances of the same flash. Although the progress of a flash of lightning through the air were absolutely instantaneous, still, if its path were in a line that would carry it further from the ear in one place than in another, there would be a corresponding difference in the times at which the sound generated in different portions of the path would reach the ear. Herschel observes that if (as is almost always the case) the flash be zigzag, and composed of broken rectilinear and curvilinear portions, some concave, some convex to the ear,—and especially if the principal trunk separates into many branches, each breaking its own way through the air, and each becoming a separate source of thunder,—all the varieties of that awful sound are easily accounted for.‡

547. The *Speaking Trumpet* has been supposed by most writers on sound, to owe its peculiar properties to its multiplying sound by numerous reflexions. Hence is suggested the form of a parabolic conoid, or a tube the section of which is a parabola, the place of the mouth being at the focus of the parabola. The vibrations emanating from the mouth would then be reflected into straight lines parallel with the axis of the trumpet, and

* Edinburgh Encyc., Art. *Acoustics.* † Herschel.

‡ Herschel on Sound, Encyc. Metrop., II, 754.

would thus go forward in a collected body to a distant point.* And, since such a form is also favorable for collecting distinct sounds into one point, the same figure is proposed as the most suitable for the *Ear Trumpet.* But the sound of these instruments may be regarded as merely the longitudinal vibration (Art. 525) of a body of air, to which momentum is given in the direction of the axis, not by reflexion from the sides, but by the direct impulse of the mouth.† The ancients were acquainted with the speaking trumpet. Alexander the Great is said to have had a horn, by means of which he could give orders to his whole army at once.‡

548. Sound may be conveyed to a much greater distance by being confined, during its whole transmission, within a pipe. Pipes used for this purpose are called *Acoustic Tubes.* Such tubes are frequently employed in public houses for conveying orders to the attendants. Dr. Herschel employed a similar tube, attached to his forty feet telescope, for communicating his observations to an assistant who sat in a small house near the instrument, and thus, under cover, noted them down, and the particu lar time in which they were made. Acoustic tubes are commonly of a cylindrical form, and have at each extremity a mouthpipe like that of a speaking trumpet, to which either the mouth or ear is applied, according as the person is speaking or listening to another. In the deception called the *Invisible Girl,* the sound of the voice is transmitted and returned through acoustic tubes.

549. *Ventriloquism* does not, as is frequently supposed, depend on the reflexion of sound, but wholly on the inaccuracy with which the ear judges of the direction from which sound proceeds,—enabling the performer, by a variation of his tone of voice, and by seeming not to move his lips, to persuade the spectators that the sound proceeds from some object to which he has directed their attention. The imitations of different sounds by which the ventriloquist is able to personate a variety of characters, and to represent them as engaged in an animated dialogue with each other, are usually limited to a comparatively small number, which have been acquired and rendered very familiar by long practice. Hence, like the performer on a musical instrument, he makes his transitions from one sound to another with a facility which can be acquired only by force of habit.

550. *Sounding Boards* were formerly constructed over the desks of public speakers, particularly in churches, with the view of aiding the powers of the voice. Their efficacy depended on

* Dr. Young, Nat. Phil., I, 375. † Ed. Encyc. II, 118.

‡ Enfield's Scient. Rec., p. 157

the reflexion of the sound; for being near the speaker, the echo or reflected sound, uniting itself with the direct sound, would augment its force or loudness. In stringed instruments, however, as the violin, the sounding board acts by receiving vibrations from the string. Thus by impelling the air with a greater surface, it produces a more powerful sound than the string alone. Hence, if some weight, (called a *mute*,) as a penknife partly open, is attached to the bridge of a violin, the sound is greatly deadened, the vibrations of the string being thus prevented from extending to the sounding board.*

The concave, undulating, and perfectly polished surface of many sea shells, fits them to catch, to concentrate, and to return the pulses of all sounds that happen to be trembling about them, so as to produce that curious resonance from within, which resembles the distant murmur of the ocean.† The organs of speech and of hearing have a mechanical structure most skilfully adapted to the peculiar nature of sound.

CHAPTER IV.

OF THE PHILOSOPHICAL PRINCIPLES OF MUSIC.

551. On this subject, we have room for only a few leading principles.

When separate sounds are repeated with a certain degree of frequency, the ear loses the power of distinguishing the intervals, and they appear united in one continued sound. By this means also, sounds harsh and dissonant in themselves, form a soft and agreeable tone. Any sound whatever, repeated not less than thirty or forty times in a second, excites in the hearer the sensation of a musical note. Nothing is more unlike a musical sound than that of a quill drawn slowly across the teeth of a coarse comb; but when the quill is applied to the teeth of a wheel whirling at such a rate that 720 teeth pass under the quill in a second, a very soft, clear note is heard.‡ In like manner the vibrations of a long harp-string, while it is very slack, are separately visible, and the pulses produced by it in the air are separately audible; but as it is gradually tightened, its vibrations quicken, and the eye soon sees, when it is moving, only a broad shadowy plain; the distinct sounds which the ear lately perceived, run together, owing to the shortness of the intervals, and are heard as one uniform continued tone, which constitutes the note or sound proper to the string.§

Nature presents us with numerous examples of a musical

* Ed. Encyc., II, 119.
† Arnott.
‡ Robison's Mech. Phil. IV, 404.
§ Arnott.

sound produced by the rapid succession of an individual sound, not at all musical in itself. The hum of winged insects, produced by the frequent motion of their wings—the murmur of a forest, occasioned by the agitation of the leaves and boughs—and the sublime roar of the ocean, constituted of the separate sounds produced by innumerable waves, are familiar examples of the operations of this principle.

552. *Musical intervals*, or sounds differing from each other in pitch by a certain interval, are found by experience to be peculiarly agreeable to the human ear; a fact for which we can assign no reason, except that such is the constitution of the mind.*

Musical sounds have certain *ratios* to one another, and are thus brought within the province of Mathematics, because the number of vibrations which produce one musical note, has a constant ratio to the number which produces another musical note. Thus, if we diminish the length of a musical string one half, we double the number of its vibrations in a given time, (Art. 524,) and it gives a sound eight notes higher in the scale than that given by the whole string. Therefore, these sounds are represented by the numbers 2 and 1, and are said to be in the ratio of 2 to 1. The upper note is said to be the *octave* of the lower; and from its great resemblance to the *fundamental* note, or that afforded by the whole string, it is considered as the commencement of a repetition of the same series; so that all audible sounds are considered as repetitions of a series contained within the interval of an octave.†

553. The length of the entire string being called 1, the respective lengths of the strings which sound the eight notes, are $\frac{8}{9}$, $\frac{4}{5}$, $\frac{3}{4}$, $\frac{2}{3}$, $\frac{3}{5}$, $\frac{8}{15}$, $\frac{1}{2}$. The sound given by the whole string, which is denoted by 1, is called the *key* note, and the other notes are called, respectively, the second, third, fourth, fifth, sixth, seventh, and eighth, and the fractions denote the relation of each note in the scale to the key note. Since the number of vibrations is inversely as the length of the string, (Art. 524,) these fractions inverted will express the number of vibrations which produce the several notes of the scale respectively. Thus, $\frac{8}{9}$ denotes that the string which sounds the next note above the key note vibrates 9 times, while the whole string vibrates 8 times. Hence, the series expressing the number of vibrations which produce the notes of the scale, are 1, $\frac{9}{8}$, $\frac{5}{4}$, $\frac{4}{3}$, $\frac{3}{2}$, $\frac{5}{3}$, $\frac{15}{8}$, 2.

But, on reducing these numbers to a common denominator, and taking their numerators, (which express the ratios of the fractions,)‡ we have the following series, 24, 27, 30, 32, 36, 40, 45, 48.

* Playfair's Outlines, I, 274. † Young, II, 393. ‡ Day's Algebra, Art. 360, Cor. 1.

Hence we have the following proposition:

If a string be divided, so that the number of vibrations performed by each part in a given time, shall be in the ratio respectively of the numbers 24, 27, 30, 32, 36, 40, 45, 48, *the sounds of the first seven will be perceived as increasing in acuteness one above another, from the first to the last, and will yield the notes from the combinations of which all musical effects are produced.**

554. By inspecting the last series of numbers, namely, that which expresses the relation between the successive notes of the diatonic scale, we shall perceive that the ratios between two successive numbers, and of course the intervals between the several notes of the scale, are not all equal to each other.

1.	The ratio of	27 to 24	is that of	9 : 8
2.	"	30 to 27	"	10 : 9
3.	"	32 to 30	"	16 : 15
4.	"	36 to 32	"	9 : 8
5.	"	40 to 36	"	10 : 9
6.	"	45 to 40	"	9 : 8
7.	"	48 to 45	"	16 : 15

Hence it appears that there are in the musical scale three sorts of intervals, of which three bear to the fundamental or key note the ratio of 9 to 8, two that of 10 to 9, and two more that of 16 to 15. The first of these intervals being the largest, is denominated the *major tone*, the second the *minor tone*, and the third the *semitone*. The scale therefore is made up of three major, two minor, and two semitones, as represented in the table.

After ascending through the first seven notes of the scale, we arrive, as has been already intimated, (Art. 552,) at a note which seems to be only a repetition of the first; hence it commences a new series of seven notes analogous to the former series, each note being an *octave* above the corresponding note in that series, and therefore implying vibrations twice as rapid. A third series is constituted in the same manner, called the *double octave*, in which the lengths of the string are $\frac{1}{4}$ of those in the first part of the scale.

All musical sounds are computed to be contained between *ten octaves;* so that the number of vibrations in a given time that yields the gravest note, is to that which yields the most acute, as 1 to 2^{10}, that is, as 1 : 1024.†

555. When the vibrations are less numerous than about 16 per second, the ear loses the impression of a continued sound, and perceives, first, a fluttering noise, then a quick rattle, then a succession of distinct sounds capable of being counted. On the other hand, when the frequency of the vibrations exceeds a cer-

* Playfair's *Outlines*, I, 273. † Ib. 274.

tain limit, all sense of *pitch* is lost; a shrill squeak, or chirp, only is heard; and what is very remarkable, many individuals, no way inclined to deafness, are altogether insensible to very acute sounds, even such as painfully affect others. This singular observation is due to Doctor Wollaston.* Nothing can be more surprising than to see two persons, neither of them deaf, the one complaining of the penetrating shrillness of a sound, while the other maintains that there is no sound at all.† Few musical instruments comprehend more than six octaves, and the human voice has only from one to three, the male voice being in pitch an octave lower than the female.‡

556. The intervals of the diatonic scale are denoted by the first seven letters of the alphabet, A, B, C, D, E, F, G; which are repeated usually in small letters, *a*, *b*, *c*, &c., in the higher series.

A succession of single musical sounds constitutes *melody*; the combination of such sounds, at proper intervals, forms *chords*; and a succession of chords constitutes *harmony*. Two notes produced by an equal number of vibrations in a given time, and of course giving the same sound, are said to be in *unison*. The relation between a note and its octave is, next after that of the unison, the most perfect in nature; and when the two notes are sounded at the same time, they almost entirely unite.§ The fifth (Art. 553) constitutes the next most perfect chord, while the second and the seventh are peculiarly harsh discords. By examining the scale of vibrations in Art. 553, we shall perceive that the chords are characterized by frequent *coincidences of vibration*, while in the discords such coincidences are more rare. Thus in the unison, the vibrations are perfectly isochronous; in the octave the two coincide at the end of every vibration of the longer string, the shorter meanwhile performing just two vibrations; and in the fifth, they coincide at the end of every two vibrations of the longer string, the shorter vibrating three times in the same period. But in the second, the longer and shorter vibrations can coincide only after eight of the longer and nine of the shorter, and in the seventh, only after eight of the longer and fifteen of the shorter. Hence the concord is more perfect as the common period is shorter.‖

Musical intervals, therefore, are divided into *chords* and *discords*. The octave, the major fifth, the major and minor thirds, the major and minor sixths, are concords, and are pleasing in themselves. The seconds, the sevenths, the minor fifths and major fourths, are discords. The chord consisting of the fundamental note with its third and fifth, and called the harmonic

---

* Phil. Transac., 1820. † Herschel. ‡ Arnott, El. Phys. I, 481

§ Ed. Encyc., Art. Music. ‖ Young's N. Phil. I, 391.

triad, forms the most perfect harmony, and contains the constituent parts of the most simple and natural melodies.*

Discords, however, are employed in musical composition; but their use is limited by special rules. Their use does not consist in the excess or defect of intervals, which when false produce jargon, not music; but in the warrantable and artful use of such combinations as, though too disagreeable for the ear to dwell upon, or to finish a musical period, yet so necessary are they to modern counterpoint, and modern ears, that harmony without their relief, would satiate, and lose many of its pleasing effects.†

557. When a long string is made to vibrate, there are heard not only the note belonging to the whole length of the string, but also more feebly the subordinate notes belonging to its half, its third, its fourth, &c., thus giving to single sounds the effect of harmony. Hence such subordinate sounds are called, with respect to the principal sound, its *harmonics*. Often the subordinate sounds swell with such force as to overpower for a time the fundamental note; and then if the string be carefully examined, it will be found to be vibrating, not as a whole, but in two, three, or four distinct portions, with points of rest between them.‡ The sounds thus belonging to a single string, and produced by its spontaneous division into different numbers of equal parts, constitute, when heard together or in succession, the simple music of Nature herself. It is produced in the most perfect manner by the *Æolian Harp*.

558. Hence arises what is denominated the *sympathy of sounds*. If two strings equally stretched, and in all other respects similar, but one only half, one third, or some other aliquot part of the length of the other, be placed side by side, and the shorter be struck or sounded, the vibration will be communicated to the longer by the intervention of the air, which will thus at once be thrown into a mode of vibration in which the whole length is divided into segments, each equal to the shorter string. Here the vibrations imparted to the string that is struck, are communicated to the aërial pulsations, which will impress on any body *capable of vibrating in their own time*, an actual vibratory motion; and if a body is susceptible of a number of modes of vibration performed in different times, that mode only will be excited which is *synchronous with the aërial pulsations*. All other motions, though they may be excited for a moment by one pulsation, will be extinguished by a subsequent one. Hence, if two strings have any mode of vibration in common, that mode may be excited by sympathy in either of them when the other is sound-

* Young. † Burney.

‡ Arnott, El. Phys. I, 478; Young's Nat. Phil. I, 382; Haüy's Nat. Phil. I. 316

ed, and that only. For example, if the length of one string is to that of the other as 2 : 3, and if either be set vibrating, the mode of vibration, corresponding to a division of the former into two, and of the latter into three segments, will, if it exists in the one, be communicated by sympathy to the other. In the vibrations of strings, which, from their small surface, can receive nothing but a trifling impulse from the air, the sounds and motions excited by this sort of sympathetic communication are feeble; but in vibrating bodies which present a large surface, they become very great. It is a pretty well authenticated feat, performed by persons of a clear and powerful voice, to break a drinking glass by singing its proper fundamental note close to it. Looking-glasses also are said to have been occasionally broken by music, the excursions of their molecules in the vibrations into which they are thrown being so great as to strain them beyond the limits of their cohesion.*

559. The theory of *Musical Instruments* will be readily understood from the principles already explained. It will be seen that they all owe their power of producing musical sounds to their susceptibility of vibrations; that the force or loudness of the sounds they afford depends on the *length* of the vibrations, and the graveness or acuteness of the sound, in other words the pitch, on their *slowness* or *frequency;* and that their chords depend, in general, upon *frequency of coincidence in the vibrations* that afford the several sounds of the concord.

The nature of stringed instruments may be learned from the *violin.* Here the strings are of the same length, but differ in weight and tension; those designed to afford the lower notes being heavier and less strained, and those for the higher notes being lighter and more tense. The lengths, moreover, are altered by applying the fingers. The several strings are usually so adjusted to each other, that is, so *tuned,* that any two contiguous strings make a *fifth.* Hence the fourth or highest stop on one string brings it into unison with the string above; and the third stop on any string forms an octave with the open string next below. On account of this power of altering the effective lengths of the strings at pleasure, of developing the harmonic sounds by a skilful application of the fingers, and of varying constantly the degrees of fulness or force in each sound by a dexterous use of the bow, the violin becomes, in the hands of an accomplished performer, an instrument of great power and compass, while it is capable of greater variety than any other musical instrument.

The *flute* affords an example of wind instruments. Here the vibrating body is a column of air, to which different lengths are given by means of the stops which are opened and closed by the

* Herschel on Sound.

fingers. The rapidity of the vibrations, and consequently the pitch, is also changed a whole octave by the management of the breath.

560. In mixed wind instruments, the vibrations or alternations of solid bodies, are made to coöperate with the vibrations of a given portion of air. Thus, in the trumpet, and in horns of various kinds, the force of inflation, and perhaps the degree of tension of the lips, determines the number of parts into which the tube is divided, and the harmonic which is produced. The hautboy and clarionet have mouth-pieces of different forms, made of reeds or canes; and the reed pipes of an organ, of various constructions, are furnished with an elastic plate of metal, which vibrates in unison with the column of air which they contain. An organ generally consists of a number of different series of pipes, so arranged, that by means of registers, the air proceeding from the bellows may be admitted to supply each series, or may be excluded from it at pleasure; and a valve is opened when the proper key is touched, which causes all the pipes belonging to the note in those series of which the registers are open, to sound at once. These pipes are not only such as are in unison, but frequently also one or more octaves above and below the principal note, and sometimes also twelfths and seventeenths, imitating the series of natural harmonics.*

For the further elucidation of this interesting subject, we are compelled to refer to more extensive treatises, as *Smith's Harmonics*, *Herschel on Sound*, and particularly, the late work of *Professor Pierce on Sound.*

---

* Young's Lectures, I, 402.

PART VI.—ELECTRICITY.

---

561. ELECTRICITY is a term derived from ἤλεκτρον, the Greek word for *amber*,* that being the substance in which a property of the agent now denominated Electricity was first observed.

The ancient Greek philosophers were acquainted with the fact that amber, when rubbed, acquires the property of attracting light bodies; hence the effect was denominated *electrical;* and in later times, the term *electricity* has been used to denote both the unknown *cause* of electrical phenomena, and the *science* which treats of electrical phenomena and their causes.

The science of electricity is hardly more remarkable on account of its surprising and beautiful phenomena, than it is curious in its *history.* The first observation recorded of it was made by Thales of Miletus,† who ascribed it to the functions of some hidden animal.‡ Theophrastus,§ the natural historian, mentions a stone called *lyncurium,* (supposed to be the *tourmalin* of modern mineralogists,) possessing the property of attraction as well as amber. He observes that it is said not only to attract straws and small pieces of sticks, but even copper and iron, if they be finely divided.‖ This is nearly the amount of what was known of electricity by the ancients; nor, so far as appears, was there a single important fact added to the science for the period of nineteen centuries.

562. In the year 1600, Dr. Gilbert, an English philosopher, published a work on Magnetism, comprising also many observations on Electricity. He knew nothing more of this agent, however, than as a power of attraction. Little was added to the knowledge of *Gilbert* on this subject until the latter part of the same century, when, after the establishment of the Royal Society

---

* Amber is a resinous substance having the appearance of indurated honey. It sometimes naturally exhibits the shape of water-worn pebbles. When heated, it exhales a highly agreeable odor. From its scarcity it bears a high price. Much of the amber found in the market is brought from Prussia, where it is found in mines, or loosely scattered along the sea coast; and it is found in other countries, imbedded in a peculiar kind of sand and gravel.

† Sometimes styled the "father of Grecian philosophy." Flourished 600 years before the Christian era.

‡ Priestley's History of Electricity, p. 1.

§ Lived at Athens, 300 years B. C.

‖ Τὸ λυγκούριον ἕλκει γὰρ ὥσπερ τὸ ἤλεκτρον. οἱ δέ φασιν οὐ μόνον κάρφη καὶ ξύλον, ἀλλὰ καὶ χαλκὸν καὶ σίδηρον, ἐὰν ᾖ λεπτός. ὥσπερ καὶ Διοκλῆς ἔλεγε.—Theophrastus περὶ τῶν λίθων.

of London, and of the Academy of Sciences at Paris, philosophical experiments began to be prosecuted with a zeal before unknown. *Boyle** discovered a number of interesting facts in electricity, and *Otto Guericke*† constructed the first electrical machine, using a globe of sulphur, instead of the glass cylinder, at present employed.

But the first sixty years of the eighteenth century, may be remembered as the period when the greatest discoveries in electricity were made. *Grey*,‡ in England, *Du Fay*,§ in France, and *Franklin*,‖ in America, are the names most distinguished in the history of this period. Each of these individuals made numerous and important discoveries; and the last two severally proposed hypotheses to account for the phenomena of electricity, hypotheses which have ever since divided the opinions of electricians.

For the sake of convenience, the term *electric fluid* is employed, without, however, implying any thing more than the *unknown cause* of electrical phenomena, whatever that cause may be.

---

* Honorable Robert Boyle, an English philosopher, lived in the reign of Charles the Second, and flourished about the year 1670. He was one of the founders of the Royal Society of London, and was a very zealous and diligent experimenter, and distinguished for his virtues and piety. Though the facts discovered by Boyle were valuable contributions to the science, yet it may serve to show the absurd notions which prevailed at that time on points of theory, to recite his views of electrical attraction He supposed that an excited body emitted a glutinous effluvium, which laid hold of small bodies in its way, and, in its return to the body which emitted it, carried them back with it.—*Priestley's Hist. Elec.* p. 7.

† Otto Guericke, of Magdeburg in Germany, better known as the inventor of the air-pump. He was contemporary with Boyle, and united an inventive talent with a taste for philosophical experiments. His electrical machine consisted of a globe of sulphur, made by melting that substance in a hollow globe of glass, and then removing the glass by breaking it. This globe he mounted upon an axis, and whirled it in a wooden frame, rubbing it at the same time with his hand. Guericke first observed the electric spark.

‡ Stephen Grey, a pensioner of the British government—flourished about the year 1730—made numerous discoveries, the most important of which was the division of bodies into *conductors* and *non-conductors*.

§ Du Fay was a member of the Academy of Sciences at Paris—flourished about the year 1733—he discovered, among other things, the influence of *moisture* upon the conducting power of bodies—the fact that *electrified* attract *unelectrified* bodies—and the *two different kinds of electricity*, the vitreous and resinous, or positive and negative.

‖ Dr. Franklin commenced his labors in electricity in 1747. The results of his experiments and observations were communicated in several letters addressed to Peter Collinson, Esq., of London, Fellow of the Royal Society, written at different times from 1747 to 1754. "Nothing," says Dr. Priestley, (*Hist. Elec.* p. 159,) "was ever written upon the subject of electricity, which was more generally read and admired in all parts of Europe, than these letters. There is hardly any European language into which they have not been translated; and, as if this was not sufficient to make them properly known, a translation of them has lately been made into Latin. It is not easy to say, whether we are most pleased with the simplicity and perspicuity with which these letters are written, the modesty with which the author proposes every hypothesis of his own, or the noble frankness with which he relates his mistakes, when they were corrected by subsequent experiments."

CHAPTER I.

OF THE GENERAL PRINCIPLES OF THE SCIENCE.

563. The most general effect by which the presence of electricity is manifested, is *attraction*. Thus, when a glass tube is rubbed with a dry silk or woolen cloth, it acquires the property of attracting light bodies, as cotton, feathers, &c. When by any process a body is made to give signs of electricity, it is said to be *excited*. When a body receives the electric fluid from an excited body, it is said to be *electrified*. Since there is found to be a great difference in bodies in regard to the power of transmitting electricity, all bodies are divided into two classes, CONDUCTORS and NON-CONDUCTORS. *Conductors* are bodies through which the electric fluid passes readily; *non-conductors* are bodies through which the electric fluid either does not pass at all, or but very slowly. The latter bodies are also denominated *electrics*, because it is by the friction of bodies of this class that electricity is usually excited. An electrified body is said to be *insulated*, when its connection with other bodies is formed by means of non-conductors, so that its electricity is prevented from escaping. Instruments employed to detect the presence of electricity are denominated *electroscopes;* such as are employed to estimate its comparative quantity, are called *electrometers*. This distinction, however, is neglected by some writers, and, to avoid the unnecessary multiplication of terms, it will be neglected in the present treatise, instruments of either kind being called *electrometers*.

Fig. 211

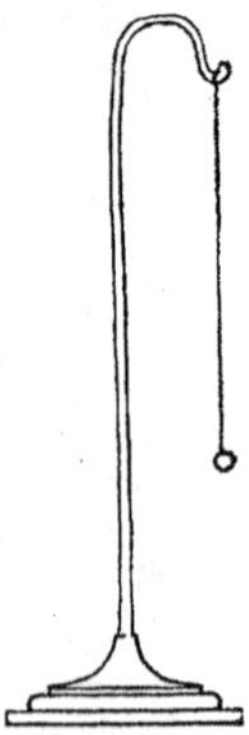

564. The *Pendulum Electrometer*, is formed by suspending some light conducting substance by a non-conducting thread. Thus, *a small ball of the pith of elder hung by a silk thread*, constitutes a very convenient instrument for detecting the presence and examining the kind of electricity. Fig. 211 represents a pendulum electrometer, consisting of a glass rod fixed in a stand, and bent at the top so as to form a hook. From this hook hangs a thread of raw silk, to the bottom of which is attached a small pith ball, made smooth and round, and weighing only a small part of a grain. The attenuated thread of silk, unwound from the ball of the silk-worm, forms a very delicate insulator; but for ordinary purposes, a common thread of silk may be untwisted, and a single filament taken for the suspending thread. For the pur-

poses of the learner, it may even be sufficient to suspend, by a thread of silk, a ball of cork, or a lock of cotton, or a feather.

565. The *Gold Leaf Electrometer*, represented in Fig. 212, consists of two strips of gold leaf, suspended from the metallic cover of a small glass cylinder. By this arrangement, the pieces of gold leaf are insulated, they are protected from agitation by the air, and electricity is easily conveyed to them by bringing an electrified body into contact with the cover. The approach of an electrified body causes the leaves to separate, or when previously separated, to collapse, according to principles to be explained presently.

Fig. 212.

566. *Coulomb's Electrometer*, Fig. 213, is an apparatus of still greater delicacy and perfection than either of the preceding instruments. It consists of a cylindrical glass vessel, having also a lid of glass, in the center of which a small hole is drilled. Through this hole passes an untwisted raw silk thread four inches long, and fixed at the top to a micrometer, by means of which it may be turned round any number of degrees at pleasure. To the silk thread is attached a very fine thread of lac, H, having at each extremity a small pith ball. This lac needle with its knobs weighs only one fourth of a grain. A small hole is drilled in the side of a vessel at A, through which passes a fine wire, terminated at both extremities by a knob. When an excited body is placed in contact with the knob at A, the knob at the other extremity will acquire the same electricity as the excited body. This electricity it will communicate to the knob of the lac needle, suspended by the silk thread, which was previously almost in contact, and the two knobs will repel each other. The movable knob attached by the silk thread, will separate from the other, and the quantity of electricity will be proportional to the distance to which it recedes.*

Fig. 213

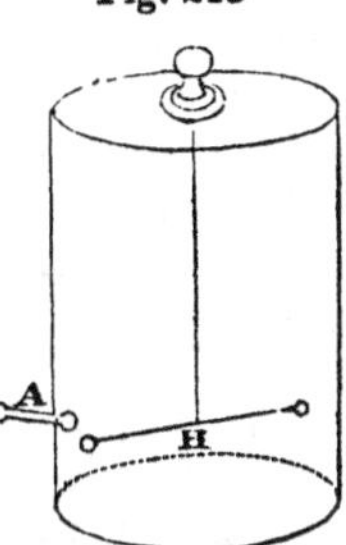

By the aid of the foregoing instruments, or even by means of the pendulum electrometer alone, we may ascertain the following LEADING FACTS, which are so many fundamental truths, in the science of electricity.

567. PROP. I. *Electricity is produced by the friction of all bodies.*

Although friction is the most common, and by far the most extensive means of exciting bodies, yet it is not the only means. Electricity is manifested during the *changes of state* in bodies,

* Thomson's Outlines of Heat and Electricity, p. 374.

such as liquefaction and congelation, evaporation and condensation. Some bodies even are excited by mere *pressure ;* others by the *contact* or *separation of different surfaces.* Most *chemical combinations* and *decompositions* are also attended by the evolution of electricity, which manifests its presence to delicate electrometers.

If we rub a piece of amber, sealing wax, or any other resin ous substance, on dry woolen cloth, or fur, or silk, and bring it toward an electrometer, it will give signs of electricity. A glass tube may be excited in a similar manner. Moreover, if we bring the excited tube near the face, it imparts a sensation resembling that produced by a cobweb. If the tube is strongly excited, it will afford a spark to the knuckle, accompanied by a snapping noise. A sheet of white paper, first dried by the fire, and then laid on a table and rubbed with India rubber, will become so highly excited as to adhere to the wall of the room, or any other surface to which it is applied. Indeed, friction is so constantly attended by electricity, that in favorable weather the fluid is abundantly indicated on brushing our clothes, which thus are made to attract the light downy particles that are floating in the air.

568. Our proposition asserts, that electricity is produced by the friction of *all* bodies, whereas, if we hold in the hand a metallic substance, a plate of brass or iron, for example, and subject it to friction, we shall not discover the least sign of electrical excitement. In such cases, however, the electricity is prevented from accumulating in consequence of the substance being a *good conductor,* and thus conveying the fluid to the hand, which is another good conductor, by which means it is lost as fast as it is excited. But if we insulate a metallic body, or any other conducting substance, then, on being rubbed, it gives signs of electricity, like electrics.

Liquids and gases, by friction against solid bodies, excite electricity. Thus, quicksilver rapidly agitated in a glass tube electrifies it, and the blast of a bellows against the projecting knob of Coulomb's electrometer, (see Fig. 213,) puts the needle in motion. Even a slight puff with the mouth, directed upon the knob, will produce a sensible degree of excitation.

569. PROP. II. *The electricity which is excited from* GLASS *and a numerous class of bodies, exhibits different properties from that which is excited from* AMBER, *or sealing wax, and a class of bodies equally numerous with the other.*

The kind of fluid excited from glass and analogous bodies is called *vitreous,* and that from amber and analogous bodies, *resinous* electricity. The term *positive* is also used instead of vitreous, and *negative* instead of resinous.

In order to understand the applications of the preceding terms, *vitreous* and *resinous*, *positive* and *negative*, it is necessary to know something of the two hypotheses upon which these terms are respectively founded. The first hypothesis is that proposed by *Du Fay*.* It ascribes all electrical phenomena to the agency of *two* fluids, specifically different from each other, and pervading all bodies. In unelectrified bodies, these two fluids exist in combination, and exactly neutralize each other. By the separation of the two fluids it is that bodies are electrified; and it is by the reunion of the two fluids, that the electricity is discharged. or bodies cease to be excited. The second hypothesis was proposed by Dr. Franklin. It ascribes all electrical phenomena to the agency of *one* fluid, which, as in the other case, is supposed to pervade all bodies, being naturally in a state of equilibrium. It is only when this equilibrium is destroyed that bodies become electrified, and it is by the restoration of the equilibrium that the electricity is discharged, or bodies cease to be excited. But a body is electrified when it has either more or less of the fluid than its natural share; in the former case it is *positively*, in the latter case *negatively*, electrified; positive electricity, therefore, implies a redundancy, and negative electricity, a deficiency of the fluid.

This much being sufficient for the understanding of the terms, and of the general principles of these two celebrated hypotheses, we shall postpone all discussions respecting them, until the learner has become acquainted with a sufficient number of electrical phenomena, to enable him to understand and to judge of the evidence adduced in support of each hypothesis.

570. Prop. III. *Bodies electrified in different ways attract, and in the same way repel each other.*

Thus if an insulated pith ball, (Art. 564,) or a lock of cotton, be electrified by touching it with an excited glass tube, it will immediately recede from the tube, and from all other bodies which afford the vitreous electricity, while it will be attracted by excited sealing wax, and by all other bodies which afford the resinous electricity. If a lock of fine long hair be held at one end, and brushed with a dry brush, the separate hairs will become electrified, and will repel each other. In like manner, two insulated pith balls, or any other light bodies, will repel each other when they are electrified the same way, and attract each other when they are electrified different ways.

Hence it is easy to determine *whether the electricity afforded by a given body is vitreous or resinous;* for, having electrified the electrometer by excited glass, then all those bodies, which when

* This is sometimes called the hypothesis of Symmer, after an English electrician of that name, who matured and illustrated the principle first suggested by Du Fay. (See Phil. Trans. 1759.)

excited *attract* the ball, afford the resinous, while all those which *repel* the ball, afford the vitreous electricity.

571. PROP. IV. *The two kinds of electricity are produced simultaneously; the one kind in the body rubbed, the other in the rubber.*

For example, if we rub a glass tube with a silk or woolen cloth, the glass becomes positive and the cloth negative. The foregoing law holds true universally; but the kind of electricity which each substance acquires, depends upon the substance against which it is rubbed. If we rub dry woolen cloth against *smooth* glass, it acquires the resinous, and the glass the vitreous electricity; but if we rub the same cloth against *rough* glass, it becomes positively, while the glass becomes negatively electrified.* The following table contains a number of electric substances, arranged in such a way that when they are rubbed against each other, any substance in the list before another becomes positively, and any substance below it negatively electrified:

1. Fur of a Cat,	6. Paper,
2. Smooth Glass,	7. Silk,
3. Woolen Cloth,	8. Lac,
4. Feathers,	9. Rough Glass,
5. Wool,	10. Sulphur.

The fur of a cat, when rubbed against any of the bodies in the table, always affords the vitreous, and the sulphur always the resinous electricity. Feathers become negative when rubbed against the fur of a cat, smooth glass, or woolen cloth; but positive when rubbed against wool, paper, silk, lac, rough glass, or sulphur.†

572. PROP. V. *Electricity passes through some bodies with the greatest facility; through others with the greatest apparent difficulty, or scarcely at all; and others have a conducting power intermediate between the two.*

Metals and charcoal, water and all liquids, (oils excepted,) are good conductors. *Melted* wax and tallow are good conductors; but these bodies while solid conduct very badly. Glass, resins, gums, sealing wax, silk, sulphur, precious stones, oxides, and all gases, are non-conductors, or at least very bad conductors.‡ Atmospheric air is a non-conductor of the highest class, when perfectly dry; but it becomes a conductor either when moist or when rarefied. The electric fluid easily pervades the vacuum

* The cloth should be attached to a glass handle to insulate it.

† When black stockings are worn over white, numerous sparks are frequently observed on pulling off the outer pair. The same appearances occur when a silk garment is worn over flannel. (See an interesting account of Symmer's experiments on this subject, in Priestley's History of Electricity, p. 267.)

‡ Thomson.

of an air-pump, or of the Torricellian tube, (Art. 460 ;) but these are imperfect vacuums: it is said that electricity cannot pass through a perfect vacuum.*

The conducting powers of most bodies are influenced by changes of temperature, and also by changes of form. Water, in its natural state, is a good conductor; but its conducting power is increased by heat and diminished by cold. Steam and ice are each inferior, in conducting power, to pure water; and ice below the temperature of $-13°$ Fah. becomes an electric of the highest class. Snow, when cold and dry, is a bad conductor. During a dry snow storm the air frequently becomes highly electrical.

The same body frequently exhibits great changes in conducting power by changes of state or chemical constitution. Thus, green wood is a conductor, dry baked wood a non-conductor; charcoal a conductor, ashes a non-conductor.

573. Strictly speaking, there is no substance known that is entirely impervious to electricity; for the intensity of that agent may be so increased as to force it, for a greater or less distance, through all bodies. Neither is there any body in which the conducting power is perfect. The following table presents a catalogue of bodies arranged in the order of their conducting powers:

CONDUCTORS.

Metals, the more perfect, or least oxidable, the better.

Charcoal, better when prepared from hard wood and well burned.

Plumbago.

Charcoal in fine powder.

Pure Water.

Snow, better when moist, worse when dry.

Living Vegetables.

Living Animals.

Flame, Smoke, Steam.

Rarefied Air.

NON-CONDUCTORS OR ELECTRICS.

Lac,† *Amber, Resins.*

Sulphur.

Wax.

Fat.

---

* Lib. Use. Knowl., Art. Electricity, p. 5.

† Lac, which is placed at the head of non-conductors, is a species of resin, sold by the druggists, and is a substance deposited upon a tree in India, by a certain species of insect.—*Shell lac*, the most common form employed in electrical experiments, is nothing more than lac in its purest form.—*Sealing wax* is substituted for lac in electrical experiments, being made chiefly of that substance.—*Varnishes* also, which are employed to coat the surfaces of electrical apparatus, owe their efficacy to lac, of which they are chiefly composed.

Glass, Gems, Precious Stones.
Silk, Wool.
Hair, Feathers.
Cotton, Paper.
Dry Atmospheric Air, and other gases.
Baked Wood.
India Rubber.

It is particularly important to remember that Metals, Water, and all moist substances, Animal substances, as the human body, and the Earth itself, are *conductors;* while the Air, when dry, and all Resinous and Vitreous substances are *non-conductors.* These bodies are those which are chiefly concerned in making experiments with electrical apparatus.

574. Prop. VI. *Insulation is effected in various degrees of perfection, according to the state of the atmosphere, and the nature of the substances employed as insulators.*

If the air were a conductor, it is not easy to see how the electric fluid could be confined so as to be accumulated. It is, moreover, only when the air is *dry* that it is capable of insulating well; hence, in damp, foggy, and rainy weather, electrical apparatus will not work well unless the air is dried artificially by operating in a close room highly heated by a stove.*

Lac, drawn into fine threads, is the most perfect insulator. Compared with silk thread, such a filament is ten times more effectual in preventing the loss of the fluid. Fine silk thread, however, when perfectly dry, is among the best insulators, and where great delicacy is required, a single filament of silk as it comes from the ball of the silk-worm is employed. Its conducting power is somewhat influenced by its color,—black being the worst, and a gold yellow the best color for insulating. Glass is much used as an insulator, especially when great strength is required, as in supports to various kinds of electrical apparatus. Glass, however, is liable to acquire moisture on its surface, in consequence of which its properties as an insulator are materially impaired. This inconvenience is obviated by giving it a thick coat of varnish. Fine hair is a good and convenient substance in some cases of insulation.

In certain instances, conducting or uninsulating threads are required. Then fine silver wires, or linen threads first steeped in a solution of salt and dried, are used.

575. The *sphere of communication* is the space within which a spark may pass from an electrified body, in any direction from it. It is sometimes called the striking distance. The *sphere of*

* We have been able to hold public lectures on electricity, illustrated by numerous experiments, in the most unfavorable weather, by keeping the room highly heated by close stoves.

influence is the space within which the power of attraction of an electrified body extends in every way, beyond the sphere of communication. A glass tube strongly excited will exert an influence upon the gold leaf electrometer at the distance of ten or even twenty feet, although a spark could not pass from the tube to the cap of the electrometer at a greater distance than a few inches.

The electricity which a body manifests by being brought near to an excited body, without receiving a spark from it, is said to be acquired by *Induction.*

When an insulated conductor, unelectrified, is brought into the neighborhood of an insulated charged conductor, its electricity undergoes a new arrangement. The end of it next to the excited conductor, assumes a state of electricity opposite to that of the excited conductor; while the farther extremity assumes the same kind of electricity. Suppose the excited conductor is electrified positively. The end of the insulated conductor next to it becomes negative, and the remoter end positive; and intermediate between these two points, there occurs a place where neither positive nor negative electricity can be perceived. This place is called the *neutral point.*

The reason why unelectrified bodies are attracted by excited electrics is, that they are put into the opposite state by induction, and then attracted upon the general principle laid down in Prop. III. When they come into the sphere of communication of the excited body, they immediately acquire the same kind of electricity, and are repelled. If they come into contact with uninsulated bodies, they lose the electricity they have acquired, are again put into the opposite state by induction, again attracted and again repelled. This process will go on until the electricity of the insulated conductor is all conveyed away.*

A body which has been electrified by induction, returns to its natural state instantaneously when the electrifying body is suddenly withdrawn. This is called the *return stroke.* Thunder clouds sometimes put objects beneath them under so powerful an influence of induction, that on the return to the natural state the shock is so violent as to destroy life.

The foregoing general principles may be verified with very simple apparatus, such as pith balls, a glass tube, and a stick of sealing wax. But the same facts may be exhibited in a much more striking and impressive manner by the electrical machine and its appendages, and our attention will therefore be now turned to the consideration of the subject of electrical apparatus.

* Thomson's Outlines of Heat and Electricity, p. 362.

CHAPTER II.

OF ELECTRICAL APPARATUS.

576. The object of the electrical machine is to *accumulate* electricity. It is made of several different forms, but two of these forms are predominant, which it will be sufficient for our present purpose to describe; of these, one is called the Cylinder, the other, the Plate Machine. The Cylinder Machine is represented in Fig. 214. The principal parts belonging to it, are the cylinder, the frame, the rubber, and the prime conductor. The *cylinder* (A) is of glass, from eight to twelve inches in diameter, and from twelve to twenty-four inches long. It should be perfectly cylindrical, otherwise it will not press the cushion or rubber evenly when turned. It must be as smooth as possible, for rough glass becomes a partial conductor; the former only is suitable for affording positive electricity. The cylinder should be so mounted on the frame as to revolve without waddling, for such a motion would prevent its being in uniform contact with the rubber. The *frame* (B, B) is made of wood, which must be close-grained, well seasoned, and baked in an oven, and finally coated with varnish, the object of all this preparation being to

Fig. 214.

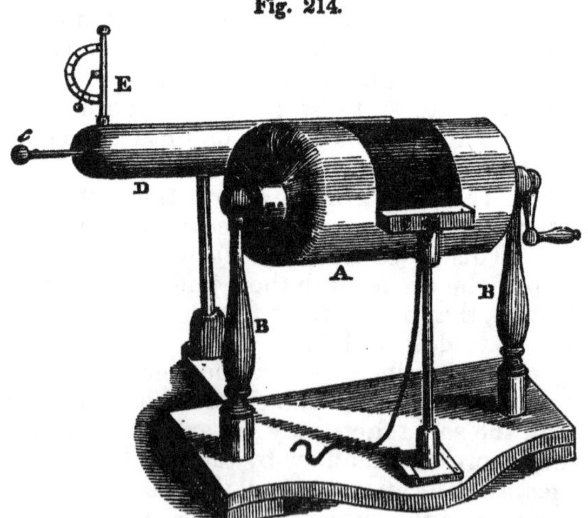

diminish its conducting powers, and thus prevent its wasting the electricity of the cylinder. The *rubber* (C) consists of a leathern cushion, stuffed with hair like the padding of a saddle. This is covered with a black silk cloth, having a flap which extends from the cushion over the top of the cylinder to the distance of an inch from the points connected with the prime conductor, to be mentioned presently. The rubber is coated with an amalgam,* made of mercury, zinc, and tin, which preparation has been found, by experience, to produce a high degree of electrical excitement, when subjected to the friction of glass. The rubber is insulated by placing it on a solid glass pillar, and it is made to fit closely to the cylinder by means of a spring worked by a screw.

The *prime conductor* (D) is usually a hollow brass cylinder with hemispherical ends. It is mounted on a solid glass pillar, with a broad and heavy foot, made of wood, to keep it steady. The cylinder is perforated with small holes, for the reception of wires (*c*) with brass knobs.

It is important to the construction of an electrical machine, that the work should be smooth and free from points and sharp edges, since these have a tendency to dissipate the fluid, as will be more fully understood hereafter. For a similar reason, the machine should be kept free from dust, the particles of which act like points, and dissipate the electricity.

577. The Plate Machine, (Fig. 215,) consists of a circular plate of glass from eighteen to twenty-four inches or more in

Fig. 215.

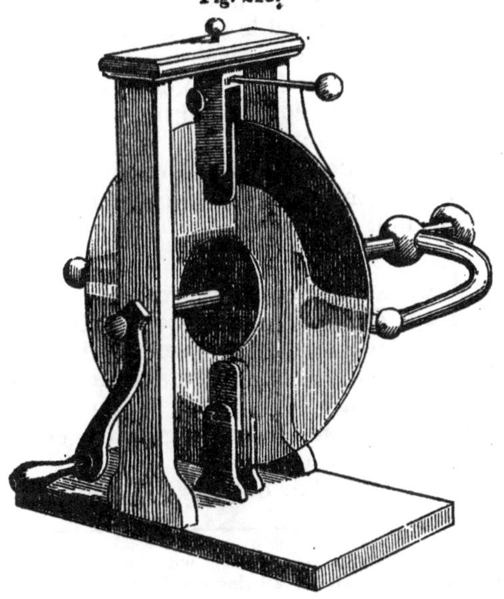

* The amalgam recommended by Singer, one of the ablest practical electricians, is composed of zinc two ounces, of tin one ounce, and of mercury six ounces. The zinc and tin may be melted together in a ladle or crucible, and poured into a mortar previously heated, to prevent the sudden congelation of the melted metals. As soon

diameter, turning vertically on an axis that passes through its center. The frame is composed of materials similar to those which compose the frame of the cylindrical machine. This machine is furnished with two pairs of rubbers, attached to the top and bottom of the plate. The prime conductor consists of a brass cylinder, proceeding from the center in a line with the axis, and having two branches, which serve to increase its surface, and at the same time to connect it with the opposite sides of the plate, so as to receive the electricity as it is evolved from each cushion.

It is not agreed which of these two machines affords the greatest quantity of electricity from the same surface ; but the cylinder is less expensive than the plate, and less liable to break, and is more convenient for common use.

578. The principles of the electrical machine, will be readily comprehended from what has gone before. It differs from the glass tube, only in affording a more convenient and effectual mode of producing friction. By the friction of the glass cylinder or plate against the rubber, electricity is evolved, which is immediately transferred to the prime conductor, and may be taken from the latter by the knuckle, or any other conducting substance. If the glass and the rubber both remain insulated, the quantity of electricity which they are capable of affording, will soon be exhausted. Hence, a chain or wire is hung to the rubber and suffered to fall upon the table or floor, which, communicating as it does with the walls of the building, and finally with the earth, supplies an inexhaustible quantity of the fluid to the rubber. In cases where very great quantities of electricity are required, a metallic communication may be formed immediately between the rubber and the ground.

579. In order to indicate the degree of excitation in the prime conductor, the *Quadrant Electrometer* is attached to it, as is represented at E in Fig. 214. This electrometer is formed of a semicircle, usually of ivory, divided into degrees and minutes, from 0° to 180°,* the graduation beginning at the bottom of the arc. The index consists of a straw, moving on the center of the disk, and carrying, at the other extremity, a small pith ball. The perpendicular support is a pillar of brass, or some conducting

---

as they are introduced, they must be rapidly stirred with the pestle, during which process the mercury may be added, and the stirring continued, until the amalgam is cold, when it will be in the form of paste or fine powder. A little lard is added, to give the amalgam the proper consistence ; but, if when applied it be warmed a little, but a small proportion of lard need be used. In hot weather, less quicksilver is to be employed.

* Sometimes the division is carried only to 90°, which is all that is necessary

substance. When this instrument is in a perpendicular position, and not electrified, the index hangs by the side of the pillar, perpendicularly to the horizon; but when the prime conductor is electrified, it imparts the same kind of electricity to the index, repels it, and causes it to rise on the scale toward an angle of 90°, or to a position at right angles with the pillar. It is obvious that the index can never rise higher than 90°, since the knob which terminates the brass pillar is electrified to the same degree as the prime conductor, and repels the index with equal force.* Nor is the angle at which the index remains suspended to be regarded as the true measure of the repulsive force. It has been demonstrated, that, in order to estimate this force truly, the arc of the electrometer should be divided according to a scale of arcs, the tangents of which are in arithmetical progression.†

580. When an electrical machine is skillfully fitted up, and works well, on turning it, circles of light surround the cylinder or plate, and brushes or pencils of light emanate copiously from the cushion and other parts of the machine. The circles of light consist of electric sparks, which discharge themselves between the excited surface and the rubber, their passage being so rapid as to appear like a continued line, like that of a small stick ignited at the end and whirled in the air. The brushes of light arise from the facility with which the fluid escapes from points or thin edges.

The experiments which were previously performed on electrical attractions and repulsions, (Arts. 567—575,) may now be repeated in a much more striking manner, and various other experiments added, which can be shown only when electricity is accumulated.

---

* Sometimes when the knob is so small that the electricity escapes from it as from a point, it does not repel the index with the same force as the prime conductor; in which case the index rises above 90°.

† Partington's Manual, Nat. Phil., II, 157.—As electrical machines are expensive, and not always easily procured by the private learner, it may be useful to suggest a mode of fitting up a cheap apparatus. A large tincture bottle may be procured of the apothecary, for the cylinder. A cover of wood may be cemented to each end, to the center of which, next to the bottom, is screwed a projecting knob for one end of the axis, while the part of the axis to which the handle is attached, is screwed into the center of the cover of wood next to the nozzle. Thus prepared, it may be mounted on such a frame of hard dry wood as every joiner or cabinet maker can construct. A tinner can make the prime conductor, and several other appendages to be described hereafter. Junk bottles or long phials serve well as insulators. Ingenious students of electricity frequently amuse themselves with making machines of this description, some of which have answered nearly every purpose of the most expensive kinds of apparatus.

A *cement*, for electrical purposes, may be made by melting together five ounces of resin, one ounce of beeswax, one ounce of Spanish brown, and a teaspoonful of plaster of Paris, or brick dust.

581. We proceed to enumerate a few of the effects of electricity, as they are exhibited by the electrical machine, confining ourselves for the present to those experiments which relate to attraction and repulsion, and the passage of the spark, reserving such as relate to light and heat to future sections. The following effects may be observed with a machine of moderate powers, the rationale of which the learner will readily supply from the propositions given in Art. 567, &c.

(1.) When the machine is turned, a downy feather, or a lock of cotton held in the hand by a conducting thread,* will be strongly attracted toward the excited surface.

(2.) A skein of thread, or lock of fine hair, looped, and suspended by the loop from the prime conductor, will exhibit strong repulsions between the threads or hairs. Lamp-wick, of light and spongy cotton, furnishes the best threads for this experiment.

(3.) The quadrant electrometer (Art. 579) being attached to the prime conductor, the conducting powers of different substances may be readily tried. Thus, an iron rod held in the hand, and applied to the prime conductor, will cause the index of the electrometer to fall instantly; and the same effect will follow the application of any metallic rod. A wooden rod of the same dimensions, will cause the index to descend more slowly; and a glass rod will hardly move it at all. These experiments show that iron is a perfect, wood an imperfect conductor, and glass a non-conductor. In the same manner the conducting powers of a stick of sealing wax, a roll of silk, or cloth, and of various other bodies, may be illustrated.

(4.) If a pith ball, or feather, or any other light body, held by a silk thread, be presented to the prime conductor, it will first be attracted and then repelled, and it cannot again be brought into contact with the electrical conductor, until its electricity is discharged by communicating with the finger, or some unelectrified conductor.

(5.) By placing light bodies between an electrified conductor and an uninsulated body, they may be made to move with great rapidity backward and forward, from one surface to the other, being alternately attracted and repelled by the electrified surface. By this means are performed electrical dances, the ringing of bells, and a variety of interesting and amusing experiments.†

(6.) If the rubber be *insulated* while the machine is turned, the rubber and the glass cylinder, or plate, will be found to be in different electrical states; an insulated body attracted by the one will be repelled by the other.

Bodies are electrified positively by connecting them with the

* The conducting power of linen or cotton threads is improved by moistening them with the breath.

† See *Singer's Elements of Electricity*, for a good selection of these experiments.

glass, by means of the prime conductor, and negatively by connecting them with the rubber, the latter being insulated, and the prime conductor uninsulated.

(7.) An electrified body frequently exhibits a tendency to separate into minute parts, these parts being endued with the power of mutual repulsion. Thus, a lock of cotton, when electrified, is separated into its minutest fibres. Melted sealing wax, when attached by a wire to the prime conductor, is divided into filaments so small as to resemble red wool. Water, dropping from a capillary syphon tube, on being electrified, is made to run out in a great number of exceedingly fine streams. Water, spouting from an air fountain, (Art. 458,) is divided into a number of rays, presenting the appearance of a brush.

(8.) A portion of electrified air, in consequence of the mutual repulsion between its particles, expands, and when at liberty to escape, becomes rarefied. Thus, a current of air may be set in motion from an electrified point, or small ball, or be made to issue from the neck of a bottle.

Such are some of the leading experiments which may be performed with the common electrical machines, in addition to those which are connected with light and heat, to be more particularlv described hereafter.

TORSION BALANCE.

582. The instrument called the Torsion Balance, invented by Coulomb,* exceeds all others in delicacy and the power of measuring small forces; and in the skillful hands of the inventor, it furnished the means of very refined investigations into the most hidden laws of electricity. The same instrument was also applied to similar researches in several other branches of physics, affording in each case an example of the most refined experimental analysis.

The force employed to estimate any given power of electric attraction, is the force of *torsion;* that is, the effort made by a twisted thread or wire to untwist itself. Since the thread may be small to any extent, and may be of any length, (and the force of torsion is found to be inversely as the length, and directly as the fourth power of the thickness,†) the degrees by which this force is increased as the thread is turned, may differ from each other by the smallest conceivable quantity, and yet be separated by spaces far enough asunder to be susceptible of being measured

* Charles Augustus Coulomb, was a very distinguished member of the French Academy, and remarkable for his assiduity and precision in experimental researches. He flourished during the latter part of the last century. His experiments on electricity, magnetism, friction, and the resistance of fluids, are among the finest in natural philosophy.

† Biot, Précis El. tome I, 339.

with the utmost precision; and thus any force, as that of electrical attraction, required to hold the successive degrees of the force of torsion in equilibrium, may be exactly ascertained. If, by a fine thread, (which may be either the smallest filament of silk, or the finest silver wire,) we suspend a horizontal needle, as in the electrometer represented in Fig. 213, the least conceivable force applied at the extremities of the needle, will put it in motion. A lever an inch long, suspended by a fibre of silk four inches in length, requires a force only the sixty thousandth part of a grain, to twist it three hundred and sixty degrees.

Fig. 216.

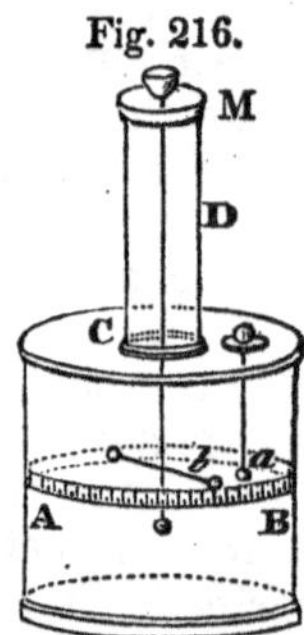

583. The construction of the instrument is as follows. In order to guard the suspended needle from the agitations of the air, it is protected by a glass cylinder, AB, having a movable lid, C, from the center of which rises a smaller glass cylinder, D, which covers the suspending thread; this latter cylinder is surmounted by a graduated circle, M, upon which moves a pointer or index, connected at the center with the suspending thread, which is twisted when the index is turned. The lid, C, is perforated with a hole to allow access to the pith ball of the needle. In the figure, this opening is represented as closed by the handle of a movable rod of glass or lac, which insulates the ball *a*, by which electricity is conveyed to the ball *b* of the needle. On a level with the needle is a circular band, graduated into degrees and minutes. It is usually made of paper, and pasted around the cylinder.

To prepare the apparatus for experiments, the index on M is set opposite to zero, and then the circle, conveying the index along with it, is turned, until the ball of the needle rests opposite to zero, on its graduated circle. In this situation, the suspending thread is entirely untwisted, or free from torsion. Now let the ball *a* be electrified, by receiving a spark from the prime conductor, and let it be introduced to the level of the needle. The ball *b* of the needle being unelectrified, is first attracted to the electrified ball, imbibes the same kind of electricity, and is then repelled to a greater or less distance, according to the intensity of the electricity. On account of the extreme delicacy of the instrument, only a very small charge must be applied; otherwise the agitation of the needle will be in danger of breaking the thread,* or the arc described by the needle will be inconveniently large. The charge is therefore applied from a pin's head, the pin itself being concealed in sealing wax. The pin's head being electrified, it is touched by the ball *a*, by means of which, the charge is introduced into the cylinder and made to communicate

* The filament used by Coulomb in some of his experiments, was a silver wire, a foot of which weighed only one sixteenth of a grain.

with the ball b of the needle. Suppose the force of repulsion between the two balls to be such, that the needle will finally settle at the distance of 36° from zero, or the point where it was quiescent, it would describe a greater arc in that direction, were not its motion counteracted by the force of torsion, exerted by the suspending wire.

584. Our object, it will be recollected, is to estimate the force of this repulsion at different distances from the electrified ball a. This is done by finding the relative forces of torsion required to bring those respective forces of repulsion to an equilibrium. We therefore turn the index upon the circle M in a direction opposite to that in which the needle moved, and observe the number of degrees through which the index must be turned, in order to make the ball b approach to any given distance of the ball a. Coulomb proceeded as follows. The ball b being electrified by contact with a, receded from it, describing an arc of 36°. The index on the circle M was then turned in the opposite direction, until the needle was carried back to the distance of 18°, which required the index to be turned over 126°. Again the index was turned until the needle was brought to the distance of 8½°, which required it to be turned over 567°. Let abd represent the circle in which these movements were performed, c being its center. Take ab equal to 36°, then b will be the position of the needle after the first repulsion. The index which carries the thread, being now turned backward 126°, the ball b, were it free to move, would be carried over the same arc to d', 126° beyond a, but on account of the repulsion of the ball a, it stops short at b', at the distance of 18° from a. Therefore, the force of repulsion of the two balls is 126°+18°=144°. In the third case, where the index was turned 567°, and the needle brought to the distance of 8½° of a, were it not for the repulsion between the balls, the needle would have been carried 567° beyond a to d, but stops short of a 8½°; therefore, that repulsion is equal to 567°+8½=575½°. Hence, the respective forces of repulsion exerted at the several distances, were as follows:

Fig. 217.

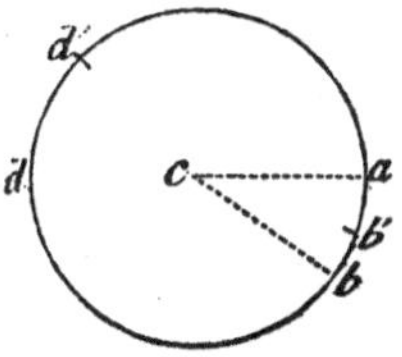

36°	- - - - - 36	which are	1 : 1.
18°	- - - - - 144	" "	½ : 4.
8½°	- - - - - 575½	" "	¼ : 16, nearly.

It appears that the distances are to one another nearly in the ratio of the numbers 1, ½, ¼, while the corresponding forces are as 1, 4, 16; that is, *the force of repulsion between two electrified bodies, at different distances, varies inversely as the square of the distance.**

* Biot, Précis Elém. tome I, 482.

The same law, therefore, governs the electrical forces as that which prevails among the bodies of the solar system.

585. Analogous experiments prove that attraction obeys the same law. Some practical difficulty was experienced by Coulomb, in his experiments on attraction, since, when the balls are differently electrified, as they must of course be in experiments on attraction, they will come together if brought within moderate distances of each other. But the law was satisfactorily shown to hold good, at such distances as were susceptible of measurement, and the law was further established by a process totally different from the preceding. It consisted in bringing the suspended needle near to an insulated electrified sphere, by which it is made to oscillate with greater or less rapidity, according to its degree of proximity. The number of oscillations, in a given time, is a measure of the force of attraction, as the number of oscillations of the pendulum measures the force of gravity, being universally as the square root of the forces. (Art. 183.) The proposition may therefore be stated in general terms—

The force of electrical attraction or repulsion, at different distances from an electrified body, varies inversely as the square of the distance.

RATE AT WHICH CHARGED BODIES LOSE THEIR ELECTRICITY.

586. It is a well-known fact, that when an insulated conductor, charged with electricity, is suffered to remain untouched for a certain time, it will gradually lose its charge. Now since, in some of the delicate researches of Coulomb, a considerable time was necessarily occupied, the electrified bodies under examination might change their degree of excitement during the experiments, and thus give a fallacious result. It became important, therefore, to ascertain the law according to which this dissipation or loss of electricity took place, and to make suitable allowance for it.

Three causes chiefly operate in depriving a body under these circumstances of its electricity:—first, the imperfection of bodies employed as insulators; secondly, the contact of successive portions of air, every particle of which carries off a certain quantity of the fluid; thirdly, the presence of moisture, which increases the conducting powers of all surfaces. (Art. 573.) No substance is actually impervious to electricity; that is, there is no substance known, of which any portion, however small, will insulate perfectly any charge however great. Still, by diminishing the intensity of the charge, or by increasing the length of the substance it has to traverse, a degree of insulation may be obtained in which the escape of the fluid is imperceptible. This tendency of electricity to escape from charged bodies, is independent of the chemical nature of those bodies, being the same, under similar circumstances, for balls of wax, copper, elder pith, and various other substances. The same tendency is equally independent of the shape and magnitude of bodies, unless when the intensity of the charge is high; in which case, a figure that involves points and edges favors the dissipation of the fluid. When bodies are highly charged, the electricity is lost with comparative rapidity; more slowly as the charge is less; and the air being dry, and the insulator of a proper length, a certain charge will be retained without further loss.*

But the chief source of dissipation of the electric charge, arises from moisture, either existing in the air, or settling upon the surface of the insulating supports, or imbibed into the fibres of insulating threads.

DISTRIBUTION OF ELECTRICITY.

587. Does electricity reside only at the surfaces of bodies, or is it expanded throughout the whole of their substance? Coating a conductor with some non-conducting substance, (as a wire with sealing-wax, leaving the ends naked,) does not in the least impede the passage of fluid through it. Indeed, every conductor may be considered as really in this situation, being in contact with a stratum of air on every side, which, when dry, is a good non-conductor. The conclusion from this fact is, that the passage of the fluid is not confined to the surface, mathematically considered, but must, at least, occupy the exterior *stratum* of the conductor. It was found, however, by Coulomb, that if, of two bodies of equal surface and similar form, as two equal spheres, one be electrified, and the other be brought into contact with it, the electricity will be equally divided between them, and that this takes place when one sphere is solid and the other hollow, equally as when both spheres are solid. Hence it is inferred, that electricity resides only at or very near the surfaces of bodies.†

Fig. 218.

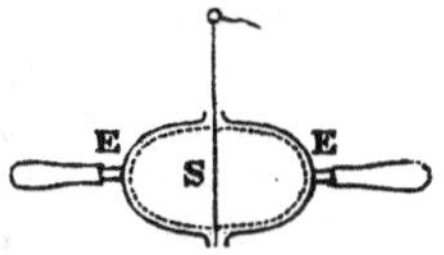

588. This fact is strikingly illustrated by an experiment, proposed by M. Biot.‡ Let S, (Fig. 218,) represent any spheroid of conducting matter, suspended by a thread of some perfectly insulating substance. Let E, E, be two caps formed of gilt paper, tinfoil, or any other conductor, and such that when united, they accu-

* Lunn, Encyc. Metrop.

† Although electricity resides only at the *surfaces* of bodies, yet the conducting power of a body, that is, its power to transmit a charge, is proportioned to the mass, or quantity of matter. (Faraday.)

‡ Précis Elém. tome I, p. 498.

rately fit the surface of the spheroid. An insulating handle of lac is also attached to each of the caps. Now let there be communicated to the ball S, any degree of electricity, and then carefully apply to it the two caps, holding them by their insulating handles. Upon removing these caps, it will be found that every particle of electricity has been abstracted from the spheroid, so that it will no longer affect the most delicate electrometer; while the two caps will be found, upon accurate trial, to have acquired precisely the same quantity of electricity that before resided upon the body S.

A proof of this point, equally conclusive, and applicable to bodies of every form, was devised by Coulomb. An insulated, solid conductor, of any figure, being provided, cavities were dug in it, to different depths below the surface, and in several different places, and the body was electrified. A *proof plane*, as it was called, consisting of a small circle of gilt paper, to which was attached an insulating handle of lac, was introduced into these various cavities at different depths. It was then withdrawn, and tested by the electrometer, and not the slightest trace of electricity was indicated. In these experiments, care was taken to introduce the proof plane, in such a way as not to touch the edges of the cavities, or any part of the surface, the object being to ascertain whether signs of electricity were exhibited at any depth below the surface. The conclusion was, that there were none, and consequently that the electricity of excited bodies resides wholly at the surface.*

An experiment, which may be easily repeated, shows how much the intensity of an electric charge is affected by the extent of surface which it pervades. Let a sheet of tinfoil be wrapped several times around an insulated cylinder, which is mounted so as to turn horizontally on an axis. Upon unwinding the metallic sheet, and thus increasing the extent of electrified surface, an electrometer connected with the cylinder will indicate a decline in the intensity of the charge, at every successive enlargement of surface.

589. Although electricity resides at the surface of an electrified body, yet it is not distributed *uniformly* over that surface, except the body be a perfect sphere, but is unequally accumulated, in different parts of the surface, in a manner depending on the figure of the body. The principle may be enunciated in general terms, thus:—

In conductors of an elongated figure, the electricity is accumulated toward the two ends, and withdrawn more or less from the central parts.

Coulomb, in his investigations on this subject, employed the

* Biot, Précis El. I, p. 500

proof plane, (Art. 588,) the circle of gilt paper being so small as to bear no considerable ratio to the surface of the electrified body under examination. By touching this plane to different points of the surface, the plane imbibes the charge belonging to that point, and may be made to transfer it to the balls of the electrical balance. (Fig. 216.) Then the amount of torsion required to bring the balls to the same given distance of each other, will be a measure of the charge communicated to the balls in each case; that is, the torsions will indicate the ratios existing between the different charges of electricity, at different points in the surface of the body under examination.

In this manner, Coulomb determined the distribution of electricity upon a steel plate, eleven inches long, one inch broad, and half a line thick, insulated and electrified. In order to cover the breadth of the plate, the gilt paper was made an inch long, but very narrow. First, the proof plane was applied to the center of the plate, and at one inch from the extremity; the latter charge was to the former as 1.2 to 1, and therefore nearly equal. Secondly, on applying the plane quite at the extremity, the charge was to that at the center as 2 to 1. Thirdly, the plane was applied, at one end, to the extreme edge, so as to be in contact with both surfaces; in which case, the charge was double that of each extreme surface, and, of course, four times that of the central parts.

590. Hence it appears, that the electricity of a conductor, analogous to the steel plate employed in the foregoing experiments, is nearly uniform on all parts of the surface, except the two ends, where it becomes twice as great as in the other parts. The rapid increase of electricity toward the extremities, appears also in other bodies of an elongated figure; and the augmentation is the more rapid, as the length is greater in respect to the diameter; and when the extremity becomes elongated, like the point of a cone, the accumulation at that extremity becomes so great, that the resistance of the air is not sufficient to retain it, and it escapes, producing the electric spark. Hence the reason why points, connected with an electrified conductor, dissipate the fluid so rapidly.

The limited extent of this work, does not permit us to give a more particular account of the researches of Coulomb, carried on by the aid of the torsion balance; but we would recommend these researches, as detailed by Biot,* to the student of natural philosophy, as examples of the most refined, ingenious, and conclusive experiments.

---

* Précis Elémentaire de Physique, tome I.

CHAPTER III.

OF THE LEYDEN JAR.

591. This instrument, which is a very important and interesting article of electrical apparatus, consists of a glass jar, coated on both sides with tinfoil, except a space on the upper end, within two or three inches of the top, which is either left bare, or is covered with a coating of varnish, or a thin layer of sealing wax. To the mouth of the jar is fitted a cover of hard baked wood, through the center of which passes a perpendicular wire,

Fig. 219. Fig. 220.

terminating above in a knob, and below in a fine chain, that rests upon the bottom of the jar. On presenting the knob of the jar near to the prime conductor of an electrical machine, while the latter is in operation, a series of sparks passes between the conductor and the jar, which will gradually grow more and more feeble, until they cease altogether. The jar is then said to be *charged*. If now we take the *discharging rod*, (which is a crooked wire, armed at each end with knobs, and insulated by a glass handle, as in Fig. 220,) and apply one of the knobs to the outer coating of the jar, and bring the other to the knob of the jar, a flash of intense brightness, accompanied by a loud report, immediately ensues. On applying the discharging rod a second time, a feeble spark passes, being the *residuary charge*, after which all signs of electricity disappear, and the jar is said to be *discharged*.

If, instead of the discharging rod, we apply one hand to the outside of the charged jar, and bring a knuckle of the other hand to the knob of the jar, a sudden and surprising *shock* is felt, convulsing the arms, and, when sufficiently powerful, passing through the breast.

592. The Leyden jar derives its name from the place of its discovery. In the year 1746, while some philosophers of Leyden

were performing electrical experiments, one of them happened to hold, in his hand, a tumbler partly filled with water, to a wire connected with the prime conductor of an electrical machine. When the water was supposed to be sufficiently electrified, he attempted, with the other hand, to detach the wire from the machine; but as soon as he touched it, he received the electric shock. It was by imitating this arrangement, that the Leyden jar was constructed; for here was a glass cylinder, having good conductors on both sides, viz. the hand on the outside, and water on the inside, which were prevented from communicating with each other by the non-conducting powers of the glass. A metallic coating, as tinfoil or sheet lead, was substituted for the two conductors, and a jar for the tumbler, and thus the electrical jar was constructed.

593. Those who first received the electric shock from the Leyden jar, gave the most extravagant account of its effects. M. Muschenbroeck, a philosopher of Leyden, of much eminence, said, that "he felt himself struck in his arms, shoulders, and breast, so that he lost his breath, and it was two days before he recovered from the effects of the blow and the terror; adding, that he would not take a second shock for the kingdom of France." M. Winkler, of Leipsic, testified, that "the first time he tried the Leyden experiment he found great convulsions by it in his body; and that it put his blood into great agitation, so that he was afraid of an ardent fever, and was obliged to use refrigerating medicines. He also felt a heaviness in his head, as if a stone lay upon it, and twice it gave him a bleeding at the nose."

In an age less enlightened than the present, and less familiar with the wonders of philosophy and chemistry, the striking and truly surprising effects of electricity, as exhibited by the Leyden jar, would naturally excite great admiration and astonishment. Accordingly, showmen travelled with this apparatus through the principal cities of Europe, and probably no object of philosophical curiosity ever drew together greater crowds of spectators. It was this astonishing experiment, (says Dr. Priestley,) that gave eclat to electricity. From this time it became the subject of general conversation. Everybody was eager to see, and, notwithstanding the terrible account that was reported of it, to *feel* the experiment; and in the same year in which it was discovered, numbers of persons, in almost every country in Europe, got a livelihood by going about and showing it. All the electricians of Europe, also, were immediately employed in repeating this great experiment, and in attending to the circumstances of it.* With similar assiduity, and unequalled success,

* Priestley's Hist. Elec., p. 84.

Dr. Franklin betook himself to experiments on the Leyden jar. He effectually investigated all its properties, by very diversified and ingenious experiments, and gave the first rational explanation of the cause of its phenomena. The following experiments may be easily repeated.

594. (1.) *The jar is charged by bringing the knob near to the prime conductor, while the machine is in operation.* One mode of charging the jar has been already mentioned in Art. 591. It may, however, either be held in the hand, or placed on the table, or on any conducting support: the only circumstance to be attended to is, that the outside shall be uninsulated. A jar, while charging, will sometimes discharge itself spontaneously. This effect will be more likely to happen, if the uncoated interval is very clean and dry, and may be prevented altogether, by previously breathing on the uncoated part.*

(2.) *The opposite sides of a charged jar are in different electrical states, the one positive and the other negative.* Thus, if a pith ball, suspended by a silk thread, be applied to the knob, it will first be attracted to it, and then repelled; but it will now be attracted by the outside coating, until it becomes electrified in the same way, and then repelled, and so on.

(3.) *In order to receive the charge, the outside of the jar must be uninsulated.* If we attach a string to the knob of the jar, and suspend the jar in the air, to the prime conductor, and put the machine in operation, no charge will be communicated to the jar. The same result will follow, if the jar stands on an insulated stand,† or is insulated by any other method. An insulated jar, however, may be charged by connecting its knob with the positive conductor, and its outer coating with the rubber.‡

(4.) *A second jar may be charged, by communication with the outside of the first, while the latter is receiving its charge.* The charge communicated to the second jar, is of the same kind as that of the first, and nearly of the same degree of intensity, provided the capacity of the two jars be the same. Moreover, if a third, a fourth, or any number of jars of the same size, be connected in a similar manner with each other, namely, having the knob of each in communication with the outside coating of the next preceding,—then all the jars will be charged with the same kind of electricity, but the degree of intensity will decline a little in the successive jars. If the charge be derived, through the

* Singer, El. Elec., p. 101.

† An insulating stand is any flat support, insulated by a pillar of glass. The pillar s usually a solid cylinder of glass, from six to twelve inches long, varnished so as to protect it from moisture. A junk bottle, surmounted by a circular piece of polished wood, dry and varnished, makes a very good insulating support.

‡ Singer, El. Elec., p. 106.

prime conductor, from the cylinder or plate, as is usually the case, it will be the positive, or vitreous electricity.

(5.) *A jar may be charged negatively, by receiving the electricity of the rubber,*—the rubber being insulated, and the prime conductor uninsulated. For this purpose, the chain usually attached to the rubber may be transferred to the prime conductor. Also, a jar may be charged negatively, by grasping the jar by the knob, and receiving the electricity of the prime conductor on the outside. It must be set down on an insulated support, else the operator will receive a shock.

(6.) *When two jars are charged, the one positively and the other negatively, on forming a communication between the insides of both, by connecting the two knobs, no discharge will take place, unless the outsides be in conducting communication.* Thus, if two jars be charged, the one from the prime conductor and the other from the rubber,* and placed at the distance of a few inches from each other, on insulated supports, on connecting the two knobs by the discharging rod, no discharge will follow; but, let a wire be laid across the supports, touching the outside of each jar; then, on applying the discharging rod to the two knobs, an explosion will immediately ensue.

By means of two jars differently charged, and placed as above, with their outsides in conducting communication, the experiment may be exhibited, which is called the *Electrical Spider.* It consists of a small piece of cork, so fashioned as to represent the body of a spider, and blackened with ink, having a number of black linen threads drawn through it to represent the legs. This is suspended by a silk thread, half way between the knobs of the two jars, and vibrates for a long time from one knob to the other, until both jars are discharged. The rationale will be obvious on a little reflection.

(7.) *The charge of any jar may be divided into definite parts;* that is, the half, the fourth, or any aliquot part of the charge may be taken.† This may be done by connecting the inner and outer coating of the charged jar, with the inner and outer coating of an unelectrified jar, of the same size and thickness. The respective charges will be measured by the quadrant electrometer.‡ (Fig. 214.)

(8.) *The electricity is accumulated on the surface of the glass, and the coatings serve merely as conductors of the charge.* This is proved by the fact that when the coatings are movable, so that

* And both may be thus charged at the same time, by connecting one with the insulated rubber, and the other with the insulated prime conductor, the jars themselves being uninsulated.

† Singer, p. 110.

‡ It is essential, however, that the electrometer should be graduated, not by equal divisions, but according to a scale of arcs, the tangents of which are in arithmetical progression

they can be taken off from the jar after it is charged, neither of them exhibits the least sign of electricity; while if another pair of coatings is substituted, which have not been electrified, on forming the communication between the inside and outside, the usual discharge takes place, showing that the whole of the charge was retained on the glass surfaces of the jar.*

(9.) *The charge of a Leyden jar may be retained for a long time.* If the surfaces are well separated from each other, the charge remains for many days, or even weeks. The charge is usually dissipated by the motion of particles of dust, or other conducting substances in the atmosphere, from one of the coatings to the other, or by the uncoated interval becoming moist and losing its insulating power; consequently a jar will retain its charge longer in dry than in damp weather. Covering the uncoated part of the jar with melted sealing wax or varnish, prevents the deposition of moisture upon it, and consequently tends also materially to prevent the dissipation of its charge.†

(10.) *A pane of glass, a plate of air, or any other similar electric, may be charged to a greater or less degree in a manner analogous to that of the Leyden jar.*—If a pane of glass is coated on both sides with a sheet of tinfoil, leaving an uncoated interval all round the edges for the space of two inches;—and if we then hold the pane by one corner and apply the knuckle to the outer coating, and bring the inner coating to the prime conductor, the pane will be charged, and may be discharged, by applying the knobs of the discharging rod to the opposite metallic coatings. A plate of air may be charged in the same manner as a plate of glass; but as air is more readily displaced by electricity, in consequence of the mobility of its particles, a thicker stratum of it must be employed. The usual form of the experiment is to employ two circular disks of wood covered with tinfoil, and well rounded at the edges, having a diameter of from two to four feet. One of the boards is to be placed flat upon a table, and the other being suspended by a silk cord from the ceiling, is adjusted so as to hang parallel over its surface, and at the distance of an inch or an inch and a half from it. The upper insulated board being connected with an electrical machine, the stratum of air between the boards becomes charged, and will communicate a shock if the upper and lower one be touched at the same time with opposite hands. The shock produced in this way is considerably less violent than that from an equal surface of coated glass; for the distance of the coatings is of necessity much greater, and the medium between them less perfectly insulating; and this last circumstance operates so rapidly when the charge is high, that its maximum of effect cannot be obtained but by making the discharge while the machine is in action. If the

* Singer, p. 112.

† Ib. 116.

discharge is not made, spontaneous explosions from one disk to the other, through the intervening plate of air, will occur at intervals, as long as the electrization of the upper disk is continued.

(11.) If a coated pane of glass be held vertically, with two of its edges parallel with the horizon, and to the upper edges of the metallic coating two threads be attached directly opposite to each other; on communicating a spark to one of the coatings, the two threads both rise, forming equal angles with the surface of the glass. On applying a conductor, as the finger, to one of the coatings, the thread on that side immediately falls, while the other thread doubles its angle of elevation; so that the angles intercepted between the two threads, is a constant quantity.*

Before the learner is qualified to understand the explanation of the foregoing experiments, he must become more fully acquainted with the law of induction, (Art. 575,) upon which the theory of the Leyden jar depends.

LAW OF INDUCTION.†

595. Active electricity, existing in any substance, tends always to induce the opposite electrical state in the bodies that are near to it. It is our object, in this section, to exhibit this important principle more fully than has yet been done in the preceding pages.

Fig. 221.

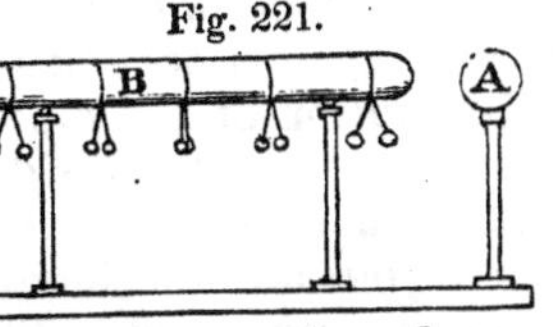

Let A (Fig. 221,) represent an electrical glass globe, and B a metallic cylinder, placed on insulating supports, near to the glass globe, but not near enough for a spark to pass. To the cylinder, let five pairs of pith balls be suspended, by conducting threads, viz. one pair near each end, one near the center, and one about half way between the center and either extremity. We shall find that every pair of pith balls, except those situated at a particular part of the cylinder not far from the center, will immediately diverge, indicating the electrical state of the part from which they are suspended. Those at either extremity diverge most; and the divergence diminishes as we approach the central parts to a certain point, where the pith balls suffer no effect, and where, consequently, the body is in its natural state. By means of the electrometer, we may ascertain that the species of electricity is negative, or opposite to that of the glass globe, in all those parts of the cylinder which are nearer to the globe than the before mentioned neutral point; and that it is positive in all parts of the cylinder more distant

* Encyc. Metropol., Elec., p. 92.
† Biot, Précis Elém. tome I., or Library of Useful Knowledge, Art. *Electricity*.

than this point. We may ascertain with much greater accuracy these electrical states, by the employment of the proof plane and electrometer of Coulomb, (Art. 589,) than by pith balls; and the results are then found to correspond with the results of theory, to be stated hereafter.

596. These effects, it should be remarked, are simply the result of electrical action at a distance; for they depend upon no other circumstance. They take place in an equal degree, whatever substance is interposed between the bodies which are exerting this action on one another, provided the interposed substance undergoes no change in its own electrical state; a condition which is fulfilled in electrics, or non-conducting bodies only. Thus, induction will take place just as effectually through a plate of glass, as if no such substance had intervened.

Let us now suppose that the acting body A is not glass, or any electric, but a conducting body, a sphere of copper, for example, charged with positive electricity, and insulated on a glass support. The primary effects of this sphere on the cylinder will be the same as in the former case; but the electrical state which the cylinder has acquired at the end adjacent to the globe, will *react* upon the electricity of the globe, tending to put it into a state still further opposite to its own, that is, to render the nearer parts of the globe positive in a higher degree than they were before. This can be done only at the expense of the other side of the globe, which thus becomes less positive than before. But this new distribution of the electric fluid in the globe, by increasing the positive state of the side next to the cylinder, tends to augment its inductive influence upon the fluid in the cylinder; that is, to drive out an additional quantity of the fluid from the negative to the positive end. This is followed in its turn by a corresponding reaction on the globe, and so on, constituting a series of smaller adjustments, until a perfect equilibrium is established in every part. When this has been attained, the electrical states will, it is evident, be of the same kind as those consequent upon the *immediate* actions, though somewhat increased in intensity by the series of reactions.

597. The following experiment is a practical illustration of the preceding remarks. Furnish the copper globe with a pair of pith balls on each of two opposite sides. When the globe is insulated and alone, any electricity communicated to it will diffuse itself equally over the surface, and both pairs of balls will diverge equally. But on bringing near to it a conducting body, the balls on the remoter side will immediately begin to collapse, while those at the nearer side diverge to a greater degree than before; thus showing the nature of the reflex operation of the induced

electricity of the conductor, upon the body from which the induction originated.

It should be recollected, that in all the changes we have thus traced as the effects of induction, there has been no *transfer* of electricity from either of the bodies to the other; as might be inferred from their taking place equally well when a plate of glass is interposed. Another proof is afforded by the circumstance, that the mere removal of the bodies to a distance from one another, is sufficient to restore each of them to its original state. The globe remains as positively electrified as before; the cylinder returns to its condition of perfect neutrality; nothing has been lost and nothing gained on either side. The experiment may be repeated as often as we please, without any variation of the phenomena. But this would not be the case if the cylinder were divided in the middle, and one or both of the parts were removed separately, while they still remained under the influence of the globe. The return of the electric fluid from the positive to the negative end being thus prevented, each part will retain, after its separation, the electricity which had been induced upon it. The nearer portion will remain negative; the remoter portion positive. If the division had been in three parts, the middle part only would have been neutral. The experiment may be made by joining two or more conductors endwise, similar to B, (Fig. 221,) so that they may act as a single conductor when placed near to the electrified globe, and, after induction has thus been produced, removing them separately, and examining their electrical states. If the number of conductors be three, the first will be found negative, the third positive, and the second neutral.

598. Another modification of effect will take place when an insulated conductor, rendered electrical at both ends by induction, is made to communicate with another insulated conductor. Let us first suppose that a long metallic conductor is brought into contact with the remote end of the first cylinder B, (Fig. 221,) which has been rendered positive by induction. The fluid accumulated at this end will now pass into the conductor, and will remove to the most distant part of it. The transit will take place before actual contact, and will be manifested by the appearance of a spark, when the bodies are brought within the striking distance. The removal of this portion of fluid to a greater distance, will occasion a disturbance in the equilibrium that had before been established. The repulsion which that fluid had excited, and which had contributed to prevent any more fluid from being repelled from the negative end, is now considerably weakened by the greater distance at which it acts; and more fluid will leave the negative end, which end will consequently become more highly negative. This change of distribution will again occasion a further effect, by its reaction on the fluid in the globe whence the action origi-

nally proceeded; and another series of changes and adjustments will follow, until a new condition of equilibrium takes place and then the fluid will be at rest.

599. Thus we learn that the effects of induction in a conductor are augmented by increasing its length; they would therefore be greatest of all, if we could give it infinite length; but the same condition is attainable by placing the conductor in communication with the earth, which will carry off all the fluid which the electrified body is capable of expelling from the nearest end. Accordingly, if we touch with the finger, or with a metallic rod held in the hand, the remote end of an insulated conductor under the influence of induction, we obtain a spark more or less vivid according to the intensity of the electricity so induced; and the conductor so touched has now only one kind of electricity, namely, the one opposite to that of the electrified body which is acting upon it. The part touched is brought into a state in which it appears to be neutral, as long as it remains in the vicinity of the electrified body; because the actions of the redundant fluid and unsaturated matter in the two bodies, exactly balance one another. But it all the while really contains less fluid than its natural share, in consequence of the repulsive tendency of the fluid in the body which produces the induction; and this negative state will readily become active if the conductor that has been touched be again insulated, and then removed from the influence of the former. This peculiar condition of a body, in which its parts are really undercharged or overcharged with fluid, although, from the action of electrical forces derived from bodies in its vicinity, a state of equilibrium is established, and no visible effect results, has been denominated by Biot *disguised electricity.*

600. We have hitherto supposed the acting body to be positively electrified; but precisely the same effects would happen with regard to degree, although opposite as to the species of electricity, if it had been negatively electrified: and the same explanations will in every respect apply, with the requisite substitution of the terms negative for positive, and of attraction for repulsion, and *vice versa.* A little reflection will also show the application of the theory of double electricities to explain the same phenomena. Calling the electricity of the globe *vitreous* instead of positive, and substituting the term *resinous* for negative, we then say that the vitreous electricity of the globe drives off the similar electricity from the contiguous end of the cylinder, and attracts to it the resinous fluid. This again attracts the vitreous fluid from the remoter parts of the globe to the nearest surface; and thus, the vitreous and resinous, instead of the positive and negative fluids, act and react on each other.

601. Another consequence of the induction of electricity must not be overlooked, namely, that the bodies between which it takes place, necessarily attract one another: for the mutual actions between the contiguous surfaces of the globe and the cylinder, (Fig. 221,) which are in opposite electrical states, exceed that of the remoter surfaces of those two bodies which are in the same electrical state, because the latter surfaces are more distant from each other than the former, and the force of electrical action is inversely as the square of the distance. Hence the attractive force always exceeds the repulsive. We have already seen (Art. 575,) that this circumstance sufficiently explains the fact, that conducting bodies previously neutral, are attracted by electrified bodies. Another fact, which appears more singular, and which cannot be accounted for on any other principle, is also a direct consequence of the law of induction. If a small insulated body, weakly electrified, be placed at a distance from another and larger body more highly charged with the same species of electricity, it will, as usual, be repelled; but there is a certain distance, within which if it be brought, attraction will take place instead of repulsion. This happens in consequence of the inductive influence producing so great a change in the distribution of electricity, as to give a preponderance to the attractive forces of the adjacent parts of the two bodies, over the repulsive forces that take place in the other parts, and which would have acted alone if the fluid had been immovable.

602. From the foregoing principles it will be easy to understand how induction may operate through a succession of conductors, which are all of them insulated except the last; and which are separated from each other by distances greater than that at which a transfer of electricity would take place. If, under such circumstances, the first be electrified, alternate states of opposite electricities will be produced in the two ends of each conductor in succession. In all the ends nearest the first body, the electricity will be of the opposite kind to that with which the first has been charged; in the other ends it will be of the same kind as that of the first body. The vicinity of these opposite electricities will tend powerfully to retain them in that condition, and will diminish their electric action on surrounding bodies. A large portion of the electricities so arranged and retained, is therefore in the condition designated by the term *disguised electricity*.* (Art. 599.)

The principles of induction developed in the preceding articles, serve to explain a number of the most curious and intricate phenomena of electricity, among which are those of the Leyden jar; to this instrument, therefore, let us now return.

* Lib. of Use. Knowl., Art. *Electricity*

THEORY OF THE LEYDEN JAR.

603. Upon what principle does this instrument receive and retain such an accumulation of the electric fluid? The answer is, because *the two surfaces of the jar mutually augment each other's capacities, upon the principle of induction.* To trace the operation of this principle a little more particularly, let us observe what takes place while a jar is charging from the prime conductor of the electrical machine. And first, suppose the jar is insulated; a spark passes to the inner surface, and electrifies it positively. The inner surface now stands in the same relation to the outer, that the globe in Fig. 221 stands to the cylinder; that is, it tends to drive off the electricity of the same kind, and, in the same proportion, to attract the electricity of the opposite kind. But as the fluid cannot escape from the outer surface, (the jar being insulated,) it of course remains to oppose the further accumulation of the similar fluid on the inner surface. But secondly, suppose the jar uninsulated, its outer coating having free communication with the earth. A spark passes to the inside as before, and electrifies positively the inner coating. This repels the similar electricity from the outer coating, and renders the outside negative. Being negative it reacts by induction, (as the nearer surface of the cylinder, in Art. 595,) on the inside, and attracts to it a still greater charge, which is supplied by the prime conductor. This additional charge, acting in the same manner on the outside, renders it more highly negative than before, in consequence of which it attracts to the inside a still further charge of electricity from the machine. This series of actions and reactions between the two surfaces of the jar, proceeds in a diminishing series, until each surface becomes too feeble to exert any further influence on the other, and the jar is then *charged.*

Substituting the terms vitreous and resinous, for positive and negative, as in Art. 600, we may easily make the foregoing explanation conform to the supposition of two fluids.

604. For the purpose of making the theory of the Leyden jar familiar, we may now recur to the experiments mentioned in Art. 594, and attempt the explanation of them.

In the structure of the jar, we recognise the operation of the principle of *induction.* Here, an unelectrified body (the outer surface) is brought very near to an electrified body, (the inner surface,) without the possibility of communicating with each other, on account of the non-conducting properties of the glass. The nearer the two surfaces can be brought to each other, the more powerful is the effect of induction, that effect being inversely as the square of the distance. Accordingly, the thinner the jar, the more powerful is the charge it will receive; but the danger of breaking prevents our employing such as are very thin.*

To trace the process of charging a jar a little more minutely, let us suppose the jar connected with the prime conductor of an electrical machine, from which a spark is communicated to the inner coating. This, according to the principle of induction, expels a similar quantity of the same fluid from the opposite unelectrified surface, and renders that negative, in the same degree as the inside is positive. Being negative, it increases the attraction of the inner surface for the opposite species of fluid, and another spark is received, which again expels an additional quantity of the same species of fluid from the outside, and thus the two surfaces continue to act upon each other reciprocally, though with constantly diminishing power, until the jar is charged.

The reason also is plain, why the outside of the jar must be uninsulated; since it is only in such case, that the foregoing process of induction can take place; and we readily see why a series of jars may be charged, from the portion of electricity which is expelled from the outside of the first jar.

605. When a jar is charged negatively from the rubber, just the opposite process in all respects takes place, the outside becoming positive by induction, and reacting upon the inside. The case mentioned in Art. 594, (6,) where two jars differently charged, cannot be discharged unless their outer surfaces be in conducting communication, will be readily understood; for it is impossible for the equilibrium to be restored by the union of the electricities on the inside, while the outside remains electrified. If we could suppose this to take place for a moment, and the electricity within to be restored to its natural state, it would again be immediately decomposed by the inductive influence of the electrified coating without.

606. The phenomena of the Leyden jar, may be equally well explained, by substituting the terms vitreous and resinous, instead of positive and negative, on the supposition of two fluids, since the principles of induction apply equally well to both hypotheses. Thus, it is as easy to suppose that the resinous electricity is induced upon the outside by the attraction of the vitreous electricity within, as it is to suppose that the outside becomes negative by the loss of a portion of its natural share; and the necessity of the outer surface being uninsulated, is as apparent in the one case as in the other. But we reserve the discussion of the comparative merits of these remarkable hypotheses, un-

* The writer of this Treatise had a large jar constructed of very thin glass; it took an extraordinary charge, and when discharged gave a report like that of an ordinary battery; but it was fractured by the first experiment.

til the learner shall have become familiar with a great variety of electrical phenomena.

CHAPTER IV.

OF ELECTRICAL LIGHT.

607. Light, we have seen, is not a constant attendant of electrical phenomena. Indeed, until noticed by Otto Guericke, it it was not known to have any relation to electricity.

Electrical light appears whenever the fluid is discharged, in considerable quantity, through a resisting medium.

Accordingly, no light is perceived when electricity flows freely through good conductors; but if such conductors suffer any interruption, as by the intervention of a space of air, or even of an imperfect conductor, then the attendant light becomes manifest.

608. We shall best learn the properties of the electrical spark, by attending to a variety of experiments in which it is exhibited.*

A glass tube rubbed with black silk, which has been smeared with a little electrical amalgam, will yield copious sparks, and flashes of light. The tube should be warm, dry, and smooth, and of a size not less than two feet in length, and three fourths of an inch in diameter.

The electrical machine, when in vigorous action, affords brilliant circles and streams of light. In order to render the light afforded by turning the machine abundant, several practical expedients are necessary. All parts of the machine must be dry and warm, (but not hot.) It is useful to rub very freely the glass plate or cylinder, with an old silk handkerchief. Black spots, or lines that collect on the glass, especially when the amalgam is new, are to be carefully rubbed off, and should dust or down collect on the amalgam of the rubber, this must be removed. The action of the cylinder will be increased by the following process: rub a little tallow on the palm of the hand, and apply it to the bottom of the cylinder; then turn the machine until the tallow is all taken up by the rubber and flap. The pores of the flap will then become filled with tallow, it will apply itself more closely to the cylinder, and the supply of electricity will become more copious. A convenient method of recruiting the action of the machine, is to coat a circular disk of pasteboard or leather with

---

* In experiments on electrical light, the room is supposed to be dark. They appear to the best advantage in the night.

amalgam, and to apply it to the glass plate or cylinder while the machine is turning.

If the chain be removed from the rubber to the prime conductor, so that the former shall be insulated and the latter uninsulated, on bringing the ends of the fingers near the rubber, a stream of diluted light will pass between the fingers and the rubber.

609. *The length, color, and form of the electric spark, varies with the nature of the conductors between which it passes, and with that of the medium interposed between them.*

Electrical sparks are more brilliant in proportion as the sub stances between which they occur are better conductors. A spark received from the prime conductor upon a large metallic ball, is short, straight, and white; on a small ball it is longer, and zigzag; received on the knuckle, a less perfect conductor, it is purplish or reddish; on wood, or ice, or a wet plant, or water, it is red. Moreover, a longer spark can be obtained from a small ball, attached to the prime conductor by a wire of five or six inches long, than from the prime conductor itself; and the longest and most zigzag spark is obtained, when the knob of a Leyden jar is presented to a similar brass ball attached to the prime conductor. From a point positively electrified, the fluid passes in the form of a brush or pencil of rays; a point connected with the negative side, exhibits a luminous star.

A metallic chain connected with the prime conductor, becomes illuminated at the points where two links join, and at other points where the conducting powers of the metal are impaired by rust, or where roughnesses occur. If the chain has been previously corroded, artificially, by dipping it into a solution of salt, or a strong acid, and suffering it to remain until the outside has become rusty, the experiment will be more striking. When the chain is so good a conductor as to afford a ready passage to the fluid, the light will be produced more abundantly if the remoter end of the chain be held by the discharging rod, so as to insulate it; or it may be attached to any other insulating support.

610. *The electric spark passes, with increased facility, through rarefied air; and the distance to which it will pass between two conductors, is augmented as the rarefaction is made more complete.*

Instead of the distance of five or six inches, which is the limit of the spark from the prime conductor of an ordinary machine in the open air, the spark will pass through the space of four feet or more, in an exhausted tube. If a pointed wire, terminating in a knob above, be introduced into the top of a tall receiver, and the receiver be placed on the plate of the air-pump, on connecting the knob of the wire with the prime conductor and turning the machine, a brush of light only will appear at the extremity of the wire; but, on exhausting the air, this brush will enlarge, varying its appearance, and becoming more diffused as the air becomes more rarefied, until at length the whole receiver is pervaded by a beautiful bluish light, changing its color with the intensity of the transmitted electricity, and producing an effect which, with an air-pump of considerable power, is pleasing in the highest degree.*

When a charged jar is placed under the receiver of an air-pump, as the exhaustion proceeds, a luminous current flows over the edge of the jar from the positive to the negative side, until the equilibrium is restored.†

611. Electric light exhibits a very beautiful appearance as it passes or *flows* through the *Torricellian Vacuum.* The color is of a very delicate bluish or purple tinge, and the light pervades the entire space. But the most pleasing exhibitions of this kind are made by forming an artificial atmosphere of vapor in the Torricellian tube. Ether, or alcohol, passes into the state of vapor when the pressure of the atmosphere is removed; and, accordingly, on introducing a drop of one of these fluids into the Torricellian vacuum, it immediately evaporates and fills the void. If, now, a strong spark be passed from the prime conductor through this vapor, the spark will exhibit various colors; in ether, it is an emerald green, or mingled red and green; in alcohol, it is red or blue; but the colors vary somewhat with the distances at which they are seen.‡

612. Sir Humphry Davy performed a number of experiments, on the passage of electricity through a vacuum, of which an account is given in the Philosophical Transactions for 1822. He succeeded in forming a Torricellian vacuum quite free from air, but in such cases, a small portion of the mercury itself is converted into vapor, and from this he could not free the empty space. In all cases where the mercurial vacuum was perfectly free from air, it was permeable to electricity, and was rendered luminous by either the common spark, or the discharge from a Leyden jar. But the degree of the intensity of this phenomenon depended upon the temperature. When the tube was very hot, the electric light appeared in the vapor of a bright green color, and of great density. As the temperature diminished, it lost its vividness, and when it was artificially cooled to 20° below zero, it was so faint as to be visible only in the dark. In all cases, where the minutest quantity of rarefied air was introduced into

* Singer. † Ib.

‡ A Torricellian tube may be prepared for this purpose, by closing the upper end, next the vacuum, with a cork, having a large pin passing through it, the size of the pin-head being enlarged by a small ball of pith or cork.

the mercurial vacuum, the electrical light changed from green to sea green, and by increasing the quantity of air, it changed to blue and purple. Also when the temperature was low, the vacuum became a much better conductor.

A more perfect vacuum was formed by means of melted metals, as tin, of a more fixed nature than mercury, and therefore not liable to impair the vacuum by vapor of their own. A vacuum being made by means of fused tin, the electric light, at temperatures below zero, was yellow, and of the palest phosphorescent kind, requiring almost absolute darkness to be perceived; nor was it perceptibly increased by heat. When the temperature was diminished, the electrical light (transmitted through vapor of mercury) diminished also, till the temperature was reduced to 20°; but between 20° and —20° it seemed stationary.

Unless the electrical machine was very active, no light was visible during the transmission of electricity; but that the electricity passed was evident, from the luminous appearance of the rarefied air, in other parts of the tube.

From these and various similar experiments related by Davy, it seems demonstrated, that electricity is capable of passing through a perfect vacuum, but that the light emitted depends upon the vapor or air through which it passes, and that if the vacuum were perfect, no light whatever would appear.*

In condensed air, on the contrary, the spark passes with greater difficulty than ordinary. In such case, also, its whiteness and brilliancy are augmented, and its course is zigzag. These appearances are even exhibited by passing the spark through *confined air*, of only the ordinary density.

613. The colors of the spark are pleasingly varied by passing it, in a condensed form, as in the Leyden jar, through media of different kinds. The experiment is performed by making the given body form a part of the circuit of communication, between the inside and outside of the Leyden jar. A ball of ivory in this situation exhibits a beautiful crimson; an egg, a similar color, but somewhat lighter; a lump of sugar gives a very white light, which remains for some time after the spark has passed; and fluor spar exhibits an emerald green light, or, in some cases, a purple light, which also continues to glow in the dark for some seconds. The great intensity of the light is shown by the strong illumination which the sparks in the jar communicate to bodies slightly transparent. Thus an egg has its transparency greatly increased; and if the thumb be placed over the space which separates the two conducting wires that communicate with the two sides of the jar respectively, the illumination is so powerful, that the blood-vessels and interior structure of the organ may be distinctly seen.

* *Phil. Trans.* 1822, or *Thomson*, Outlines, p. 470.

614. Metallic conductors, if of sufficient size, transmit electricity without any luminous appearance, provided they are perfectly continuous; but if they are separated in the slightest degree, a spark will occur at every separation. On this principle, various devices are formed, by pasting a narrow band of tinfoil on glass, in the required form, and cutting it across with a penknife, where we wish sparks to appear. If an interrupted conductor of this kind be pasted round a glass tube in a spiral direction, and one end of the tube be held in the hand, and the other be presented to an electrified conductor, a brilliant line of light surrounds the tube, which has been called the spiral tube, or diamond necklace. By inclosing the spiral tube in a larger cylinder of colored glass, the sapphire, topaz, emerald, and other gems may be imitated. Words, flowers, and other complicated forms, are also produced nearly in the same manner, by a proper disposition of an interrupted line of metal, on a flat piece of glass.

615. *The light of the electric spark, is not a constituent part of the electricity, but arises from the sudden compression of the air, or other medium through which it passes.*

It is well known, that air is capable of affording a spark by sudden compression. There is a kind of match constructed on this principle, in which a small portion of air contained in a close cylinder, being suddenly compressed by forcing down a piston, yields a spark sufficient to light a quantity of tinder at the bottom of the cylinder. Now it is found by actual experiment, that electricity has the power of condensing air. This fact is shown by means of a small instrument called *Kinnersley's Air Thermometer*. It consists of a glass tube, closed air-tight at the two ends by brass caps, through each of which passes a movable wire, terminated within by a small ball. Through the lower cap is inserted a small glass tube, open at both extremities, and turned upward parallel to the cylinder. Into this tube is introduced a quantity of water sufficient to cover the bottom of the cylinder, and of course to rise a little way into the tube. The two balls being set at some distance from each other, and a spark from the Leyden jar being passed between them, the air within is suddenly rarefied, and the water ascends in the tube, and again descends, when the explosion is over. This sudden rarefaction of a portion of air before the electric spark, must cause a sudden and powerful compression in the portions of air immediately adjacent. The immense velocity of the spark must greatly increase the resistance, and of course the force of compression. This appears to be an adequate cause for the production of the

Fig. 222.

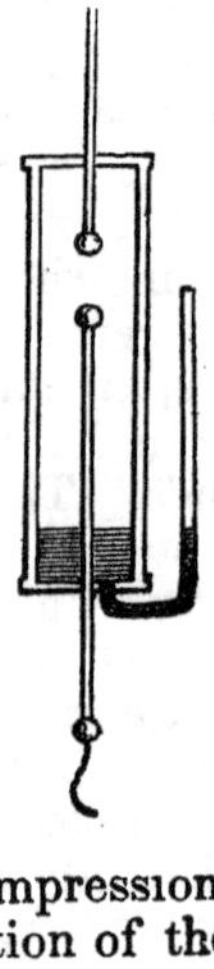

light that accompanies the electric discharge, and hence we conclude, that light is not inherent in the fluid itself. The greater density and brilliancy of the spark in condensed air, and its feebleness and diffuseness in a rarefied medium, are facts which accord well with the supposed origin; and the *zig-zag* form of the spark when long, or when passing through condensed air, is well explained by the same theory. For the electric fluid in its passage through the air, condenses the air before it, and thus meets with a resistance which turns it off laterally; in this direction it is again condensed, and has its course again changed; and so on, until it reaches the conductor toward which it is aiming. The zigzag form of lightning is accounted for on this principle.*

Electrical light is found, by optical experiments, to have in general the same nature with the light of the sun,—being like this resolved into various colors by the prism, and possessing other properties, to be described under the head of Optics, which identify it with solar light.

CHAPTER V.

OF THE ELECTRIC BATTERY.—MECHANICAL AND CHEMICAL AGENCIES, AND MOTIONS OF ELECTRICITY.—EFFECTS OF ELECTRICITY ON ANIMALS.

616. *An electric battery consists of a number of Leyden jars so combined, that the whole may be either charged or discharged at once.*

Very large jars cannot be obtained: it is rare to find one more than two feet high, by one and a half in diameter. Yet some of the mechanical effects of electricity to be described hereafter, require a much greater accumulation of the fluid than can be obtained from any single jar. The battery is constructed as follows. Large jars, twelve or fourteen inches high by five or six inches in diameter, are coated like ordinary Leyden jars. Twelve of these constitute a battery sufficiently powerful for most purposes, but the power of the battery may be carried to an indefinite extent by increasing the number of jars. When the number is twelve, they are placed four in a row in a box, the bottom of which is coated with tinfoil, by means of which the outsides of the jars are all in conducting communication. Each jar is separated from the rest by a slight partition of wood. To connect the insides of the jars, their knobs are joined by large

* Biot. Traité de Phys., tome 2.—Encyc. Metropol., Art. *Electricity*

brass wires. It is obvious, therefore, that the battery is equivalent to a single jar of enormous size, comprehending the same number of square feet.

The object of the battery is to accumulate a great *quantity* of the electric fluid, which is in proportion to the extent of surface: the *intensity* or elastic force, as indicated by the quadrant electrometer, is no greater in the battery when charged, than in a single charged jar. The battery, like the common jar, is charged by bringing the inside into communication with the prime conductor of an active and powerful electrical machine:* it is discharged as usual, by forming a connection between the inside and outside, commonly by means of the discharging rod.

Electrical batteries indicate only the *intensity* of the accumulated electricity, that is, its deviation from a state of natural distribution; the *quantity* can be inferred only from the comparative extent of the charged surface, or estimated by an examination of its effects, and is therefore by no means accurately appreciable.

617. The largest machine and battery hitherto constructed, were made for the Teylerian museum, at Haarlem. It consists of two circular plates of glass, each five feet five inches in diameter. The prime conductor consists of several pieces, and is supported by three glass pillars, nearly five feet in length. The force of two men is required to work the machine; and when it is required to be put in action for any length of time, four are necessary.†

At its first construction nine batteries were applied to it, each having fifteen jars, every one of which contained a square foot of coated glass; so that the grand battery formed by the combination of all these, contained one hundred and thirty-five feet. As examples of the great power of the Teylerian machine, we may mention the following: it charged a Leyden jar by turning the handle half round,—a charge which the jar would receive and lose by discharging itself spontaneously eighty times in a minute. A single spark from the conductor melted a considerable length

---

* As the process of charging a large battery is tedious and laborious, it has been proposed to charge each jar successively, after that which is immediately connected to the prime conductor, by means of the electricity expelled from the outside of the first, as is explained in article 594, (4.)—Encyc. Metrop., Art. *Electricity*.

† *Wright's Electrical Machine* is one of the largest hitherto constructed in this country. It has four large cylinders, all made to turn simultaneously, and to deliver their electricity to the same prime conductor. The prime conductor is made of sheet brass, of which it contains more than thirty square feet, and is mounted horizontally on glass pillars immediately over the cylinders. The spark from this conductor, when all the cylinders are in full action, received on the knuckle, gives a painful sensation. The machine charges a battery consisting of twelve gallon jars in thirty seconds, and causes a half gallon jar to explode spontaneously twelve times in a minute. T elegant piece of apparatus was constructed by Mr. Caleb Wright, and has recently been presented by that gentleman to Yale College.

of gold leaf. A spark or zigzag stream of fire would dart from the prime conductor to a neighboring conductor, to the distance of ten feet. A wire three eighths of an inch in diameter was found to be insufficient to transmit the whole charge of the prime conductor, but the wire would give small sparks to a conductor brought near it. The sphere of influence (Art. 575) extended to the distance of forty feet, so as sensibly to affect the pith ball electrometer. The *spider web* sensation, (or that peculiar sensation resembling that of the spider's web,) which is experienced by holding an excited glass tube to the face, was felt by bystanders to a great distance from the machine.*

MECHANICAL EFFECTS OF ELECTRICITY.

618. *The sound produced by an electric discharge is ascribed to the sudden collapse of the air, which has been displaced by the passage of the electric fluid.*

Hence the sound is greater in proportion to the quantity and intensity of the charge. A battery, when fully charged, gives a loud explosion.

619. *Imperfectly conducting substances, through which a powerful electric charge is passed, are torn asunder with more or less violence.*

A large Leyden jar is sufficient for exhibiting some of these mechanical effects: others require the power of the battery. When the charge is passed through a thick card, or the cover of a book, a hole is torn through it, which presents the rough appearance of a bur on each side. By means of the battery, a quire of strong paper may be perforated in the same manner; and such is the velocity with which the fluid moves, that if the paper be freely suspended, not the least motion is communicated to it.† (See Art. 231.) Pieces of hard wood, of loaf sugar, of stones, and many other brittle non-conductors, are broken, or even torn asunder with violence, by a powerful charge from the battery.

The expansion of *fluids* by electricity is very remarkable, and productive of some singular results. When the charge is strong, no glass vessel can resist the sudden impulse. Beccaria inserted a drop of water between two wires in the center of a solid glass ball of two inches diameter: on passing a shock through the drop of water, the ball was dispersed with great violence. In like manner, by the sudden expansion of a small body of confined air, strongly electrified, explosions may be produced, and bodies that resist its expansion are projected with violence. Even good conductors, when minutely divided, are expanded by electricity.

* Cavallo, Complete Treatise, Vol. II.

† Singer.

Thus mercury, confined in a capillary glass tube, will be expanded with a force sufficient to splinter the tube.*

CHEMICAL EFFECTS OF ELECTRICITY.

620. *By means of electricity, more or less accumulated, a variety of chemical effects may be produced; such as the combustion of inflammable bodies, the oxidation, fusion, and even combustion of metals, the separation of compounds into their elements, or the union of elements into compounds.*†

Ether and alcohol may be inflamed by passing the electric spark through them; nor is the effect diminished by communicating the spark by means of a piece of ice or any other cold medium. The finger may be conveniently employed to inflame these substances. Phosphorus, resin, and other solid combustible bodies, may be set on fire by the same means; gunpowder and the fulminating powders may be exploded, and a candle may be lighted. Gold leaf and fine iron wire may be burned, by a charge from the battery. Wires of lead, tin, zinc, copper, platina, silver, and gold, when subjected to the charge of a very large battery, burn with explosion, and are converted into oxides.‡

The same agent, moreover, is capable of reviving these oxides; that is, restoring them to the state of pure metals. By a singular contrariety of properties, water is decomposed into its gaseous elements, and the same elements are reunited to form water; and the constituent gases of atmospheric air, are, by passing a great number of electric charges through a confined portion of air, converted into nitric acid.

621. *The velocity of the electric fluid is apparently instantaneous.* A circuit of four miles has been formed, by means of wire, between the inside and outside of a Leyden jar, and no perceptible interval was occupied during the discharge. Analogy, however, would lead us to believe that electricity, like light, is progressive in its motions, but that it moves with a velocity too great to be measured, except for intervals of immense extent.

The velocity of light appears to be instantaneous, for such distances as four miles; but when such intervals are taken as the diameter of the earth's orbit, light is found to have a pro-

* The *illuminating powers* of the spark from a large battery, produce very striking appearances. The scintillations of an iron chain 20 or 30 feet long, when made to form the medium of communication between the inside and outside of a charged battery, are peculiarly beautiful. The chain may conveniently be hung in catenary curves from the ceiling of the room, supported by threads of silk.

† The chemical agencies of electricity, however, are much more powerful and extensive as exhibited by the galvanic apparatus, than by the common electrical machine.

‡ Singer, p. 185.

gressive velocity of 192,500 miles per second. If, therefore, electricity actually moves with a progressive velocity like that of light, still the time occupied in traversing the space of four miles would be inappreciable, since it would equal only about the fifty thousandth part of a second. Experiments are in progress for conveying *telegraphic dispatches* by means of electricity.

622. *The electric fluid, in its route, selects the best conductors.* The Leyden jar may be discharged with a wire held in the hand, without the insulating handle used in the discharging rod, since metallic wire is a better conductor than the hand, and the fluid will take its route through that in preference to the hand. But if a wooden discharger be substituted for the wire, the shock will be felt, since animal substances are better conductors than wood. It is necessary to remark, however, that when the charge is very intense or the quantity great, as in the battery, then some portion of the fluid will escape from the discharging wire and pass through the hand. In such cases, therefore, it is prudent to make use of the discharging rod.

Lightning, in striking a building, usually takes a course which indicates the preference of the fluid for the best conductors.

623. *The electric fluid will sometimes take a shorter route through a worse conductor, in preference to a longer route through a better conductor.* The sparks will pass through a short space of air, instead of following a small wire thirty or forty feet. The preference of the shorter route is sometimes indicated in taking the electric shock. While one person is receiving the shock from the Leyden jar, another may grasp his arm without feeling the least effect from the charge.

624. *The course of the charge is frequently determined by the influence of points, either in dissipating or in receiving the fluid.* Sharp points, connected with the best conductors, greatly favor the dispersion of the fluid during its passage, and sharp pointed conductors determine the charge toward them, from a great distance around. The finest needle, held in the hand toward the knob of one of the jars of a charged battery, will silently discharge it, in a few seconds; and if we apply one hand to the outside of a Leyden jar, and with the other bring a fine needle slowly to the knob of the jar, only a comparatively feeble shock will be felt, the charge being rapidly dissipated while the needle is approaching the knob.

EFFECTS OF ELECTRICITY UPON ANIMALS.

625. We have already several times incidentally adverted to the shock communicated to the animal system, when it is brought

into the electric circuit, so that the charge passes through it. We now propose to consider this interesting part of our science more particularly.

626. *The electric shock is received, whenever the animal system is made a part of the conducting communication between the inside and outside of a charged Leyden-jar.* A convenient method of administering the shock, is to place the charged jar on a table, resting immediately on a metallic plate,* as a plate of tin, lead, or copper; then grasping a metallic rod in each hand, touch one of them to the plate and the other to the knob of the jar, and a sudden convulsion of the limbs or the breast will be experienced, more or less violent according to the strength of the charge. The effect is greatly heightened by feelings of dread or apprehension, and it may be resisted to a considerable degree by a voluntary effort. A slight charge affects only the fingers or the wrists; a stronger charge convulses the large muscles above the arm-pits; a still greater charge passes through the breast, and becomes in some degree painful. Electricians, however, have frequently ventured upon charges sufficiently powerful to convulse the whole frame.

627. *The shock may be communicated to any number of persons at once.* This is usually effected by their joining hands, while the first in the series holds one of the metallic rods, (Art. 653,) with which he touches the plate or outside of the jar, and the last in the series holds the other rod, with which he touches the knob of the jar, at which instant the whole number receive the shock at the same moment, and that however extensive the circle of persons may be. The Abbe Nollet, a celebrated French electrician, gave the shock, at once, to one hundred and eighty of the king's guards, and to all the members of a convent, who formed a large community. All gave a spring at the same moment. The strength of the shock, however, is somewhat diminished by passing through a long circuit, some portion of the fluid being dissipated on the way.† The connection, instead of being made by taking hold of hands, may be formed between any number of persons A, B, C, D, &c. as follows: A may touch his foot to the foot of B; B may take the hand of C, who may touch the foot of D; then each of the company will feel the shock in one arm and one leg, showing that the fluid pursues the most direct course, agreeably to Art. 650.

628. If the discharge from two square feet of coated surface be made to pass through the region of the diaphragm, a sudden

* It is safer to employ such a plate than to bring the conducting rod immediately into contact with the outside coating of the jar; for in such case, persons unaccustomed to receive the shock, are apt to overturn the jar and break it.

† Priestley, p. 97.

convulsive action of the lungs produces a loud shout. A smaller charge produces a violent fit of laughter, even in the gravest persons. A very strong charge, passed through the diaphragm, produces involuntary sighing and tears, and sometimes brings on a fainting fit.* The charge of a large battery is sufficient to destroy human life, especially if it be received through the head. By standing on the *Insulating Stool*, which is a stool with glass feet, a person becomes an insulated conductor, and may be electrified like any other insulated conductor. A communication being made with the machine, the fluid pervades the system, but excites hardly any sensation except a prickling of the hair, which at the same time rises and stands erect; for the hairs, being similarly electrified, mutually repel each other.

While in this situation, the human system exhibits the same phenomena as the prime conductor when charged; that is, it attracts light bodies, gives a spark to conductors brought near it, and communicates a slight shock to another person who receives the spark from it. Indeed, the same shock is felt by both parties.

By means of the insulating stool, the most delicate shocks may be given; for the charge may be drawn off from any part, by imperfect conductors. Thus, a pointed piece of wood will draw off the charge from the eye, in a manner so gentle, as to secure that tender organ against any possibility of injury. By a variety of conductors, of different powers, and by points and balls, the sensations may be accommodated, with much delicacy, to the state of the patient, or to the nature of the affected part

Fig. 230.

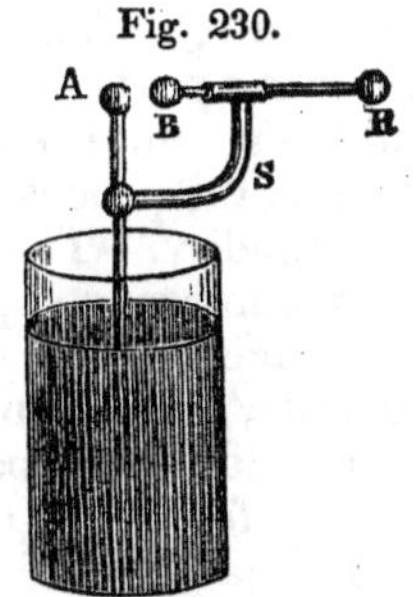

629. The shock may be communicated directly to any individual part of the system, without affecting the other parts, by making that part form a portion of the electric circuit, between the inside and outside of a Leyden jar. Thus, let it be required to electrify an arm. Two *directors* (consisting of wires terminating in brass knobs, and insulated by glass handles) are connected by chains with the knob, and the outside coating of a charged jar; then, on applying one of the directors to the hand, and the other to the naked shoulder, the arm is convulsed. In cases where the patient requires only a moderate shock, the charge is regulated by a contrivance attached to the jar, called *Lane's Discharging Electrometer*, represented in Fig. 223. S, is a stick of solid glass; B, R, two brass knobs, connected by a wire, which slides back and forth in such a way, that it may be

* Encyc. Metrop.—Morgan's Lectures.—Singer's Elements.

set at any required distance from the knob of the jar. Suppose that the outside of the jar is grasped with the left hand, and the knob R with the right, while the knob A is constantly receiving sparks from the prime conductor.* When the knob B is in close contact with A, no shock will be received, because the two sides of the jar are in conducting communication. But when B is removed a little way from A, then the shock will be felt when the charge has become sufficiently intense for the spark to pass from A to B ; and the shock will be greater or less according as the knobs are nearer to each other, or further apart, until the distance becomes too great for the jar to discharge itself. When the sliding wire is withdrawn a small distance from the knob of the jar, a constant succession of slight shocks are experienced, which are sometimes called *vibrations*.

630. It has already been mentioned, that life may be destroyed by strong electrical charges. Experiments have been made with the view of investigating the nature of this destructive action. Dr. Van Marum, of Haarlem, selected for this purpose eels, which, as is well known, retain signs of irritability, when cut into three, four, or six parts, and even when deprived of their heads. The eels employed in these experiments, were a foot and a half in length, and the shock was conveyed through the whole body. They were instantly killed, and never moved afterward. They were immediately skinned, and trial was made by pinching, pricking, &c., but no traces of irritability remained. When the shock was made to pass through individual parts, for example the head, these alone lost their irritability, while the rest retained it. When the head was kept free from the shock, the remaining parts only were paralyzed.†

It had been remarked, that whenever animals had been killed by lightning, the process of spontaneous putrefaction ensued with unusual rapidity. This subject was examined by M. Achard of Berlin, by numerous experiments. From these it appeared that electricity accelerates putrefaction, since it was found that animals recently killed, and animal substances, such as raw beef, became putrid much sooner when electrified. General credit is given to the foregoing experiments, but it seems easy to account for the increased tendency of milk to sour, and of meat to become putrid, during a thunder storm, from the effects of heat and moisture, which are known and adequate causes of these phenomena.

631. Soon after the discovery of the Leyden jar, commenced

* It will be convenient to attach one chain to the knob at R, and another to the outside of the jar, near the bottom, and to hold both in the two hands, standing on the insulated stool.

† Nicholson's Jour. VIII, 319.—Encyc. Metrop., Art. Electricity.

the application of electricity to *medicine;* and Medical Electricity became thenceforth a distinct branch of the science. The first cure said to have been effected by this agent, was upon a paralytic. Electricity shortly became very celebrated for the cure of this disorder, and patients flocked in great numbers to the practitioners of this branch of the profession. As usual, the effects of this new remedy were greatly exaggerated, and it was widely extolled, not only for the cure of palsy, but of all other diseases.* It was even pretended, that the virtues of the most valuable medicines might be conveyed into the system through the medium of electricity, preserving their specific properties in the same manner as when taken by the way of the stomach. Preparations of this kind were called *Medicated Tubes.* Pavati, an Italian, and Winkler, a German, were especially celebrated for this species of practice. The mode was, to enclose the medicines in a glass tube, then to excite the tube, and with it to electrify the patient. In this way, it was said, the healing virtues of the medicine were communicated to the system in a manner at once efficacious and agreeable.†

Pretensions so extravagant could not long be sustained, and the natural consequence was, that the use of electricity in medicine soon fell into great neglect, and has remained in this situation to the present time. There are, however, certain properties inherent in this agent, which deserve the attention of the enlightened physician, and inspire the hope that, in judicious hands, it may still be auxiliary to the healing art. First, the great activity of the agent, particularly the facility and energy with which it can be made to act upon the nervous system, indicate that it has naturally important relations to medicine. The power of being applied, locally, to any part of the system, renders it a convenient application in cases where other local remedies cannot be administered. Secondly, the acknowledged property of electricity to promote the circulation of fluids through capillary tubes, Art. 581, (7,) suggests the probability of its being efficacious in promoting the circulation of the fluids of the animal system, and in increasing the quantity of insensible perspiration. Thirdly, in the history of medical electricity are recorded well-attested cures, effected by means of electricity, of such diseases as palsy, rheumatism, gout, indolent tumors, deafness, and a variety of other disorders.‡

---

* Priestley, p. 409. † Ibid., 146.

‡ See Priestley's History of Electricity, pp. 146 and 408—Singer's Elements, p. 292—Phil. Transactions, passim—Encyc. Metropolitana, Elec. 105—Cavallo, Complete Treatise, Vol. II.

CHAPTER VI.

OF THE CAUSE OF ELECTRICAL PHENOMENA.

632. For the sake of convenience, and for the purpose of avoiding repetition and circumlocution, we have made frequent use of the phrase *electric fluid.* It may be proper now to inquire, whether there are any just grounds for supposing such a fluid or fluids to be present in electrical phenomena.

There are two modes by which the existence of such a fluid may be rendered probable: the first is, by showing that such a supposition is conformable to the analogy of nature; the second is, by proving that the agent of electrical phenomena exhibits the properties of a fluid.

633. First, *there are some reasons derived from analogy for believing in the existence of an electric fluid.* (1.) The reasons in favor of supposing that light and heat are caused by the agency of peculiar fluids, (arguments, however, that we cannot discuss here,) which have induced a general belief, are for the most part equally applicable to electricity. (2.) In the present state of our knowledge, the most subtile of all fluids, indeed the most attenuated form of matter, is hydrogen gas, of which one hundred cubic inches weigh only two and a quarter grains, being nearly fourteen times lighter than common air. But at no distant period, means had not been devised by mankind for proving the materiality of common air, nor even of identifying the existence of the other gases, which now bear so conspicuous a part in experimental philosophy. But as knowledge and experimental researches have advanced, a series of fluids still more subtle than air have come to light, until we have reached a body nearly fourteen times lighter than air, at which, at present, the series stops. Is it probable, however, that nature stops in her processes of attenuation precisely at the point where, for want of more delicate instruments, or more refined and powerful organs of sensation, our methods of investigation and powers of discrimination come to their limit? An examination of the general analogies of nature will lead us to think otherwise. The subordination which exists among the different classes of bodies that compose the other departments of nature, is endless, or at least, indefinite. In the animal creation, for example, beginning with the mammoth or the elephant, we descend through numerous tribes to the insect which is barely visible in the sunbeam. Before human ingenuity had devised means of aiding the powers of vision, the naturalist might

have fixed this as the limit of the animal creation. But the invention of the microscope has carried the range of human vision immeasurably further; and at each successive improvement in that instrument, new tribes of insects or animalcules have been revealed to the eye, still more and more attenuated. A similar subordination might be found in the vegetable kingdom, and in the organic structure of both animals and vegetables.

634. To apply this analogy to the case before us, we begin the series of inorganic bodies with platinum, and descend through classes of bodies constantly diminishing in density, until we come to ether, the lightest of liquids, and on the confines of those bodies which are invisible to the eye, and manifested only by the effects which they produce. By modern discoveries the series has been extended to hydrogen, a body two hundred and forty-seven thousand times lighter than platinum. Here for the present we pause, standing in the same relation with respect to any fluids that may lie beyond, that the ancients stood with respect to common air, and all the other aëriform fluids.

Considerations of this nature lead us to believe that there are in nature fluids more subtle than hydrogen; and such being the fact, we can hardly resist the belief, that heat, light, and electricity, are bodies of this class,—bodies which make themselves known to us by the most palpable and energetic effects, although their own constitution is too subtle and refined for our organs to recognise, or our instruments to identify them as material.

635. Secondly, in addition to the foregoing presumptions in favor of the supposition that electricity is a peculiar fluid, *it exhibits in itself the properties of a fluid.* The rapidity of its motions, the power of being accumulated, as in the Leyden jar, its unequal distribution over the surfaces of bodies, (Art. 589,) its power of being confined to the surfaces of bodies by the pressure of the atmosphere, its attractions and repulsions, are severally properties which we can hardly ascribe to any thing else than an elastic fluid of the greatest tenuity.

But granting the presence of an elastic fluid in electrical phenomena, it remains to be determined whether, according to the hypothesis of Franklin, these phenomena are to be ascribed to the agency of a single fluid, or whether, according to that of Du Fay, they imply the existence of two distinct fluids. The numerous facts with which the learner has been made acquainted in the preceding pages, will fit him to appreciate the evidence offered in favor of, or against these hypotheses respectively.

636. The principles of each hypothesis have been already explained, (see Art. 569,) and they have been rendered familiar by repeated application It will be recollected that they concur in

supposing that all bodies are endued with a certain portion of electricity, called their *natural share*, in which the fluid, whether single or compound, is in a state of perfect equilibrium; and that in the process of excitation, this equilibrium is destroyed. But here the two views begin to diverge: the one supposes that this equilibrium is destroyed in consequence of the separation of *two fluids*, which, like an acid and an alkali combining to form a neutral salt, exactly neutralize each other by mutual saturation, but which, when separated, exhibit their individual properties; the other, that the equilibrium is destroyed, like that of a portion of atmospheric air, by greater or less exhaustion on the one side, or condensation on the other. In the former case, moreover, the equilibrium is restored by the reunion of the two constituent fluids; in the latter, by the movement of the redundant portion to supply the deficient, as air rushes into the exhausted receiver of an air-pump.

It is a remarkable fact, that nearly every electrical phenomenon may be perfectly explained in accordance with either hypothesis; nor is it agreed, that an *experimentum crucis** has yet been found.†

637. One of the latest advocates of the hypothesis of a single fluid is Mr. Singer,‡ an able practical electrician, and the most distinguished defender of the doctrine of two fluids is M. Biot.§ In support of the former doctrine, are offered such arguments as the following. (1.) Its greater *simplicity*. It is supposed to be more conformable to the Newtonian rule of philosophizing, "to assign no more causes than are just sufficient to account for the phenomena." The known frugality of nature in all her operations, might lead us to suppose that she would not employ two agents to effect a given purpose, when a single agent would be competent to its production. This argument, however, cannot be applied, either where one cause is *not* sufficient to account for the phenomena, or where there is direct proof of the existence of more agents than one. (2.) The appearance of a *current*, circulating from the positive to the negative surface, analogous to the passage of air of greater density into a rarefied space. This point is much insisted on by Singer, and numerous examples are brought forward where the progress of such a current is manifest to the senses. Thus, the flame of a candle, brought into the circuit between the inside and outside of a Ley-

* The "experimentum crucis," is a phrase introduced by Lord Bacon, implying a fact which can be explained on one of two opposite hypotheses, and not on the other. The figure is derived from a cross set up where two roads meet, to tell the traveller which road to take.

† Lib. Useful Knowledge.

‡ Elements of Electricity and Electro-Chemistry, by George John Singer, London, 1814.

§ Traité de Physique, tome II.

den jar, is, on the discharge of the jar, bent toward the negative side; a pith ball, under similar circumstances, moves in the same direction when a charged jar is placed under the receiver of an air-pump, and the air is exhausted, a luminous cloud flows from the positive to the negative side, in whichever way the jar is electrified. None of these arguments, however, are found to be conclusive; for the mechanical effects, which are here ascribed to an elastic fluid, that is, the electric fluid, flowing toward the negative side, can all be accounted for, either upon the principles of attraction and repulsion, common to both hypotheses, or from the mechanical impulse of a current of air, which is known to be repelled from a point positively electrified. The electric spark passing instantaneously, or at least with a velocity entirely inappreciable, it is impossible to determine its direction.

638. The fact that bodies *negatively electrified repel each other*, (Art. 570,) is a strong argument against the truth of the hypothesis under consideration. It is not difficult to conceive that a self-repellent fluid should communicate the same property to two pith balls in which it resided; but that the mere *deficiency* of the fluid should produce the same effect is incredible. This fact drove Æpinus (a celebrated German electrician, who brought this hypo thesis to the test of mathematical demonstration) to the necessity of supposing that *unelectrified matter is self-repellent*—a supposition which is not only destitute of proof, but which is inconsistent with the general laws of nature, from which it appears that attraction and not repulsion exists mutually between all kinds of bodies. In the distribution of electricity upon surfaces differing in shape and dimensions, the fluid is found to arrange itself in strict accordance with hydrostatic principles, and that too in bodies negatively as well as positively electrified. Now that the privation, or mere absence of a fluid, should exhibit such properties of a present fluid, is inconceivable.

639. In favor of the doctrine of two fluids, the following arguments are urged. (1.) *Two opposite currents* are supposed to be sometimes indicated. Thus, (Art. 619,) a card perforated by a strong electric discharge, exhibits burs or protrusions on both sides. The appearance of the *electric spark* passing between two knobs, is supposed by some writers to indicate the meeting of two fluids from opposite parts. When the spark is short, the whole distance between the two knobs through which it passes, is illuminated. But when the spark is long, those portions of it which are nearest to the knobs, are much brighter than the central portions. Near the knobs the color is white, but toward the center of the spark it is purplish. Indeed, if the spark is very long, the middle part of it is not illuminated at all, or only

very slightly. Now this imperfectly illuminated part, is obviously the spot where the two electricities unite, and it is in consequence of this union, that the light is so imperfect.* (2.) The two electricities are characterized by *specific differences.* The light afforded by the vitreous surface is different from that of the resinous; when the two opposite portions of the spark meet, as above, the place of meeting is only half the distance from the negative that it is from the positive side; the bur protruded from the card is larger in the direction of the vitreous than in that of the resinous fluid; and the two severally produce certain chemical effects in bodies which are peculiar to each. (3.) But the most conclusive argument in favor of two fluids, is the perfect manner in which this supposition accounts for the *distribution of electricity* on bodies of different dimensions. (See Arts. 587—590.) On the hypothesis, that electrical phenomena are owing to the agencies of *two fluids, both perfectly incompressible, the particles of which possess perfect mobility, and mutually repel each other, while they attract those of the opposite fluid, with forces varying in the inverse ratio of the square of the distance,*—on this hypothesis, M. Poisson, a celebrated mathematician of France, applied the exhaustless resources of the calculus, to determine the various conditions which electricity would assume in distributing itself over spheres, spheroids, and bodies of various figures. The results at which he arrived were such as accord in a very remarkable degree with experiment, and leave little doubt that the hypothesis on which they were built must be true. Nor is any supposition involved in the hypothesis itself inconsistent with established facts. (4.) Finally, authority is, at the present day, almost wholly on the side of the doctrine of two fluids—an opinion which has constantly gained new adherents with every new discovery in the science of electricity, particularly in the department of galvanism.

CHAPTER VII.

OF ATMOSPHERICAL ELECTRICITY—THUNDER STORMS—LIGHTNING RODS.

640. Having learned the laws of electricity from a great variety of experiments, the student is now prepared to look upon the works of Nature, and to study the phenomena which the same agent produces there on a most extensive scale.

* Thomson.

The atmosphere is always more or less electrified. This fact is ascertained by several different forms of apparatus. For the lower regions, it is sufficient to elevate a *metallic rod* a few feet in length, pointed at the top, and insulated at the bottom. With the lower extremity is connected an electrometer, which indicates the presence and intensity of the electricity. For experiments on the electricity of the upper regions, a kite is employed, not unlike a boy's kite, with the string of which is intertwined a fine metallic wire. The lower end of the string is insulated by fastening it to a support of glass, or by a cord of silk. But as experiments of this kind involve some personal hazard, we subjoin, from an excellent treatise on practical electricity,* a few directions for the construction of this apparatus.

641. An electric kite should be constructed in the most simple manner, for it is an apparatus very liable to be injured or lost; its size should be moderate, as there is not often sufficient wind to raise one that is very large, which is besides on several other accounts very troublesome to manage. An ordinary paper kite, about four feet in height, and two feet wide, varnished with drying oil to defend it from the rain, is sufficiently well adapted to this purpose. The string must be made with a thin copper or silver thread, (such as is used for gilt lace,) interwoven with the twine of which it is formed, through its whole length. When the kite is raised, the string is insulated by attaching it to a silk cord, whose opposite extremity may be fastened to a rail, or any fixed or heavy body. The end of the metallic string is to be connected with an insulated conductor, and at two inches from the extremity of this conductor, a brass ball, well connected with the ground, or the nearest water, is to be placed; so that when the electricity becomes sufficiently intense to pass an interval of two inches, it will be conducted safely away without injury to the experimenter, who should be cautious, in such cases, not to approach the insulated conductor; but if he has occasion to remove any apparatus to or from it, to do so by the aid of long insulating handles or forceps.†

642. A few facts may be mentioned to show the hazard attending this class of experiments. Cavallo, on one occasion, had raised a kite, the string of which was insulated by silk lace. A cloud was over head, and the electricity began to be abundant, with which he charged a pair of Leyden jars. In order to prevent any accident which might arise from too great an accumulation of the fluid, he wished to take off the insulating silk, and connect the string immediately with the ground. For this purpose, he took hold of the string and detached it from its support.

* Singer's Elements. † Singer, p. 267.

"While I effected this, (says he,) which took up less than half a minute of time, I received about a dozen or fifteen very strong shocks, which I felt all along my arms, in my breast, and legs, shaking me in such a manner, that I had hardly power enough to effect my purpose, and to warn the people in the room to keep their distance."* Professor Richman, of Petersburgh, a distinguished devotee of our science, fell a victim to his temerity. He had constructed an apparatus for observations on atmospherical electricity, which was entirely insulated, and had no contrivance for discharging it when electrified too strongly. On the 6th of August, 1753, he was examining the electricity of this apparatus in company with a friend; while attending to an experiment, his head accidentally approached the insulated rod, when his attendant observed a globe of blue fire, as he called it, as big as his fist, jump from the rod to the head of the professor, which, at that instant, was about a foot from it. M. Richman was killed instantly; a red spot was left on his forehead, his shoe was burst open, and part of his waistcoat singed; his companion was benumbed, and rendered senseless for some time; and the door-case of the room was split, and the door torn off its hinges.

643. The most powerful apparatus ever employed for atmospheric electricity, was constructed in France by M. de Romas. He procured a kite seven feet long and three feet wide, and elevated it to the height of five hundred and fifty feet. A cloud coming over, the most striking and powerful electrical phenomena presented themselves. Light straws that happened to be on the ground near the string of the kite, began to erect themselves, and to perform a dance between the apparatus and the ground, after the manner of dancing images, as exhibited in ordinary electrical experiments. Art. 581, (5). At length streams of fire began to dart to the ground, some of which were an inch in diameter, and ten feet long, exhibiting the most terrific appearance.

The foregoing facts evince the abundance of electricity in the atmosphere at particular periods; but experiments of a less formidable kind have been instituted, to ascertain the electrical changes of the air. For this purpose, Mr. Canton, an English philosopher, constructed an ingenious apparatus, which warned him of the presence of any unusual quantity of electricity, by causing it to ring a bell connected with the lower extremity of the apparatus.

644. Obvious as is the connection between the phenomena of common electrical apparatus, and those exhibited in the heavens during a thunder storm, yet the identity of lightning with the electric spark, was not dreamed of by the earlier electricians.

* Cavallo's Complete Treatise on Electricity, II, 22

To Dr. Franklin, is universally conceded the merit of having established this fact, first by reasoning on just principles of analogy, and afterward by actually bringing down the lightning from the skies. The resemblances between the appearances of lightning and electricity, were thus enumerated.

(1.) The zigzag form of lightning corresponds exactly in appearance with a powerful electric spark, that passes through a considerable interval of air.

(2.) Lightning most frequently strikes such bodies as are high and prominent, as the summits of hills, the masts of ships, high trees, towers and spires. So the electric fluid, when striking from one body to another, always passes through the most prominent parts.

(3.) Lightning is observed to strike most frequently into those substances that are good conductors of electricity, such as metals, water, and moist substances: and to avoid those that are non-conductors.

(4.) Lightning inflames combustible bodies; the same is effected by electricity.

(5.) Metals are melted by a powerful charge of electricity: this phenomenon is one of the most common effects of a stroke of lightning.

(6.) The same may be observed of the fracture of brittle bodies.

(7.) Lightning has been known to strike people blind: Dr. Franklin found, that the same effect is produced on animals, by a strong electric charge.

(8.) Lightning destroys animal life; Dr. Franklin killed turkeys of about ten pounds weight, by a powerful electric shock.

(9.) The magnetic needle is affected in the same way by lightning and by electricity, and iron may be rendered magnetic by both causes. The phenomena therefore are strictly analogous, and differ only in degree; but if an electrified gun-barrel will give a spark, and produce a loud report at two inches distance, what effect may not be expected from 10,000 acres of electrified cloud? But (said Franklin) to ascertain the accuracy of these ideas, let us have recourse to experiment. Pointed bodies receive and transmit electricity with facility; let therefore a pointed metal rod be elevated into the atmosphere and insulated; if lightning is caused by the electricity of the clouds, such an insulated rod will be electrified whenever a cloud passes over it; this electricity may be then compared with that obtained in our experiments.*

645. Such were the suggestions of this admirable philosopher; they soon excited the attention of the electricians of Europe, and

* Singer.

having attracted the notice of the King of France, the approbation he expressed excited in several members of the French Academy, a desire to perform the experiment proposed by Franklin, and several insulated metallic rods were erected for that purpose. On the 10th of May, 1752, one of these, a bar of iron forty feet high situated in a garden at Marly, became electrified during the passage of a stormy cloud over it; and during a quarter of an hour it afforded sparks, by which jars were charged and other electrical experiments performed. During the passage of the cloud, a loud clap of thunder was heard, so that the identity of these phenomena was thus completely proved. Similar experiments were made by several electricians in England.

Doctor Franklin had not heard of these experiments, and was waiting the erection of a spire at Philadelphia to admit an opportunity of sufficient elevation for his insulated rod, when it occurred to him that a kite would obtain more ready access to the regions of thunder than any elevated building. He accordingly adjusted a silk handkerchief to two light strips of cedar, placed crosswise; and having thus formed a kite, with a tail and loop, at the approach of the first storm, he repaired to a field, accompanied by his son. Having launched his kite, with a pointed wire fixed to it, he waited its elevation to a proper height, and then fastened a key to the end of the hempen cord, and attached this by means of a silk lace (which served to insulate the whole apparatus) to a post. The first sign of electricity which he perceived, was the separation of the loose fibres of the hempen cord: a dense cloud passed over the apparatus, and some rain falling, the string of the kite became wet; the electricity was then collected by it more copiously, and a knuckle being presented to the key, a stream of acute and brilliant sparks was obtained. With these sparks, spirits were fired, jars charged, and the usual electrical experiments performed. Thus was the identity of lightning and electricity, which had been indicated by so many analogies, now established by the most decisive experiment.

646. It is a matter of much importance to the science of Meteorology, to ascertain from what *source* atmospherical electricity originates. Among the known sources of this agent none seems so probable, as the evaporation and condensation of watery vapor. We have the authority of two of the most able and accurate philosophers, Lavoisier and La Place, for stating that *bodies in passing from the solid or liquid state to that of vapor, and, conversely in returning from the aëriform condition to the liquid or solid state, give unequivocal signs of either positive or negative electricity.**

* Dr. Thomson, in his Outlines of Electricity, makes the following note.—M. Pouillet has lately published a set of experiments, which seems to overturn Volta's theory of the evolution of electricity by evaporation. He has shown that no electricity is evolved by evaporation, unless some chemical combination takes place at the same

Combustion is also attended with the evolution of electricity and even the *friction* of opposite currents of wind, or of a high wind against opposing objects, probably generates more or less of the same agent. The production of electricity during evaporation and condensation, may be rendered evident by Coulomb's electrical balance; as may that evolved during the friction of air. If the stem of a tobacco pipe be heated red hot, and a drop of water be introduced by way of the bowl, the jet of steam falling on the brass ball (Fig. 216, *a*,) of the balance will electrify it, so that it will set the index of the balance in motion.

It is obvious, that a cause which produces only very feeble signs of electricity in so small a quantity of vapor as that which arises from a single drop of water, may still be sufficient to occasion a vast accumulation of the same agent, in such a quantity of vapor as that which is daily ascending into the atmosphere; for it has been calculated, that more than two thousand millions of hogsheads of water are evaporated from the Mediterranean alone in one summer's day.*

THUNDER STORMS.†

647. The following are the *leading facts* respecting the electricity of the atmosphere in relation to this subject, and these are facts which have been established by numerous observers, of the most accurate and diligent class. Beccaria, an Italian electrician, continued his observations on the electricity of the atmosphere for fifteen years with the greatest assiduity; and Cavallo, Read, Saussure, and others, prosecuted the same inquiries with similar zeal.

(1.) Thunder clouds are, of all atmospheric bodies, the most highly charged with electricity; but all single, detached, or insulated clouds, are electrified in greater or less degrees, sometimes positively and sometimes negatively. When, however, the sky is completely overcast with a uniform stratum of clouds, the electricity is much feebler, than in the single detached masses before mentioned. And, since fogs are only clouds near the surface of the earth, they are subject to the same conditions: a driving fog, of limited extent, is often highly electrified.‡

---

time. But it follows from his experiments, that electricity is evolved abundantly during *combustion*, the burning body giving out resinous, and the oxygen vitreous electricity. In like manner, the carbonic acid emitted by vegetables, is charged with resinous electricity, and the oxygen (probably) charged with vitreous electricity.—*Thomson's Outlines*, p. 440. But we shall be slow to reject the results of experiments performed by such experimenters as Lavoisier and La Place, especially when confirmed by the testimony of Volta and Saussure.

* Singer.

† The most complete enumeration of facts hitherto made respecting thunder storms, was published by M. Arago in the *Annuaire* for 1838.

‡ Ed. Encyc., VIII, 310.

(2.) The electricity of the atmosphere is strongest when hot weather succeeds a series of rainy days, or when wet weather succeeds a series of dry days; and during any single day, the air is most electrical when the dew falls before sunset, or when it begins to exhale before sunrise.

(3.) In clear, steady weather, the electricity generally remains positive; but in falling or stormy weather, it is constantly changing from positive to negative, or from negative to positive.*

648. Such are the circumstances of atmospheric electricity in general; next, let us attend to the peculiar phenomena of thunder storms, chiefly as they are exhibited in our own climate.

(1.) In thunder storms there is usually a singular and powerful combination of all the elements,—of darkness, rain, thunder and lightning, and sometimes hail.

(2.) They occur chiefly in the hottest season of the year, and after mid-day; and are more frequent and violent in warm than in cold countries.

(3.) Thunder storms never occur beyond 75° of latitude—seldom beyond 65°.†

(4.) In this state, (Connecticut,) thunder storms usually come from the west, either directly, or from the northwest or southwest; but occasionally from the east.

(5.) Violent thunder and lightning are frequently seen in volcanoes and water spouts.

(6.) Thunder storms sometimes descend almost to the surface of the sea, and fall upon the sides of mountains; in which case, they are extremely violent.

(7.) We occasionally observe the following circumstances succeed each other in regular order: first, a vivid flash of lightning,—then a loud peal of thunder,—and, after a short interval, a sudden fall of rain, which sometimes stops as suddenly as it began.‡

649. There are in thunder storms, evidently, two distinct classes of phenomena to be accounted for. The first class consists of the common elements of a storm,—clouds, wind, and rain; the second, of thunder and lightning. The following proposition embraces, in our view, the true explanation of both these classes of phenomena:

The storm itself, including every thing except the electrical appearances, is produced in the same manner as other storms of wind

---

* Singer, p. 273. † Arago, *Annuaire*, 1838, p. 389.

‡ Morgan's *Lectures on Electricity* contain an excellent view of the natural agencies of electricity.

*and rain; and the electricity, and of course the thunder and lightning, is owing to the rapid condensation of watery vapor.**

We do not, therefore, consider electricity as the *cause*, but as the *consequence* of the storm; or as a concomitant of the clouds, wind, and rain.

A sudden and copious deposition of condensed vapor, is an essential preliminary, or concomitant, of a thunder storm, since when the process of condensation is slow, too much of the electricity evolved would escape to allow of the requisite accumulation; and the amount of vapor condensed must be copious, else the quantity of the electric fluid produced would not be sufficient to cause the violent phenomena of a thunder storm. These conditions imply, first, that the air is extremely humid, or the dew-point (Art. 483) very high, so that a slight reduction of temperature will precipitate vapor; and, secondly, that the air which affords the vapor, or materials of the storm, is suddenly cooled. The cooling may be conceived to take place in several different ways,—as the meeting of hot and cold bodies of air by opposite winds, or the sudden transference of hot and humid air from the surface of the earth to the region of congelation by the agency of tornadoes or whirlwinds. All that we require is that the reduction of temperature should be great and sudden.†

650. The earth itself, in its natural state, is a vast conductor, where any excess of the electric fluid may readily discharge itself. Accordingly, where a cloud highly charged comes near to the earth, it puts the latter in the opposite electrical state by induction, and a discharge takes place between the earth and the cloud. When the electricity which is expelled from the earth by the approach of a cloud, returns to it, it sometimes produces a violent shock, known by the name of the *returning stroke.*‡ (Art. 575.) Indeed, in some instances, lightning is supposed to take a circuitous route in its way from one cloud to another, first darting to the earth and thence to the opposite cloud, the distance of the clouds from each other being too great to permit the discharge through the intervening space of air. And since electricity passes quietly without light or noise, when it makes its way through good conductors, and manifests its splendors and mechanical energies only when its path is obstructed by imperfect conductors, it is reasonably inferred that the light-

* Other causes, such as friction, change of temperature, &c., may have some influence, but the condensation of vapor, producing electricity, which is accumulated in *insulated* clouds, (thunder clouds being insulated by the circumambient air,) is to be regarded as the chief source of the electricity of thunder storms.

† The great amount of electricity which is found to be evolved from the cloud of steam suddenly condensed from the pipe of a locomotive, favors the explanation here given of the origin of the electricity of thunder storms. (See Experiments of Dr. Pattersen and others, *Franklin Journal*, Feb. 1842.)

‡ Mahon's (Earl Stanhope's) "Principles of Electricity."

ning and thunder have an origin extrinsic to the fluid itself; that the lightning is produced by the sudden and powerful *condensation* which the air experiences when compressed before the fluid, (a known cause of heat and light,) and that the thunder is produced by the *collapsing* of the air, filling the sudden void, occasioned by the passage of the fluid, (a known cause of sound.)* The zigzag appearance of lightning is well explained by supposing the air so much condensed before it, as to turn its course in another direction, where the same resistance is again experienced and another change encountered. This explanation is rendered the more probable by experiments, which show that the zigzag appearance is very much increased when the electric spark is passed through condensed air, but disappears entirely when it is passed through a vacuum. (Arts. 610, 613.)

651. If we now apply these principles to the facts before enumerated, (Art. 647,) we shall find them capable of a clear and satisfactory explanation.

All insulated clouds are electrical in a greater or less degree, because their very formation implies a condensation of watery vapor, and the state of insulation prevents the escape of the electric fluid, that is thus evolved. The electricity is stronger in such insulated detached clouds, some of which are positive, and some negative, than in a sky uniformly overcast, because in the latter case the opposite electricities are neutralized, while in the former they are kept separate. The electricity of the atmosphere is strongest when hot weather succeeds a series of rainy days, and when wet weather succeeds a series of dry days, because then, in both cases, the evaporation is most sudden and abundant; and on a single day, the signs of electricity are strongest at the rising and falling of the dew, that being the very moment when the evaporation in the morning, and the condensation in the evening, are most copious. Thunder showers are most frequent and violent in hot climates, and during the hottest seasons of the year, for in such places and at such times, the causes supposed are in most active operation.† Electricity, if evolved at all by slower processes of evaporation and condensation, finds its equilibrium before it can accumulate in sufficient quantity to produce the phenomena of a thunder storm. Thunder storms usually occur after mid-day, because it is chiefly during the hottest part of the day, or a little after it, that the meeting of those opposite currents occurs, which generate the storm; since it is at the places of greatest rarefaction that this concourse of winds takes place; and therefore during the heat of summer,

* Cavallo, Complete Treatise, p. 274.

† See a tabular view of the thunder storms of different climates, Arago, *Annuaire* 1838, p. 403.

the sun is sometimes followed round the globe by a succession of thunder storms.

In volcanoes, the most vivid lightnings and the heaviest thunders are produced, because here an immense quantity of heated vapor is thrown out, which, on reaching the cold regions of the atmosphere, is suddenly condensed into thick clouds; and the same phenomena are often terrific in water spouts, because here the sudden formation of clouds and rain, occasions a vast evolution of electricity. Violent thunder storms sometimes fall upon the sides of mountains, or upon the surface of the sea, for here, on account of the proximity of the clouds, the discharges are made toward the earth or sea, which, in ordinary cases, are made from cloud to cloud.

652. All the foregoing facts appear to admit of a clear explanation, in conformity with the supposition, that the storm itself, including all the phenomena except the electrical, is produced like other storms of wind and rain, by the sudden cooling of heated air, charged with watery vapor, (Art. 487,) and that the electrical phenomena are produced by the condensation of the vapor itself into clouds and rain. But the last fact mentioned appears to present greater difficulties. We refer to that quick succession of events, Art. 648, (7,) occurring in the following order; namely, first, a vivid flash of lightning—then a loud peal of thunder—and, after a little interval, a sudden fall of rain, which frequently stops as suddenly as it commenced. At first view, it would seem that the rain which follows the electrical discharge is produced by it; whereas, according to the foregoing views, the lightning is not the cause, but rather a consequence of the formation of the rain. (Art. 649.) But suppose that the events were to take place as required by our principles; that drops of rain were suddenly to coalesce, forming a shower, and that the attendant lightning and thunder were produced by this process; let us see in what order the notice of these events would reach the earth. The passage of light being nearly instantaneous, the flash would be seen the instant of the explosion; but sound is a comparatively slow traveller, and would take its own time to reach the ear; and rain, a slower traveller still, would arrive much later than the other two. To submit these successive events to something like mathematical calculation, we will suppose the cloud to be one fourth of a mile high, and that the precipitation of the rain, and the evolution of the electricity, which causes the explosion, are cotemporaneous events. First, the *flash* would reach us without any perceptible interval. Secondly, the *sound* travelling at the rate of 1130 feet per second, would require 1.15 seconds to reach the ear. Thirdly, the *rain*, descending like any other falling body, we may calculate its time accordingly. The times being as the square roots of the

spaces, $\sqrt{16.1} : 1 :: \sqrt{1320} : 9$ *seconds*. The time would be considerably more than this, on account of the resistance of the air. Our principles, therefore, require that the flash, the report, and the shower, should succeed each other in the order in which they actually occur.

LIGHTNING RODS.

653. Dr. Franklin had no sooner satisfied himself of the identity of electricity and lightning, than, with his usual sagacity, he conceived the idea of applying the knowledge acquired of the properties of the electric fluid, so as to provide against the dangers of thunder storms. The conducting powers of metals, and the influence of pointed bodies, to collect and transmit the fluid, naturally suggested the structure of the Lightning Rod. The experiment was tried, and has proved completely successful; and probably no single application of scientific knowledge ever secured more celebrity to its author.

654. Lightning rods are at present usually constructed of wrought iron, about three fourths of an inch in diameter. The parts may be made separate, but, when the rod is in its place, they should be joined together so as to fit closely, and to make a continuous surface, since the fluid experiences much resistance in passing through links and other interrupted joints. At the bottom, the rod should terminate in two or three branches, going off in a direction from the building. The depth to which it enters the earth should not be less than five feet; but the necessary depth will depend somewhat on the nature of the soil: wet soils require a less, and dry soils a greater depth. In dry sand it must not be less than ten feet; and in such situations, it would be better still to connect, by a convenient conducting communication, the lower end of the rod with a well or spring of water. It is useful to fill up the space around the part of the rod that enters the ground, with coarsely powdered charcoal, which at once furnishes a good conductor, and preserves the metal from corrosion. The rod should ascend above the ridge of the building to a height determined by the following principle: that *it will protect a space in every direction from it, whose radius is equal to twice its height.* It is best, when practicable, to attach it to the chimney, which needs peculiar protection, both on account of its prominence, and because the products of the combustion, smoke, watery vapor, &c. are conductors of electricity. For a similar reason a kitchen chimney, being that in which the fire is kept during the season of thunder storms, requires to be especially protected. The rod is terminated above in three forks, each of which ends in a sharp point. As these points are liable to have their conducting power impaired by rust, they are

protected from corrosion by being covered with gold leaf; or they may be made of solid silver or platina. Black paint, being made of charcoal, forms a better coating for the rod than paints made of other colors, the bases of which are worse conductors. The rod may be attached to the building by *wooden* stays. Iron stays are sometimes employed, and in most cases they would be safe, since electricity pursues the most direct route, (Art. 623;) but in case of an extraordinary charge, there is danger that it will divide itself, a part passing into the building through the bolt, especially if this terminates in a point. Buildings furnished with lightning rods have occasionally been struck with lightning; but on examination it has generally, if not always, been found that the structure of the rod was defective; or that too much space was allotted for it to protect. When the foregoing rules are observed, the most entire confidence may be reposed in this method of securing safety in thunder storms.

CHAPTER VIII.

PRECAUTIONS FOR SAFETY DURING THUNDER STORMS.—ANIMAL ELECTRICITY.—CONCLUDING REMARKS.

655. THE great number of pointed objects that rise above the general level, in a large city, have the effect to dissipate the electricity of a thunder cloud, and to prevent its charge from being concentrated on any single object. Hence, damage done by lightning is less frequent in a populous town, than in solitary buildings. For similar reasons, a great number of ships, lying at the docks, disarm the lightning of its power, and thus avert the injury to which the form of their masts would otherwise expose them. A solitary ship on the ocean, unprotected by conductors, would appear to be peculiarly in danger from lightning; but, while the greater number of ships that traverse the ocean are wholly unprotected, accidents of this kind are comparatively rare. The reason probably is, that water being a better conductor than wood, the course of the discharge toward the water is not easily diverted, and will not take the mast in its way unless the latter lies almost directly in its course. Barns are peculiarly liable to be struck with lightning, and to be set on fire; and as this occurs at a season when they are usually filled with hay and grain, the damage is more serious, for the quantity of combustible matter they contain, is such as to render the fire unmanageable. Professor Silliman ascribes this liability of barns to be struck with lightning, to the influence of the evaporation that proceeds from

the fresh hay, &c., which is supposed to furnish a conducting medium like the smoke of a chimney.*

656. Silk dresses are sometimes worn with the view of protection, by means of the insulation they afford. They cannot, however, be deemed very effectual unless they completely envelop the person: for if the head and the extremities of the limbs are exposed, they will furnish so many avenues to the fluid as to render the insulation of the other parts of the system of little avail. The same remark applies to the supposed security that is obtained by sleeping on a feather bed. Were the person situated *within* the bed, so as to be entirely enveloped by the feathers, they would afford some protection; but if the person be extended on the surface of the bed, in the usual posture, with the head and feet nearly in contact with the bedsted, he would rather lose than gain by the non-conducting properties of the bed; since being a better conductor than the bed, the charge would pass through him in preference to that.† The horizontal posture, however, is safer than the erect; and if any advantage on the whole is gained by lying in bed during a thunder storm, it probably arises from this source. The same principle suggests a reason why men or animals are so frequently struck with lightning when they take shelter under a tree during a thunder storm. The fluid first strikes the tree, in consequence of its being an elevated and pointed object, but it deserts the tree on reaching the level of the man or animal, because the latter is a better conductor than the tree.

Tall trees situated near a dwelling house, furnish a partial protection to the building, being both better conductors than the materials of the house, and having the advantage of superior elevation.

657. The protection of chimneys is of particular importance, for to these a discharge is frequently determined. When a fire is burning in the chimney, the vapor, smoke and hot air, which ascend from it, (as has been intimated in article 654,) furnish a conducting medium for the fluid; but even when no fire is burning, the soot that lines the interior of a chimney is a good conductor, and facilitates the passage of the discharge.

It is quite essential, during a thunder storm, to avoid every considerable mass of water, and even the streamlets that have resulted from a recent shower; for these are all excellent con-

---

* American Journal of Science, Vol. III, p. 345.

† Security to the person might be obtained by an entire covering of either a very bad or a very good conductor. In the former case, the electricity would not approach the system; in the latter case it would confine itself to the covering. Clothes when very wet have been supposed to furnish a protection on this principle. (See an interesting case stated by Professor Hitchcock, in the American Journal of Science.)

ductors, and the height of a human being, when connected with them, is very likely to determine the course of an electric discharge. The partial conductors, through which the lightning directs its course, when it enters a building, are usually the appendages of the walls and partitions; the most secure situation is therefore the middle of the room, and this situation may be rendered still more secure by standing on a glass legged stool, a hair mattress, or even a thick woolen rug. The part of every building least liable to receive injury is the middle story, as the lightning does not always pass from the clouds to the earth, but is occasionally discharged from the earth to the clouds. Hence it is absurd to take refuge in a cellar, or in the lowest story of a house; and many instances are on record in which the basement story has been the only part of the building that has sustained severe injury. Whatever situation is chosen, any approach to the fire-place should be particularly avoided.* An open door or window is an unsafe situation, because the lightning is apt to traverse the large timbers that compose the frame of the house, and would be determined towards the animal system on account of its being a better conductor. In a carriage the passenger is safer in the central part than next to the walls; but a carriage may be effectually protected by attaching to its upper surface metallic strips connected with the wheel tire. The fillets of silver plating which are frequently bound round the carriage, may be brought into the conducting circuit.

ANIMAL ELECTRICITY.

658. Of the natural agencies of electricity, one of the most remarkable, is that exhibited by certain species of fish, especially the *Torpedo* and the *Gymnotus*. This peculiar property of the Torpedo was known to the ancient naturalists, and is accurately described by Aristotle, and by Pliny. Aristotle says that this fish causes or produces a torpor upon those fishes it is about to seize, and having by that means got them into its mouth it feeds upon them. Pliny says that this fish if touched by a rod or spear, even at a distance, paralyzes the strongest muscles.

The fact, however, that this extraordinary power depends upon electricity, was not known, until about the year 1773, when it was ascertained by Mr. Walsh, that the Torpedo is capable of giving shocks to the animal system, analogous to those of the Leyden jar. Though this property is regarded as establishing the identity of the power with the electric fluid, yet this power, as developed in the Torpedo, has never been made to afford a spark, nor to produce the least effect upon the most deli-

* Singer.

cate electrometer.* As late as the year 1828, experiments were made upon the Torpedo, by Sir Humphry Davy, and the conclusions at which he arrived, were that the electricity resides in this animal in a form suited exclusively to the purpose of communicating shocks to the animal system, while it has little or nothing else in common with the properties of electricity, as developed in various artificial arrangements.†

659. The Torpedo is a flat fish, seldom twenty inches in length, but one found on the British coast was four and a half feet long The electricity of the Torpedo has the same relation as commo: electricity to bodies in respect to their conducting power, being readily transmitted through metals, water, and other conductors, and not being transmitted through glass, and other non-conductors.

The electric organs of the Torpedo are two in number, and placed one on each side of the cranium and gills. The length of each organ is somewhat less than one third part of the length of the whole animal. Each organ consists of perpendicular columns reaching from the under to the upper surface of the body, and varying in length according to the various thickness of the flesh in different parts. The number of these columns is not constant, varying not only in different Torpedoes, but likewise in different ages of the animal, new ones seeming to be produced as the animal grows. In a very large Torpedo, one electric organ has been found to consist of one thousand one hundred and eighty-two columns. The diameter of a column is about one fifth of an inch. Each column is divided by horizontal partitions, consisting of transparent membranes, placed over each other at very small distances, and forming numerous interstices, which appear to contain a fluid. The number of partitions contained in a column one inch in length, has been found in some instances not less than one hundred and fifty. By this arrangement the amount of electrified *surface* is exceedingly great; equivalent in one instance to one thousand and sixty-four feet of coated glass. Hence, the effects of the electricity of the Torpedo are such as correspond to those, which, in artificial arrangements, are produced by diffusing a given quantity of fluid over a great surface, by which its intensity is much diminished.‡

---

* Humboldt.

† Phil. Trans. 1829. A reflection naturally suggested by this fact is, that the fluid which is excited in the various species of electrical apparatus, both the common and Voltaic, is a compound, embracing several distinct substances.

‡ In the Philosophical Transactions for 1832, Dr. Davy has given an interesting series of observations on the Torpedo. He ascertained that the animal has the power of giving magnetic polarity to iron, and of affecting chemical decompositions in a slight degree. The shocks were very powerful. By giving repeated shocks, the older fish were rendered languid and died soon.

660. The *Gymnotus*, or Surinam eel, is found in the rivers of South America. Its ordinary length is from three to four feet; but it is said to be sometimes twenty feet long, and to give a shock that is instantly fatal. The electrical organs of the Gymnotus, constitute more than one third part of the whole animal; they consist of two pairs, of different sizes and placed on different sides. The shock communicated to fishes instantly paralyzes them, so that they become the prey of the Gymnotus. By irritating the animal with one hand, while the other is held at some distance in the water, a shock is received as severe as that of the Leyden jar.

Unlike the Torpedo, the Gymnotus gives a small but perceptible *spark*, affording additional proof of the identity of the power with that of electricity.

M. Humboldt, in his travels in South America, describes a singular method of catching the Gymnotus, by driving wild horses into a lake which abounds with them. The fish are wearied or exhausted by their efforts against the horses, and then taken; but such is the violence of the charge which they give, that some of the horses are drowned before they can recover from the paralyzing shocks of the eels.

The *Silurus electricus*, is a fish found in some of the rivers of Africa. Its electrical powers are inferior to those of the Torpedo and Gymnotus, but they are still sufficient to give a distinct shock to the human system.

661. Certain furred animals, particularly the cat, become spontaneously electrified. This is more especially observable on cold windy nights, when the state of the air is favorable to insulation. At such times a cat's back will frequently afford electrical sparks. Ancient historians mention a number of very remarkable occurrences, of good or evil omen, which are due to the electricity of the atmosphere. Herodotus informs us that the Thracians disarmed the sky of its thunder by throwing their arms into the air; and that the Hyperboreans produced the same effect by launching among the clouds darts armed with points of iron. Cæsar, in his Commentaries, says that in the African war, after a tremendous storm which threw the whole of the Roman army into great disorder, the points of the darts of a great number of the soldiers shone with a spontaneous light. In the month of February (says he) about the second watch of the night, there suddenly arose a great cloud, followed by a dreadful storm of hail, and in the same night the points of the darts of the fifth legion appeared on fire.*

During a dry snow storm, when electricity is evolved in great quantities, and, on account of the dry state of the air, is partly

* Ed. Encyc. VIII, 311. Arago, *Annuaire*, 1838, p. 375.

insulated on conducting bodies, similar appearances are exhibited. Thus, the ears of horses, and various pointed bodies, emit faint streams of light. These phenomena are sometimes exhibited in a most striking manner in a storm at sea, when the masts of a ship, yard-arms, and everv other pointed object, are tipped with lightning.*

CONCLUDING REMARKS.

662. From the energy which electricity displays in our experiments, and much more in thunder storms, there can be no question that it holds an important rank among the ultimate causes of natural phenomena. Its actual agencies, however, are liable to be misinterpreted, and that they have been so in fact, is too manifest from the history of the science. After the splendid experiments with the Leyden jar, and more especially after the identity of electricity with lightning had been proved, electricians fancied that they had discovered the clue which would conduct them safely through the labyrinth of nature. Every thing not before satisfactorily accounted for, was now ascribed to electricity. They saw in it, not only the cause of thunder storms, but of storms in general; of rain, snow, and hail; of whirlwinds and water spouts; of meteors and the aurora borealis; and finally, of tides and comets, and the motions of the heavenly bodies.† Later electricians have found in the same agent the main spring of animal and vegetable life, and the grand catholicon which cures all diseases. Recent attempts have been made to establish the very identity of galvanic electricity and the nervous influence, by which the most important functions of animal life are controlled.‡

Among the most important of the agencies of electricity in the economy of nature, is that which, according to the views of Sir Humphry Davy, it sustains in relation to the chemical agencies of bodies. Chemical and electrical attractions, he supposes, are one and the same thing, or at least dependent on the same cause, the attractions between the elements of a compound arising solely from their being naturally in opposite electrical states. But the discussion of this hypothesis belongs more appropriately to galvanism, a branch of our subject which, on account of its peculiarities, especially in the mode of excitation, has been constituted a separate department of science.

It is a remark of Lord Bacon, that things appear usually to better advantage and more important in their *relations*, than in their individualities. "The contrary, (says he,) has made many

---

* See Jones's Sketches of Naval Life, I, 199.

† Encyc. Brit., *Electricity*.

‡ *Wilson Philip*, Phil. Trans. Tilloch's Phil. Mag. XXX, 488.

particular sciences to become barren, shallow, and erroneous, while they have not been nourished and maintained from a common fountain." The truth of this remark is strikingly exhibited in respect to electricity. If in its original form it is an interesting and wonderful agent, still more astonishing and important are the properties it has disclosed in its relations to certain chemical agents, to magnetism, to heat, and to light, giving rise to the different sciences of Galvanism, Electro-Magnetism, Thermo-Electricity, and to the art of gilding and copying by the Electrotype process. Much, we believe, remains to be discovered respecting the useful purposes which this mysterious agent is destined to perform for man; but enough has been revealed to assure us that electricity is the agent by which are to be achieved the most refined and delicate performances of art, and by which man is to acquire his most perfect mastery over nature.*

* In the distribution of subjects in Yale College, Galvanism and its kindred subjects are assigned to the chemical department.

The most extensive and complete treatise hitherto published on the subject of Electricity in all its relations, is the work of *Becquerel*, in French, consisting of seven volumes 8vo. *Pouillet*, in the first volume of his *Elémens de Physique*, gives a full and able view of these subjects.

PART VII.—MAGNETISM.

663. MAGNETISM *is the science which treats of the properties and effects of the magnet.* The same term is also used to denote the unknown cause of magnetic phenomena; as when we speak of magnetism as excited, imparted, and so on.

Magnets are bodies, either natural or artificial, which have the property of attracting iron, and the power, when freely suspended, of taking a direction toward the poles of the earth. The natural magnet is sometimes called the *loadstone*.* It is an oxide of iron of a peculiar character, found occasionally in beds of iron ore. Though commonly met with in irregular masses only a few inches in diameter, yet it is sometimes found of a much larger size. One recently brought from Moscow to London, weighed one hundred and twenty-five pounds, and supported more than two hundred pounds of iron.†

664. The *attractive* powers of the loadstone have been known from a high antiquity, and are mentioned by Homer, Pythagoras, and Aristotle. But the *directive* powers were not known in Europe until the twelfth century, though some writers have endeavored to trace the history of the compass needle to a remoter period, and some have strenuously maintained, that the Chinese were in possession of it many centuries before it was known to the Europeans.‡

Magnetism is the most recent of all the physical sciences, and notwithstanding the numerous discoveries achieved in it within a few years, and the remarkable precision with which its laws have been ascertained, yet it is still to be regarded as a science quite in its infancy, although it is rapidly progressive.

665. If a magnet be rolled in iron filings, it will attract them to itself. This effect takes place especially at two opposite points, where a much greater quantity of the filings will be collected than in any other parts of the body. The two opposite points in a magnet, where its attractive powers appear chiefly

* Said to be derived from *lædan*, a Saxon word which signifies *to guide*
† Partington's Manual, II, 243.
‡ Cavallo *on Magnetism*; Barlow, *Encyc. Metrop.*; Klaproth, *Amer Jour.* xl., 242

to reside, are called its *poles*. The straight line which joins the poles, is called the *axis*. (See Fig. 224.)

Fig. 224.

Fig. 225.

If a large sewing needle, or a small bar of steel be rubbed on the loadstone, one extremity on one pole, and the other extremity on the other, the needle or bar will itself become a magnet, capable of exhibiting all the properties of a loadstone. Without staying at present to describe more minutely the process of making artificial magnets, we will suppose ourselves provided with several magnetic needles and bars, and we may proceed with them to study the leading facts of the science of magnetism. By attaching a fine thread to the middle of a needle, and suspending it so as to move freely in a horizontal plane, or by resting it on a point, as is represented in Fig. 225, we shall have a simple and convenient apparatus for numerous experiments. The needle thus suspended will place itself in a direction nearly, though not exactly, north and south. If the needle is drawn out of the position it assumes when at rest, it will vibrate on either side of that position until it finally settles in the same line as before, one pole always returning toward the north, and the other toward the south. Hence the two poles are denominated respectively *north and south poles.* In magnets prepared for experiments, these poles are marked either by the letters N and S, or by a line drawn across the magnet near one end, which denotes that the adjacent pole is the north pole.

666. By means of the foregoing apparatus, we may ascertain that the magnet has the following general properties, viz:

First, powers of attraction and repulsion.

Secondly, the power of communicating magnetism to iron or steel by induction.

Thirdly, polarity, or the power of taking a direction toward the poles of the earth.

Fourthly, the power of inclining itself toward a point below the horizon, usually denominated the *dip of the needle.*

The further development of these properties will constitute the subjects of the following chapters.

CHAPTER I.

OF MAGNETIC ATTRACTION.

667. When *either pole of a magnet is brought near to a piece of iron, a mutual attraction takes place between them.*

Thus, when the ends of a magnetic bar or needle are dipped into a mass of iron filings, these adhere in a cluster to either pole. A bar of soft iron, or a piece of iron wire, resting on a cork, and floating on the surface of water or quicksilver, may be led in any direction by bringing near to it one of the poles of a magnet. This action is moreover *reciprocal;* that is, the iron attracts the magnet with the same force that the magnet attracts the iron. If the two bodies be placed on separate corks and floated, they will approach each other with equal momenta; or if the iron be held fast, the magnet will move toward it.

668. Two other metals besides iron, namely, nickel and cobalt, are susceptible of magnetic attraction. These metals, however, exist in nature only in comparatively small quantities, and therefore by magnetic bodies, are usually intended such as are ferruginous. Even iron, in some of its combinations with other bodies, loses its magnetic properties; indeed, only a few of the numerous ores of iron are attracted by the magnet. But soft metallic iron and some of the ores of the same metal, affect the needle even when existing in exceedingly small quantities, so that the magnet becomes a very delicate test of the presence of iron. Compass needles are sometimes said to be disturbed by the minute particles of steel left in the dial plate by the graver;* and the proportion of iron in some minerals may be exactly estimated by the power they exert upon the needle.†

669. *In the action of magnets on each other, poles of the same name repel, while those of different names attract each other.*

Thus, the north pole of one magnet will repel the north pole of the other, and attract its south pole. The south pole of one will repel the south pole of the other, and attract its north pole. These effects, it will be perceived, are analogous to those produced by the two species of electricity; and they equally imply two species of magnetism or two magnetic fluids, (as it is convenient to call them,) namely, the northern and the southern, or as they are now denominated, the *boreal* and the *austral* fluids.

---

* Eaton, *Am. Jour. Science.* xiv, 15.

† Biot.

A very simple piece of apparatus will serve to exhibit the foregoing property. The accompanying figure represents two sewing-needles magnetized and suspended by fine threads. On approaching the north pole of a magnetic bar to the north poles of the needles, they are forcibly repelled; but on applying the south pole of a bar as in figure 226, the north poles of the needles are attracted toward it.

Fig. 226.

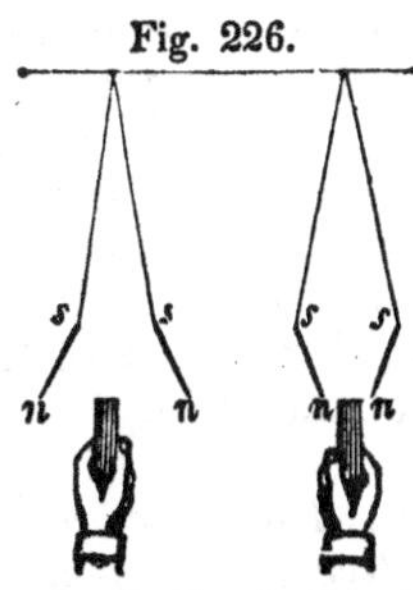

670. *By bringing a magnet near to iron or steel, the latter is rendered magnetic by induction.*

Thus, let the north pole of a magnetic bar A, (Fig. 227,) be brought near to one end of an unmagnetized bar of soft iron B: the iron will immediately become itself a magnet, capable of attracting iron filings, having polarity when suspended, and possessing the power of communicating the same properties to other pieces of iron. It is, however, only while the iron remains in the vicinity of the magnet, that it is endued with these properties; for let the magnet be withdrawn, and it loses at once all the foregoing powers. This, it will be remarked, is asserted of *soft iron;* for steel and hardened iron are differently affected by induced magnetism.

Fig. 227.

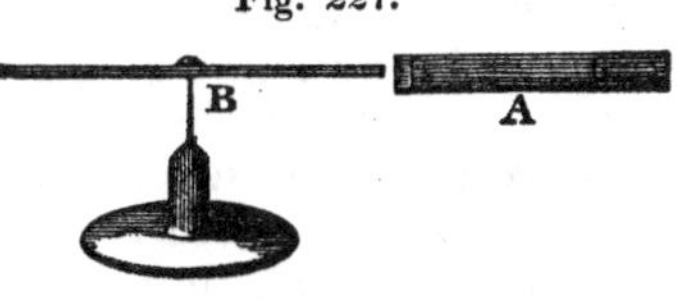

On examining the kind of magnetism induced upon the two ends of the iron bar B, (Fig. 227,) which we may easily do by bringing it near to the poles of the needle, (Fig. 225,) we shall find that the nearer end has south, and the remoter end north polarity. This effect also is analogous to that produced by electrical induction. (See Arts. 595, 596.) A corresponding effect would have taken place, had the south, instead of the north pole of the magnet been presented to the bar of iron; in which case, the nearer end would have exhibited the northern, and the remoter end southern polarity. Or, to express this important proposition in general terms,

Each pole of a magnet induces the opposite kind of polarity in that end of the iron which is nearest to it, and the same kind in that end which is most remote.

671. It is not essential to the success of these experiments, that the bars of iron which receive magnetism by induction, should be placed in a straight line with the magnet; they may be at right angles to it, or inclined at any other angle, the only essential condition being, that the end of the bar should be brought near to the pole of the magnet. Indeed, the effect is increased,

that is, the magnetism of the iron bar is rendered stronger, when the bar is inclined toward the magnet, as in Fig. 228, and is the strongest of all when it is placed parallel to the magnet; for it will be seen that in these two latter positions, both poles of the magnet conspire in their action upon the iron bar.

Fig 228.

672. *The power of a magnet is increased by the exertion of its inductive power upon a piece of iron, in its neighborhood.*

The end of the piece of iron contiguous to the pole of the magnet, is no sooner endued with the opposite polarity, than it reacts upon the magnet and increases its intensity; and a series of actions and reactions take place between the two bodies, similar to what occurs in electrical induction. (See Art. 596, &c.) On this account the powers of a magnet are increased by action, and impaired or even lost by long disuse. By adding, from time to time, small pieces of iron to the weight taken up by a magnet, its powers may be augmented greatly beyond their original amount. Hence, the force of attraction of the dissimilar poles of two magnets, is greater than the force of repulsion of the similar poles: because, when the poles are unlike, each contributes to enhance the power of the other, but when they are alike, the influence which they reciprocally exert, tends to make them unlike, and of course to impair their repulsive energies.

Hence, also, a strong magnet has the power of reversing the poles of a weak one. Suppose the north pole of the weaker body to be brought into contact with the north pole of the stronger; the latter will expel north polarity, or the boreal fluid, and attract the austral, a change which in certain cases will be permanent.

673. If the north pole of a magnetic bar be placed upon the middle of an iron bar, the two ends of the latter will each have north polarity, while the part of the bar immediately in contact with the magnet receives south polarity; and if the same north pole be placed on the center of a circular piece of iron, all parts of the circumference will be endued with north polarity, while the plate will have a south pole in the center. By cutting the plate into the form of a star, each extremity of the radii becomes a weak north pole when the north pole of a magnet is placed in the center of the star. If an iron bar is placed between the dissimilar poles of two magnetic bars, (all being in one straight line,) both of the magnets will conspire to increase the intensity of each pole of the bar, and the magnetism imparted to the bar will be considerably stronger than from either magnet alone; but if the same bar be placed between the two similar poles, the opposite polarity will be imparted to each end, while the same polarity

is given to the center of the bar. Thus if the bar be placed between the north poles of two magnets, each end of the bar will become a south pole and the center a north pole. When one end of a magnetic bar is applied to the ends of two or more wires or sewing needles, the latter arrange themselves in radii diverging from the magnetic pole. This effect is in consequence of their remoter ends becoming endued with similar polarity, and repelling each other. A like effect is observable among the filaments of iron filings, that form a tuft on the ends of a magnetic bar.

674. The foregoing experiments are sufficient to show that when a piece of iron is attracted by the magnet, it is first itself converted into a magnet by the inductive influence of the magnetizing body. Each of the iron filings which compose the tuft at the pole of a magnetic bar or needle, is itself a magnet, and in consequence of being such, induces the same property in the next particle of iron, and that in the next, and so on to the last. Hence magnetic attraction does not exist, strictly speaking, between a magnet and iron, but only between the opposite poles of magnets; for the iron must first become a magnet before it is capable of magnetic influence.

675. *Soft iron readily acquires magnetism and as readily loses it; hardened steel acquires it more slowly, but retains it permanently.*

In the preceding example, the magnetism acquired by a bar of iron, by the process of induction, is retained only so long as the magnetizing body acts upon it. Soon after the two bodies are separated, the bar loses all magnetic properties.

When a bar of steel is placed very near a strong magnet, the action of the magnet commences immediately upon the end of the bar nearest to it, the north pole for example communicating south polarity to the contiguous extremity of the bar. According to our previous experience, we should expect to find the remote end of the bar a north pole; but such is not the *immediate* result; a sensible time is required before the north polarity is fully imparted to the remote extremity. Indeed, if the bar be a long one, it sometimes happens that the north polarity never reaches the farthest end, but stops short of it at some intermediate point. This north pole is succeeded by a second south pole, that by another north pole, and thus several alternations between the two poles occur before reaching the end of the bar.

676. The process of magnetizing a steel bar or needle, is accelerated by any cause which excites a tremulous or vibratory motion among the particles of the steel. Striking on the bar with a hammer promotes the process in a remarkable degree, especially

if it occasions a ringing sound, which indicates that the particles are thrown into a vibratory motion. The passage of an electric discharge through a steel bar under the influence of a magnet, produces permanent magnetism. Heat also greatly facilitates the introduction of the magnetic fluid into steel. The greatest possible degree of magnetism that can be imparted to a steel bar, is communicated by first heating the steel to redness, and while it is under the influence of a strong magnet, quenching it suddenly with cold water.

A magnet, however, loses its virtues by the same means as, during the process of induction, were used to promote their acquisition. Accordingly, any mechanical concussion or rough usage, impairs or destroys the powers of a magnet. By falling on a hard floor, or by being struck with a hammer, it is greatly injured. Heat produces a similar effect. A boiling heat weakens and a red heat totally destroys the power of a needle. On the other hand, cold augments the powers of the magnet; indeed, they improve with every reduction of temperature hitherto applied to them.*

As iron and steel are found of various degrees of hardness, so the power of acquiring and of losing magnetism, is very various in different ferruginous bodies. It is in general true, that this power is in proportion to the hardness. Thus, the attraction of soft malleable iron for the magnet being 100, that of hard cast steel is only 49, and that of cast iron only 48.†

677. *If a steel bar rendered magnetic by induction, be divided into any two parts, each part will be a complete magnet, having two opposite poles.*

We here meet with a remarkable distinction between magnetic and electric induction. When a body electrified by induction, is divided into two equal parts, the individual electricities alone remain in each part respectively; but in the case of magnetic induction, although no appearance of polarity be exhibited except at the two ends, yet wherever a fracture is made, the two ends separated by the fracture immediately exhibit opposite polarities, each being of an opposite name to that of the original pole at the other end of the fragment. If each of the two fragments be again divided into any number of parts, each of these parts is a magnet perfect in itself, having two opposite poles.

In magnetism, therefore, there is never, as in electricity, any *transfer* of properties, but only the excitation of such as were already inherent in the body acted upon. Magnetism never passes out of one body into another; nor can we ever obtain a piece of iron or steel, that contains exclusively either northern or southern polarity.

* Christie, Phil. Trans., 1825.

† Barlow.

678. *The force of attraction, or of repulsion, exerted upon each other by the poles of two magnets, placed at different distances, varies inversely as the square of the distance.*

This law was ascertained by Coulomb, by means of the torsion balance, in a manner similar to that adopted in investigating the law of electrical attraction. (See Art. 583.) The same law therefore which governs the attraction of gravitation, likewise controls electrical and magnetic attractions. It is the most extensive law of the physical world. Nor is this action at a distance prevented, or even impaired, by the interposition of other bodies not themselves magnetic.

679. *The magnetic power of iron resides wholly on its* SURFACE, *and is independent of the mass.*

Thus a hollow globe of iron of a given surface will have the same effect on the needle as though it were solid throughout.* In this fact we again meet with a striking analogy between magnetism and electricity, a similar property having before been shown to belong to the electric fluid. This is one of the most recent discoveries in magnetism, and was made by Professor Barlow of the Military Academy at Woolwich, (Eng.) to whose ingenious and assiduous labors are due many of the latest and most important investigations in this science.

CHAPTER II.

OF THE DIRECTIVE PROPERTIES OF THE MAGNET.

680. *If a small needle be placed near one of the poles of a magnet, with its center in the axis of the magnet, it will take a direction in a line with that axis.*

Thus, let S N be a large magnetic bar, and *s n* a small needle placed near the north pole of the magnet with its center in the axis; it will be seen that the action of the pole of the magnet is such as to bring the needle into a line with the magnet. The action of the bar upon the needle tending to give it this direction, is, since it repels *n* and attracts *s*, equal to the sum of its actions upon both poles.

Fig. 229.

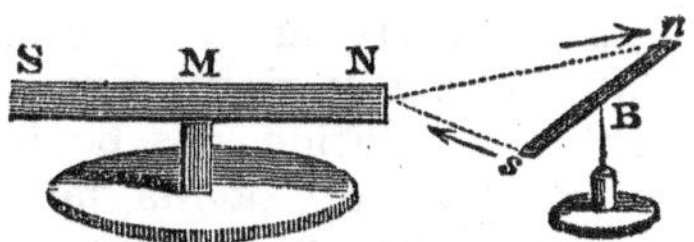

681. *If the needle be placed at right angles to the bar, with one of its poles directed toward the center of the bar, it will take a direction parallel to the bar.*

* It appears, however, that a certain thickness is necessary to the maximum effect, although that thickness is very small. (See Phil. Trans., 1831, p. 81.)

By supposing B (Fig. 229,) to be placed as indicated in the above proposition, it will be seen, that the actions of both poles of the magnet would conspire in relation to each pole of the needle, and that these forces can be in equilibrium only when the needle is parallel with the bar. The needle in this situation has a tendency to move toward the magnet, because the attraction being exerted on the nearer, and the repulsion on the remoter pole, the sum of the attractions exceeds that of the repulsions.

682. *Iron filings or other ferruginous bodies, which are free to obey the action of a magnetic bar, naturally arrange themselves in curve lines from one pole of the magnet to the other.*

Thus, if we place a sheet of white paper on a magnetic bar laid on the table, and sprinkle iron filings on the paper, the filings will arrange themselves in curves around the poles of the magnet. A small sewing needle suspended horizontally by a slender string, on being brought near to different parts of the magnet, will take directions corresponding to the part of the curve in which it happens to be placed. At the poles it will be in a line with the axis of the magnet; opposite the center of the bar it will be parallel to it; and between these two points it will take intermediate directions, as is represented in Fig. 231.

Fig. 230.

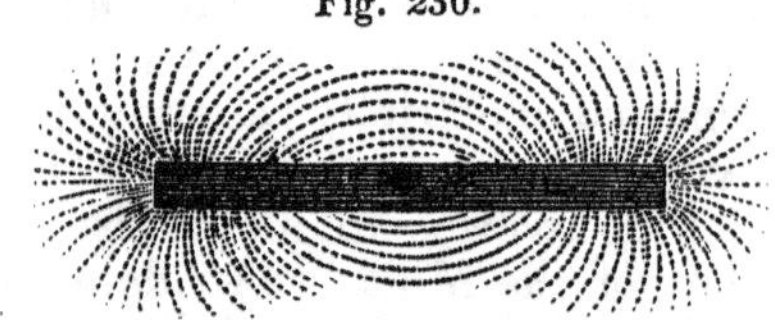

Fig. 231.

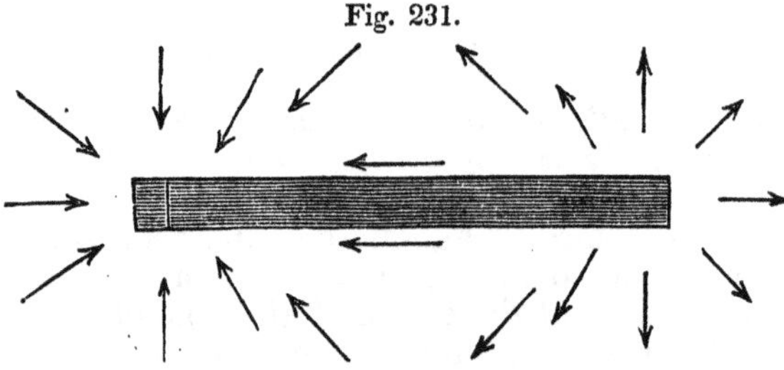

These curves have given rise to the most fanciful theories of magnetism, having been assumed as the traces of an invisible fluid perpetually circulating between the poles of the magnet; and this circulation has been afterwards employed for illustrating every variety of magnetic phenomena, but in such a way as to leave the subject involved in greater mystery than at first. These causes are nothing more than the necessary results of forces like those described in the foregoing propositions.*

The curves which iron filings describe when thus arranged

* Barlow.

are called *magnetic curves*. They present several curious properties, which have been investigated by mathematicians; but we must refer the student to more extensive treatises than the present for a full development of this subject.*

683. *By different methods, nearly all bodies may be made to affect the magnetic needle.*

Coulomb, by the assistance of the extremely delicate apparatus he employed, detected, as he thought, some slight portion at least of magnetism in various metals besides those to which it had been exclusively ascribed, and hence announced that *all bodies whatsoever are subject to the magnetic influence.* The possibility, however, that the different bodies tried by Coulomb might contain small portions of iron, threw some doubt on his conclusions; but within a few years the doctrine has been confirmed, that magnetism, in some way or other, pervades all bodies.

M. Arago, of France, found that the oscillations which a needle would make when set in motion, were affected by its being placed above or below plates of different substances, as metal, ice, or water. The *number* of oscillations in a given time, remained the same whatever substance was employed for the plate; but the horizontal *range* or amplitude of the needle was much greater in some cases than in others. Copper, for example, would diminish the range four times as much as lead, and a hundred times as much as antimony. This fact led to the discovery, that if plates of copper and various other substances are put into rapid *rotation* below a horizontal needle freely suspended, the needle will be made to revolve with great velocity. The experiment has been varied in many different ways. Thus, a powerful horseshoe magnet itself, placed vertically, has been made to revolve, having a plate of metal or other substance suspended immediately over it. When the motion of the magnet acquires a certain velocity, the suspended plate will likewise revolve in the same direction, and sometimes with so great a speed that the eye cannot distinguish it.

Mr. Faraday, of England, has recently asserted the doctrine, supported by numerous experiments, that all the metals are magnetic at a certain temperature, and that at a certain other temperature, they all lose this power. Even iron loses all magnetic properties at an orange heat, and nickel at a heat still lower.

684. *The magnetic needle, when freely suspended, seldom points directly to the pole of the earth, but its deviation from that pole is called the* DECLINATION *or the* VARIATION *of the needle.*

A vertical circle drawn through the line in which the needle naturally places itself, is called the *magnetic meridian.* A plane,

* Journal of the Royal Institution, Feb. 1831.—Leslie's Geometrical Analysis.

passing at right angles to the magnetic meridian, through the center of the needle, is called its *magnetic equator*. A line drawn on the surface of the earth, passing through the places where the needle points directly to the north pole, and where of course the geographical and magnetic meridians coincide, is called the *line of no variation.*

The discovery of the variation of the needle has been commonly ascribed to Columbus. His son Ferdinand states, that on the 14th of September, 1492, his father first discovered the variation, and that in consequence, his crew mutinied, supposing that the needle had lost its polarity, and that they would not be able to find their way back to Europe. It appears, however that the same phenomenon had been discovered about two hundred years before that period, though it had not become generally known to navigators.*

685. *The declination of the needle is not constant, but is subject to a small annual change, which carries it to a certain limit on one side of the pole of the earth, when it becomes stationary for a time, and then returns to the pole, and proceeds to a certain limit on the other side of it, occupying a period of many years, during each vibration.*

At London, in the year 1580, the needle pointed $11\frac{1}{4}$ degrees to the east of north ; in 1657, it pointed directly to the pole ; after which period, it continued to move west for one hundred and fifty-seven years, until the year 1814, when its western declination was nearly $24\frac{1}{2}$ degrees ; since 1814, it has been moving slowly eastward. If it takes as many years to return as it did to move from the pole to its western limit, it will reach the pole again in the year 1971 ; and should it proceed as far eastward as it did westward, and occupy as long a time, it will reach its eastern limit in 2128. The total arc of declination will be 48° 35′ 48″, and the period occupied in passing over it, three hundred and fourteen years. This would be an average of 9′ 17″ annually. But the annual variation is much smaller than this toward its eastern and western limits, but much greater when the needle is in the vicinity of the line of no variation. Thus, during the nine years that elapsed between 1814 and 1823, the progress eastward was only 11′ 22″, or only 1′ 1.6″ annually, while from 1657 to 1672, a period of fifteen years, the declination west amounted to 2° 30′, or 10′ annually ; and between 1692 and 1722, the annual increase of declination was 16′ 40″. It performed half the amount of its western declination in fifty-seven years, while to complete the other half, occupied one hundred years.†

The variation of the needle, however, is not the same at the

* Cavallo, Treatise on Magnetism, Supplement.—Barlow, Phil. Trans. 1833.
† Thomson, Outlines of Heat, Elec., and Mag., p. 545.

same time in all parts of the earth, but every part has its particular declination. For instance, if we sail from the Straits of Gibraltar to the West Indies, in proportion as we recede from Europe and approach America, the compass will point nearer and nearer due north; and when we draw near the American coast, it will point exactly north. But if we sail from Great Britain to the southern coast of Greenland, we shall find the needle deviate further and further from the north, as we approach Greenland, where the deviation will not be less than 50°.* In some parts of Baffin's Bay the needle points due west.

686. The *line of no variation* encompasses the globe, but its course is subject to numerous irregularities. Commencing at the magnetic pole in Lat. 70°, Lon. 90° 30′, it runs a few degrees east of south, through Hudson's Bay, New South Wales, Lake Erie, the western part of the state of Pennsylvania, passing a little westward of Washington City, and enters the Atlantic Ocean near Newbern in North Carolina. After leaving the United States, it veers a little more to the east, running a few degrees eastward of the West India Islands, and meeting the eastern coast of South America, near Maranham, on the northeast coast of Brazil. Crossing the great eastern promontory of South America, it pursues a regular course in a southeasterly direction toward the south polar regions. In the eastern hemisphere, the line of no variation presents greater anomalies. Proceeding from south to north, it passes in a northerly direction through New Holland, somewhat westward of the eenter. Thence its course changes to the west, and it sweeps in a great curve through the Indian Ocean, around Hindostan, returning through China to nearly the same longitude as it had in New Holland, where it turns to the north again. Places lying westward of either of these lines, have, within certain limits, easterly variations, and those lying eastward of those lines, have westerly variations.† In Europe, Africa, and the western parts of Asia, together with the greater part of the Atlantic Ocean, the variation is to the west. Throughout the greater part of the western hemisphere, the variation is to the east.

687. At *New Haven*, the variation of the needle is at present about 6 degrees, and is on the increase. In 1820, it was 4° 35′ 10″;‡ and in 1835, it was 5° 40′ 34″.§ On this subject, Professor Loomis remarks: From the time of the earliest observations, (1673,) down to about the commencement of the present century, the westerly variation was decreasing, and the easterly in-

* Thomson, Outlines of Heat, Elec., and Mag., p. 543.
† Barlow, Phil. Trans., 1833.
‡ According to observations made by Professor Fisher. See Am. Jour., XVI, 60
§ Professor Loomis, Ib. XXX, 224.

creasing, in every part of the United States. Since that period. the movement of the needle has been in the opposite direction. At present, therefore, *the westerly variation is increasing, and the easterly diminishing.* This change commenced between the years 1793 and 1819, though probably not everywhere simultaneously. The present *annual change of variation,* is about 2 minutes in the southern and western states, from 3 to 4 minutes in the middle states, and from 5 to 7 minutes in New England.

The variation for several prominent points is as follows:

Cambridge, Mass. - - -	in 1835,	8° 51′ W.
New Haven, Conn. - - -	1836,	5° 55′ "
New York City, - - -	1837,	5° 40′ "
Albany, - - - - -	1836,	6° 47′ "
Philadelphia, - - - -	1837,	3° 52′ "
Washington City, - - -	1809,	0° 52′ "
Charlottesville, Va. - - -	1835,	0° 0′ "
Newbern, N. C. - - -	1806,	2° 0′ E.
Athens, Geo. - - - -	1837,	4° 31′ "
St. Louis, Mo. - - - -	1819,	10° 47′ "

688. *Besides the annual variation, the magnetic needle is subject to daily changes called the* DIURNAL VARIATION.

According to the observations of Professor Loomis, made at Yale College in 1835 and 1836, the north end of the needle has in the morning a motion eastward, amounting to from one to three minutes, when the declination is usually less than at any other hour of the day, and may therefore be called the minimum. During winter, this minimum is attained about 8 o'clock, but as early as 7 o'clock during summer. After reaching its minimum position, it gradually moves to the west, and attains its maximum declination about 3 o'clock in winter, and 1 o'clock in summer. From this time the needle again returns eastward. The whole amount of the diurnal variation rarely exceeds 12 minutes, and is commonly much less than that. These changes of declination during the day are connected with changes of *temperature,* being from May to October, inclusive, 11′ 56″, and from November to April, 5′ 11″.

689. *A needle first balanced horizontally on its center of gravity and then magnetized, no longer retains its level, but its north pole spontaneously takes a direction to a point below the horizon, called the* DIP OF THE NEEDLE.

The *Dipping Needle* is represented in figure 232. When used, it is to be placed in the magnetic meridian, and the stand which supports it rendered perfectly level, by means of the adjusting screws attached.

The dip of the needle is very different in different parts of the globe being in general least in the equatorial and greatest in the polar regions. At certain places on the globe the needle has no dip, that is, it becomes perfectly horizontal, and a line uniting all such places is called the *magnetic equator of the earth.* Again in the polar regions, the dipping needle sometimes becomes nearly perpendicular to the horizon. In the middle latitudes, the dip is greater or less, but does not correspond exactly to the latitude.

Fig. 232.

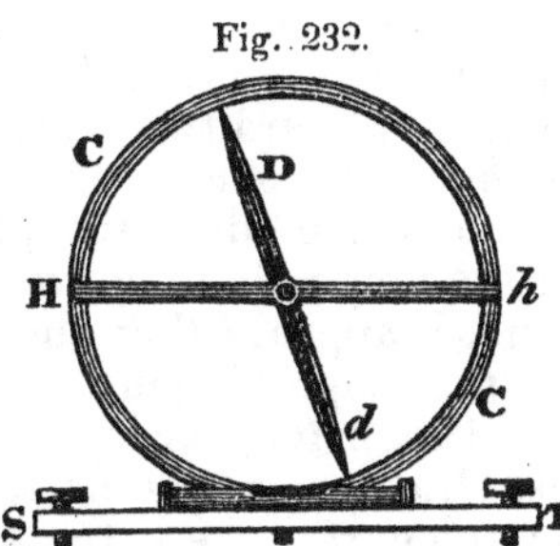

If the magnetic meridian coincided with the geographical, the magnetic equator would coincide with the earth's equator ; but such is not the fact. We may consider the magnetic equator, in general, as a great circle encompassing the earth, and inclined to its equator at an angle of about 12 degrees. It not only crosses the equator at two points diametrically opposite to each other, as a regular great circle would do, but crosses it also in one or perhaps two intermediate points.

The dip of the needle, like the declination, is not constant at the same place, but undergoes a slight variation from year to year. In the course of two hundred and forty-five years, it has varied at London more than 5°. Its present amount is about 70°, and the variation is from two to three minutes annually. The following table exhibits the dip of the magnetic needle for the year 1839, at the places annexed.*

Hudson, Ohio, ...	Lat. 41° 15′ N.	72° 48′.4
Buffalo, ...	42° 53′	74° 40′.8
Schenectady, ...	42° 48′	74° 36′.1
Albany, ...	42° 39′	74° 51′.3
New York City, ...	40° 43′	72° 52′.2
New Haven, Conn., ...	41° 18′	73° 26′.7
Cambridge, Mass., ...	42° 22′	74° 20′.1
Princeton, N. J., ...	40° 22′	72° 47′.1
Washington City, ...	38° 53′	71° 21′.4

690. *The force exerted by the magnetism of the earth varies in different places: its comparative estimate for any given place, is called the* MAGNETIC INTENSITY *for that place.*

As in the case of the pendulum in its relation to the force of gravity, the magnetic intensity may be measured by the *number of oscillations* (Art. 183) which a needle drawn a given number of degrees from its point of rest, performs in a certain time, as a minute for example, the force being as the square of the number

* Professor Loomis, Amer. Phil. Trs., 1839.

of oscillations. In general it is well ascertained that the magnetic intensity is least in the equatorial regions, and increases as we advance toward the poles. It is probably at its maximum at the magnetic poles. By ascertaining, from actual observation, a unmber of different places on the surface of the earth, where the magnetic intensities are equal, and connecting them by a line, it appears that they arrange themselves in a curve around the magnetic pole. These lines are called *isodynamic curves*. Extensive journeys have been undertaken by Humboldt, Sabine, Hansteen, and others, to ascertain the point on the surface of the earth where the magnetic intensities are equal, for the purpose of describing these curves. The earlier results indicated the position of the magnetic pole to be in the northeastern part of Hudson's Bay, lat. 60° N., lon. 80° W. ;* but the directions of these curves presented such anomalies as to suggest the idea of a second magnetic pole in the opposite hemisphere : with the view of ascertaining this point, Professor Hansteen, of Christiana, several years since, undertook a journey into Siberia, at the expense of the King of Sweden, and has fully confirmed the fact, that there exists a second magnetic pole to the north of Siberia, around which the isodynamic curves arrange themselves in regular order.† From experiments made in deep mines, and in the upper regions of the atmosphere by aëronauts, it appears, that in both these situations, the magnetic intensity is the same as at the corresponding places on the surface of the earth.

691. *The effects produced by the earth on a magnetic needle, correspond to those produced on it by a powerful magnet, and hence the earth itself may be considered as such a magnet.*

The magnetism of the earth has been supposed by some to result from a great magnet lying in the central parts of the earth ;‡ by others,§ to be nothing more than the *resultant* of all the smaller magnetic forces scattered through various parts of the terrestrial sphere ; and by others, to be excited on the surface of the earth by the action of the solar rays.

The supposition of a great magnet in the interior of the earth, to which all the phenomena of terrestrial magnetism are to be ascribed, is the earliest hypothesis, and is adequate to explain most of the facts of the science. But such a supposition is inconsistent with the recent discovery of two north poles, (Art. 690,) implying the existence of four magnetic poles of the earth.

* In the year 1832, Commander James Ross, of the British navy, supposed that he had reached the true magnetic pole in N. lat. 70°, W. lon. 96° 30′, in a region lying northward of Hudson's Bay, and westward of Baffin's Bay. The best chart hitherto published of the magnetic lines of equal variation on the earth's surface, has recently been constructed by Professor Barlow, from an immense number of observations, and published in the Philosophical Transactions for 1833.

† *Sabine*, Amer. Jour. XVII, 145.

‡ Gilbert.

§ Humboldt and Biot.

The opinion of Biot, that terrestrial magnetism is only the aggregate, or resultant, of all the individual magnetic forces residing in different parts of the earth, appears to be no improbable supposition, and accords well with the general doctrine of the composition of forces.

692. In the year 1813, Dr. Morichini, of Rome, announced that the violet rays of the solar spectrum have the property of rendering iron magnetic. In 1825, these experiments were repeated and extended by Mrs. Somerville,* and resulted in proving, that the magnetizing power is not confined to the violet rays, but extends to the indigo, blue, and green rays. The probable conclusion is, that a class of rays emanate from the sun, which have the property of producing magnetism, and are distinct from those which afford light and heat, and produce chemical changes. Hence, in the solar beam there are at least four distinct kinds of rays, denominated, respectively, *colorific*, *calorific*, *chemical* and *magnetizing* rays.†

693. *Electricity and magnetism are, in some of their properties, remarkably alike, but in others strikingly dissimilar.*

Several of these analogies have been already incidentally mentioned; but it will be useful to the student to consider them in connection. Electricity and magnetism agree in the following particulars. (1.) Each consists of two species, the vitreous and resinous electricities, and the austral and boreal magnetisms. (2.) In both cases, those of the same name repel, and those of opposite names attract each other. (3.) The laws of induction in both are very analogous. (4.) The force, in each, varies inversely as the square of the distance. (5.) The power, in both cases, resides at the surface of bodies, and is independent of their mass.

But electricity and magnetism are as remarkably unlike in the following particulars. (1′.) Electricity is capable of being excited in all bodies and of being imparted to all: magnetism resides almost exclusively in iron in its different forms, and, with a few exceptions, cannot be excited in any other than ferruginous bodies. (2′.) Electricity may be *transferred* from one body to another: magnetism is incapable of such transference; magnets communicate their properties merely by *induction*, a process in which no portion of the fluid is withdrawn from the magnetizing body. (3′.) When a body of an elongated figure is electrified by induction, on being divided near the middle, the two parts possess respectively the kind of electricity only which each had before the separation; but when a bar of steel or a needle magnetized by induction, is broken into any number of parts, each part has both polarities and becomes a perfect magnet. (4′.) The directive

* Phil. Trans. 1826.

† See Brewster's Optics, p. 92

properties and the various consequences that result from it, the declination, annual and diurnal variations, the dip, and the different intensities in different parts of the earth, are all peculiar to the magnet, and do not appertain to electrified bodies.

694. *The phenomena of magnetism are explained on the hypothesis of two fluids, residing naturally in iron and all ferruginous bodies, which, when united, neutralize each other's effects, but which, when separated, exhibit the respective properties of boreal and austral magnetism.*

Nearly all the arguments alleged in favor of the hypothesis of two fluids in electricity, apply equally well to magnetism. It is necessary however to assume, that the two magnetic fluids are separated from each other only at distances extremely small, for otherwise it is impossible to account for the fact, that when a magnet is divided into minute fragments, each piece contains both fluids, being a perfect magnet with two opposite poles. This hypothesis, like the corresponding one in electricity, has been submitted by Poisson to the most rigorous mathematical analysis, and all the deductions made from it are found to accord exactly with the facts ascertained by experiment. Hence this doctrine is generally received, and has nearly superseded the explanation formerly given by Æpinus, who accounted for magnetic phenomena on the supposition of a single fluid, similar to Franklin's hypothesis of electricity.

According to the foregoing hypothesis, iron differs from nearly every other natural substance, in containing a certain portion of the compound magnetic fluid. This usually maintains its equilibrium, and therefore is latent or insensible ; but various causes disturb this state of equilibrium, and then the separate fluids exhibit their peculiar properties. When once separated, they have the power of producing on the magnetic fluid of other masses of iron a similar separation, each repelling the similar, and attracting the dissimilar species. Hence one magnet affords the means of making another ; and the process of magnetizing consists not in imparting any thing from the magnetizing body, but merely in decomposing the fluid before residing in the body magnetized, that is, separating it into its constituent fluids. Indeed, so far from losing by the process of magnetizing, the original magnet itself gains by the reaction of the new magnet it has formed, which tends still more fully to develope or separate its own constituent fluids. By this means, what was originally a very weak, may become a strong and powerful magnet, without any other aid, than what contributes to separate more fully the two fluids naturally inherent in it.*

* Pouillet, Elémens de Physique, I, 325

695. The facility with which soft iron acquires and loses magnetism, (Art. 675,) is conceived to depend on the ease with which the magnetic fluids pervade a mass of loose texture, in which the particles have comparative freedom of motion; while the greater fixedness of the particles of hard steel, creates an obstruction to the motions of the same fluids. Thus a magnet loses its powers by exposure to a white heat, (Art. 676,) because the separate fluids, having freedom of motion, combine and neutralize each other; and the method of imparting magnetism to iron by magnetizing it while softened by heat and suddenly cooling it, is so effectual, because in this way the two fluids are first easily separated by induction, and afterward are prevented from combining by the increased obstruction created by hardening the metal. The development of magnetism in an iron bar by percussion, (Art. 676,) is supposed to be owing likewise to the greater freedom of motion secured to the magnetic fluids by the vibration of the particles of iron, thus enabling these fluids to separate from one another, while as soon as the vibration ceases, that freedom of motion is lost, and the fluids are prevented from reuniting. That such a vibration is favorable to the effect produced, is inferred from the fact that blows which produce a *ringing* sound are peculiarly efficacious in developing magnetism. The same explanation is applied to the case where magnetism is lost by percussion; since here, the vibrations would enable the separate fluids to combine.

The periodical changes in the situation of the magnetic poles of the earth, upon which the direction of the needle depends, including the annual and diurnal variations, the dip, and the intensity of the force, result from causes which have hitherto eluded discovery.

METHODS OF MAKING ARTIFICIAL MAGNETS.

696. If the learner has made himself acquainted with the principles expounded in the preceding propositions, he will be qualified to proceed, with interest and intelligence, to an explanation of the leading methods practised in the manufacture of artificial magnets. These methods also, by involving a practical application of those principles, will serve to impress them on the memory and to render the knowledge of them familiar.

It will be recollected that magnets are made from other magnets; that this is done, not by any *transference* of a portion of the power of the magnetizing body, but by the development of the powers naturally residing in the body to be magnetized; that this development is effected wholly on the principle of induction; that the original magnet gains instead of losing by its action on other bodies; that this power may be induced on iron by the agency of an artificial magnet, or of the loadstone, or of the earth, which is itself a weak magnet, and acts upon the same principles as any other magnet. It must also be kept clearly in mind, that soft iron or steel readily acquires and as readily loses the magnetism induced upon it, and that hardened iron or steel receives it slowly and with much difficulty, but retains it permanently. As the earth itself may be supposed to have been the original source of magnetism in all other bodies in which it is found, we shall begin by describing the methods of magnetizing from the earth, without the aid of either a loadstone or an artificial magnet.

697. *A certain degree of magnetism may be given to steel bars by hammering them while in a vertical position.*

Bars of steel prepared for this purpose are of a prismatic form with rectangular sides, the length being ten times the breadth, and twenty times the thickness. Six or eight bars of equal size are to be provided, and being held in a vertical position they are to be struck with a few blows of the hammer, when they will be found to have acquired a feeble degree of magnetism, which is indicated by their exhibiting polarity and having the power of attracting iron filings. This effect will be much greater if the bars, while receiving the blows, be placed near to a mass of iron, so as to experience its inductive influence. A pair of tongs may be used for this purpose; during the process, the tongs themselves become magnetic, and by their reaction greatly increase the magnetism of the bars.

698. *A needle may be magnetized by simply suffering it to remain in contact with the pole of a strong magnet; or better, between the opposite poles of two magnets.*

The effect produced by two magnets is much more than double that of one magnet, as may be inferred from article 673. But if the needle be of considerable length, several intermediate sets of poles are sometimes developed, as will be seen by applying iron filings. It adds much to the power of the two magnetic bars between which the needle is placed, if to the extremity of the bar most remote from the needle, a mass of soft iron is placed. (See Art. 673.) The iron, in this case, acts and reacts by induction: and hence whenever magnets are not in use, they require to be connected with iron to prevent the loss of their powers. Pieces of soft iron thus connected with magnets for the purpose of augmenting their powers by induction, are called *armatures*. Thus A is the armature of the horse-shoe magnet represented in Fig. 234.

699. But it must be recollected, (Art. 693,) that the two species of magnetism are not, like those of electricity, separated to a distance from each other, so that one kind may be wholly collected at one end of the bar and the other kind at the other end,

but that the two are separated only at a minute distance, remaining in the immediate vicinity of each other throughout the whole length of the bar. Hence, in order to give the magnetizing pole its full effect, it becomes necessary to apply it successively to every part of the bar from one end to the other.

A more effectual method of magnetizing a needle is the following:—Place two magnetizing bars, A, B, parallel to each other, with their dissimilar poles adjacent; unite the poles at one end by a piece of soft iron, R, and apply the poles at the other end to the needle, as is represented in figure 233. Upon this principle, that is, the increased energy with which the two poles

Fig. 233. Fig. 234.

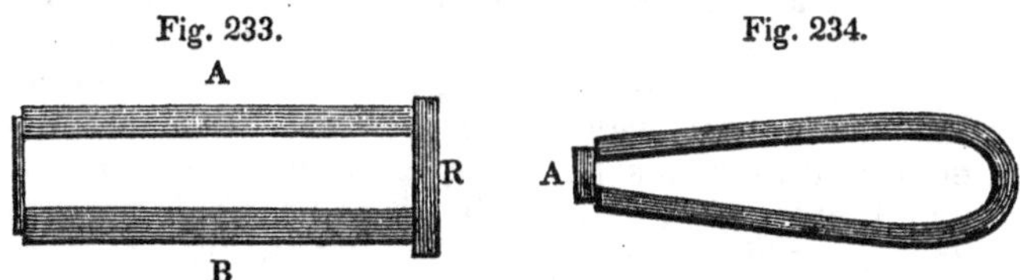

act together, is formed what is called the horse-shoe magnet, which derives its name from its peculiar figure, (Fig. 234.) Bars of this form are converted into magnets upon the same principles as straight bars, the magnetizing bar being made to follow the curvature always in the same direction. A very efficacious mode of making horse-shoe magnets is thus described by Professor Barlow. Two horse-shoe bars may be united at their ends in such a manner, that the poles which are to be of opposite names shall be in contact. They are then to be rubbed with another strong horse-shoe magnet, placing the latter so that its north pole is next to the south pole of one of the new magnets, and consequently its south pole next to the north pole of the same, carrying the movable magnet round and round always in the same direction. This is esteemed one of the most eligible modes of making powerful magnets.

The horse-shoe magnet is itself very convenient for imparting magnetism to other bodies. Place the poles near the center of the needle; move them along its surface backward and forward, taking care to pass over each half of it an equal number of times; repeat the same operation on the other side; and the needle will become speedily and effectually magnetized.

700. The best mode of making magnetic needles in general, is expressed in the following rule, given, as the result of very extensive and accurate experiments, by Capt. Kater.

Place the needle in the magnetic meridian; join the opposite poles of a pair of bar magnets, (the magnets being in the same line,) and lay the magnets so joined flat upon the needle, with their poles upon its center; then having elevated the distant extremities of the *magnets, so that they may form an angle of about two or three degrees with the needle, draw them from the center of the needle to the extremities, carefully preserving the same inclination ; and having joined the poles of the magnets at a distance from the needle, repeat the operation ten or twelve times on each surface.**

In connection with the foregoing rule, Capt. Kater gives the following summary of principles, established with respect to the compass needle. 1. That the best *material* for compass needles is a clock spring; but care must be taken, in forming the needle, to expose it as seldom as possible to heat, otherwise its capability of receiving magnetism will be much diminished. 2. That the best *form* of a compass needle is a pierced rhombus, (Fig. 236,) in the proportion of about five inches in length to two in width, this form being found susceptible of the greatest directive force. 3. That the best method for *tempering*, is first to harden the needle at a red heat, and then to soften it from the middle to about an inch from each extremity, by exposing it to heat sufficient to cause the blue color which arises, again to disappear. 4. That in the same plate of steel, of the size of a few square inches only, portions are found *varying considerably in their capability of receiving magnetism*, though not apparently differing in any other respect. 5. That *polishing* the needle has no apparent effect on its magnetism. 6. That in needles from five to eight inches in length, their weights being equal, *the directive forces are nearly as the lengths*. 7. That the directive force does not depend upon extent of surface, but, in needles of the same length and form, it is as the *mass*. 8. The deviation of a compass needle, occasioned by the attraction of soft iron, depends on *extent of surface* and is wholly independent of the mass, except a certain thickness of iron, amounting to about two tenths of an inch, which is requisite for the complete development of its attractive energy.

701. The reasons on which the preceding rule and the annexed principles are founded, will for the most part be understood from what has gone before. A needle to be magnetized is placed in the magnetic meridian, because (the earth being considered as a magnet) the needle has its axis then parallel to that of the magnet, a position in which, Art. 671, it receives the greatest effect from induction. The opposite poles are joined, because acting reciprocally upon each other by induction, they augment each other's powers. The bars thus joined are placed on the center of the needle and drawn in opposite directions, for, by this means, upon that part of the needle which lies between them, the forces of the two poles conspire. Upon the part which lies between each bar and the adjacent extremity of the needle, these two poles are opposed to each other; but as the poles are more remote

* Phil. Trans.

from the parts where their actions are opposed to each other, than from those whose actions conspire, they on the whole tend to augment each other's effects. The bars are first laid flatwise, and afterward elevated by as small an angle as will serve the purpose of drawing them asunder, with their poles only in contact with the needle, because the effect of induction is strongest when the magnetizing bars are nearest to a parallelism with the body to be magnetized; and the same angle of inclination is carefully preserved, for it is only in this way that both sides of the needle will have precisely the same strength, a condition essential to its perfection. In renewing the application of the bars, they are removed to a distance before their poles are joined again, because it is important to secure the magnetism which the needle has already acquired, against those partial disturbances which might arise from the irregular action of the magnetic bar.

702. *Magnets are liable to lose their power, to prevent which, certain precautions are necessary.*

If a single magnet is kept out of its natural direction, it grows gradually weaker, and this loss of power is most rapid when its position is the reverse of the natural one, that is, when its north pole is turned toward the south. Under these circumstances, indeed, unless the magnet is made of the hardest steel, it will in no long time lose the whole of its magnetic power. Two magnets may also very much weaken each other if they are kept, even for a short time, with their similar poles fronting each other. The polarity of the weaker magnet, especially, is rapidly impaired, and sometimes found to be actually reversed. More frequently, however, there arises from this opposition of powers, considerable irregularity and confusion in the poles of both magnets.

Since heat also impairs the powers of the magnet, (Art. 675,) this instrument should never be exposed to a high temperature. We should likewise be very cautious to avoid all rough and violent treatment; for its virtues are speedily impaired by concussion, or whatever occasions a vibration among its particles. It must not, therefore, be suffered to fall on the floor, or be rubbed with coarse powders, or be ground with the view of altering its form. The loadstone has its powers impaired by similar means; hence, we should attempt to alter its natural form as little as possible; and when it is necessary to do so, it should be effected very rapidly by cutting it on a lapidary's wheel.

Although the loadstone retains its magnetic virtue more tenaciously than any artificial magnet that can be constructed, yet even this body requires a certain management for the permanent preservation of its powers. For this purpose it should be *armed*, as it is called; that is, a piece of soft iron should be constantly kept in contact with the two poles. In order to do this most effectually, we must first ascertain the situation of the poles of the loadstone, which we may do by rolling it in iron filings; and then cutting off all the superfluous parts, we may give it the shape of a parallelopiped, having the poles in the middle of two opposite surfaces, and at the same time taking care to preserve the axis, which passes through the poles, of as great a length as can be obtained; for it has been observed that any curtailment of the magnet in the direction of this line, deprives it of force in a greater degree than when shortened in any other direction. Plates of soft iron are next attached to the two sides containing the poles, which are made to project a little way below the bottom of the loadstone, terminating in two bars, like the poles of a horse-shoe magnet, to which bars a short bar of soft iron is attached, upon which the whole force of both poles acts simultaneously. This action exerted upon the iron bar, is sufficient to preserve the powers of the loadstone from decay. (See Fig. 235.) A similar piece of iron is applied by way of armature, to the two poles of a horse-shoe magnet. Bar magnets also, when laid aside, should be placed with the north pole of one in contact with the south pole of another, or what is better, two bars may be placed parallel, at a little distance from each other, with their like poles in opposite directions, and having their dissimilar poles united by short pieces of iron, so as to form, with the bars, a parallelogram. Magnets should be polished, because they are then less liable to contract rust.*

Fig. 235.

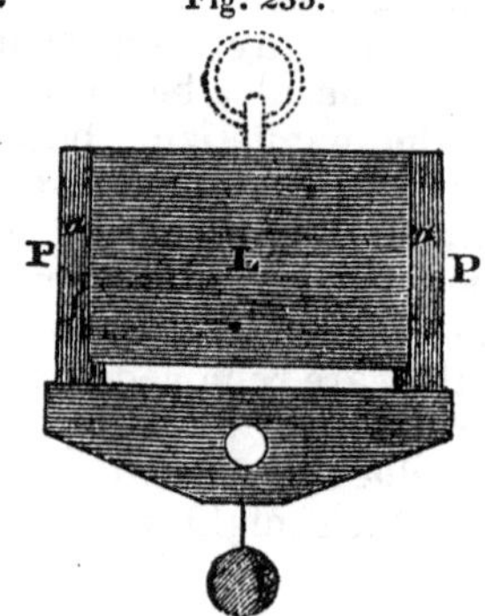

703. The compass (the importance of which to mankind, has attached to the subject of magnetism its principal value) is of many different forms, but the chief varieties are the land compass, the Mariner's compass, the Azimuth compass, and the Variation compass. The needle, in all these varieties, is usually a thin flat plate of steel, tapering at the extremities; but, as we have already mentioned, (Art. 700,) a more eligible form has been proposed by Capt. Kater, consisting of four narrow strips of steel, united in the form of a hollow rhombus, (Fig. 236.) It is found advantageous to concentrate the powers of the needle as much as possible in the two extremities, and to avoid all inequalities, arising from intermediate poles, or

Fig. 236.

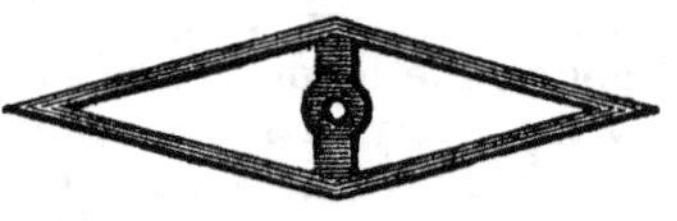

* Lib Useful Knowledge, *Magnetism*, p. 54.

from a difference of strength in different parts. The needle is secured at the point of suspension, and furnished with a conical cap of brass which rests on a perpendicular pin; and still further to diminish friction, the point which rests on the extremity of the pin is made of agate, one of the hardest mineral substances. Since, if the needle is magnetized after having been balanced on its center of gravity, it would no longer remain horizontal, the equipoise is restored by attaching a small weight to the elevated side.

704. The compass, in its simplest form, consists of a needle like the foregoing, enclosed in a suitable box covered with glass. This is all that is essential when it is required merely to know the direction of the meridian, or the north and south points. But, for most purposes, the compass is furnished with a graduated circular card, divided into degrees and minutes; and in the mariner's compass, the card is also divided into thirty-two equal parts, called rhumbs. The card thus divided is fastened to the needle itself, and turns with it.

Thin, slender needles, have the greatest directive powers, and are most sensible, since they undergo less friction than those which are heavier, but due regard to strength requires them to be made of a certain degree of thickness; an increase of length is attended with an increase of directive power; but when the thickness remains the same, the weight, and consequently the friction, increases in the very same ratio; no advantage, therefore, as to directive power, can be obtained by any increase of length. Moreover, needles which exceed a very moderate length, are liable to have several sets of poles, a circumstance which is attended with a great diminution of directive force. On this account, short needles, made exceedingly hard, are generally preferable.

705. The great importance of the Mariner's compass, has made its construction an object of much attention, and the best artists have tried their skill upon it. The compass is suspended in its box in such a manner as to remain in a horizontal position, notwithstanding all the motions of the ship. This is effected by means of *gimbals*. This contrivance consists of a hoop, CD, usually of brass, (Fig. 237,) fastened horizontally to the box by two pivots placed opposite to each other, and constituting the axis on which the hoop turns up and down. At an equal distance from the pivots on each side, that is, at the distance of 90° from each pivot, two other pivots are attached to the ring at right angles to the former, on which the inner box that contains a card is hung. Of course when it turns on these pivots, its motion is at right angles with that of the hoop. Therefore all the motions of which the compass box is capable, are performed

Fig. 237

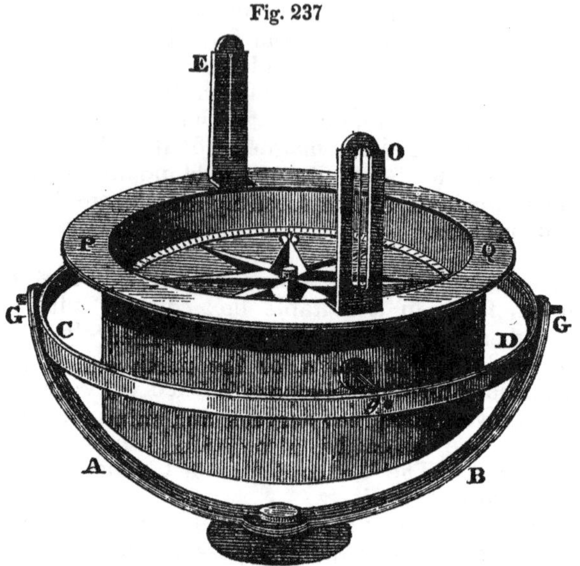

around two axes which intersect each other at right angles; consequently, the point of intersection, being in both axes, will not move at all. But the needle and the attached card rest upon this point, and are connected with the compass box in no other point. Hence they remain constantly horizontal in every position of the box.

The Azimuth compass* differs from the common mariner's compass only in having sights attached, by which the bearing of any object with the meridian may be ascertained. The Surveyor's compass is a variety of the azimuth compass.

LOCAL ATTRACTION OF VESSELS.

706. A few years since it was observed, for the first time, that the needle of the mariner's compass, on board of a ship, does not continue to point constantly in the same direction, but alters its direction as the ship heads toward different parts. Changing the position of the ship from north or south to east or west, sometimes changes the direction of the needle 20° or 30°. Indeed, in one instance mentioned by Capt. Sabine, shifting the ship's head from east to west, produced a change in the direction of the needle amounting to 50°. Such irregularities are found to be greatest in the polar seas. This effect is caused by the attraction which the large quantity of iron on board a ship exerts upon the

---

* *Azimuth*, as applied to a star or any celestial object, is an arc of the horizon intercepted between the meridian and a vertical circle passing through the object.

needle, consisting of the guns on board of a man of war, of the masses of iron sometimes employed as ballast, of the iron tanks recently substituted for water casks, and of the various bolts, bars, nails, &c. which enter more or less into the construction of every sort of vessel. Indeed, on account of the greater quantity of iron now employed, the mariner's compass, at present, is more subject to irregularities from this cause than it was formerly.*

In order to investigate the laws by which these effects were controlled, and to devise a remedy for them, Professor Barlow, of the Military Academy of Woolwich, instituted a great number of experiments, which resulted in the discovery of a method of obviating completely, every difficulty, by neutralizing the effect of the iron of the ship, and leaving the needle free to obey the impulse of terrestrial magnetism alone. It is easy to understand that all the various forces exerted by the iron in different parts of the vessel, will have a single *resultant*, equivalent to the whole; and that, if we can discover the amount of this resultant, it will be only necessary to make a suitable correction, to be either added or subtracted, according as the indications of the needle are too small or too great.

707. Mr. Barlow procured a solid ball of iron, thirteen inches in diameter, and two hundred and eighteen pounds in weight. When the compass was *above* the ball, he found that the north end of the needle was attracted toward it; and that when it was *below* the ball, the south end was attracted toward it: and that, in traversing the interval between these two positions, it always passed through a point in which the ball had no effect on the needle. Instead, however, of moving the compass through these different positions, the compass was suffered to remain stationary, and the ball, suspended by means of pulleys, was raised or lowered at pleasure, and thus easily brought into any required position with respect to the compass. The experiment showed, that all those points in which the ball exerts no influence on the needle, are in the same plane, and that this plane is inclined to the horizon toward the south, making an angle with the horizon equal to the complement of the dip; and of course the direction of this plane is at right angles to the direction of the dip. This plane, therefore, in reference to the iron sphere, constitutes its magnetic equator. It is at right angles to the magnetic meridian, and cuts the horizon in the magnetic east and west points. A compass needle, whose center is anywhere in this plane, will not have its action disturbed in the least by the influence of the ball. Hence, this plane is denominated the *plane of no attraction*, or the plane of neutrality. Nor is the existence of such a plane confined to masses of iron of a globular shape; it extends equal-

* Barlow, Phil. Trans. 1833.

ly to masses of the most irregular form, and even to an assemblage of detached masses, like those disposed through the different parts of a ship.

708. The actual amount of deviation produced in the ship's compass by its local attraction, will, of course, be different in different vessels. With an easterly or westerly course, it has been observed in the latitude of London to vary from five to twelve or fourteen degrees : it is of greater amount as the ship is in higher latitudes; and diminishes, without vanishing, at the equator; and again increases as we approach the south pole. Experiments were made on eight different men of war in the British harbors, and in all of them very considerable deviations were detected from the local cause under consideration, and an average deviation in the whole of 8° 44′. The Gloucester, one of these ships, was invariably drawn to the southward of her intended place, notwithstanding the greatest care was taken in steering her. Had it not been ascertained, by taking an observation, that this error was altogether the effect of local attraction, it would probably have been ascribed to the influence of an unknown current. The real deviation, estimated in distance, would occasion the vessel, after running ten miles, to be more than a mile and a half to the southward of her reckoning, and so on as the distance increased. An error of this magnitude, occurring in a narrow channel, and in a dark night, were it unknown or disregarded, might lead to the most disastrous consequences; and shipwrecks have been traced, with much probability, to this source of error in the reckoning. The loss of the Thames, Indiaman, a few years since, was ascribed to this cause. This vessel, besides the usual supply of guns, had a cargo of more than four hundred tons of iron and steel. The influence of such an enormous magnetic mass, would alone be quite sufficient to explain the otherwise unaccountable circumstances, that after leaving Beachey Head in sight at six o'clock in the evening, the ship was wrecked upon the same spot, between one and two o'clock in the morning, without the least apprehension of being near the shore.*

709. The *Correcting Plate* of Professor Barlow, affords an effectual remedy for these errors. It consists of a double plate, formed of two thin disks of iron, screwed together in such a manner as to combine any strong irregular power of one plate, with a corresponding weak part of another; by which means a more uniform action is obtained. These plates are of a circular form, twelve or thirteen inches in diameter. Now, it being ascertained from actual experiment, (comparing the direction of the

* Barlow.

compass on board with its direction on the shore,) what is the amount of deviation occasioned by the iron of the ship, it is evidently possible, by bringing a small quantity of iron *near* to the needle, to produce in it a deviation of the same amount, and of course to double the error in question. In Fig. 238, A represents a vertical stand or log of wood, turning horizontally on its base, to the top of which the compass B is firmly attached. The correcting plate C, is supported by a pin passing through its center, and entering a hole made in the side of the stand. Of these holes there are several, so as to admit of shifting the positions of the plate. To ascertain the local attraction of a ship, the direction of the needle is first observed on board, free from the influence of the plate. The apparatus represented in Fig. 238, is then removed on shore, and the bearing of the compass observed before applying the plate. The difference in direction on board and on shore, shows the effect of the ship's iron. The plate is now applied, and the log, together with the plate, is turned round, until the direction of the needle is the same as on board. In order to make it the same, it may be necessary to shift the position of the plate, by inserting the pin in a different place. We have now ascertained the position of the plate, with respect to the needle, required in order to indicate the local attraction of the ship. If the apparatus is placed again on board, (the compass and the plate retaining the same relative position,) the whole deviation of the needle from its true place will be doubled; that occasioned by the iron of the ship being equal in amount to that made by the plate. Hence, in any given case, we have only to observe the effect of the plate upon the needle, in order to learn the amount of the local attraction.

Fig. 238.

In order to bring the efficacy of the correcting plate to the test of experience, several of the ships of the royal navy of Great Britain were furnished with it, and trials were instituted with it in various parts of the world, from the arctic to the antarctic circle, and with the most satisfactory results. This expedient, therefore, is at present held to be a most effectual corrective of the errors from the local attraction of vessels.

710. Chronometers, also, which are carried on board of ships for the purpose of finding the longitude, are liable to have their rate of going affected by the magnetic action of the iron of the ship. Although a sudden alteration in the rates of chronometers at sea had frequently been observed, yet the cause was not detected until as late as the year 1818. It appeared on examination, that the effect was produced by the magnetic action of the

ship's iron upon the inner rim of the balance of the chronometer, which is made of steel. A similar influence was perceptible on placing magnets in the neighborhood of the chronometer. Mr. Barlow applied himself to experiments on this subject, and found that masses of iron wholly destitute of permanent magnetism, occasioned an alteration in the rates of chronometers, placed near them in different positions. Sometimes they were accelerated, and sometimes retarded. Hence, it is recommended to keep the chronometer, on board of any ship, out of the vicinity of any large mass or surface of iron. The method proposed for rectifying this error is the same as that for correcting the compass, viz. by first ascertaining what the effect of the ship's iron upon the chronometer is, and then applying the correcting plate upon the same principles as in the case of the compass.

The late voyages to the northern seas, undertaken by the British government, however they may have failed of gaining their principal object, namely, the discovery of a northwest passage, still achieved many valuable results in experimental science, but in none, perhaps, more than in the science of magnetism. Among the rest, they made numerous observations on the local attraction of vessels; on the magnetic effect of the ship's iron upon the rate of chronometers; upon the position of the magnetic poles; upon the phenomena of the dipping needle; and upon the magnetic intensities of different places on the earth's surface.

MAGNETIC CHARTS.

711. The great importance of the mariner's compass to the art of navigation, has induced the British government, at different times, to send abroad men of science to make observations on magnetism in different countries, with the view of reducing the principles on which the variation of the compass depends, to settled laws. The first great enterprise of this kind was undertaken about the year 1680, by Dr. Halley, one of the most distinguished and zealous philosophers of that age. For the purpose of ascertaining the law of the variation of the compass, Dr. Halley was invested with the command of a national ship, in which he traversed the Atlantic ocean in various parts, extending his voyage to the fiftieth degree of south latitude. After he had collected a great number of observations made by others, and compared them with his own, he published, in the year 1700, a synopsis of them in the form of a chart, in which the ocean was represented as crossed by a number of lines passing through those places where the compass had the same deviation. Thus, in every point of one line there was, in the year 1701, no variation; in any point of another line, the compass had twenty de-

grees of east variation; and in every point of a third line, it had twenty degrees of west variation.*

But though Halley's chart was constructed with all possible care, and presented a comprehensive view of all that was then known of the subject, yet it could not be of much permanent utility, since the lines of which it is composed are themselves continually changing their relation to one another. Among the recent magnetic charts which have been published, that of Professor Barlow is the most extensive and useful.† Professor Loomis has published a valuable chart of the United States, in the 39th volume of the American Journal of Science.

The great and constant irregularities of all the lines described on magnetic charts, whether they relate to the variation of the compass, or to the magnetic dip and intensity, are such as almost to preclude the hope of reducing the phenomena of terrestrial magnetism to laws so definite, as to afford rules of calculating these particulars for any given place, independently of experiment.

712. The presence of the *Aurora Borealis,* is found to have a remarkable effect on the magnetic needle. A brilliant Aurora will sometimes cause an enormous deflexion in the needle. Indeed this deflexion has been found, in certain cases, to remain to a greater or less degree permanent. In the great Aurora of November 14th, 1837, the needle often moved thirty minutes in three seconds, and its entire range of motion was no less than six degrees.‡ The luminous columns, also, which frequently appear during this phenomenon are generally, if not always, parallel to the magnetic meridian;§ and the *Corona,* or luminous circle, which, in some great exhibitions of the Aurora, is formed southeast of the zenith, has its center at the point toward which the upper end of the dipping-needle is directed, that is, toward the *pole* of the dipping-needle. These facts evidently show that the matter that forms the Aurora has magnetic properties, but they give no information respecting its *origin.* This is believed to be independent of the earth, and the discussion of it therefore belongs to astronomy.

* Robison's Mech. Phil. IV., 358.

† Phil. Transactions for 1833.

‡ Mr. E. C. Herrick.

§ See Dalton's Met. Essays.—Am. Jour. xxx, 227.

PART VIII.—OPTICS.

CHAPTER I.

PRELIMINARY DEFINITIONS AND OBSERVATIONS.

713. OPTICS *is that branch of Natural Philosophy which treats of Light and Vision.*

More particularly, it is the object of this science to investigate the *nature* of the agent on which the phenomena of vision depend; to treat of the *motions* of light in respect to its direction, its velocity, and its reflexion from the surface of bodies; to trace its change of direction, and the various other modifications it undergoes by passing through different transparent media; to explain the *phenomena of nature* which depend upon the properties of light, embracing the doctrine of *color;* to trace the relation between light and the structure of the eye, comprehending the subject of *vision;* and finally, to describe the various *instruments* to which a knowledge of the principles of Optics has given birth, disclosing many new and wonderful properties of light, and extending the range of human vision, on the one hand, to myriads of objects too minute, and on the other, to numberless worlds too remote, to be seen by the unassisted eye.

714. Luminous bodies are naturally of two kinds, such as shine by their own light, as a lamp or the sun, and such as shine by borrowed light, as the moon, and most of the visible objects in nature.

A *ray* is a line of light; or it is the line which may be conceived to be described by a particle of light. In a more general sense, the term is applied to denote the smallest portion of light which can be separately subjected to experiment. A *beam* is a collection of parallel rays. A *pencil* is a collection of converging or diverging rays. A *medium* is any space through which light passes. When a space is a perfect void, so as to offer no obstruction to the passage of light, it is said to be a *free medium;* when the space intercepts a portion only of the light, it constitutes a *transparent medium.* Transparency, however, may exist in different degrees. When the medium itself is invisible, as portions of air, it is said to be *perfectly transparent;* when the medium is visible, but objects are seen distinctly through it, as in the clearest specimens of glass and crystal, it is said to be, simply, *transparent;* when objects are indistinctly seen through it, it is *semi-transparent;* and when a mere glimmering of light passes through, without representing the figure of objects, it is *trans lucent.* Bodies that transmit no light are said to be *opake.*

715. *Rays of light, while they continue in the same uniform medium, proceed in straight lines.*

For objects cannot be seen through bent tubes; the shadows of bodies are terminated by straight lines; and all conclusions drawn from this supposition, are found by experience to be true. If two bodies with plane surfaces, as two disks of metal, be held between the eye and some luminous point, as a star, on bringing the two planes gradually toward each other, the star may be seen through the intervening space until the planes come completely into contact; but if one of the surfaces is convex and the other concave, the light is intercepted before the surfaces have met.* In consequence of the rectilinear motion of light, it forms angles, triangles, cylinders, cones, &c., and thus its affections fall within the province of geometry, the principles of which are applied with great effect to the development of the properties and laws of light, after a few fundamental properties are established by experiment.

From every point in a luminous object, an inconceivable number of rays of light emanate in every direction, when not prevented by obstacles that intercept it. Thus, from every point in the flame of a candle, as seen by night, light diffuses itself, pervading an immense sphere, and filling every part of the space so perfectly, that not the minutest point can be found destitute of some portion of its rays. Any luminous body of this kind is called a *radiant.* The pencil of light which proceeds from a radiant is a cone, the sections of which made by any plane correspond to the figures called conic sections. If any portion of the pencil be intercepted by a rectilateral figure, that portion constitutes a pyramid of which the figure is the base, and the luminous point itself is the vertex.

716. *Light has a progressive motion of about one hundred and ninety-two thousand five hundred miles per second.*

The estimation of the velocity of light, (which may be classed among the greatest achievements of the human mind,) has been effected in two different ways. The first method is by means of the eclipses of Jupiter's satellites. To render this mode intelligible to those who have not studied astronomy, it may be premised, that the planet Jupiter is attended by four moons which revolve about their primary, as our moon revolves about the earth.

* Biot.

These small bodies are observed, by the telescope, to undergo frequent eclipses by falling into the shadow which the planet casts in a direction opposite to the sun. The exact moment when the satellite passes into the shadow, or comes out of it, as would be seen by a spectator at the mean distance of the earth from the sun, is calculated by astronomers. But sometimes the earth and Jupiter are on the same side, and sometimes on opposite sides of the sun; consequently, the earth is, in the former case, the whole diameter of its orbit, or about one hundred and ninety millions of miles nearer to Jupiter than in the latter Now it is found by observation, that an eclipse of one of the satellites is seen about sixteen minutes and a half sooner when the earth is nearest to Jupiter, than when it is most remote from it, and consequently, the light must occupy this time in passing through the diameter of the earth's orbit, and must therefore travel at the rate of about one hundred and ninety-two thousand miles per second.* Another method of estimating the velocity of light, wholly independent of the preceding, is derived from what is called the *aberration of the fixed stars.* The full explanation of this method must be referred to astronomy; but it may be understood in general, that the apparent place of a fixed star is altered from the effect of the motion of its light, combined with the motion of the earth in its orbit. It will be remarked that the place of a luminous object is determined by the direction in which its light meets the eye. But in the case of light coming from the stars, the apparent direction is altered in consequence of the motion of the earth in its orbit, being intermediate between the actual direction of the earth and of the light of the star; and the velocity of the earth in its orbit being known, that of light may be computed from the proportional part of the effect produced by it in causing the aberration. The velocity of light, as deduced from this method, comes out very nearly the same as by the other.† Hence it is inferred, that the velocity of light is uniform.

717. *The intensity of light, at different distances from the radiant, varies inversely as the square of the distance.*

This proposition is proved in the same manner as that respecting gravity, (Art. 7, p. 21,) the reasoning in which, applies to all emanations from a center.

Although the intensity of light decreases rapidly as we recede from the radiant, yet the *brightness* of the object suffers little diminution by increase of distance. Thus, a candle appears nearly as bright at the distance of a mile as when close to the

* $\frac{190000000}{16.5 \times 60} = 192000$ *nearly.*

† See Herschel on Light, *Encyc. Metrop.*

eye. If, while the observer remains stationary, the light which was before spread over a given area, should be all collected into a space half as large, the brightness would obviously be twice as great as before; or, in general, the brightness, the quantity of light being given, is inversely as the area; that is, inversely as the square of the diameter. Now, as we recede from an object, its area is apparently diminished, and on this account, its brightness is increased in the same ratio as it is diminished by the cause operating according to the foregoing proposition. The brightness therefore remains constant.*

This is to be understood however, only of light passing through a *free medium;* by traversing the air, the brightness is diminished according to the following law.

718. *The effect of a transparent medium of uniform density, is to diminish the intensity of light in a geometrical ratio.*

For, imagine that the medium, a piece of glass, for example, is divided into equal laminæ, of such thickness, that the first lamina shall stop $\frac{1}{n}$th part of the rays that fall upon it. Then there will issue from the lamina $1-\frac{1}{n}=\frac{n-1}{n}$ rays. The second lamina, in like manner, will stop $\frac{1}{n}$th part of the light which falls upon it; that is, $\frac{1}{n}$ of $\frac{n-1}{n}=\frac{n-1}{n^2}$. There will, therefore, issue from the second lamina, $\frac{n-1}{n}-\frac{n-1}{n^2}=\frac{(n-1)^2}{n^2}$. In the same manner it may be shown, that there will issue from the third lamina, $\frac{(n-1)^3}{n^3}$. Hence, the series expressing the decreasing quantities of light, is $\frac{n-1}{n}, \frac{(n-1)^2}{n^2}, \frac{(n-1)^3}{n^3}$, &c. which is evidently a series in geometrical progression.†

719. The shadow of a globe that is illuminated by an equal globe, is cylindrical, and indefinitely long. The shadow of a less globe, illuminated by a greater, (as of the earth, or of the moon, illuminated by the sun,) is a cone of finite length, whose dimensions may be easily computed when the diameters and distances of the globes are known. And, lastly, the shadow of a globe, illuminated by one that is smaller, extends itself indefi-

* Herschel on Light.

† Barlow.

nitely in a truncated cone, perpetually enlarging. These several truths will be readily understood by referring to Fig. 239.

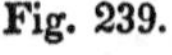
Fig. 239.

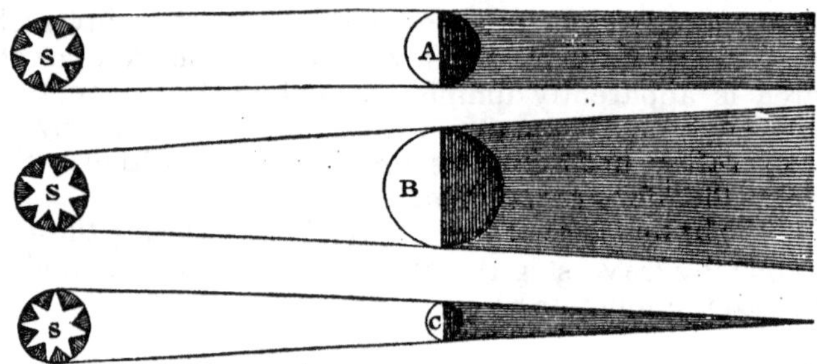

Light, when it impinges on smooth surfaces, is *reflected* back into the same medium, and when it passes out of one medium into another, it is bent out of its former course, or *refracted.* The laws of reflexion and refraction constitute, severally, important departments of the science of Optics, and to these our attention will now be directed.

CHAPTER II.

OF THE REFLEXION OF LIGHT.

620. Light *is said to be reflected when, on impinging upon any surface, it is turned back into the same medium.*

Instruments employed as reflectors are divided into *mirrors* and *speculums.* The name mirror is applied to reflectors made of *glass* and coated with quicksilver, as common looking-glasses: the word speculum, is applied to a *metallic* reflector, such as those made of silver, steel, tin, or of a peculiar alloy, called speculum metal. As the light which falls on glass mirrors, is intercepted by the glass before it is reflected from the quicksilvered surface, a speculum, or a reflector of polished metal, is that supposed to be employed in optical experiments, unless the contrary is specified. Such a surface, indeed, is to be understood where the word mirror is used without distinction.

The surface of the mirror or speculum, may be either plane, concave, or convex, and the reflector is denominated accordingly.

* That part of Optics which treats of reflected light, is sometimes denominated *Catoptrics,* (Κατὰ,) and that part which treats of refracted light, *Dioptrics,* Διὰ.

A ray of light before reflexion, is called the *incident* ray. The angle made by an incident ray, at the surface of the reflector, with a perpendicular to that surface, is called the *angle of incidence*; the angle made by the reflected ray, with the same perpendicular, is called the *angle of reflexion*. Thus, in Fig. 240, if MN represents the reflecting surface, DC a perpendicular to it at the point C, AC the incident, and BC the reflected ray; then ACD will be the angle of incidence, and BCD the angle of reflexion.

Fig. 240.

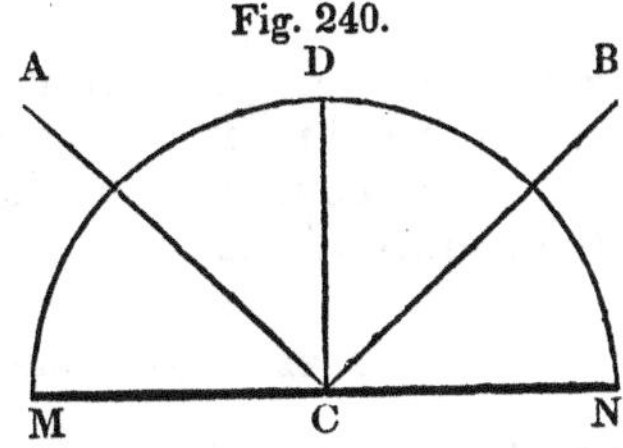

721. Experiments on light are usually conducted in a room which can be made dark with close shutters, one of which is perforated with a circular hole, an inch or two in diameter, for admitting a beam of light. This opening is rendered smaller to any required degree by covering it with a piece of board or metallic sheet, having a smaller aperture. And as the sun may not shine directly into the shutter at the time required, a mirror is sometimes attached to the outside of the shutter, so contrived that, by means of adjusting screws, it may be made to turn the rays of the sun into the opening, and to give them a horizontal or any other required direction. The course of the rays is rendered palpable to the eye, by the illuminated particles of dust that are floating in the air.

722. *The angles of incidence and reflexion are in the same plane, and are equal to each other.*

Let a ray of light AC, (Fig. 240,) admitted into a dark chamber as above, be incident upon a horizontal speculum MN at the point C, to which the line CD is perpendicular, and let CB be the reflected ray. Then if the plane surface of a board or a metallic plate, be made to coincide with the incident ray and the perpendicular, it will be found to coincide also with the reflected ray, showing that the three rays are in the same plane. Again, if, from the point C, with the radius CA, a circle be described, on measuring the arcs subtended by the angles of incidence and reflexion, they will be found to be exactly equal to each other.* The following corollaries will be evident on consideration: That the complements of the angles of incidence and reflexion, are also equal; that the reflected ray may be taken for the incident ray, and vice versa; and that, if the incident ray be perpendicular to the reflecting surface, it will be reflected back in the same

* An ingenious apparatus is described by Biot, (*Précis Elém.* tome II, 136,) by which this experiment may be performed with the utmost degree of precision; the results are as enunciated in the proposition.

line. The angles of incidence and reflexion are also equal when the reflexion takes place from a concave or convex surface; for the reflexion being from a *point*, the curve and tangent plane at that point coincide, and have both the same perpendicular, viz. the radius of the curve.

REFLEXION OF LIGHT FROM PLANE MIRRORS.

723. *When rays of light are reflected from a plane surface, the reflected rays have the same inclination to one another, as their corresponding incident rays.*

Case 1. *Parallel rays.*—Let RS (Fig. 241,) be the reflecting surface; AB, CD, the incident, BG, DH, the reflected rays. Then the angle CDR=ABR: but CDR=HDS, and ABR=GBS. Therefore, HDS=GBS, and hence BG and DH are parallel.

Fig. 241.

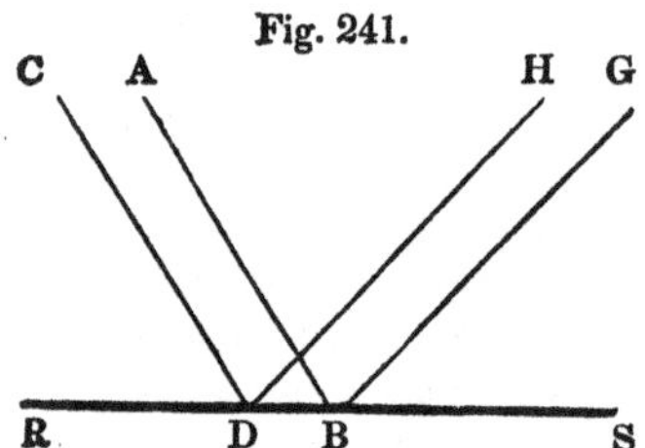

Fig. 242.

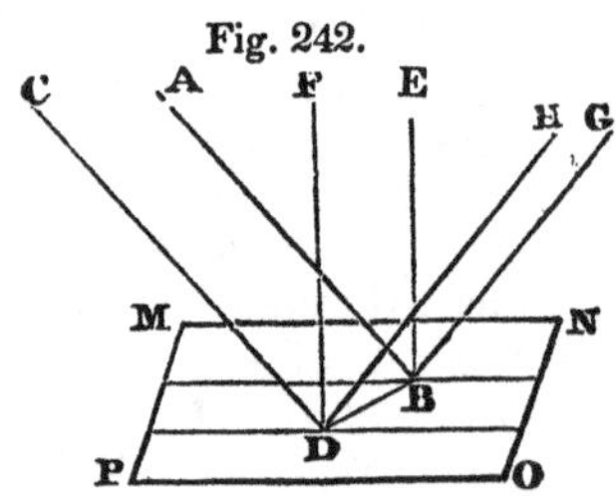

In the foregoing example, the angles of incidence are supposed to be in the same plane; but where these angles are in different planes, let AB, CD, (Fig. 242,) be two parallel rays incident upon the plane mirror MNOP, having their angles of incidence in different planes; from their points of incidence, B, D, draw the perpendiculars BE, DF, and let BG, in the plane that passes through AB and BE, be the reflected ray; join BD, and let DH be the intersection of the two planes, CDH and GBDH. Since BE, DF, are both drawn perpendicular to the same plane, they are parallel;* and as AB and CD are parallel by supposition, the angles of incidence, ABE, CDF, are equal.† Because EB, FD, and AB, CD, are parallel, the planes ABG, CDH are also parallel,‡ and they are intersected by the plane GBDH; consequently DH is parallel to BG,§ and EBG=FDH. But EBG=ABE=CDF ∴ FDH=CDF, and hence DH is the reflected ray, and it was before proved to be parallel to BG, the other reflected ray.

Case 2. *Diverging Rays.*—Let RAB (Fig. 243,) be a pencil of diverging rays, incident upon the plane mirror PB; and from R draw RF perpendicular to PAB, and cutting the mirror in P. Let AD be the reflexion of an incident ray RA, and produce

---

* Euc. 6, 2, Sup. † 9, 2, Sup. ‡ 13, 2, Sup. § 14, 2, Sup.

Fig. 243.

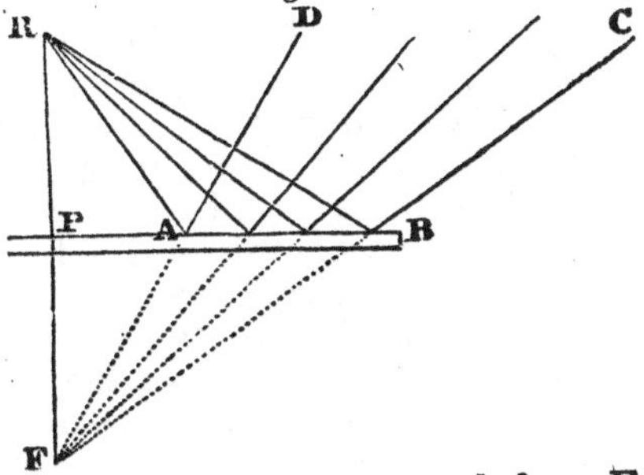

DA backward to F. Then PAR=BAD=PAF; consequently, in the right-angled triangles PAR, PAF, the angles are all equal, and PA common; hence RP=PF, that is, the reflected ray proceeds as if it came from a point F, on the other side of the mirror, and from the same distance from it as R. In like manner it may be shown, that all the other rays will proceed as if they diverged from F, and therefore F is the *virtual focus*, or imaginary radiant, of all the reflected rays. Since PRA=PFA, it may be shown in the same way that PRB=PFB; hence, taking equals from equals, the remainder AFB=ARB, that is, the rays after reflexion have the same inclination as before.

Case 3. *Converging Rays.*—If DA, CB, (Fig. 243,) constitute a pencil of incident rays converging to the point F, it follows from the above reasoning that they will converge to the focus R after reflexion.

724. *Parallel rays, incident upon a concave mirror, and near its axis, are reflected to a focus equidistant from the surface and the center of the mirror.*

Fig. 244.

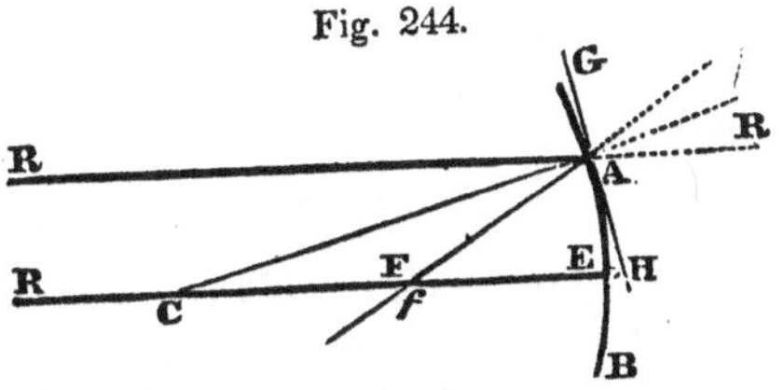

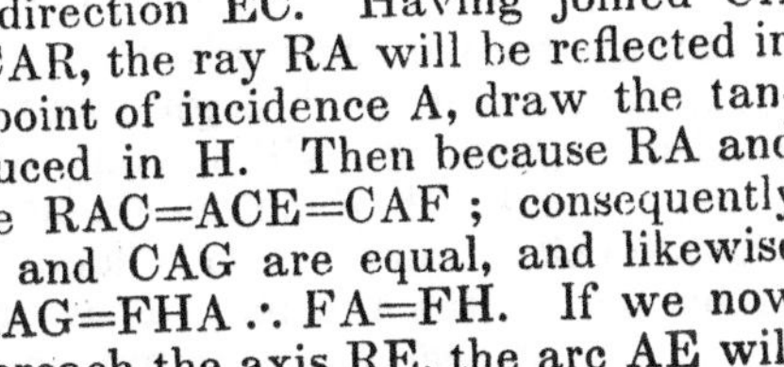

Let RA, RE, (Fig. 244,) be parallel rays incident upon the spherical mirror AEB, whose center is C. The ray RE, passing through the center C, and therefore falling perpendicularly on the mirror at E, will be reflected in the direction EC. Having joined CA and made the angle CAF=CAR, the ray RA will be reflected in the direction AF. At the point of incidence A, draw the tangent GH, cutting CE produced in H. Then because RA and RE are parallel, the angle RAC=ACE=CAF; consequently CF=FA. But since CAH and CAG are equal, and likewise CAF and CAR, ∴ FAH=RAG=FHA ∴ FA=FH. If we now suppose the ray RA to approach the axis RE, the arc AE will diminish, and its secant CH will ultimately become equal to the radius CE, and then FH will be equal to FE, and of course FA or FC will equal FE.

The foregoing proposition is applicable to such rays only as are exceedingly near to the axis of the mirror CE. As the parallel rays are more remote from the axis, the focus F approaches nearer and nearer to the point E, until, when the arc EA becomes equal to 60°, F coincides with E; for then the angle CAF and ACF being each equal to 60°, the remaining angle of the triangle ACF must also be equal to 60°; consequently, CF must equal CA, and of course the point F will coincide with E.

If several beams of parallel rays be incident nearly perpendicularly upon a spherical mirror, the foci of the reflected rays will be in a spherical surface *concentric* with that mirror. For since each focus (Fig. 244) is, by the proposition, equidistant from the center and from the surface of the mirror, the distances of all the foci from the mirror must be exactly the same; that is, they must be in a surface concentric with that of the mirror.

Rays falling on any part of a concave mirror parallel to its axis, will all be brought to a focus at the same point, if the curvature of the mirror be that of a *parabola*. For then, according to a property of the parabola, all diameters, or lines parallel to the axis, and a line drawn from the focus to the point where the diameters meet the curve, make equal angles with the tangents at those points.* But these equal angles are the complements of the angles of incidence and reflexion, which are also equal. Wherefore rays, parallel to the axis, will be reflected into the lines which all meet at one and the same focus.

725. Diverging Rays, *incident upon a concave mirror, are collected into a focus, which changes its situation as the distance of the radiant from the mirror is changed, conformably to the law, that the angle of incidence is equal to the angle of reflexion made with the radius of concavity.*†

If the radiant point be further from the mirror than the center, as at A, (Fig. 245,) then the focus will be between the center and the mirror; if the radiant be at the center, the rays will return to the center again; if the radiant comes still nearer to the mirror, the focus will pass to the other side of the center and continue to recede from it, until the radiant has arrived at the place of the focus of parallel rays, when the focus on the other side of the center will be thrown to an infinite distance; and finally, if the radiant be brought nearer to the mirror than the principal focus, the rays will go out diverging, and will never come to a focus;—all which is evident from the general law of reflexion, the situation of each reflected ray being easily determined by that of the incident ray with respect to the perpendicular, that is, the radius of the mirror. Thus, the rays emitted from A will be collected in *a*; those from C will return to C again; those from *a* will be collected in A; those from F, the

* Conic Sections.

† The several cases will be the more easily remembered, by keeping in mind the situation of the incident ray relatively to the perpendicular; that is, the radius of concavity.

focus of parallel rays, will be reflected into the parallel lines *cf*, *cf*; and those from D into the diverging lines *cd*, *cd*, which will appear to proceed from A′. Again, if the radiant is first placed near the mirror, and removed from it by successive steps, just the converse effects will follow. Hence, the radiant and its corresponding focus are denominated *conjugate foci*. In the foregoing experiment, the conjugate foci approach one another—meet in the center of concavity—pass to different sides of that center—and afterward recede from one another, until the focus nearest to the mirror arrives at the focus of parallel rays, when the two conjugate foci are separated to the greatest possible distance from each other.

Fig. 245.

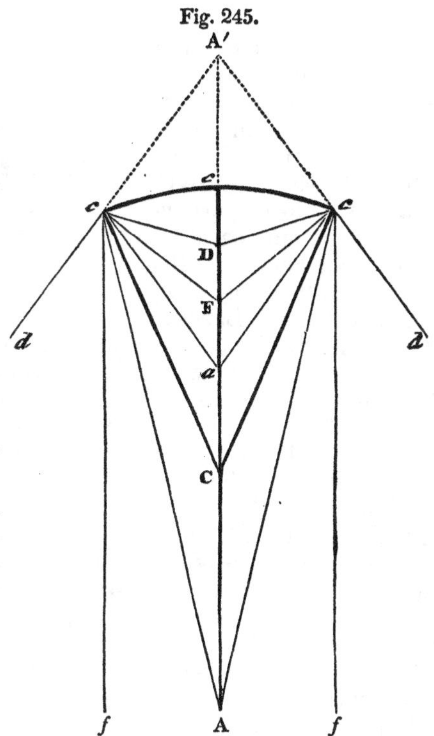

726. *Parallel rays incident upon a* CONVEX MIRROR, *are made to diverge as from a point behind the mirror.*

Let MN (Fig. 246) be a convex mirror whose center is C, and let AM, AD, AN, be parallel rays falling upon it. Continue the lines CM and CN to E, E, and ME, NE, will be perpendicular to the surface of the mirror at the points M and N. The rays AM, AN, will therefore be reflected in the directions MB, NB, the angles of reflexion EMB, ENB, being equal to the angles of incidence EMA, ENA. By continuing the reflected rays BM, BN backward, they will be found to meet behind the mirror at F, their virtual focus.

Fig. 246.

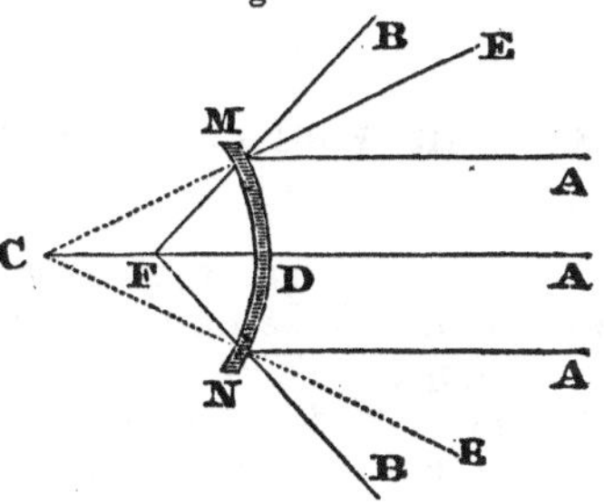

727. *Diverging rays incident upon a convex mirror, are made to diverge as from a point behind the mirror, and nearer to it than the virtual focus of parallel rays.*

Let MN (Fig. 247,, be a convex mirror, C its center of convexity, and AM, AN, rays diverging from A, which fall upon the mirror at the points M, N. The lines CME and CNE, will be, as before, perpendicular to the mirror at M and N; and, consequently, if we make the angles of reflexion EMB, ENB, equal to the angles of incidence EMA, ENA, then MB, NB, will be the reflected rays, which, when continued backward, will meet at F, their virtual focus behind the mirror. By comparing Figs. 246 and 247, it will be obvious that the ray AM in Fig 247, is further from ME than in Fig. 246, and consequently, the reflected ray MB must also be further from it. Hence, as the same is true of the ray NB, the point F, where these rays meet, must be nearer D in Fig. 247, than in Fig. 246; that is, in the reflexion of diverging rays, the virtual focal distance DF, is less than for parallel rays. For the same reason, if we suppose the radiant point A to approach the mirror, the virtual focus F will approach it; and when A arrives at D, F will also arrive at D. In like manner, if A recedes from the mirror, F will recede from it; and when A is infinitely distant, or when the rays become parallel, as in Fig. 246, F will reach the place of the virtual focus of parallel rays.

Fig. 247.

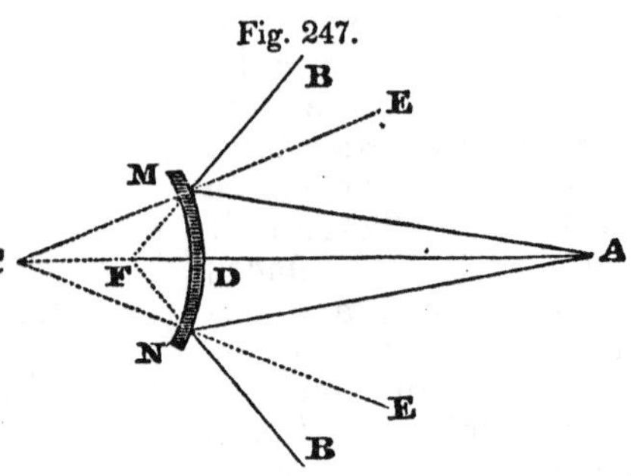

CHAPTER III.

OF IMAGES FORMED BY PLANE, CONCAVE, AND CONVEX MIRRORS.

728. *In all mirrors, plane or spherical, the place of the imaginary radiant, is the intersection of the perpendicular from the radiant point of the object to the mirror, with any reflected ray.*

All the rays which diverge from any point in the object before reflexion, appear, after reflexion, to diverge from one and the same point, namely, from the imaginary radiant. (See Fig. 243.) And since the perpendicular ray RP is reflected back to R, the imaginary radiant must be in that line produced; and since the imaginary radiant must likewise be in any other reflected ray as AD, produced, it must be in the intersection of the two lines.

The same reasoning would apply to a concave or a convex surface, since the reflexion at any point of such a surface, is the same as it would be from a plane surface which is a tangent to the curve at that point.

729. *When any object is placed before a* PLANE *mirror, the image of it appears at the same distance behind it, of the same magnitude, and equally inclined to it.*

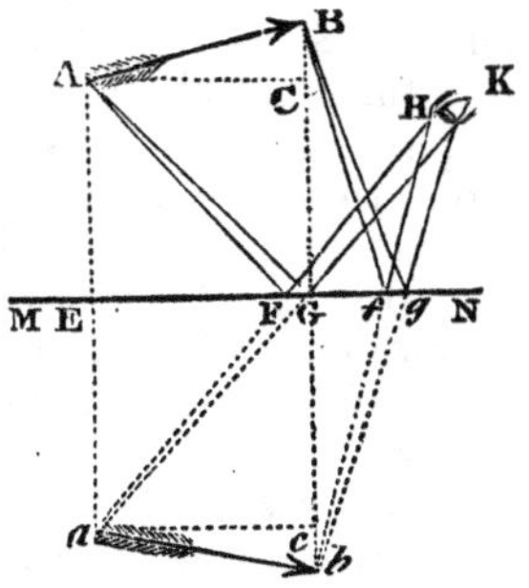

Let MN (Fig. 248,) be a plane mirror, and AB an object placed before it, and let the position of the object be such that the reflected rays may enter an eye placed at H. From A and B let fall upon the mirror the perpendiculars A*a*, B*b*. Then the rays AF, AG, diverging from A, will be reflected in the lines FH, GK, as if they came from the point *a*, so situated that EA=E*a*, (Art. 723;) and hence the point A will be seen at *a*, *as far behind the mirror as* A *is before it.* In like manner, it may be shown that the point B of the object, will appear at *b*, so situated that GB=G*b*. By taking any other rays at pleasure, divergent from any other point of the object AB, it may, in a similar manner, be shown, that they will have their foci in points of the line *ab*, formed by drawing perpendiculars from the given points of the object. Now, since GB=G*b* and *a*G*b*=BGK=AGB, G*a*=GA, ∴ AB=*ab*. That is, the *magnitude* of the image equals that of the object. From A and *a* draw the perpendiculars AC, *ac*; then the angle BAC=*bac*, that is, the object and the image are equally *inclined* to the mirror.

Fig. 248.

Hence objects that are perpendicular to the horizon, seen in a plane mirror, appear inverted, the highest point of the object being the lowest point of the image.

If a plane mirror is inclined to the horizon at an angle of 45°, an object parallel to the horizon will appear erect, and perpendicular objects will appear horizontal. For, since the image has the same inclination to the mirror as the object has, when the angle made by the object is 45°, that made by the image must be 45° also, and both together must make 90°.

730. *If the image of an object is seen by reflexion, from two plane mirrors, the reflexion being in a plane perpendicular to their common section, the angular deviation of the image from the object, will be equal to twice the inclination of the reflecting mirrors.*

Let AB, CD, (Fig. 249,) be two plane mirrors, inclined to each other in any angle. Produce AB and CD until they meet in G, then AGC will be the *angle of inclination of the mirrors.* Let SB be a ray of light proceeding from any distant object, as a star, and reflected from AB to CD, and from CD to the eye at H; then the image of S will be at O in the line HD produced. Also produce SB to H, then SHO will be the *angular deviation of the image from the object,* the line of common section of the mirrors passing through G, perpendicular to the plane AGC. It is required to prove that the angle SHO is double the angle AGC. Because HBG=ABS=GBD ∴ HBD=2GBD. In like manner BDO=2BDC.

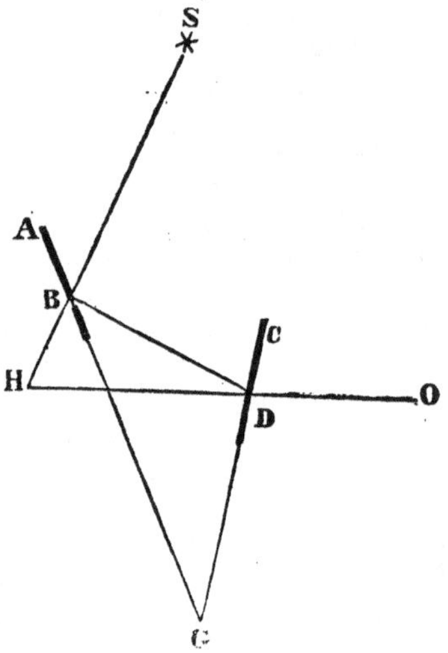

Fig. 249.

But, SHO=BDO−HBD=2BDC −2GBD=2BGD.

Therefore, SHO=2BGD.

Hence, when a plane mirror revolves on an axis, the angular velocity of the reflected ray is double that of the mirror. Therefore, by turning a mirror through 45°, the image is carried through 90°, so that a mirror set at an angle of 45° with the horizon, represents horizontal objects in a perpendicular position, and perpendicular objects on a horizontal level, agreeably to the last article

Upon the foregoing proposition depends a principle employed in Hadley's Quadrant, in which two mirrors, inclined to one another, measure the angular distance between two objects, by indicating the arc through which the image of one of them must be made to pass, in order to carry it over that distance.* Thus, if in order to make the image of a star descend to the horizon, the mirror that reflects it must be turned 20°, the altitude of the star is 40°. Hence, an *octant* only is required to measure a quadrant, or an angle of 90°.

731. *When the object is parallel to a plane mirror, the length or breadth of that part of the mirror upon which the image appears, is to the length or breadth of the object, as any reflected ray is to the sum of the incident and reflected rays.*

If the object DE (Fig. 250,) is parallel to the mirror AB, and the image LM is seen by the eye at C, then FN, the length of that part of the mirror which is taken up by the image, subtends the angle LCM, under which the image appears. Now the length of the image LM is equal to the length of the object DE. (Art. 729.) And because FN is parallel to LM,

* See Day's Navigation and Surveying, Art. 91.

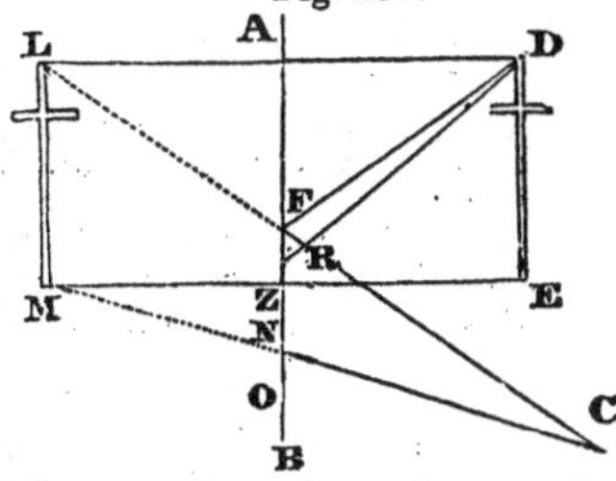

Fig. 250.

∴ FN : LM : : CF : CL. But CF is the reflected ray, and CL is equal to the sum of the incident and the reflected rays.

Hence, an object which is not wholly visible when the eye is at a certain distance from the mirror, may become so by bringing the eye nearer to the mirror; for in proportion as the ratio of the reflected ray to the sum of the incident and reflected rays is diminished, in the same proportion is the part of the mirror required to comprehend the entire image diminished.

If a spectator sees himself entirely in a plane mirror placed parallel to him, the mirror must be half as long as himself. For then the incident and reflected rays will be equal, and consequently the latter equal to half the sum of the two, and hence the mirror must be half the length of the object.

732. *When an object is placed between two parallel plane reflectors, a row of images is formed in each mirror, appearing in a straight line behind each to an indefinite extent.*

Let there be two plane reflectors, parallel to each other; and let an object, a candle for example, be placed between them. An image of the candle will be formed in each mirror, as far behind it as the object is before it. Again, each of these images becomes in its turn a new object to the opposite mirror,

Fig. 251.

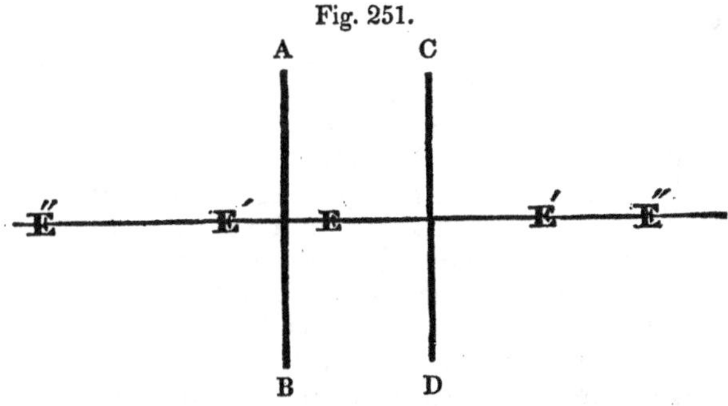

and forms a corresponding image as far behind that mirror as it is itself before it, and thus the images are repeated in a right line until the light becomes too feeble to be visible. Thus let AB, CD, (Fig. 251,) be two plane mirrors, and E an object between them; two images will be formed of E at E′ and E′, from which *virtual foci* the light will emanate, and will strike upon the opposite mirrors, respectively, in the same manner as if it came from luminous bodies placed at those points; hence two images of E' and E' will be formed at E'' and E''; and thus a succession of images will arise to an indefinite extent; but since a part of the light is lost at every reflexion, each succeeding image is fainter than the preceding. The *Endless Gallery* is formed on this principle. It consists of a box, in the opposite sides of which are placed two parallel reflectors, and between them a number of objects are placed, which are repeated in an endless succession.

733. *If an object be placed between two plane reflectors inclined to each other, the images formed will lie in the circumference of a circle, whose center is in the intersection of the two planes, and whose radius is the distance of the object from that point.*

Fig. 252.

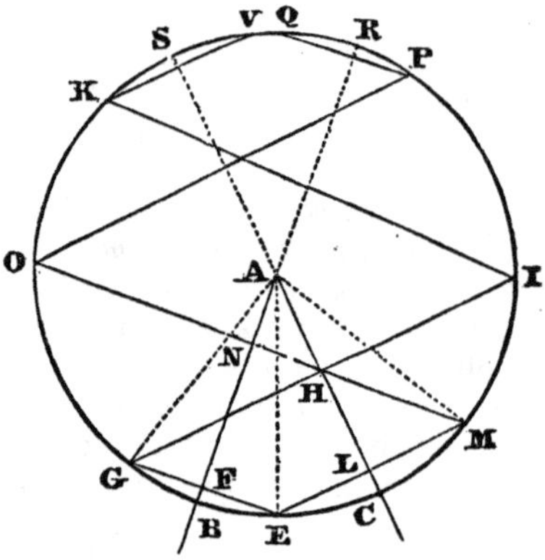

Let AB, AC, (Fig. 252,) be two plane reflectors inclined at the angle BAC, and E an object placed between them. Draw EF perpendicular to AB, and produce it to G, making FG=EF; then the rays which diverge from E and fall upon AB will, after reflexion, diverge from G; or G will be an image of E.* From G, draw GH perpendicular to AC, and produce it to I, making HI=GH, and I will be a second image of E, &c. Again, draw ELM perpendicular to AC, and make LM=EL; also draw MNO perpendicular to AB, and make NO=MN, &c. Therefore, the successive images formed, beginning on the side of AB, are G, I, K, V; and those on the side of AC, are M, O, P, Q. Then, since EF is equal to FG, and AF common to the triangles AFG, AFE, and the angles at F are right angles, AG is equal to AE. In the same manner it may be shown that AM, AO, AI, &c., are severally equal to AE; and of course, the points G, M, O, I, &c. are in the circumference of a circle whose center is A and radius AE.

If the angle BAC is finite, *the number of images will be limited.* For BA and CA being produced to R and S, the rays which are reflected from either surface, diverging from any point Q and between S and R, will not meet the other reflector, since it is not before either reflector, but behind both.

* The eye is supposed to be situated between the planes of the mirrors. The learner may find some difficulty in conceiving how rays of light can proceed from G, a point behind the mirror; but, it must be recollected that this is only an *imaginary* point, from which the rays that are reflected from the mirror into the line GH *seem* to emanate.

If we consider the whole angular opening of the mirrors, namely, the *sector* ABC, as the object, images of it will be formed in a circle as of any other object.

If the inclination of the mirrors gradually diminishes, the magnitude of the sectors will also be diminished, and the number of repetitions of them increased in the same proportion. The number of images of any object placed between the mirrors, will be in like manner increased as the inclination of the mirrors is diminished; and since, when the angle of inclination is very small, the mirrors approach the situation of parallel mirrors, so the number of images approach to infinity.

734. As the learner sometimes experiences a difficulty, in conceiving clearly the course which the rays take in forming the successive images, we subjoin a brief illustration. It may be premised, that *whatever turns a ray of light may take, in passing from an object to the eye, the object will be seen in that direction in which the light finally meets the eye.* Thus in forming some of the foregoing images, a ray of light undergoes three or four reflexions, in different directions, but still the last ray that meets the eye will fix the position of the image.

Let RS, RT, (Fig. 253,) be two plane mirrors inclined to each other, and Q a luminous object between them. Let the eye be situated at any convenient point as O, and let A, B, C, D, be the several images. Draw a line from O to any one of these images, as D, and from the point where this line intersects the speculum, draw a line to the next preceding image, and from the point where this line meets the speculum, draw a line to the next preceding image, and so on back to the object Q.

Fig. 253.

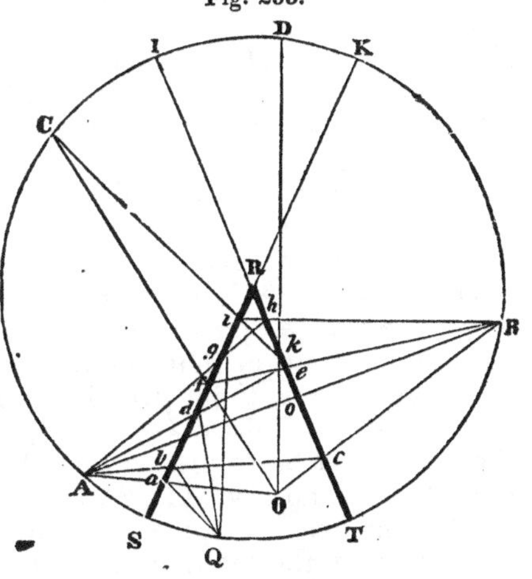

Now the image A is seen by the ray Q*a*, *a*O; B, by Q*b*, *bc*, *c*O; C, by Q*d*, *de*, *ef*, *f*O; and D, by Q*g*, *gh*, *hi*, *ik*, *k*O.

For, since S*a*Q=S*a*A=R*a*O ∴S*a*Q=R*a*O, and therefore Q*a* being the incident, *a*O must be the reflected ray; and the same may be proved in each of the several cases of reflexion.

735. It is found by experiment, that when a pencil of light is incident perpendicularly upon *water*, only 18 rays out of 1000

are reflected, while the greater part of the remaining rays are transmitted. As the angle of inclination is increased, the proportion of rays reflected is also rapidly increased, till at an angle of 75° the reflexion is 211 rays; at 85°, 501; and at 89°, 692. In *glass*, 25 out of 1000 are reflected at a perpendicular incidence; and the glass always reflects more light than water, till we reach very great angles of incidence, such as 87½°, when it reflects only 584 rays, while water reflects 614.

736. *The image is inverted, when the rays of light which come from one extremity of the object, cross those which come from the other extremity, before they meet in the corresponding points of the image.*

Fig. 254.

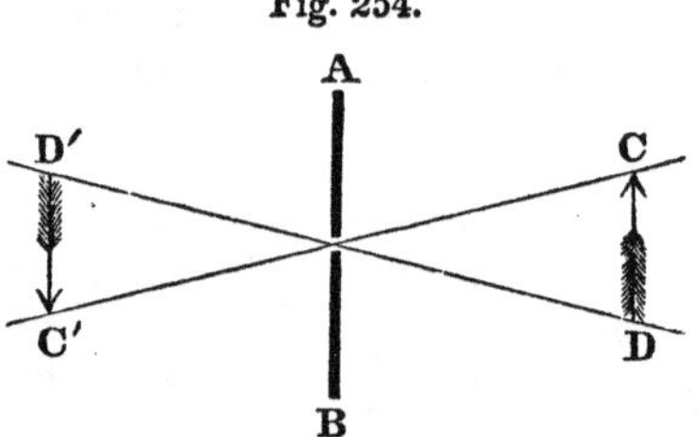

Thus, let AB (Fig. 254,) represent a board or screen having a small opening through which rays of light may pass. Since light passes in right lines, it is obvious that rays proceeding from C, the highest point in the object CD, will pass to C′, while rays from D, the bottom of the object, will pass to D′, crossing each other at the orifice, and thus forming an inverted image of the object, D′C′. In like manner rays coming from all the intermediate points, on opposite sides, will cross before they take their respective places in the image.

737. *When an object is placed before a* CONCAVE *mirror, the image of it has various magnitudes and positions, depending on the distance of the object from the mirror.**

1. When an object is *between the mirror and the focus of parallel rays*, the image appears behind the mirror, and is more distant from it, and larger than the object. Let MN (Fig. 255,) be the concave mirror, F its principal focus, C its center, and AB the object. From C draw CA*a*, CB*b*, passing through A and B, and let the object be so placed that the reflected rays will reach the eye at H. The rays AD, AG, proceeding from A, will be reflected to the eye at H, making equal angles with the perpendicular CD, and they will diverge as if they had come from a remote point *a*, situated in the intersection of those rays with the perpendicular CA*a*, (Art. 723.) In like manner, the rays B*d*, B*g*, will enter the eye at H, as if they had proceeded from *b*, a point where they intersect CB*b*. Since the rays diverge less than before reflexion, (Art. 725,) these points, *a*, *b*, will be further

* These different places of the image depend on the principles demonstrated in Art. 725, and they will be easily remembered by considering the relation of the incident rays to the perpendicular, that is, the radius of the concavity, conformably to the general laws of reflexion.

from MN than A and B are, and from similar triangles, the image ab will be greater than AB, in the ratio of Cb to CB.

Fig. 255. Fig. 256.

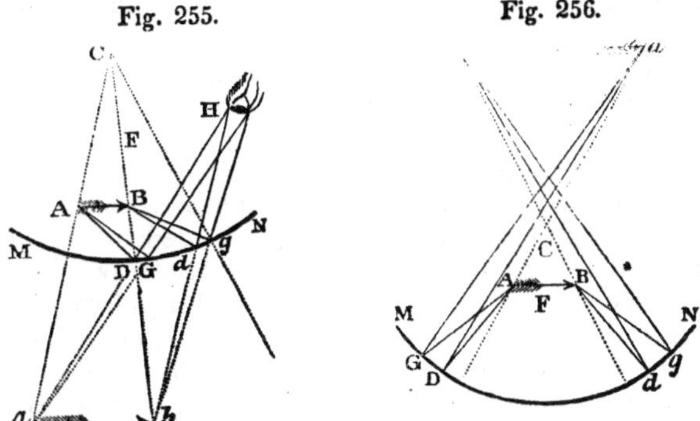

2. When the object is placed *in the principal focus*, the rays will go out parallel, and will never come together so as to form an image of themselves, nor will they proceed from any point behind the mirror, so as to form an imaginary image, like that of case 1.

3. When the object is situated *between the principal focus and the center*, the image is formed on the other side of the center, and is inverted and larger than the object. Let MN, (Fig. 256,) be the mirror, C its center, F its focus, and AB the object. Through C draw the lines CA, CB, and continue them backward to a and b. Then let AD, AG, and Bd, Bg, be two sets of rays flowing from the extremities A, B. These rays will, after reflexion in the directions Da, Ga, and db, gb, meet the perpendicular lines Ca, Cb, in the points a, b, at a greater distance from the mirror than the center C, (being reflected to the other side of the radius,) and will there form an image of those points of the object. (Art. 725.) The image is therefore more remote from C than the object is, and the size of the one will be to that of the other as aC is to AC ; that is, the image will be larger than the object.*

4. When the object is situated *beyond the center*, the image will then be formed between the center and the principal focus, and will be inverted and less than the object. This is the converse of the preceding, and will be made obvious by considering the rays as first flowing from ab and converging to AB. When the middle part of the object is placed *in the center* of the mirror, the object will coincide with the image, and the image will be inverted. That the center of the image will coincide with that of the object, may be inferred from Art. 725 ; the reflected ray being returned back in the incident ray or perpendicular ; and

* For, since the rays AD, Da, make equal angles with the radius CD, Ca is greater than CA, and consequently, from similar triangles, $ab > $AB. See Fig. 245.

rays proceeding from the extremities of the object A and B, will make equal angles with this perpendicular on the different sides of it, and therefore an inverted image will fall back upon the object.

738. The following experiments, which may be easily repeated, will serve to render familiar the different cases above demonstrated.

We will suppose a lighted candle to be placed very near to a concave mirror: it will form no image before it, because the rays go out still diverging, but we see an enlarged image of the candle behind the mirror. As the radiant is withdrawn from the mirror toward the principal focus, the image will rapidly recede on the other side of the mirror, and grow larger and larger until the radiant reaches the focus, when the image will suddenly disappear. On removing the radiant a little further, the image will be found at a distance before the mirror, and very much enlarged. As the radiant approaches the center, the image approaches it rapidly on the other side of it, constantly diminishing in size, until they both meet and coincide in the center. Removing the radiant still further, the image appears again between the center and the focus, diminished in size, and slowly approaching the focus as the radiant recedes, but never reaches it, unless when the radiant may be considered as at an infinite distance, as in the case of the heavenly bodies. By applying principles already explained, the learner will readily account for these various appearances.

739. *When any object is placed before a* CONVEX *mirror, the image of it appears nearer to the surface of the mirror than the object is, and of a less size.*

Let MN (Fig. 257) be a convex mirror whose center is C, and AB the object; and let the position of the object be such that a reflected ray may enter the eye placed at H. From C draw CA, CB, cutting the mirror MN in E and F. The rays AF, AG, will be reflected to H and K, making equal angles with the perpendicular passing from C through F and G, and will therefore enter the eye as if they came from some point as *a*, at the intersection of these rays with the perpendicular AC; consequently, the point A of the object will have its image visible at *a*. In like manner, rays B*f*, B*g*, falling upon the points *f*, *g*, will be reflected to the eye as if they came from *b*, the point where they intersect the perpendicular drawn from B to C. Now, as the reflected rays diverge more than the incident ones, the point *a* will be nearer the

Fig. 257.

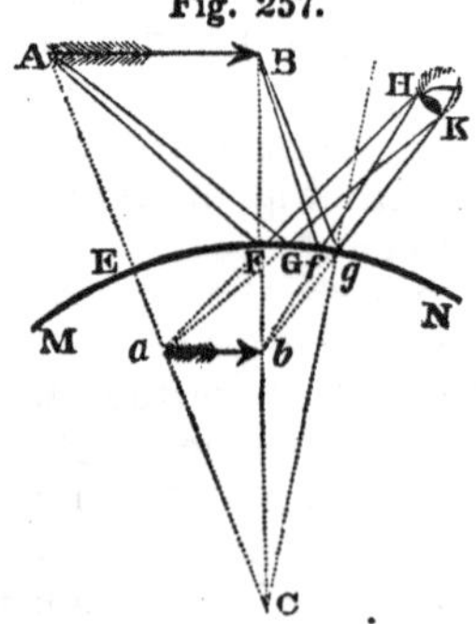

mirror than A, and the image *ab* will be less than the object AB, in the ratio of C*b* to CB.

740. *In spherical mirrors, concave or convex, the diameter of the object is to the diameter of the image, as the distance of the object from the center is to the distance of the image from the center; and also as the distance of the object from the surface is to the distance of the image from the surface.*

It is evident, from Figs. 256, 257, that the object and the image subtend each the same angle, the former at the center and the latter at the surface; and as they are parallel to one another, their lengths are as their distances from these points respectively.*

741. One who looks into a concave mirror, sees his own face varied in the following manner.

When he holds the reflector near to his face, he sees his image *distinct*, because the rays come to the eye diverging, (which is their natural state with respect to near objects,) and *enlarged*, because, as the rays diverge less than before, the image is thrown back to a greater distance behind the mirror than the object is before it, and the magnitude is as that distance, by Art. 740. As he withdraws the eye, the image grows larger and larger, until the eye reaches the focus. From the focus to the center, no distinct image is seen, because the rays come to the eye converging, a condition incompatible with distinct vision. At the center, the eye sees only its own image, since the image is reflected back to the object and coincides with it. Beyond the center, his face will be seen on the other side of the center before the mirror, (though habit may lead him to refer it to a point behind it;) and it will be *diminished*, being nearer to the mirror than the object is, (Art. 740,) and *inverted*, because the pencils of rays from the extreme points of the object, cross each other in the focus.†

742. *By the reflexion of light from concave mirrors, there are exhibited curves of a peculiar kind, called* CAUSTICS BY REFLEXION.‡

Let MBA (Fig. 258) be a concave spherical mirror whose center is C, and whose focus for parallel and central rays is F. Let RMB be a pencil of light falling on the upper half, MB, of the mirror, at the points 1, 2, 3, 4, &c. If we draw radii to all these points from the center C, and make the angles of reflexion equal to the angles of incidence, we shall obtain the directions and foci of all the incident rays. The ray R1, near the axis RB, will

---

* Euc. VI, 4.

† These phenomena may be all observed with an ordinary concave shaving glass.

‡ Called caustics, or *burning* points, because, since the rays of light or heat cross each other in the points that make up these curves, the intensity of light or heat is twice as great there as elsewhere.

have its conjugate focus at f, between F and the center C. The next ray, R2, will cut the axis nearer F, and so on with all the rest, the foci advancing from C to F. By drawing all the re-

Fig. 258.

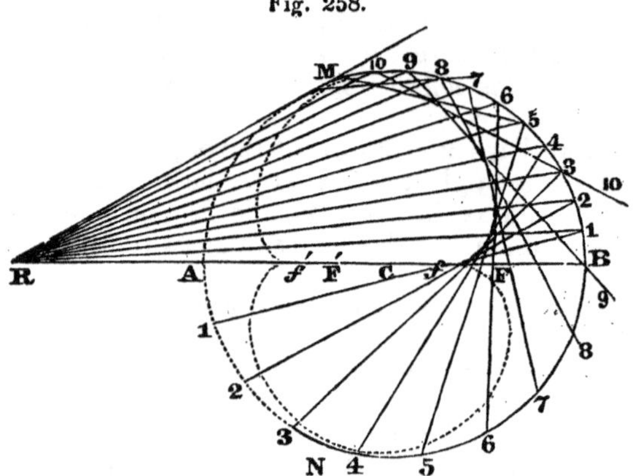

flected rays to these foci, they will be found to intersect one another, as in the figure, and to form by the intersection the *caustic curve* Mf. They are so called because, in consequence of the intersections of the rays in the points forming these curves, those points are brighter, or, where heat is reflected, hotter, than the contiguous spaces. If the light had also been incident on the lower half of the mirror, a similar caustic, shown by a dotted line, would also have been formed between N and f. If we suppose, therefore, the point of incidence to move from M to B, the conjugate focus of any two contiguous rays, or an infinitely slender pencil diverging from R, will move along the caustic from M to f.*

743. Concave mirrors, in consequence of the property they have of forming images in the air, were, in a less enlightened age than the present, frequently employed by showmen for exhibiting surprising appearances. The mirror was usually concealed behind a wall, and the object, which might be a skull, a dagger, &c., was placed between it and the wall, and strongly illuminated. The rays proceeding from the object, fell upon the mirror, and were reflected by it through an opening in the wall, and brought to a focus, so as to form an image in the same room with the spectator. If a fine transparent cloud of blue smoke is raised, by means of a chafing dish, around the focus of a large concave mirror, the image of any highly illuminated object will be depicted in the middle of it with great beauty. A dish of

* These curves may be seen on the surface of milk, placed in a white bowl or tea-cup, set in the sun.

fruit thus represented invites the spectator to taste, but the instant he reaches out his hand a drawn dagger presents itself.*

744. Concave mirrors have been used as *lighthouse reflectors*, and as *burning instruments*. When used in lighthouses, they are formed of copper plated with silver, and they are hammered into a parabolic form, and then polished with the hand. A lamp placed in the focus of the parabola, will have its divergent light thrown, after reflection, into something like a parallel beam, which will retain its intensity to a great distance.

When concave mirrors are used for burning, they are generally made spherical, and regularly ground and polished upon a tool, like the specula used in telescopes. The most celebrated of these were made by M. Villele, of Lyons, who executed five large ones. One of the best of them, which consisted of copper and tin, was very nearly four feet in diameter, and its focal length thirty-eight inches. It melted the metals, as silver and copper, and even some of the more infusible earths.

Burning mirrors, however, have sometimes been constructed on a much larger scale, by combining a great number of plane mirrors. It is supposed that it was a mirror of this kind which Archimedes employed in setting fire to the Roman fleet under Marcellus. Athanasius Kircher, who first proved the efficacy of a union of plane mirrors, went with his pupil Scheiner to Syracuse, to examine the position of the hostile fleet; and they were both satisfied that the ships of Marcellus could not have been more than *thirty* paces distant from Archimedes.

Buffon, the celebrated naturalist, constructed a burning apparatus upon this principle, which may be easily explained. He combined one hundred and sixty-eight pieces of mirror, six inches by eight, so that he could, by a little mechanism connected with each, cause them to reflect the light of the sun upon one spot. Those pieces of glass were selected which gave the smallest image of the sun at two hundred and fifty feet. With one hundred and fifty-four mirrors, he was able to fire combustibles at the distance of two hundred and fifty feet.

CHAPTER IV.

OF THE REFRACTION OF LIGHT.

745. When the rays of light pass out of one medium into another, as out of air into water, they are bent out of their previous direction; and hence,

* Hutton's Recreations

Refraction is the change of direction which light undergoes by passing out of one medium into another.

The lines which a ray describes before and after refraction are called *incident* and *refracted* rays; the angle contained between the incident ray and a perpendicular to the surface drawn from the point on which the ray falls, is called the *angle of incidence;* the angle contained between the refracted ray and the perpendicular, is called the *angle of refraction.* The angle which the refracted ray makes with its previous line of direction is called the *angle of deviation.* These several definitions the learner will easily apply to the following figure. Thus AC, (Fig. 259,) is the incident, and CE the refracted ray; ACD is the angle of incidence, ECF the angle of refraction, GCE the angle of deviation. It is a general fact, to which there are only a few exceptions, that a ray of light in passing out of a rarer into a denser medium is refracted *toward* the perpendicular to the surface; and in passing out of a denser into a rarer medium, it is refracted *from* the perpendicular. But the chemical constitution of bodies, as well as their density, sometimes affects their refracting power. Thus, inflammable bodies, as sulphur, amber, and essential oils, have a very great refracting power in comparison with other bodies; and in a given instance, a ray of light in passing out of one of these substances into another of greater density but of less refractive power, might not be turned toward, but from, the perpendicular.

Fig. 259.

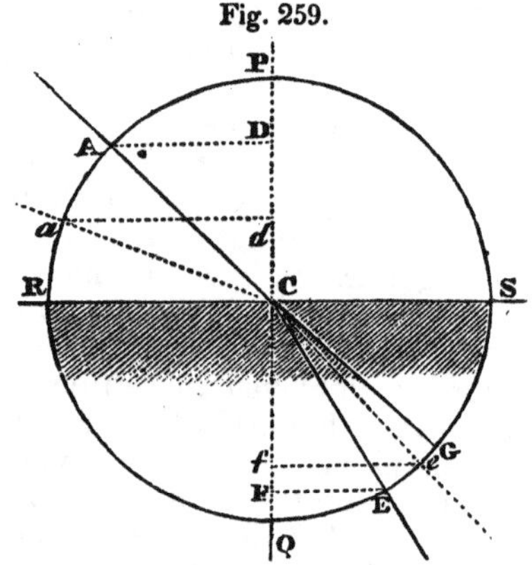

746. *When a ray of light passes from one medium into another of different density, the sines of the angle of incidence and refraction have always the same ratio to each other; and the incident and refracted rays are in the same plane.*

This proposition may be proved by experiment.* Let AC (Fig. 259,) be a ray of light incident upon the surface RS of water, or any other medium. This ray, instead of proceeding directly forward in AC produced, is bent or refracted at C into the direction CE. In like manner, another ray *a*C, incident upon the same point C, is found to be bent or refracted into the line C*e*. Through the point C draw the line PCQ perpendicular to

* For a theoretical demonstration, see Newton's Optics, or Encyc. Metropolitana.

the refracting surface RS, and upon C as a center, describe a circle APQ. If we now compare the angles of refraction with the corresponding angles of incidence, we shall perceive no particular relation between them, except that in general, one increases or diminishes with the other; but if we compare the *sines* of these angles, viz. AD with EF, and *ad* with *ef*, we shall find that the ratio of the one to the other is constant, so that AD is always to EF as *ad* to *ef*, whatever is the value of the angles of incidence or refraction. If the surface RS is that of *water*, into which a ray passes from the atmosphere, the ratio of the sines of incidence and refraction will be as 4 to 3 nearly, and this ratio will always be the same at whatever angle the ray enters the medium. From air into crown glass, the ratio is as 3 to 2; from air into *sulphur*, as 2 to 1; from air into *diamond* as 1 to $\frac{2}{5}$. (See Fig. 260.)

Fig. 260.

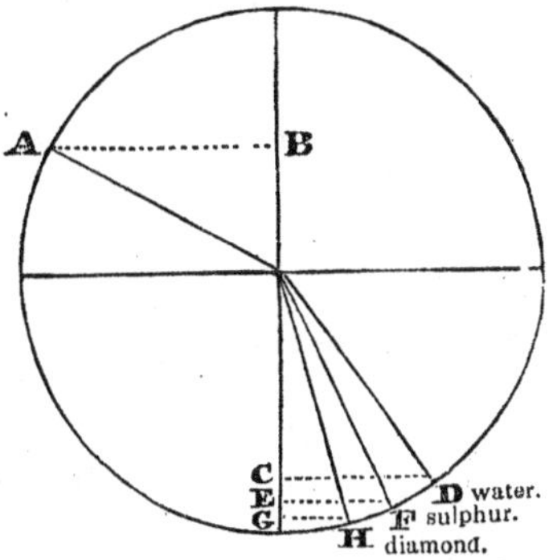

A B C E G D water. F sulphur. H diamond.

By admitting the light through a small aperture at A, (Fig. 259,) so as to pass through another aperture at C, and fall upon the bottom of the vessel at E, it will be found by experiment that the three points A, C, E, are always in the same plane, whatever be the angle of incidence ACP; that is, the incident and refracted rays are always in the same plane.

747. Supposing the sine of the angle of refraction to be always 1, then the sine of the angle of incidence will be nearly 1.33 in water, and 1.5 in glass. The sine of the angle of incidence, that of refraction being taken for unity, is called the INDEX OF REFRACTION.* Consequently it is the ratio of the sine of the angle of incidence to that of refraction. Thus the index of refraction for sulphur is 2, because when light passes out of air into sulphur, the angle of incidence is double that of refraction. Rays of light which pass perpendicularly out of one medium into another, suffer no refraction; for the sine of the angle of incidence then becomes nothing. When the ray passes in the opposite direction, that is, from a denser into a rarer medium, as from water into air, the same constant ratio is found to exist between the sines of incidence and refraction. Thus, (Fig. 259,) the light from E to C will pass into CA, and the ratio of the sines of incidence and refraction will be that of EF to AD.

We see an example of the foregoing principle in the bent ap-

* It is understood that the passage is from air into the given medium

pearance of an oar in the water, the light of the part immersed (by which it is visible) being turned from the perpendicular, and causing it to appear higher than its true place. In the same manner, the bottom of a river appears elevated, and diminishes the apparent depth of the stream. The following ancient experiment illustrates the same principle. If a small piece of silver be placed in the bottom of a bowl, and the eye be withdrawn until the piece of silver disappears, on filling up the bowl with water, the silver comes into view again.

748. *A ray of light cannot pass out of a denser into a rarer medium, when the angle of incidence is greater than that at which the sine of the angle of refraction becomes equal to radius.*

Let AC (Fig. 261,) be the ray incident upon the rarer medium RS. It will be refracted from the perpendicular DF into the direction CE, so that AD is to EF in a constant ratio. (Art. 746.) If we increase the angle ACD, the angle FCE will also increase, till the lines CE and FE coincide with the radius CS. But if beyond this position of the ray AC, the angle ACD is still further increased, it is manifest its sine is also increased; and consequently, in order that the ratio may be constant, the sine of refraction EF must also increase, which is impossible, since it is already by hypothesis equal to the radius CS. Hence it follows, that whenever the angle of incidence is greater than that at which the sine of the angle of refraction becomes equal to radius, the ray cannot be refracted consistently with the constant ratio of the sines.

Fig. 261.

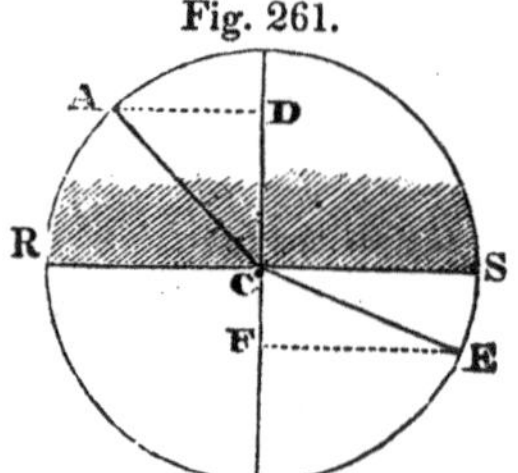

This is found to be the case by experiment; and at the angle thus indicated, all the incident rays are *reflected* from the inner surface of the denser medium, having a reflexion more brilliant than what can be produced from any metallic surface. This reflexion is then called *total reflexion.*

In *water*, whose index of refraction is 1.336, the angle of total reflexion is 48° 28′. In *glass*, whose index of refraction is 1.50, it is 41° 49′. In *sulphur*, it is 30°; and in *diamond*, it is 23° 35′

749. *Transparent bodies differ much among themselves in refracting power.*

The following table will be useful by way of reference.

TABLE OF REFRACTIVE POWERS.

	Index of Refraction
Chromate of Lead,..............................	2.974
Red Silver Ore,..............................	2.564
Diamond,..............................	2.439

	Index of Refraction
Phosphorus,	2.224
Sulphur, (melted.)	2.148
Glass, (composed of lead two parts, flint one,)	1.830
Sapphire, and other precious gems,	1.800
Sulphuret of Carbon,	1.768
Oil of Cassia,	1.641
Quartz, or Rock Crystal,	1.548
Amber,	1.547
Crown Glass,	1.530
Oil of Olives,	1.470
Alum,	1.457
Fluor Spar,	1.434
Mineral Acids,	1.410
Alcohol,	1.372
Water,	1.336
Ice,	1.309
Tabasheer,	1.111

Hence it appears, that certain salts of silver and lead, the diamond, phosphorus, and sulphur, rank highest in refracting power; next come the precious gems, and flint glass, containing a large proportion of the oxide of lead, which has a refracting power considerably higher than crown glass, containing less metallic oxide; to which succeed the aromatic oils. Among transparent solids, fluor spar is distinguished for its low refracting powers; but tabasheer, a substance formed from the concreted juice of the Indian bamboo, is more particularly remarkable for the same property. Figure 260, will convey an idea of the comparative refractive properties of several of these substances.

In the preceding table, the refractive powers of different bodies are given without any consideration of their densities or specific gravities; but it is evident, that if a body of small specific gravity has the same refractive power as another body of greater specific gravity, the former must have a greater *absolute* action upon light than the latter. Hence, in order to measure the absolute refractive powers of bodies, their specific gravities must be taken into the account. When estimated on this principle, *hydrogen* will be found to have the greatest refractive power of all bodies,—it being, according to Dr. Brewster, equal to 3.0953; and it is also the most inflammable of all bodies. It was in consequence of the high refractive properties of inflammables, that Sir Isaac Newton expressed the opinion that the diamond is a body of this class, before its chemical constitution had been discovered.*

750. The *Multiplying Glass* (Fig. 262) exhibits as many ima-

* It is now known to consist of carbon, or pure charcoal.

ges of a luminous object, as there are surfaces exposed to it. The candle at A, sends rays to each of the three surfaces of glass. Those which fall on it perpendicularly, pass directly through the glass to the eye, without change of direction, and form one image in its true place at A. But the rays which fall on the two oblique surfaces, have their directions changed both in entering and in leaving the glass, (as will be seen by following the rays in the figure) so as to meet the eye in the directions of B and C. Consequently, images of the candle are formed, also, at both these points. A multiplying glass has usually a great many surfaces inclined to one another, and the number of images it forms is proportionally great.

Fig. 262.

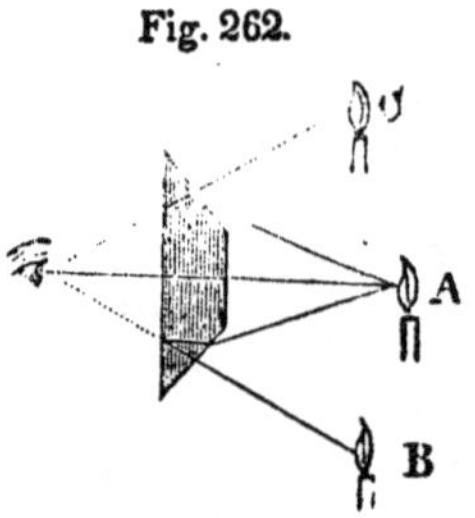

751. The PRISM is an important instrument in Optics, especially as it affords the means of decomposing light, and enters into the construction of several optical instruments. The *triangular* prism is the only one employed in experiments, and of this, nothing more is essential than barely the inclination of two plane transparent surfaces to one another. The optical prism, however, is usually understood to be a piece of solid glass, having two sides constituted of equal parallelograms, and a third side, called the *base*. The line of intersection of the two sides is called the *edge*, and the angle contained by the sides, the *refracting angle* of the prism. A straight line, passing lengthwise of the prism, through its center of gravity, and parallel to the edge, is called the *axis*. A section made by a plane perpendicular to the axis, is an isosceles triangle. Frequently, the three angles of the prism are made equal to one another, each being 60 degrees.*

Figure 263, represents a section of a prism ABC, of which AB is the *base*, and ACB the *refracting angle*. DE is a beam of the sun's light falling obliquely on the first surface AC, where one portion is reflected but another portion transmitted. The latter portion, instead of passing directly forward and forming an

Fig. 263.

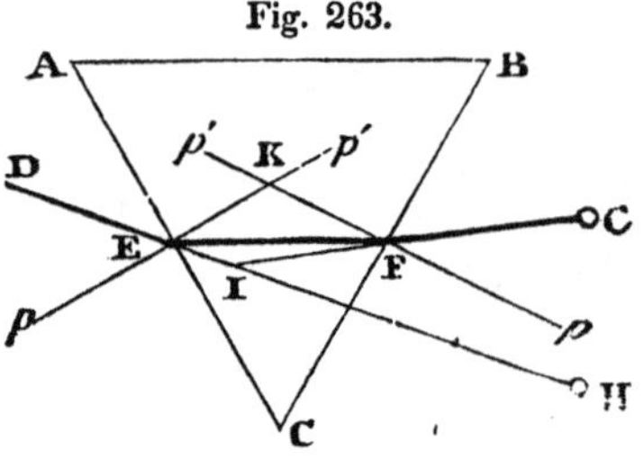

* A very convenient prism for common experiments may be constructed as follows. Select two plates of window glass of the best quality, or better, two pieces of looking-glass, from which the silvering has been removed. The plates may be five or six inches long, and one and a half or two inches broad. They are to be united at their edges at an angle of about 60°, and furnished with a tin case, which shall afford the base and the two ends, and a covering for the edge. One of the ends has an orifice with a stopper, for the convenience of filling with a fluid, which may be pure water, or better, a saturated solution of the sugar of lead, filtered perfectly clear

image of the sun at H, is turned upward toward the perpendicular pp', meeting the opposite surface CB in F, where it is again turned upward, from the perpendicular $p'p$, in the direction FG, carrying the image of the sun from H to G. If the incident and emergent rays be produced so as to meet in I, the angle FIH is called the *angle of deviation.*

752. By means of the prism, the index of refraction for different bodies may be found very conveniently from the following theorem.

The index of refraction diminished by unity, is always equal to the angle of deviation divided by the refracting angle of the prism.

In demonstrating this proposition it is necessary to premise, that when angles are small their ratio is nearly that of their sines; and since the sine of the angle of incidence is to that of refraction as the index of refraction to unity, (Art. 747,) therefore, n being the index of refraction, (see Fig. 263,)

$$p'\text{EI}(=\text{DE}p) : p'\text{EF} :: n : 1 \therefore \text{FEI} : p'\text{EF} :: n-1 : 1;$$
$$\text{also, } p'\text{FI}(=\text{GF}p) : p'\text{FE} :: n : 1 \therefore \text{EFI} : p'\text{FE} :: n-1 : 1;$$
$$\therefore \text{FEI}+\text{EFI} : p'\text{EF}+p'\text{FE} :: n-1 : 1,$$
$$\therefore \text{FIH} : p'\text{KF} :: n-1 : 1$$

But p'KF and ACB are equal, being each a supplement to four right angles in the quadrilateral figure ECFK. Therefore,

$$\text{FIH} : \text{ACB} :: n-1 : 1 \therefore n-1\times\text{ACB}=\text{FIH}.$$

$$\text{Hence,} \qquad n-1=\frac{\text{FIH}}{\text{ACB}}.$$

Now in prisms of glass, $n=\frac{3}{2}$; therefore, $\frac{\text{FIH}}{\text{ACB}}=\frac{1}{2}$, or FIH = $\frac{1}{2}$ ACB; that is, the angle of deviation equals half the refracting angle of the prism.

In order to find the index of refraction for any solid substance, the substance itself may be formed into a prism. The refracting angle of the prism being always known, and the angle of deviation easily measured, the index of refraction is readily found, by dividing the latter angle by the former, and adding one to the quotient. If the substance is of such a nature, that it cannot be fashioned into a prism, as a liquid, for example, it may then be introduced into the refracting angle of a prism formed by two plates of glass inclined to each other.

753. *When light is transmitted through a medium bounded by plane and parallel surfaces, the incident and emergent rays are parallel.*

Let AB*ba* (Fig. 264,) be the medium bounded by parallel surfaces AB, *ab*; and let DE be the incident ray refracted in the di-

Projections may be attached to the two ends to serve as handles or as an axis, by which the prism may rest on supports. Instead of the tin case, we may employ a block of hard wood, first formed into a triangular prism, and then dug out so as to admit the plates.

rection EF and emerging in the direction FG; the ray FG will be parallel to DE. Through the points E, F, draw the perpendiculars PQ, RS. Then, since PQ and RS are parallel, the angle of refraction QEF at the first surface, is equal to EFR, the angle of incidence at the second surface; but as the ratio of the sine of QEF to DEP is the same as that of EFR to SFG, (Art. 746,) the angles DEP and SFG must be equal, and, consequently, their complements AED, *b*FG; and if we add to these the equal angles AEF, *b*FE, the whole angles DEF, GFE will be equal, and consequently the rays DE, FG parallel.*

Fig. 264.

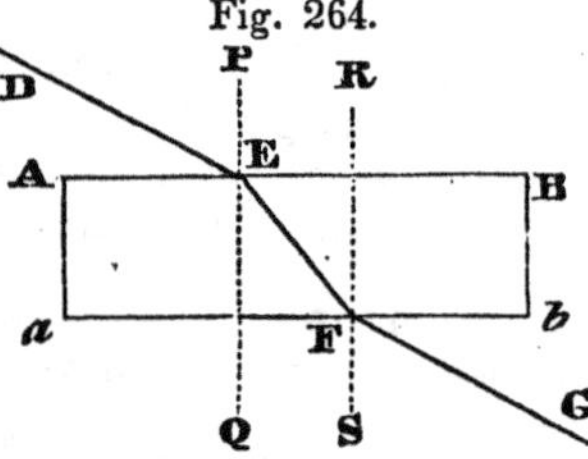

It is found by experiment that when light is transmitted through *two* contiguous mediums, bounded by plane and parallel surfaces, the incident and emergent rays are parallel to one another.†

754. *Through a plane surface, if diverging rays pass out of a rarer into a denser medium, they are made to diverge less than before: if out of a denser into a rarer medium, to diverge more.*

For since the sine of the angle of refraction is always as that of incidence, the most divergent lines in a pencil will be the most refracted, and will of course be brought nearer to a parallelism with those rays which diverge less when the refraction is *toward* the perpendicular, but will be still further separated when the refraction is *from* the perpendicular.

755. Lenses, on account of their extensive use in the construction of optical instruments, require very particular attention in the study of Optics. They are of several varieties, as is shown in the following figure.

Fig. 265.

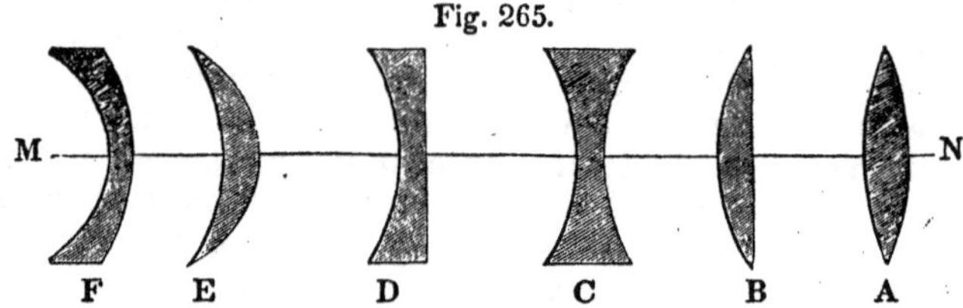

A *double convex lens* (A) is a solid formed by two segments of a sphere, base to base.‡

A *plano-convex lens* (B) is a lens having one of its sides convex and the other plane, being simply a segment of a sphere.

* Euc. I, 27. † Wood's Optics, p. 40.

‡ Though this is the most common form of the double convex lens, yet it is not essential that the two segments should be portions of the same sphere: they may be segments of different spheres, in which case the curvatures will be unequal on the two sides of the lens.

A *double concave lens* (C) is a solid bounded by two concave spherical surfaces, which may be either equally or unequally concave.

A *plano-concave lens* (D) is a lens one of whose surfaces is plane and the other concave.

A *meniscus* (E) is a lens one of whose surfaces is convex and the other concave, but the concavity being less than the convexity, it takes the form of a crescent, and has the effect of a convex lens whose convexity is equal to the difference between the sphericities of the two sides.

A *concavo-convex lens* (F) is a lens one of whose surfaces is convex and the other concave, the concavity exceeding the convexity, and the lens being therefore equivalent to a concave lens, whose sphericity is equal to the difference between the sphericities of the two sides.

A line (MN) passing through the center of a lens perpendicular to its opposite surfaces, is called the *axis*.

756. The manner in which light is refracted by passing into denser or rarer mediums bounded by spherical surfaces, may be readily understood and easily remembered, by keeping in mind the position of the incident rays with respect to the perpendicular, that is, the radius of the spherical surface. Suppose the two mediums are air and glass, and let us take first, the case of a *convex* surface of glass: then, since rays passing into the glass would be turned toward the perpendiculars, (all of which being radii, tend toward a common center,) parallel rays would be made to converge; diverging rays would become less diverging; converging rays, more converging. These are the *general* results; but let us trace the progress of diverging and converging rays a little more particularly. If the rays came from a near radiant, so as to diverge very much from each other, the effect of the glass would be simply to *diminish* their divergency; but if they came from some more distant point, so as to be less diverging, they might be turned so far toward the perpendicular as to become parallel, or even converging. But suppose the incident rays to come to the glass converging, then if they were directed toward the center of the sphere they would coincide with the radii or perpendiculars and suffer no change of direction; if they originally tended to a point more distant than the center, being turned toward the radii, they would be rendered more convergent; but if they tended toward a point nearer than the center, for the same reason they will converge less than before.

These several cases will be rendered familiar by studying the representation in Fig. 266.*

*The student is expected to make the explanation of each case from the figure, following the rays AN, &c. to GN. Thus, AN being refracted toward the perpen-

Fig. 266.

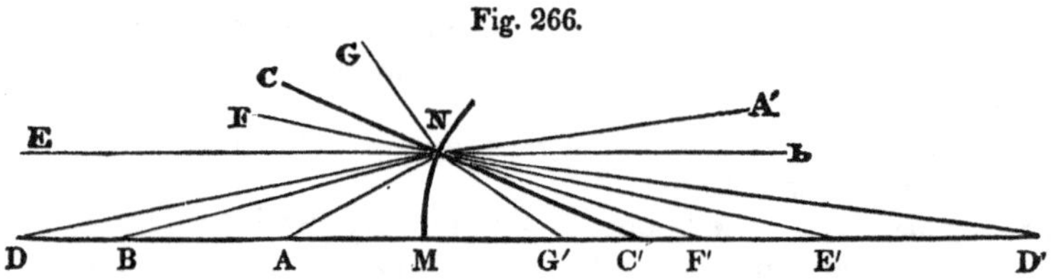

757. Secondly, let us consider the case of a *concave* surface We shall perceive, by inspecting Fig. 267, that *parallel* rays, b

Fig. 267.

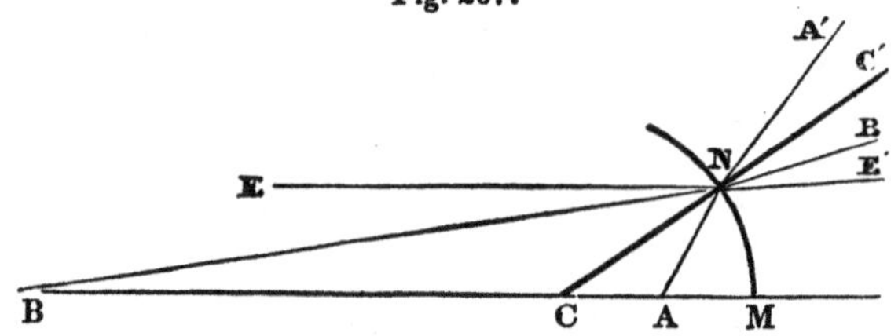

being turned toward the perpendicular, are made diverging; *diverging* rays are, in general, rendered more diverging; but when they come from the center of concavity, they suffer no refraction, and when from a point nearer the surface than the center, they diverge less than before; and *converging* rays are, in general, rendered less converging, but they may be so slightly convergent before, that the refracting power of the glass shall be sufficient to render them parallel or even divergent.

758. Thirdly, if we now trace the progress of the rays through LENSES, we shall readily follow their course by applying the foregoing principles.

1. Let AB (Fig. 268,) be a double *convex* lens, C, C′, the centers of curvature, and ED a ray of light falling upon the lens at D. According to the principles just explained, ED would be turned toward CD, the perpendicular to the refracting surface.

Fig. 268.

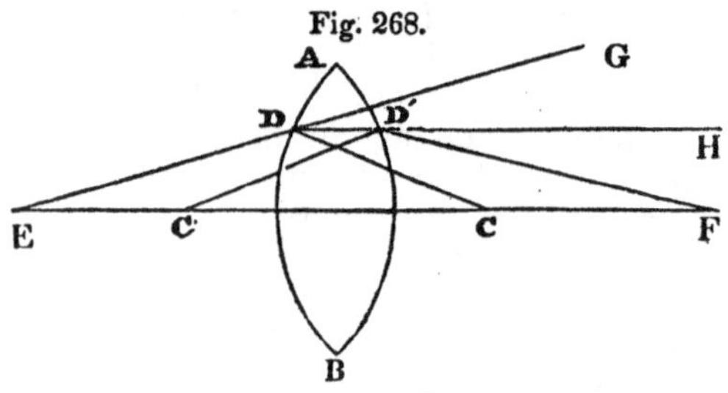

---

dicular NC′, is rendered less diverging; BN is turned so far toward NC′ as to become parallel to AM, &c.

and instead of passing onward in the same straight line EDG, it would proceed in the line DD′. Again, on passing out of the denser into the rarer medium at the second surface at D′, instead of proceeding onward in the line DD′H, it would be turned further from the perpendicular to that surface, namely C′D′, so as to proceed in the line D′F. Both surfaces of the lens, therefore, conspire to turn the ray out of its former course, and when the curvature of the two sides is the same, they contribute equally to produce this effect.

Fig. 269.

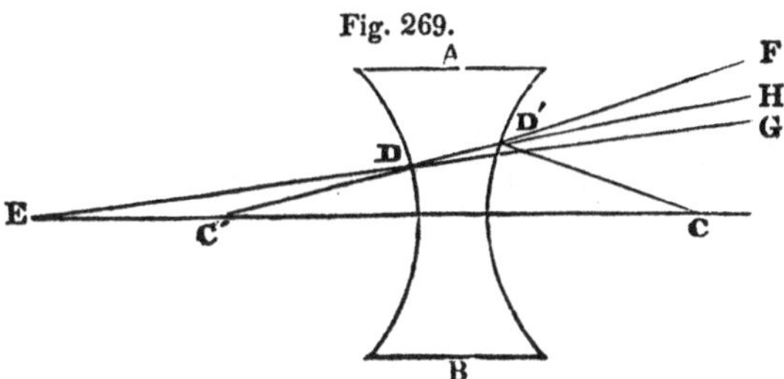

2. Let AB (Fig. 269,) be a double *concave* lens, then by tracing the progress of the ray ED, DD′, D′F, it will be seen that the effect of each surface of the lens is to cause the ray to diverge further from the axis. Thus C′D, CD′, being the radii of curvature, the ray ED, on entering the lens, is refracted into the line DH; and again, on leaving the lens, it is refracted into D′F.

759. *In a double convex, or double concave lens, there is a certain point called its center, through which every ray that passes, has its incident and emergent rays parallel.*

Fig. 270.

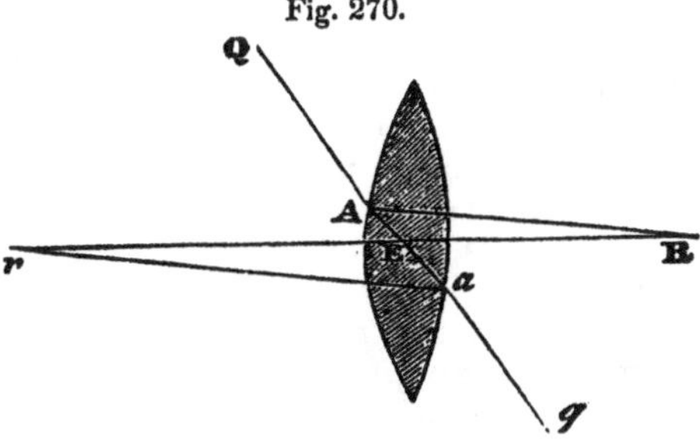

Let R, *r*, (Figs. 270, 271,) be the centers from which the surfaces of these lenses are described, and RE*r* their axis. Draw any two of their radii RA, *ra*, parallel to each other, and join A*a*; the point E, where this line intersects the axis, will be the point above described, and any ray, as Q*q*, passing through A, will have the incident ray QA, parallel to the emergent ray *aq*

Fig. 271.

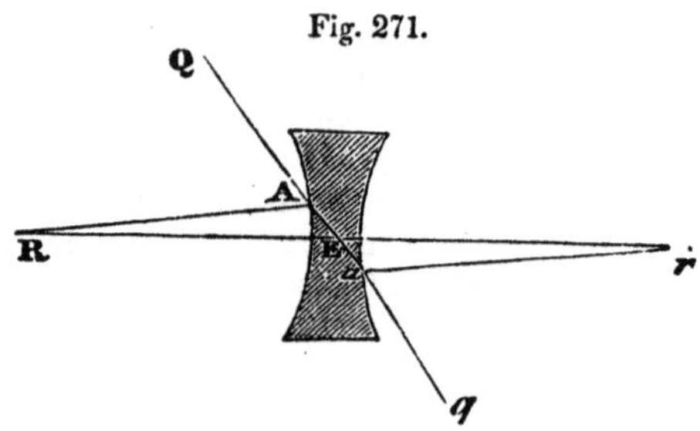

For since the triangles REA, rEa, are similar, RA : ra : : RE : rE, ∴ RA±ra : ra : : RE±rE : rE. And, as the three first terms of this proportion are invariable, the last, rE, must also be invariable. Hence it follows, that to whatever points in the surface of the lens, the parallel radii RA, ra, are drawn, the line Aa will always cut the axis Rr in the same point E. If we now suppose the ray Aa to pass both ways out of the lens, it will be refracted equally and in contrary directions; because RA, ra, being perpendiculars to the surface at A and a, the angles of incidence of the ray Aa or aA, will be equal. Consequently, AQ will be parallel to aq. When the thickness of the lens is inconsiderable, and when a ray falls nearly perpendicularly upon it, the part of the ray through E, viz. QAEaq, may be taken as a straight line, passing through the center E of the lens; for the perpendicular distance between AQ, aq, diminishes, both with the thickness of the lens and with the obliquity of the ray to the axis.

760. The office of a convex lens is to *collect* rays of light. Hence, when applied to parallel rays, it makes them converge; to diverging rays, it makes them diverge less; and to converging rays, it makes them converge more. Moreover, with regard to diverging rays, the degree of divergence may be reduced so much as to render the rays parallel, or even to make them converge, which will depend both on the position of the radiant, as illustrated in Art. 756, and on the power of the lens.

On the contrary, the office of a concave lens is to *separate* rays of light. Hence, when it is applied to parallel rays, it makes them diverge; to rays already diverging, it makes them diverge more; and to converging rays, it makes them converge less, become parallel, or even diverging.*

With these general principles in view, we may now advantageously investigate the manner in which IMAGES are formed by means of lenses.

1. If we place a radiant, as a candle, nearer to a lens than its

* A striking analogy will be remarked between the convex lens and concave mirror, and between the concave lens and convex mirror.

principal focus, then, since the rays go out diverging, (Art. 756,) no image will be formed on the other side of the lens.

2. If we place the radiant in the focus, the rays will go out parallel, but will still not be collected into a distinct image.

3. If the radiant is removed further from the lens than its principal focus, then the rays will be collected on the other side of the lens, so as to form a distinct representation of the object.

As this last case is particularly important, since it exhibits the manner in which images are formed by means of convex lenses, let us examine it with more attention.

761. *Rays of light diverging from the several points of any object, which is further from a convex lens than its principal focus, will be made to converge on the other side of the lens, to points corresponding to those from which they diverged, and will form an image.*

Let MN (Fig. 272,) be a luminous object placed before a double convex lens LL. Now every point in the radiant sends forth innumerable rays in all directions, part of which fall upon the lens LL. Each pencil may be considered as a cone of rays, having for its axis the straight line which passes through the center of the lens, which line suffers no change of direction, (Art. 757,) while those rays of the pencil which strike upon the extreme parts of the lens, form the exterior rays of the cone; all the others are of course included between these. It will be sufficient to follow the course of the central and the two extreme rays. Let ML, MC, ML, represent such a pencil. The two extreme rays will be collected by the lens and made to meet in the axis or central ray, in some point on the other side, as at *m*. For the same reason, every other point in the object will have its corresponding point in the image, and all these points of the image taken together, will form a true representation of the object. By inspecting the figure, it will be seen, that the axes of all the pencils cross each other in the center of the lens; that the part corresponding to the top of the object is carried to the bottom of the image, while that corresponding to the bottom of the object is at the top of the image, and, consequently, that the image is inverted with respect to the object. It will be further seen, that although the individual rays which make up a single pencil are made, on passing through the lens, to converge, yet the axes of all the pencils go out diverging from each other, which carries them further

Fig. 272.

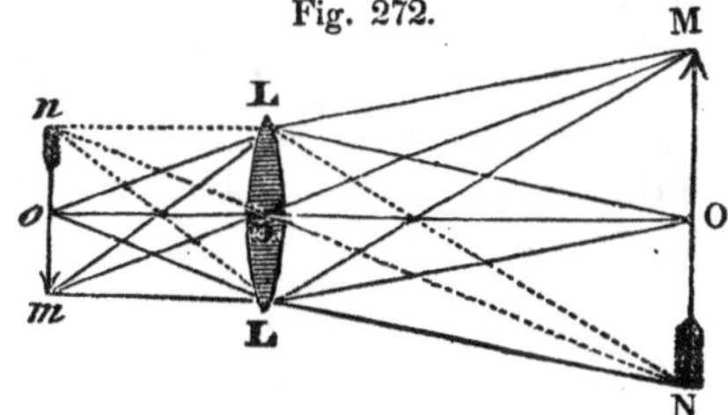

and further asunder, the further they proceed before they come to a focus.

762. *The diameter of the object is to the diameter of the image as the distance of the object from the lens is to the distance of the imag from the lens.*

For the two triangles MOC and *mo*C are similar; therefore MO : *mo* :: CO : C*o*. With a given object, the diameter of the image is as its distance from the lens. And, since the surfaces of the object and the image are similar figures, (being parallel sections of similar pyramids, or cones, whose vertices meet in the center of the lens,) the surface of the image is as the square of its distance from the lens. By bringing the object nearer to the lens, the image recedes from it on the other side, since the rays, being more divergent, are not so soon brought to a focus; therefore, by bringing the radiant very near to the focus of parallel rays, so as to throw the image very far back, the latter becomes exceedingly magnified.

The diameter of the image will not be altered by changing the area of the lens; for that diameter will be determined in all cases by the distances between the *axes* of the two pencils which come from the extremities of the object, and cross each other in the center of the lens. The size of the image, however, will be affected by changing *the convexity of the lens,* while the object remains the same and at the same place, being found nearer the lens, as the latter is more convex.

763. *Rays proceeding from any radiant point, which are refracted by the different parts of the same lens, do not meet accurately in one focus, but their points of meeting are spread over a certain space, whose diameter is called the* SPHERICAL ABERRATION *of the lens.*

Let LL (Fig. 273) be a plano-convex lens, on which are incident the parallel rays RL, RL, at the extremities, and R′L′, R′L′, near the axis: then, according to Art. 756, the axis will proceed on without any change of direction, and the rays which are very near to the axis, being also nearly perpendicular to the refracting surface, sustain only a slight change of direction, sufficient, however, to collect them into a focus at some distance from the lens in the point F. But the rays RL, RL, meeting the refracting surface more obliquely, are more turned out of their course, and are therefore collected into a focus in some point nearer to the lens than F, as at *f*. The intermediate rays refracted by the lens will have

Fig. 273.

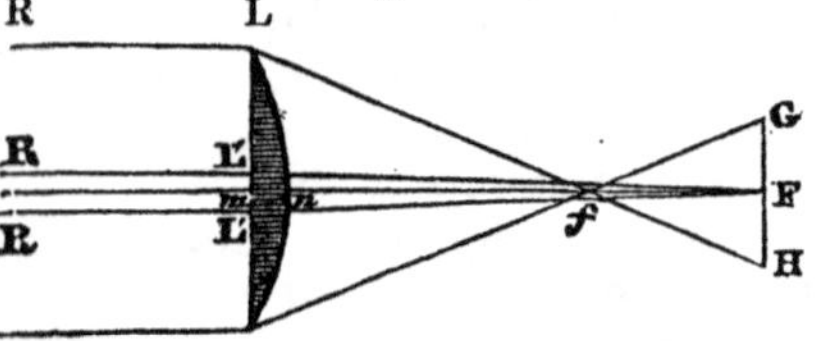

their foci between F and f. Continue the lines Lf and Lf, till they meet at G and H, a plane passing through F. The distance fF is called the *longitudinal spherical aberration*, and GH the *lateral spherical aberration.*

It is obvious that such a lens cannot form a distinct picture of any object in its focus F. If it is exposed to the sun, the central parts of the lens L′mL′, whose focus is at F, will form a pretty bright image of the sun at F; but as the rays of the sun which pass through the outer part LL of the lens have their foci at points between f and F, the rays will, after arriving at these points, pass on to the plane GH, and occupy a circle whose diameter is GH; hence the image of the sun in the focus F, will be a bright disk, surrounded and rendered indistinct by a broad halo of light, growing fainter and fainter from F to G and H. In like manner, every object seen through such a lens, and every image formed by it, will be rendered confused and indistinct by spherical aberration.

If we cover up all the exterior portions of the lens, so as to permit only those portions of the rays which lie near the axis to pass through the lens, then the rays all meet at or very near to the point F, and a much more distinct image is formed; but so much of the light is excluded by this process, that the brightness of the image is considerably diminished. The *dimensions* of the image are the same in both cases. (Art. 762.)

764. By experiments made with different kinds of lenses, the following results are obtained. In *plano-convex* lenses, placed as in Fig. 272, the greatest spherical aberration is $4\frac{1}{2}$ times mn, the thickness of the lens. In a *plano-convex* lens, with its convex side turned toward the parallel rays, the aberration is only $1\frac{17}{100}$ths of its thickness. In using the plano-convex lens, therefore, it should always be so placed, that the parallel rays may be incident upon the convex surface. In a *double convex* lens, with equal convexities, the aberration is $1\frac{67}{100}$ths of its thickness. The lens which has *the least spherical aberration*, is a double convex one, whose radii are as 1 to 6. When the face, whose radius is 1, is turned toward the parallel rays, the aberration is only $1\frac{7}{100}$ths of its thickness. Hence, the lenses employed in optical instruments are made *very thin;* and the light is suffered to pass only through *the central parts* of the lens. As the central parts of the lens LL, refract the rays too little, and the marginal parts too much, it is evident, that if we could increase the convexity at n, and diminish it gradually toward L, we should remove the spherical aberration. But the ellipse and hyperbola are curves of this kind, in which the curvature diminishes from n to L; and mathematicians have shown how spherical aberration may be entirely removed, by lenses whose sections are ellipses or hyperbolas. Of a lens of this kind we will annex one example.

765. *A lens in the form of a spheroid, (generated by the revolution of an ellipse about its major axis,) whose major axis is to the distance between its foci, as the sine of incidence to the sine of refraction, will cause parallel rays incident in the direction of its axis, to converge accurately to the remoter focus.*

Let BDK, (Fig. 274,) be the generating ellipse, H and I its foci; then, by the supposition,

DK : HI : : *sin.* Incidence : *sin.* Refraction.

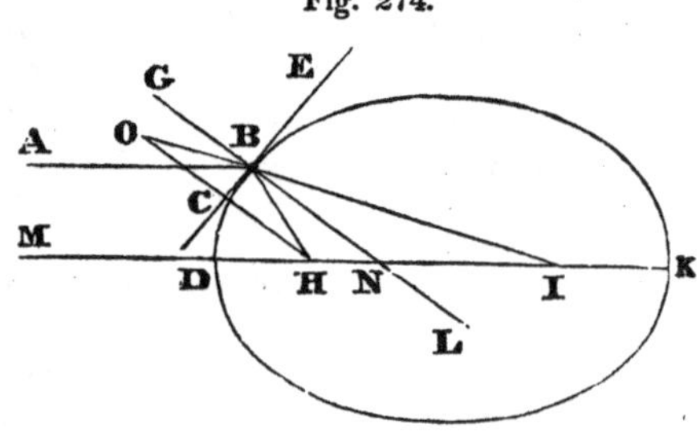

Fig. 274.

Let AB, which is parallel to DK, be a ray of light incident upon the spheroid. Join HB, IB; draw EBC, touching the generating ellipse in B; through B and H, draw GBL and HCO at right angles to EBC; let GBL meet DK in N; and produce IB till it meets HCO in O. Then, since HBC=IBE,* and OBC=IBE, therefore HBC=OBC. Also, BCH, BCO, are right angles, and BC is common to the two triangles BCH, BCO; therefore BO=BH, and IO=DK; consequently,

IO : IH : : *sin.* Incidence : *sin.* Refraction. And because BN is parallel to OH,

IB : IN : : IO : IH : : *sin.* Incid. : *sin.* Refrac.

Also, IB : IN : : *sin.* INB : *sin.* IBN : : *sin.* BNH or *sin.* ABG : *sin.* IBL; therefore, *sin.* ABG : *sin.* IBL : : *sin.* Incid. : *sin.* Refrac. And since *sin.* ABG is the sine of incidence, *sin.* IBL is the sine of refraction, LBI is the angle of refraction, and BI is the refracted ray. In the same manner it may be shown, that every other ray in the pencil will be refracted to I.

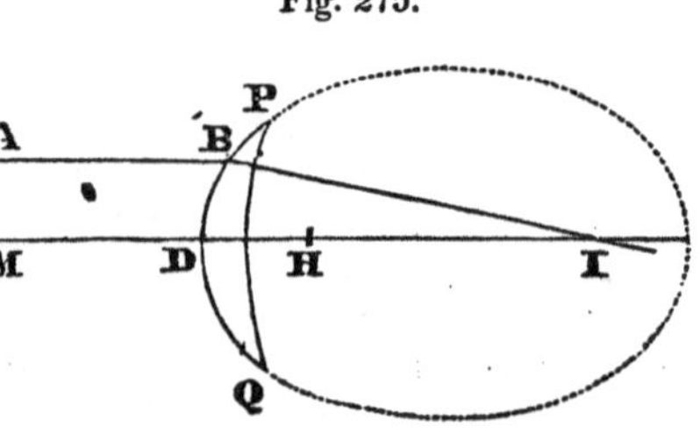

Fig. 275.

766. If from the center I, (Fig. 275,) with any radius less than ID, a circular arc PQ be described, the solid, generated by the revolution of PDQ about the axis DI, will refract all the rays incident parallel to DI, accurately to I. For, after refraction at the surface PDQ, the rays converge to I; and they suffer no refraction at the surface PQ, because they are incident perpendicularly upon it.†

Hence it follows that a *meniscus*, whose convex surface is part

* Conic Sections. † Wood's Optics, Sec. 187, 188.

of an ellipsoid, and whose concave surface is part of any spherical surface whose center is in the farther focus, may be so constructed as to have no spherical aberration, and to refract parallel rays incident on its convex surface, to the farther focus. When the foregoing properties of the ellipse were discovered, (and similar properties belong to the hyperbola,) philosophers exerted all their ingenuity in grinding and polishing lenses with elliptical and hyperbolical surfaces; and various ingenious mechanical contrivances were proposed for this purpose. These, however, have not succeeded; and the difficulty of grinding glasses of any other than a spherical curvature, is such as to prevent the use of spheroidal and other forms not subject to aberration; but other expedients have been devised for correcting this error.

Though we cannot remove or diminish the spherical aberration of *single* lenses beyond $1\frac{7}{100}$ths of their thickness, yet by combining two or more lenses, and making opposite aberrations correct each other, we can remedy this defect to a very considerable extent in some cases, and in other cases remove it altogether.* The manner in which this is effected, will be more particularly pointed out in connection with the subjects of Microscopes and Telescopes.

CHAPTER V.

OF THE DECOMPOSITION OF LIGHT AND THE SOLAR SPECTRUM.†—NATURE OF LIGHT.

767. In tracing the course of rays of light through a refracting medium, we have thus far supposed them to be homogeneous, and to be all affected in the same manner. But in nature, the fact is otherwise; that is,

The sun's light consists of rays which differ in refrangibility and in color.

The glass prism, in consequence of the strong refraction of light which it produces, (see Art. 751,) is well fitted for experiments of this kind. We procure, therefore, a triangular prism of good flint glass, and having darkened a room, admit a sunbeam obliquely, through a small round hole in the window shutter. Across this beam, near the shutter, we place the prism, with its edge parallel to the horizon, so as to receive the beam upon one of its sides. The rays, on passing through the prism, will be refracted and thrown upward, as will be rendered evident by conceiving perpendiculars drawn to the surface of the

---

* Brewster.

† That part of Optics which treats of colors, is sometimes denominated *Chromatics*.

prism at the points of incidence and emergence. If now we receive the refracted rays upon a screen, at some distance, they will form an elongated image, exhibiting the colors of the rainbow, namely, red, orange, yellow, green, blue, indigo, violet, together composing the *prismatic spectrum.* (See Fig. 276.)

Fig. 276.

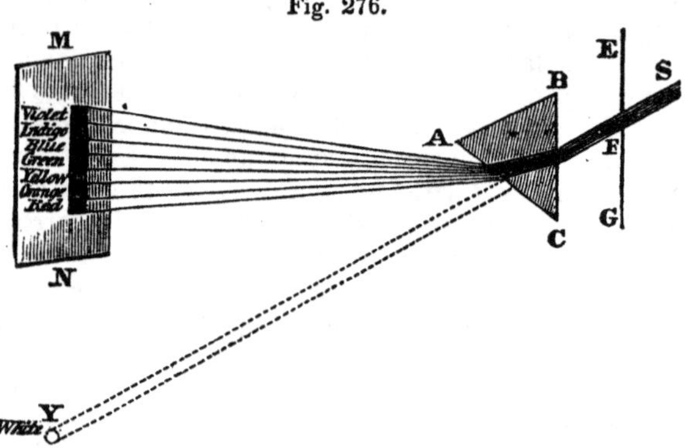

S, a sunbeam.

F, a hole in the window shutter.

ABC, the prism, having its refracting angle ACB downward.

Y, a white spot, being an image of the sun formed on the floor before the prism is introduced.

MN, the screen containing the spectrum.*

768. On viewing the spectrum attentively, we perceive that the lowest or least refracted extremity is a brilliant red, more full and vivid than can be produced by any other means, or than the color of any natural substance. This dies away, first into an orange, and then passes by imperceptible gradations into a fine pale straw-yellow, which is quickly succeeded by a pure and very intense green, which again passes into a blue, at first of less purity, being mixed with green, but afterward, as we trace it upward, deepening into the purest indigo. Meanwhile, the intensity of the illumination is diminishing, and in the upper portions of the indigo tint, it is very feeble ; but the blue is continued still beyond, and acquires a pallid cast of purplish red, a livid hue better seen than described, and which, though not to be exactly matched by any natural color, approaches most nearly to that of a fading violet.†

A pleasing way of exhibiting the separate colors of the spectrum, is to throw the prismatic beam on a distant wall or screen,

* The opposite white wall of plaster or stucco, may serve the purpose of a screen ; or the screen may be made of a large sheet of white paper ; but a convenient screen for the lecture room is made by pasting a large sheet of muslin to a frame, and attaching it to a movable stand. If the cloth is thick, it may be wet

† Herschel on Light.

so as to form a long spectrum, and into this beam, at some convenient distance from the prism, as near A, (Fig. 276,) to introduce a concave lens of a size sufficient to cover each of the different colored pencils successively. The lens will cause the rays of the same color to diverge, and to form a circular image on the screen, which will distinguish them very strikingly from the contiguous portions of the spectrum.

769. *If rays of the same color in the prismatic beam be insulated from the rest, and made to pass through a second prism, they are refracted as usual, (the amount of refraction being different for the different colored rays,) but they undergo no further change of color.*

To perform this experiment, we provide a board, perforated with a small round hole, and mounted on a stand. This screen is placed across the prismatic beam, a little way from the prism, in such a manner as to permit rays of the same color only to pass through the aperture, while the other portions of the beam are intercepted. The homogeneous light thus insulated is made to pass through a second prism, and its image is thrown on the

Fig. 277.

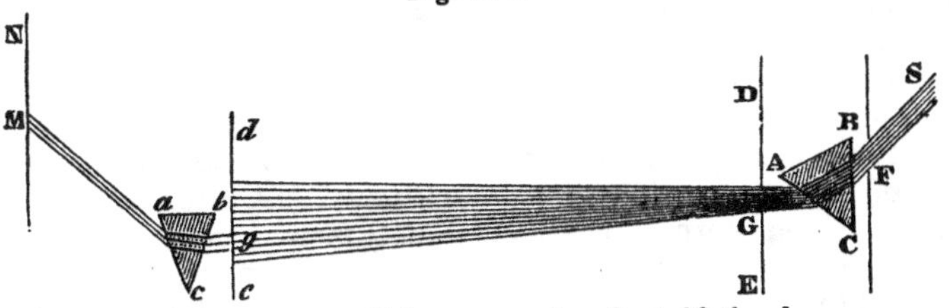

wall. The experiment will be more perfect, if the homogeneous pencil is made to pass through a second screen similar to the first, so as to let only the central rays fall upon the second prism. This second refraction produces no change of color. It will be found, however, that while all other things remain the same, the several images formed of homogeneous rays, will occupy different positions on the wall, the red being lowest, the violet highest, and the intermediate colors will be arranged between them in the order of their refrangibilities. (See Fig. 277.)

In addition to the parts of the figure enumerated in Fig. 277, DE represents the first screen, which permits only one sort of rays to pass by a small aperture at G, and *de* represents a second screen, which permits only the central rays of this pencil to pass by a small hole at *g* ; *abc* is the second prism, and M is the image of homogeneous light on the wall.

770. *The light of the sun reflected from the first surface of bodies, and also the white flames of all combustibles, whether direct or reflected, differ in color and refrangibility, like the direct light of the sun.*

The truth stated in this proposition was established by Newton, by experiments with the prism, similar to those detailed in connection with the preceding proposition.

771. *The sun's light is compounded of all the prismatic colors, mixed in due proportion.*

If we collect, by means of a convex lens, the different colored pencils in the prismatic beam, just after they have emerged from the prism, (see Fig. 276,) the image formed by the lens will be perfectly white. A concave mirror may be used instead of the lens, the image being thrown on a screen. Or the rays after they have passed the prism may be received on a second prism of the same kind, placed near the first, but with its refracting angle in the opposite direction. In this case the second prism restores the light to its usual whiteness.

That all the different colors of the spectrum are essential to the composition of white light, may be rendered evident by intercepting a portion of any one of the colors of the spectrum before they have been reunited as in the foregoing experiments. Thus, if we introduce a thread or a wire into any part of the prismatic beam between the prism and the lens, the image formed by the lens will be no longer white but discolored. If, instead of the wire, an instrument, shaped like a comb with coarse broad teeth, be introduced into the beam, the discoloration of the image is more diversified, the colors of the image being those compounded of the prismatic colors, which are not intercepted by the comb. If the teeth of the comb be passed *slowly* over the beam, a succession of different colors appears, such as red, yellow, green, blue, and purple; but if the motion of the comb be rapid, all these different hues become blended into one by the momentary continuance of each in the eye, and the sensation is that of white light.

For a similar reason, if the colors of the spectrum are painted on a top, in due intensity and proportion, and the top be set to spinning, the sensation will be that of white light. Or the colors of the spectrum may be first laid on a sheet of paper, and this may be pasted on a cylinder of wood, which may be made to revolve on the whirling tables: the result will be the same. Newton tried various experiments with different colored powders, grinding together such as corresponded as nearly as possible to the colors of the spectrum. By these means he was able to produce, from the mixture of seven different colored powders, a *grayish-white*, but could never reach a perfectly clear white, owing to the difficulty of finding powders whose colors correspond exactly to those of the spectrum.

772. *Several of the colors of the spectrum may be produced by the mixture of other colors; as green by the mixture of yellow and blue, orange by red and yellow, &c.*

Experiments were devised by Newton for thus combining the colors of two contiguous spectrums, transferring, for example, the blue of one to the yellow of the other, and forming green by their union. On causing this compound green, however, to pass through the prism, it is resolved into its original colors, yellow and blue, whereas, the green of the spectrum is not thus resolved by the prism. Hencc, Newton infers that the green of the spectrum is not a compound but a simple original color, and so of all the rest.

It has, however, been a question among opticians, since the time of Newton, *what is the number of original or fundamental colors in the spectrum?* Many years since, Mayer advanced the hypothesis that the only simple colors in the solar spectrum are *red, yellow,* and *blue,*—all the others being compounded of these; and more recently Dr. Brewster has gone far toward establishing this doctrine. According to this eminent optician, (1.) Red, yellow, and blue light exist at every point of the solar spectrum; (2.) As a certain portion of red, yellow, and blue, constitute white light, the color of every point of the spectrum may be considered as consisting of the predominating color at any point, mixed with white light. Thus in the red space, there is *more red* than is necessary to make white light with the small portions of yellow and blue which exist there; in the yellow space, there is *more yellow* than is necessary to make white light with red and blue; and in the part of the blue space which appears violet, there is more red than yellow, and hence the excess of red forms a violet with the blue.

Fig. 278.

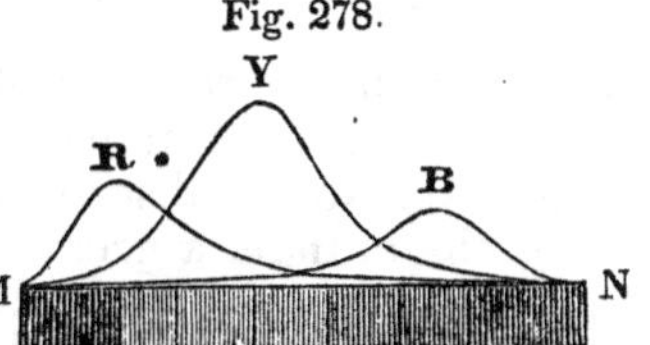

773. The mode by which these three primary colors produce by their combination the seven colors developed by the prism, is exhibited to the eye by the following diagram. MN, (Fig. 278,) is the prismatic spectrum, consisting of three primary spectra of the same length, viz. a red, a yellow, and a blue spectrum. The intensities of each color at various points of the spectrum, are represented by ordinates of different lengths, the extremities of which form the curves MRN, MYN, and MBN, corresponding to the three colors red, yellow, and blue, respectively. The red spectrum has its maximum intensity at R; and this intensity may be represented by the distance of the point R from MN. The intensity declines rapidly to M, and slowly to N, at both of which points it vanishes. The *yellow* spectrum has its maximum intensity at Y, the intensity declining to zero at M and N; and the *blue* has its maximum intensity at B, declining to nothing at M and N. The general curve which represents the total illumination at any point will be outside these three

curves, and its ordinate at any point will be equal to the sum of three ordinates at the same point. Thus, the ordinate of the general curve at the point Y, will be equal to the ordinate of the yellow curve, which may be supposed to be 10; added to that of the red curve which may be 2, and that of the blue which may be 1. Hence the general ordinate will be 13. Now if we suppose that three parts of yellow, two of red, and one of blue make white, we shall have the color at Y equal to $3+2+1=6$ parts of white mixed with seven parts of yellow; that is, the compound tint at Y will be a bright *yellow*, without any trace of red or blue. As these colors all occupy the same place in the spectrum, they cannot be separated by the prism; and if we could find a colored glass, which would absorb seven parts of the yellow, we should obtain at the point Y, a *white light*, which the prism could not decompose.*

774. The arguments on which most of these conclusions are grounded, are derived from experiments on the analysis of light by *absorption*. If, (says Dr. Brewster,) we take a piece of blue glass and transmit through it a beam of white light, the light will be of a fine deep blue. This blue is not a simple homogeneous color, like the blue or indigo of the spectrum, but is a mixture of all the colors of white light which the glass has not absorbed; and the colors which the glass has absorbed are those which the *blue wants* of white light. In order to determine what these colors are, let us transmit through the blue glass, the prismatic spectrum KL, Fig. 276; or, what is the same thing, let the observer place his eye behind the prism BAC, and look through it; he will see the spectrum on the other side of the prism, but with this remarkable change, that it will appear deficient in a certain number of its differently colored rays. A particular thickness absorbs the middle of the red space, the whole of the orange, a great part of the green, a considerable part of the blue, a little of the indigo, and a very little of the violet. The yellow space, which has not been much absorbed, has increased in breadth. It occupies part of the space formerly covered by the *orange* on one side, and part of the space formerly covered by the *green* on the other. Hence it follows, that the blue glass has absorbed the red light, which, when mixed with the yellow light, constituted *orange*, and has absorbed also the *blue* light, which, when mixed with the yellow, constituted a part of the green space next to the yellow. We have, therefore, by absorption, decomposed *green* light into yellow and blue, and *orange* light into yellow and red; and it consequently follows, that the orange and green rays of the spectrum, though they cannot be decomposed by prismatic refraction, can be decom

* Brewster's Treatise on Optics p. 73.

posed by absorption, and actually *consist of two different colors possessing the same degree of refrangibility.* Difference of color is therefore not a test of difference of refrangibility; and the conclusion deduced by Newton is no longer admissible as a general truth: "That to the same degree of refrangibility ever belongs the same color, and to the same color ever belongs the same degree of refrangibility."

By absorbing the excess of any color at any point of the spectrum above what is necessary to form white light, we may actually cause white light to appear at that point, and *this white light will possess the remarkable property of remaining white after any number of refractions, and of being decomposable only by absorption.*

FIXED LINES IN THE SPECTRUM.

775. The solar spectrum, in its greatest possible state of purity and tenuity, when received on a white screen, or when viewed by admitting it at once into the eye, is not an uninterrupted line of light, red at one end and violet at the other, and shading away by insensible gradations through every intermediate tint from one to the other, as Newton conceived it to be, and as a cursory view shows it. *It is interrupted by intervals absolutely dark;* and in those parts where it is luminous, the intensity of the light is extremely irregular and capricious, and apparently subject to no law, or to one of the utmost complexity. Consequently, if we view a spectrum formed by a narrow line of light parallel to the refracting edge of the prism, (which affords a considerable breadth of spectrum without impairing the purity of the colors, being, in fact, an assemblage of infinitely narrower linear spectra arranged side by side,) instead of a zone of equable light and graduating colors, it presents the appearance of a striped riband, being crossed in the direction of its breadth by an infinite multitude of dark, and by some totally black *bands,* distributed irregularly throughout its whole extent. This irregularity, however, is not a consequence of any casual circumstances. The bands are constantly in the same parts of the spectrum, and preserve the same order and relations to each other; the same proportional breadth and degree of obscurity, whenever and however they are examined, *provided solar light be used,* and provided the prisms employed be composed of the same material; for a difference in the latter particular, though it causes no change in the number, order, or intensity of the bands, or their places in the spectrum, as referred to the several colors of which it consists, yet causes a variation in their proportional distance from one another. By solar light must be understood, not merely the direct rays of the sun, but any rays which have the sun for their ultimate origin; the light of the clouds, or sky, for instance; of

the rainbow; of the moon, or of the planets. All these lights, when analyzed by the prism, are found deficient in the identical rays which are wanting in the solar spectrum; and the deficiency is marked by the same phenomenon, viz. by the occurrence of the same dark bands in the same situations, in spectra formed by these several lights. In the light of the stars, on the other hand, in electric light, and in that of flames, though similar bands are observed in their spectra, yet they are differently disposed; and the spectrum of each several star, and each flame, has a system of bands peculiar to itself, and characteristic of its light, which it preserves unalterably at all times, and under all circumstances.

776. Fig. 279, is a representation of the fixed lines of the spectrum, according to Fraunhofer,* the small bands observed by him (more than five hundred in number) being omitted. Of these fixed lines, he selected seven, (those marked B, C, D, E, F, G, H,) as terms of comparison, or as standard points of reference in the spectrum, on account of their distinctness, and the facility with which they may be recognised. The definiteness of these lines, and their fixed position with respect to the colors of the spectrum,—in other words, the precision of the limits of those

Fig. 279.

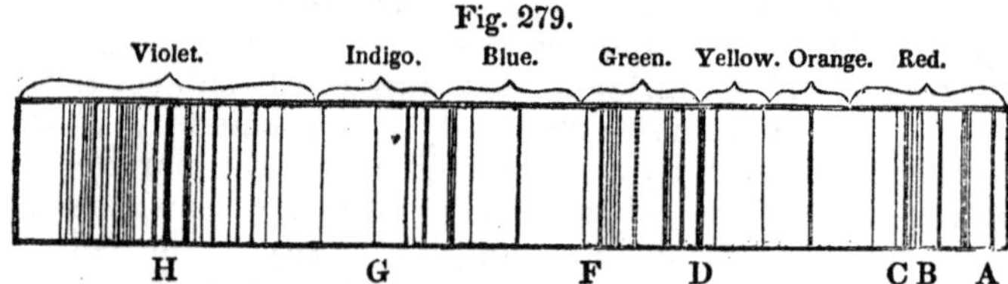

degrees of refrangibility which belong to the *deficient* rays of solar light,—renders them invaluable in optical inquiries, and enables us to give a precision hitherto unheard of to optical measurements, and to place the determination of the refractive powers of media on the several rays almost on the same footing, in respect to exactness, with astronomical observations.†

NATURE OF LIGHT.

777. *The phenomena of light may be explained, either on the supposition that light is a material fluid of extreme subtilty, or that it is produced by the undulations of an independent medium, set in motion by the luminous body.*

Opticians of great eminence, as Descartes, Huygens, Euler, and Young, have held the opinion, that light does not consist of

* A celebrated German optician, recently deceased.

† Herschel on Light.

actual emanations of material particles from the luminous body, but that such a body has merely the property of communicating a series of vibrations to a peculiar fluid that is diffused throughout the universe, which vibrations form the communication between the luminous body and the eye. The medium is conceived to be of extreme tenuity and elasticity; such, indeed, that though filling all space, it shall offer no appreciable resistance to the motions of the planets and comets, capable of disturbing them in their orbits. It is, moreover, imagined to penetrate all bodies; but in their interior, to exist in a different state of intensity and elasticity from those which belong to it in a disengaged state, and hence the refraction and reflexion of light. Newton, however, and with him the greater number of opticians, have held, that light consists of actual particles of matter sent off from luminous objects to the eye. In the former case, the fluid is only the medium of light, as air is the medium of sound; the vibrations of the medium following each other, as wave follows wave, with incredible swiftness, and thus conveying the impression from the radiant to the eye: in the latter case, the motion is simply that of a chain of particles moving in right lines with the same astonishing velocity. Thus, when the sun rises, it either sends forth luminous particles, which, entering the eye, occasion the sensation of vision; or puts in motion the peculiar fluid which is the medium of light, which motion is propagated from wave to wave till it reaches the eye.

778. It forms a strong objection against the hypothesis of undulations, that the motions of light are confined to *right lines*, a condition not essential to this species of motion; while it is a strong argument in favor of the materiality of light, that it exhibits the property of attraction, one of the most characteristic properties of matter. The motion is conformable to the laws which regulate the motions of small bodies under the same circumstances. Thus, when it meets with no impediment, it moves uniformly forward in right lines; it is affected by passing into mediums of different densities, in a manner correspondent to the law of the mutual gravitation of matter, being attracted from rarer toward denser bodies; and finally, it produces certain chemical changes in bodies which belong to none but a material agent. The rays of light, also, by passing through certain media, undergo a change, to be described hereafter under the head of *polarization*, in which the opposite sides of a ray appear to be endowed with different properties, a fact which accords ill with the idea of undulations, though it is quite consistent with the doctrine of the materiality of light. The latter hypothesis, moreover, has the advantage of leading the student to a more ready apprehension of the nature of optical phenomena. Still, the object of this science is not so much to ascertain the nature

of the agent on which the phenomena of light depend, as it is to study those phenomena themselves, and to classify them under general laws, which may be applied to the construction of optical instruments, and to the interpretation of nature.

779. To the doctrine of the materiality of light, it has been objected, first, that material particles, endued with such immense velocity, would have a momentum which nothing could resist, much less so delicate an organ as the eye; secondly, that were the rays material, so prodigious is their number scattered throughout the universe, they would interfere with one another, and, thirdly, that the sun and stars would waste away and grow dim, by such a constant and profuse expenditure of matter. But these objections severally admit of a satisfactory reply. In the first place, the momentum of a ray of light may still be inconsiderable, if the quantity of matter is small in the same proportion as the velocity is great. Though such an attenuation of matter is amazing, yet it is not incredible, but perfectly consistent with the known analogies of nature. In the second place, notwithstanding the universal diffusion of light, no interference of its particles is necessary, for it is not essential to the purpose of vision, that a ray should consist of contiguous particles of light. It is found, that the sensation continues for some time after the luminous object is removed, during an interval sufficient for light to pass through twenty-two thousand miles; consequently, particles no nearer to each other than this distance, would be competent to maintain uninterrupted vision. Thus, an ignited stick, whirled in the air, exhibits a ring of light, because the sensation continues for a longer time than the illuminated point occupies in passing round the circle. In the third place, the small danger of waste sustained by the sun in consequence of the light which it dispenses, may be inferred from the following remarks of Dr. Priestley. After relating an experiment, in which the light of the sun collected during one second, by a concave reflector of four square feet, and thrown on the arm of a delicate balance, indicated a weight *not exceeding* the 1200 millionth part of a grain, the Doctor adds: "Now the light in the above experiment was collected from a surface of four square feet, which reflecting only about half what falls upon it, the quantity of matter contained in the rays of the sun incident upon a square foot and a half of surface in one second of time, ought to be no more than the 1200 millionth part of a grain. But the density of light at the surface of the sun is greater than at the earth in the proportion of 45000 to 1; there ought, therefore, to issue from one square foot of the sun's surface, in one second of time, in order to supply the waste by light, one forty thousandth part of a grain of matter,—that is, a little more than two grains

in a day, or about 4752000 grains, which is about 670 pounds, avoirdupois, in six thousand years."*

The more recent and refined discoveries in optics have favored the theory of undulations, and the argument derived from the authority of great names is now decidedly on the side of this theory, and against that of emissions.†

CHAPTER VI.

OF COLORS IN NATURAL OBJECTS.

780. The knowledge of the composition of light, and of the properties of the solar spectrum, naturally led to an inquiry into the subject of colors, as exhibited in the phenomena of nature. The bright tints of the rainbow—the splendid hues sometimes exhibited by thin plates, as soap bubbles—and finally, the diversified colors of objects in all the kingdoms of nature, remained to be accounted for. We propose now to inquire how far this object has been effected.

THE RAINBOW.

781. The rainbow, one of the most striking and magnificent of the phenomena of nature, was long ago supposed to be owing to some modification which the light of the sun undergoes in passing into drops of rain ; but the complete development of the causes on which it depends, was reserved for the genius of Newton, and naturally followed in the train of those discoveries which he made upon the prismatic spectrum.

The rainbow, when exhibited in its more perfect forms, consists of two arches, usually seen in the east during a shower of rain, while the sun is shining in the west. These arches are denominated the outer and the inner bow, of which the inner bow is the brighter, but the outer bow is of larger dimensions every way. The succession of colors in the one is directly opposite to that of the other.

782. Drops of rain, though small, are large in comparison with the minuteness of rays of light, and are to be regarded as spheres of water, exerting the powers of refraction and reflexion in the same manner as large globes of water would do. It was,

* Priestley, Hist. Light and Colors.

† See an excellent view of the doctrine of *Undulations* in Pouillet's Traité de Physique, II. 263.

in fact, by investigating the manner in which globular glass vessels filled with water modify the solar rays, that the first hints were obtained respecting the cause of the rainbow. In the year 1611, Antonio de Dominis made a considerable advance toward the theory of the rainbow, by suspending a glass globe in the sun's light, when he found, that while he stood with his back to the sun, the colors of the rainbow were reflected to his eye in succession by the globe, as it was moved higher or lower

Let us, therefore, in the first place, follow the course of a ray of light through a globule of water. Let SA (Fig. 280) be a small beam of light from the sun, falling upon the surface of a globule of water at A. Agreeably to what is known of the laws of light in passing out of one transparent medium into another, a portion of the rays would be reflected at A, and another portion would pass into the drop and be refracted to the further surface at B. The same effect would recur here, and also at D and at F; and were the eye situated in either of the lines BC, DE, or FG, it would perceive the prismatic colors, because some of the rays which composed the beam of light that reached the eye, would be refracted more than others, and thus the different colors would be made to appear. Or if a screen were so placed as to receive these transmitted rays, a faint spectrum would be formed upon it. Such a progress of a beam of light admitted through the window shutter, and made to fall on a globular vessel of water, may be actually rendered visible by experiment.*

Fig. 280

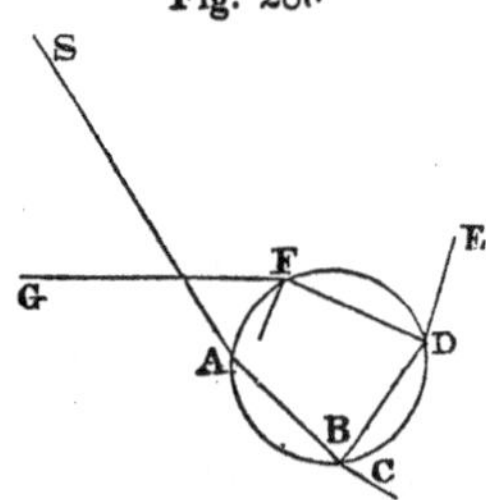

783. It may be remarked, that but a comparatively small part of the solar rays that shine upon a drop of water, are required in order to produce the mild light of the rainbow, aided as its light is by the dark ground or cloud on which it is usually projected; yet where the number of rays that enter the eye is diminished beyond a certain limit, the light becomes too feeble for distinct vision. It will also be observed, that a considerable portion of light is lost at each successive reflexion that takes place within the drop, so that a certain beam of light, conveyed to the eye after two reflexions, will be much more feeble than the same beam after one reflexion. Indeed, so much of the sun's light is dissipated at the first point of reflexion from the interior surface, added to what is transmitted at the same point, and of course never reaches the eye of the spectator, that, were it not for a great *accumulation* which the sun's rays undergo at a particular point in the drop, whence the light is reflected and con-

* Biot.

veyed to the eye, the phenomena of the rainbow would not occur. The manner in which this accumulation is effected, is now to be explained.

784. Let *fzpq* (Fig. 281) be the section of a drop of rain, *fp* a diameter, *ab*, *cd*, &c. parallel rays of the sun's light, falling upon the drop. Now *yf*, a ray coinciding with the diameter, would

Fig. 281.

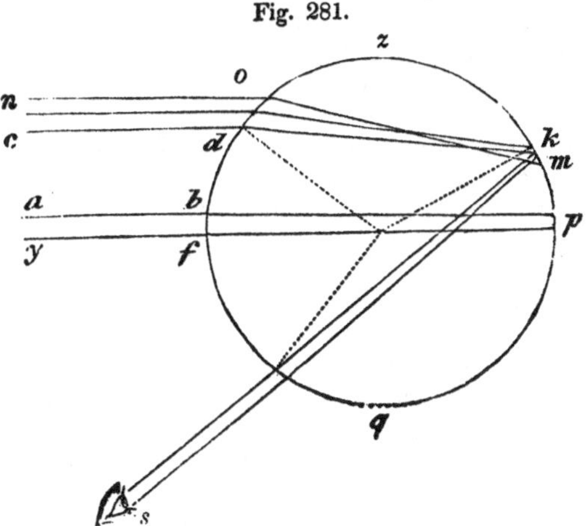

suffer no refraction; and *ab*, a ray near to *yf*, would suffer only a very small inclination toward the radius, so as to meet the remoter surface of the drop very near to *p*; but the rays which lie further from *yf*, being inclined toward the radius in a greater angle, would be more and more refracted as they were further removed from the diameter. The consequence would be, that after passing a certain limit, the rays that lay above that limit would cross those which lay below it, and meet the further surface somewhere between the diameter and the ray which passed through the said limit; that is, all the rays falling on the quadrant *fz*, would meet the circumference within the arc *kp*. But when a quantity is approaching its limit, or is beginning to deviate from it, its variations are nearly insensible. Thus, when the sun is at the tropics, being the limits to which he departs from the equator, he appears for some time to remain at the same point. In the same manner, a great number of the rays which lie contiguous to *cd*, on both sides of it, will meet in very nearly the same point on the concave surface of the drop at *km*. Consequently, a greater number of rays will be reflected from that point than from any other in the arc. Now were these rays to return in the same lines, they would emerge parallel in the lines *cd*, *no*; but if, instead of returning back in the quadrant *fz*, they are reflected on the other side of the radius, they may meet the curve at the same angle in the quadrant *fq*, and emerge parallel, coming to the eye at *s*. Hence it appears, that there is a particular point in a drop of rain, where the rays of the sun's light seem to *accumulate*, and are therefore peculiarly fitted to make an impression on the organ of vision. It is found by calculation, that the angle which the incident and emergent rays, in such cases, make with each other, is, for the *red* rays 42° 2′, and for the *violet* rays 40° 17′. These are the angles when the rays emerge after two refractions and *one* reflexion : in the case of two refractions and *two* reflexions, the angles are, for the *red* rays 50° 59′, and for the *violet* 54° 9′.

785. Let us next consider what must be the position of the spectator, in order that his eye may receive the emergent rays which make the foregoing angle with the incident rays, and which, of course, are those which cause the phenomena of the rainbow.

The spectator must stand with his back to the sun, and *a line drawn from the sun toward the bow so as to pass through his eye, will make the same angle with the emergent rays that the e make with the incident rays.* Thus, let AB (Fig. 282,) be the incident, and GI the emergent ray, and let the angle which these two rays make with each other be AKI; and let IT be a ray passing from the sun toward the bow through the eye of the spectator; then, (since the rays of the sun may be regarded as parallel,) AB and IT are parallel, and the alternate angles AKI and KIT, equal.

Fig. 282.

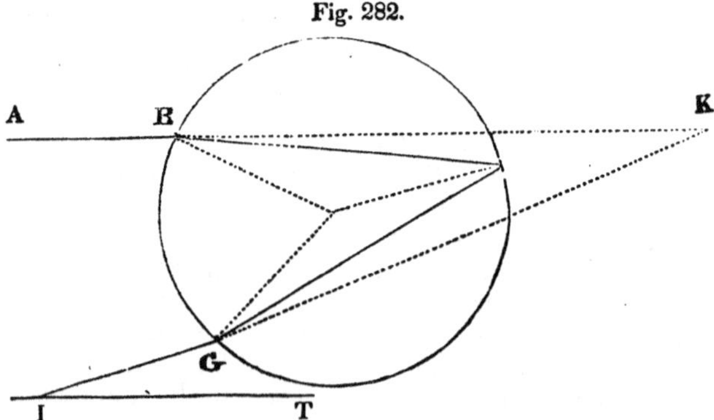

But AKI is the angle made by the incident and emergent rays, and KIT the angle made by the emergent ray, and a line drawn from the sun toward the bow through the eye of the spectator.

786. *When the sun shines upon the drops of rain as they are falling, the rays which come from those drops to the eye of the spectator after* ONE REFLEXION AND TWO REFRACTIONS, *produce the innermost or primary rainbow ; and those drops which come to the eye*

after TWO REFLEXIONS AND TWO REFRACTIONS, *produce the outermost or superior rainbow.*

Let SOC* (Fig. 283) be a straight line passing from the center of the sun through the eye of the spectator at O, toward the bow, and let SR, SV, be incident rays, which, after one reflexion and two refractions, are conveyed to the eye at O, making, (Art. 785,) with SOC, angles equal to those formed by the incident and emergent rays. If OV makes with SOC an angle of 40° 17′, and be conceived to revolve around OC, describing the surface of a cone, all the drops of rain on this surface will be precisely in the situation necessary in order that the violet rays, after two refractions and one reflexion, may emerge parallel and arrive at the eye in O, and this will not take place in the same manner on any other part of the cloud; so that, by means of this species of rays, the spectator will see on the cloud a violet colored arc, of which OC will be the axis, and C the center. He will, besides, see also an infinity of other concentric arcs exterior to the violet, each one of which will be made up of a single species of rays; and according as these rays are less refrangible, their areas will be of greater diameter, so that the largest, composed of the extreme red, will subtend an angle ROC of 42° 2′. Therefore the whole width of the colored bow will be 42° 2′—40° 17′, or 1° 45′, the red being on the outside and the violet within.

Fig. 283.

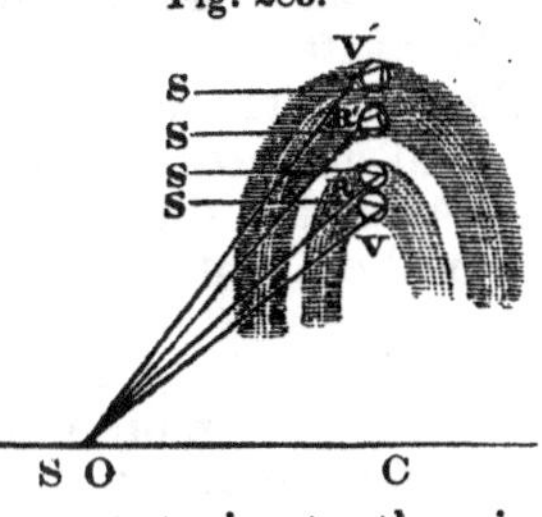

The contrary order of colors will result from *two reflexions* and two refractions. Let SV′, SR′, be the incident rays, which, after two reflexions and two refractions, converge to the eye at O, making (Art. 785,) with SOC, angles equal to those formed by the incident and emergent rays, namely, 50° 59′ and 54° 9′, and the lines R′O and V′O, as before, be conceived to revolve around SOC, they will severally meet with all the drops, which having twice refracted and twice reflected the extreme red and violet rays, can transmit them to the eye parallel to each other. Between these two arcs, there will be others exhibiting all the intermediate prismatic colors; and the whole together will form a second bow, whose breadth will be 54° 9′—50° 59′, or 3° 10′.

787. The rays, therefore, which come from all the drops which make an angle of 42° 2′ with a line passing from the sun through the eye (which may be called the *axis of vision*,) appear red; and it is obvious, that a collection of rays drawn all around this axis from the eye to drops thus situated, would form a cone, of

* It will be observed, that the line SOC is at right angles to the plane of the surface; that is, to the plane of the bows

which the drops themselves would constitute the base, and of course would form a circle. The same is true of all the other colors which emerge from drops at angles which are different for different colors, but constant for the same color. Hence, *the line which passes from the sun through the eye of the spectator, passes also to the center of the bow*, or is the axis of the cone of which the bow itself is the base. If the sun is on the horizon, this axis becomes a horizontal line; consequently, the center of the arch rests on the opposite horizon, and the bow is a semicircle, of which the highest point has an altitude above the horizon of 42° 2′. If the sun is at this altitude of 42° 2′ above the horizon, then the center of the bow will have the same depression below the opposite horizon, and the circumference, at its highest point, will just reach that horizon. When the sun is between these two points, the elevation of the bow will be the difference between the altitude of the sun and the foregoing angle.

788. When the spectator is on an eminence, as a high mountain, he may see more than half the bow, when the sun is near setting; for the axis will in that case pass to a point above the opposite horizon. Travellers who have ascended very high mountains, have occasionally observed their shadows projected on the clouds below, with their heads encircled with rainbows.* In this case, the axis passes to a point above the opposite horizon equal to or greater than the semi-diameter of the bow, so that the whole of the circumference comes into view; and the eye of the spectator being in the axis, the entire bow is projected around that as a center, upon the surface of the clouds.

COLORS OF BODIES.

789. According to the Newtonian theory, the color of a body depends on *the kind of light which it reflects*. A great number of bodies are fitted to reflect at once several kinds of rays, and consequently appear under mixed colors. It may even happen that of two bodies which should be green, for example, one may reflect the pure prismatic green, and the other the green which arises from the mixture of yellow and blue. This quality of selection, as it were, in bodies, which varies to infinity, occasions the different kinds of rays to unite in every possible manner and every possible proportion; and hence the inexhaustible variety of shades which nature, as in sport, has diffused over the surfaces of different bodies.

When a body absorbs nearly all the light that reaches it, that body appears black; it transmits to the eye so few reflected

* Amer. Jour. of Science, xii, 172.—Malté Brun's Universal Geography, vol. 1, p 363.

rays, that it is scarcely perceptible in itself, and its presence and form make no impression on us, unless as it interrupts, in a manner, the brightness of the surrounding space.

But for a body to reflect one kind of ray rather than any other kind, there must be something in that body which determines the preference. In what then does a red body differ in this respect from a yellow, a green, or a violet one? Various attempts have been made, and on various hypotheses, to resolve this question. Newton, who entered on this subject with great earnestness, has here most successfully interrogated nature by a series of experiments, of which we shall give the results.

790. Having taken two glasses of a telescope, the one plano-convex, the other slightly convex on both sides, he placed one of the faces of this upon the plane face of the former, and pressed the two glasses at first gently, and then by degrees more closely against one another. The effect of this gradual pressure was, an appearance of *colored circles* in the plate of air between the glasses, which circles had the point of contact for a common center, and which increased in number as the pressure was increased, in such a manner that the circle which appeared last always was nearest the point of contact, and on a still further pressure extended its circumference to form a kind of ring round a new circle that arose near its middle.

The pressure having been carried to a certain term, Newton stopped and observed as follows: At the point of contact was a black spot that was encompassed by several series of colors, arranged from the center outward in the following order:

First series, blue, white, yellow, red.
Second, violet, blue, green, yellow, red.
Third, purple, blue, green, yellow, red.
Fourth, green and red.
Fifth, greenish blue and red.
Sixth, greenish blue and pale red.
Seventh, greenish blue and reddish white.

Beyond this number the tints were regularly paler until the color became white.

The reason why these successive colors were arranged in rings, having the point of contact of the two lenses for their common center, is obvious, since each color was developed at a certain thickness, and the points of equal thickness being equidistant from the center, they would, of course, be arranged in the circumference of a circle.

Newton measured the diameters of the annular bands formed of these different colors, by taking the points where they had most lustre; and he found that the squares of those diameters were to one another as the terms of the ascending progression 1. 3. 5. 7. 9. 11. &c.: from which it results, that the intervals between the two glasses, relatively to the corresponding points, followed the same progression.

For, let *nam* (Fig. 284,) be a diameter taken on the surface of the plane glass, and *agf* a section of the sphere to which that part of the double convex lens that turns toward *a*, belongs. Let also *ab*, *ad*, be the semi-diameters of the two rings at the points where the colors are most vivid. Having drawn *be*, *dg*, parallel to the diameter *af*, and *eh*, *gi*, parallel to *an*, we shall have

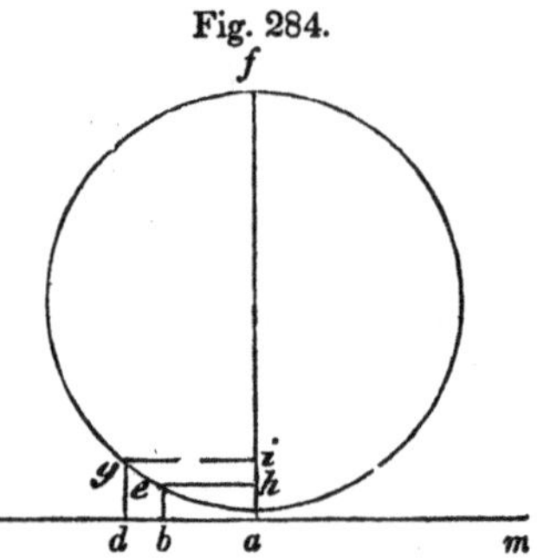

$$(eh)^2 : (gi)^2 :: ah \times hf : ai \times if.$$

But the distances between the two lenses being exceedingly small in comparison with the diameter *af*, *hf* and *if* may be taken as equal to *af*, whence, by substitution,

$$(eh)^2 : (gi)^2 :: ah \times af : ai \times af :: ah : ai :: be : dg.$$

791. From these proportions, it was merely necessary to ascertain the absolute length of a single diameter, to know the lengths of all the others, as well as the different thicknesses of the plates of air at the points where the different colors were seen. Newton drew up a table of these degrees of thickness, assigning to each color that degree at which it was developed. For example, the most intense blue makes its appearance at a thickness equal to the two millionth part of an inch, the visual ray being supposed to come to the eye perpendicularly to the two glasses. As the visual rays deviate from a perpendicular, the breadth of the rings increases, the same color requiring a greater thickness to produce it. Among other results obtained from these experiments, were the following:

Air, at and below a thickness of *half a millionth* of an inch, ceases to reflect light. At and above a thickness of *seventy-two millionths* of an inch, it reflects white; that is, all the rays of the spectrum. Between these two limits, it reflects the various orders of colors contained in the table.

Water, at and below a thickness of *three eighths of a millionth* of an inch, ceases to reflect light. At and above *fifty-eight millionths* of an inch, it reflects white; and between these two limits, it reflects the orders of the colors contained in the table.

Glass, at and below *one third of a millionth* of an inch, ceases to reflect light. At and above a thickness of the *fifty millionth* of an inch, it reflects white; and between these limits, it reflects the orders of colors contained in the table.

Newton also, having measured the diameters of the rings at the intermediate places where the colors were obscure, found that their squares were to one another as the even numbers 2, 4, 6,

8, 10, 12, &c.; and hence the intervals between the glasses, at the corresponding points, observed a similar progression.

792. Such were the phenomena which the glasses presented as seen by *reflexion;* but on looking through them to observe the effect of *refracted* light, other series of colors took the place of the preceding ones. The central spot, which was before black, now became white, and the order of colors, relatively to the different series, was this:

1. Yellowish-red, black, violet, blue.
2. White, yellow, red, violet, blue.
3. Green, yellow, red, bluish-green.
4, 5, 6. Red, bluish-green.

By comparing these colors seen by transmitted, with those seen by reflected light, it is observable, that the white answers to black, the red to blue, the yellow to violet, the green to a mixture of red and violet; that is, the part that appeared black on simply looking at the glasses, became white when the observer looked *through* them, and so of the other colors. But the tints produced by transmitted light were feeble and languishing, unless the visual ray was extremely oblique, in which case they were sufficiently vivid and brilliant.

793. Newton substituted water for air between the two glasses, and the colors instantly became fainter, and the rings contracted; that is, the ring of a particular color had its circumference nearer the center than when that color was reflected by the plate of air. The diameters of the corresponding rings were to one another nearly as 7 to 8, and consequently their squares were as 49 to 64; whence it follows, that the different thicknesses of the fluids at the places where the rings appeared, were nearly as 3 to 4; that is, in the ratio of the sine of incidence to the sine of refraction (Art. 749,) when the light passes from water into air. Newton imagined that this result might be extended to all kinds of mediums, and he therefore deduced from it this general law: that where a medium more or less dense than water is impressed between two glasses, the interval between the glasses at the place where any particular color is perceived, is to the interval which gives that color by means of air, as the sines which measure the refraction at the passage from the same medium into air. This rule may be equally applied to a thin plate, detached from any kind of body, the thickness of which we would determine by the tone of its color.*

794. The phenomena of the rings being reduced to laws extremely exact and well adapted to calculation, Newton reduced

* Haüy Nat. Philosophy, Sects. 711—720.

them all to a still simpler expression, making them depend on a physical property which he attributed to light, and of which he defined all the particulars conformably to their laws. Considering light as a matter composed of small molecules emitted by luminous bodies with very great velocities, he concluded, that since they were *reflected* from a lamina of air, at the several thicknesses corresponding to the numbers 1, 3, 5, 7, &c., and *transmitted* at the intermediate thicknesses 0, 2, 4, 6, &c., the molecules must have some peculiar modification of a periodical nature, such as to incline them alternately to be reflected and refracted after passing through certain spaces. Newton characterized this tendency to alternate reflexion and transmission, and designated the two states by the phrases *fits of easy reflexion* and *fits of easy transmission.**

795. Having defined completely all the characters of these fits, or *periodical returns of states favorable to reflexion and transmission,* Newton employed them as a simple property, not only to unite under one point of view the phenomena of the colors produced by thin plates, but also to foresee and to calculate beforehand, both as to their general tenor, and their minutest details, a crowd of analogous phenomena, observed to attend reflexion in thick plates, which, in fact, exceeded by as much as twenty or thirty thousand times those on which the calculations had been founded; moreover, applying the same reasoning to the integrant particles of material substances, which all chemical and physical phenomena show to be very minute, and to be separated even in the most solid bodies, by spaces immense in comparison of their absolute dimensions, he was able to deduce naturally from the same principles the theory of the different colors they present to us, a theory which adapts itself with surprising facility to all the observations to which these colors can be submitted. The number and importance of those applications account sufficiently for the care which Newton bestowed on his experiments on colored rings.†

796. Among the experiments of Newton on colored rings, none are more interesting than those which he instituted on *soap bubbles.* It is well known, that when these bubbles are inflated to a certain degree of thinness, very gaudy colors make their appearance, and hence these are selected as favorite objects of

* This phraseology has an air much more hypothetical than the reality, the thing signified being little more than the simple enunciation of a fact ascertained by experiment. Probably the singularity of the phrase has contributed to bring the doctrine into discredit, or even into ridicule, with those who have never looked any further into it than to read the title. The most profound opticians of modern times, have regarded these investigations of Newton, as among the most ingenious and sagacious of all his labors.

† Biot

amusement for children. But it was reserved for no less a mind than that of Newton, to make these exhibitions the means of penetrating the secrets of nature.

In preparing the bubbles for experiment, he took various ingenious precautions to form them in the most perfect manner, and preserved them for deliberate examination, by covering them with a glass receiver which protected them from the agitation of the air, and means were devised for preventing any extraneous light from mixing with that of the bubble. Things being thus arranged, and the eye placed in a favorable position, a number of concentric horizontal rings are seen, exhibiting vivid colors disposed with perfect regularity. They correspond in appearance to those exhibited by the plate of air between the lenses, (Art. 790,) but are more elegant and perfect in every respect. Similar exhibitions of color are presented in glass bubbles blown exceedingly thin; and also in thin laminæ of the mineral called mica. Analogous variations of color are seen even in the tarnish of certain metals, particularly in plates of copper and steel when they have been heated in the open air, and they appear in the plumage of birds.

797. The following propositions, several of which have already been incidentally mentioned, will present a *summary* of the Newtonian doctrine of colors.

1. The colors of natural bodies are not qualities inherent in the bodies themselves, by which they immediately affect our sight, but *are a mere consequence of that peculiar disposition of the particles of each body, by which it is enabled more copiously to reflect the rays of one particular color, and to transmit, or stifle, or, more properly, to absorb, the others.*

2. The colors of natural bodies are the colors of *thin plates*, produced by the same cause as that which produces them in thin laminæ of air, glass, &c., viz. the interval between the anterior and posterior surfaces of the atoms. The *thickness* of the atoms of a medium, and of the interstices between them, determines the color they reflect or transmit at a particular incidence, because it must depend on the thickness of any lamina, whether the light when it has reached its posterior surface is in the state favorable for transmission or for reflexion.

3. Opacity in natural bodies arises from the *multitude of reflexions* caused in their internal parts. By this means, the rays are conceived to be entangled, as it were, running their rounds from atom to atom, without a possibility of reaching the surface and escaping.

It would be inconsistent with the nature of an elementary work like the present, to enter into all the details of this remarkable hypothesis: for such disquisitions we must refer the student to Brewster's Optics, to Biot's Traité de Physique, to Herschel's

Treatise on Light, Professor Bartlett's Optics, and to various other works of great ability which have been written on these subjects within the present century.

INFLEXION OR DIFFRACTION OF LIGHT.

798. INFLEXION *or* DIFFRACTION *is a term used to denote certain phenomena which light exhibits, when, under certain circumstances, it forms parallel bands or fringes.*

For the purpose of experiments on this subject, a beam of light is admitted into a dark room, through a very small aperture, as a pin-hole made in sheet lead; or, what is better, a convex lens is placed in the window shutter, which brings the rays to a focus, and affords a divergent pencil of light. If we introduce into this pencil any opake body, as a knife-blade, for example, and observe the shadow which it casts on a white screen, we shall observe on both sides of the shadow *fringes of colored light*, the different colors succeeding each other in the following order: first fringe, *violet, indigo, pale blue, green, yellow, red;* second, *blue, yellow, red;* third, *pale blue, pale yellow, red.* The brightness of these fringes diminishes as they recede from the shadow, and the shadow itself is not quite dark, but is formed also of luminous and dark fringes, all parallel to the edges of the knife. The fringes in question are absolutely independent of the nature of the body whose shadow they surround, and the form of its edge; neither the density or rarity of the one, nor the sharpness or curvature of the other, having the least influence on their breadth, their colors, or their distance from the shadow. Thus it is indifferent whether they are formed by the edge or back of a razor, by a mass of platina, or by a bubble in a plate of glass, (which, though transparent, yet throws a shadow by dispersing away the light incident upon it;) circumstances which make it clear that their origin has no connection with the ordinary refractive powers of bodies, or with any *elective* attractions or repulsions exerted by them on light; for such forces cannot be conceived as independent of the *density* of the body exerting them, however minute we might regard the sphere of their action.*

799. If the light of the solar beam be first separated into the prismatic colors, and these severally be submitted to experiment, the fringes will in each case be of the same color as the colored pencil; but they will be broadest in *red* light, smallest in *violet,* and of intermediate sizes in the intermediate colors. If we place the screen at different distances from the interposed body which gives the shadow, it will be found that the fringes grow less and

* Herschel.—Brewster, *Life of Newton*, p. 103.

less as we approach the edge of the body from which they take their rise. On measuring the distances of the fringes from the shadow, while they are thus changing their dimensions, and connecting by a line the several points representing those distances, it is found that this line is not a straight line, but a *hyperbola*, whose vertex is at the edge of the body; so that the same fringe is not formed of the same light at all distances from the body, but resembles a caustic curve, (Art. 742,) formed by the intersection of different rays. When we consider that the fringes are largest in red, and smallest in violet light, it is easy to understand the cause of their colors in white light; for the colors seen in this case arise from the superposition of fringes of all the seven colors; that is, if the eye could receive all the differently colored fringes at once, these colors would form by their mixture the actual colors in the fringes seen by white light. Hence we see why the color of the first fringe is violet near the shadow, and red at a greater distance; and why the blending of the colors beyond the third fringe, forms white light, instead of exhibiting themselves in separate tints.

Upon measuring the proportional breadths of the fringes with great care, Newton found that they were as the numbers 1, $\sqrt{\frac{1}{3}}$, $\sqrt{\frac{1}{5}}$, $\sqrt{\frac{1}{7}}$, and their intervals in the same proportion.*

800. In the foregoing experiments, the colored fringes are supposed to be formed on the edge of the shadow of an opake body, placed in a beam or pencil of light. The same phenomena are exhibited in a more striking and beautiful manner, when we view with a magnifying glass a pencil of light as it passes through an exceedingly small aperture. Suppose, for instance, we place a sheet of lead, having a small pin-hole pierced through it, in the pencil of rays diverging from the focus of a lens. The image of the hole will be seen through the magnifier as a brilliant spot, encircled by rings of colors of great vividness, which contract and dilate, and undergo a singular and beautiful alternation of tints, as the distance of the hole from the luminous point on the one hand, or from the magnifier on the other, is changed. When the latter distance is considerable, the central spot is white, and the rings follow nearly the orders of colors of their plates. When the magnifier is brought very near to the pin-hole the central white spot contracts into a point and vanishes, and the rings gradually close in upon it in succession, so that the center assumes, successively, the most surprisingly vivid and intense hues, and the rings surrounding it undergo great and abrupt changes in their tints.†

Newton attempted to account for the inflexion of light by a supposed repulsion exerted by the edge of the interposed body

* Brewster's Optics, pp. 95—97.

† Herschel on Light, Sec. 730.

or by the edges of the circular aperture on the rays of light that are nearest to it, while they exert a less *repulsion* on such as are a little more remote. By this means, the relative direction of the rays would be so altered that they would cross one another, and their light interfere or become blended; and by following out the consequences of this interference, they are found to correspond to some of the effects actually observed to take place.*

801. A more satisfactory explanation of the inflexion of light and the formation of colored fringes, is afforded by that theory which considers light as produced, not by the *emission of luminous particles* from the radiant body, but by the *undulations of a peculiar fluid.* Phenomena of the foregoing description are accounted for on the doctrine of *interferences.* It appears by experiment, that a body already illuminated *may become less bright by the addition of more light.* Let the solar light, reflected horizontally, be admitted into a dark chamber by two small holes, which are near each other, but separated by such an interval that the conical pencils do not intermix until they have proceeded a certain distance. A little beyond the point where they intermix, let them be received on a screen; then at some points of the illumined part there will be a partial or comparative darkness. If now, one of the openings is closed, so that the light is not intermixed on the illumined part, but the screen receives light only from one hole, the partial darkness will vanish, and parts of the remaining circle will have become *brighter by this loss of light.* Here it is evident that the addition of fresh light produces darkness, and that an obscure surface becomes brighter by the removal of some of the light which shines upon it.

When the preceding experiment is made with great care, and light of *one color*, as red or blue, is admitted through two fine slits or holes near each other, the pencil produces bands, which are alternately bright and dark, exactly analogous to the bands produced in the experiments on diffraction. But when either of the apertures is closed, the bands disappear, and the space in which they were is occupied by light that is nearly uniform. Thus the stoppage of the light from one aperture removes the partial obscurity which existed between the bright spaces. This darkness, therefore, results from the concourse of the two lights meeting obliquely from the two apertures. Hence it appears that two homogeneous rays of light, emanating from the same source, may, after passing over a certain distance, come to a point under such circumstances that the brightness will be almost annihilated. This effect can be referred to nothing else but

* Fresnel, however, has shown that this phenomenon is independent of the *edges* of bodies.

the mutual action of the rays of light, and to this mutual action the term *interference* has been applied.

Something quite analogous to this occurs in the phenomena of *sound*. The transmission of sounds being by undulations, we may conceive two undulations to exist exactly similar, and to produce simultaneous impulses in the same direction, so that the effect on the ear will be double what either would have produced separately. Here one sound-wave is augmented by the addition of another. But we may also readily conceive cases in which one sound-wave may interfere with another, so that the combined effect is less than either would have produced alone. Effects of this kind, well known in music, are called *beats;* and this and numerous other analogies between the phenomena of light and those of sound, which have been traced with the most refined ingenuity by several eminent opticians, have produced a strong impression in favor of the *undulatory* theory of light.*

CHAPTER VII.

OF DOUBLE REFRACTION AND POLARIZATION.

802. Double refraction *is a modification which light undergoes in passing through certain media, by which a single pencil of light is divided into two pencils, affording two separate images of the objects.*

This phenomenon was first observed in a crystal of carbonate of lime, denominated *Iceland spar.* This substance may be seen in every cabinet of minerals, presenting the figure of a rhomb. It is a solid, bounded by six rhomboidal faces. It is colorless and highly transparent, and distinguished for its beauty in mineralogical collections; but its most remarkable property is that of rendering letters, or any other small objects placed behind it, *double.*

Fig. 285.

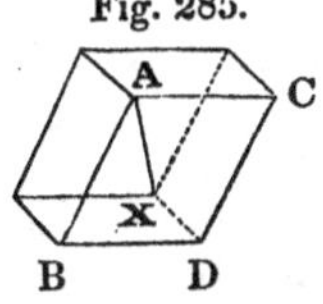

Though double refraction is exhibited by the Iceland crystal in a manner peculiarly striking, yet this phenomenon is by no means confined to that substance. It takes place in numerous transparent crystals. It also occurs in a variety of other bodies, which are more or less transparent, where there is any disposition toward a regular arrangement of the particles, such as hair, quills, and the like, and in all bodies when in a state of unequal dilatation or compression.

* See the explanation in full in Pouillet's El. de Phys., ii, 106.

Some classes of bodies possess the property of double refraction naturally and permanently; others may be made to acquire it by artificial means, but retain it only transiently.

803. The explanation of this singular effect has exercised the sagacity of the profoundest philosophers, at the head of whom are Newton and Huygens. Inquiries respecting it have of late years been associated with those respecting the polarization of light, both of which subjects have been studied with the greatest attention and zeal by some of the first philosophers of the present century; and their investigations have opened a new field of philosophical curiosity, no less ample than fertile. In a work so limited as the present, it will be impossible to give any thing more than a very slight sketch of these subjects, to serve merely for the purposes of an *introduction* to studies which, in order to be fully understood, require to be prosecuted for a length of time proportioned to their extent and intricacy. Unimportant as these researches might appear, on a superficial view, they are, nevertheless, in common with several other refined inquiries in optics, highly conducive to that full knowledge of the properties of light, which enables the artist to give perfection to such noble instruments as the microscope and the telescope.

804. If a rhomb of Iceland spar, represented in Fig. 286, be placed above a black line drawn on white paper, and viewed with the eye at R, the line will appear double, as *mn*, MN; or if we cause a pencil of light R*r*, to fall upon the surface of the rhomb, it will be separated into two pencils, *r*O, *r*E, each of which will emerge from the rhomb at *o'* and *e'* in directions O*o'*, E*e'*, parallel to R*r*. The pencil R*r* has therefore suffered *double refraction* in passing through the rhomb. This might be suspected to arise from a difference in density of the different parts, since the parts which were more dense would refract the light more than those which were less dense, and thus tend to produce separate images; but as the same effects will take place by making the pencil R*r* fall at the same incidence, and in the same direction, relative to the summit A upon any point of any of the faces, it is manifest that the double refraction cannot arise from any difference of density in different parts of the rhomb.

Fig. 286.

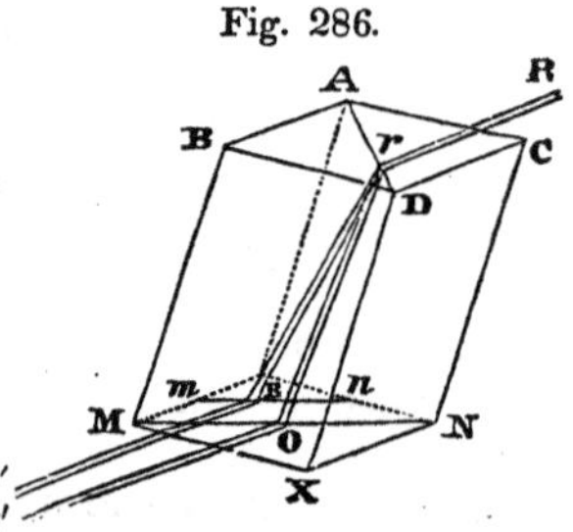

805. To illustrate the manner in which these effects take place let AMXN, (Fig. 287,) be a section of the Iceland crystal, (Fig

285,) made by a plane passing through the diagonal AX. Let PV be a perpendicular to the upper surface of the crystal at any point F. Then rays of light DF, CF, will each have two refracted rays, FO, FO′, and FE, FE′. On measuring the several angles, it will be found that the two rays FO, FO′, follow the ordinary law of refraction, making the sines of the angles of refraction as the sines of those of incidence. Each of these, therefore, is called the *ordinary ray*. But the other rays FE, FE′, will not conform to that law, but will make angles with the perpendicular, that are sometimes greater and sometimes less than would be required by it, and are thus separated from the ordinary rays. Hence, each of these is called the *extraordinary ray*. In the case of the Iceland crystal, while at a perpendicular incidence the ordinary ray undergoes no refraction, the extraordinary ray is turned from the perpendicular 6° 12′; and at angles of 10°, 20°, 30°, &c., while the ordinary ray suffers a regular refraction according to the law of the sines, the extraordinary ray suffers a refraction constantly greater than that.

Fig. 287.

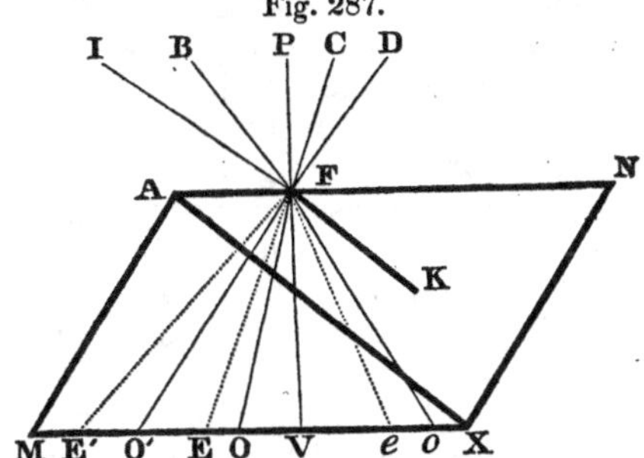

806. The line which connects the two obtuse angles of the rhomboid, as AX, (Fig. 287,) is called the *optic axis*, since when a ray enters a crystal in the direction of this line, either coincident with it or parallel to it, such a ray suffers no double refraction. Thus, the ray IF, which meets the crystal in such a direction as to be refracted into the line FK, parallel to AX, is not separated into two rays, but merely undergoes ordinary refraction. When the extraordinary ray is between the ordinary ray and the perpendicular, as Fe, the crystal is said to have a *positive* axis: when the extraordinary ray is further from the perpendicular than the ordinary ray, it is said to have a *negative* axis. Thus, quartz has a positive, and Iceland spar a negative axis.

Any plane, like that of Fig. 287, passing through the optic axis, and every plane parallel to this, is called the *principal section*. Whenever the principal section contains the ordinary ray, it also contains the extraordinary ray, which is not the case with planes inclined to that.

807. An axis of double refraction, however, is not, like the axis of the earth, a *fixed line*, within the rhomb or crystal. It is only a *fixed direction:* for if we divide, as we may do, the rhomb ABC (Fig. 285) into two or more rhombs, each of these separate

rhombs will have its axis of double refraction; but when these rhombs are again put together, their axes will all be parallel to AX. Every line, therefore, within the rhomb parallel to AX, is an axis of double refraction or optic axis; but as these lines have all one and the same direction in space, the crystal is still said to have only one optic axis.

Herschel, in his Treatise on Light, illustrates this subject by the following simile. Suppose a mass of brick work, or masonry, of great magnitude, built of bricks all laid parallel to each other. Its exterior form may be what we please; a cube, a pyramid, or any other figure. We may cut it (when hardened into a compact mass) into any shape, a sphere, a cone, a cylinder, &c., but the edges of the bricks within it, still lie parallel to each other; and their directions, as well as those of the diagonals of their surfaces, or of their solid figures, may all be regarded as so many axes, i. e. lines having (so long as the mass remains at rest) a determinate position, or rather *direction* in space, no way related to the exterior surfaces, or linear boundaries of the mass, which may cut across the edges of the bricks in any angles we please.

808. A great number of crystals have *two* axes of double refraction, or two directions inclined to each other, along which the double refraction is nothing. In crystals with one axis, the axis has the same position, whatever be the color of the pencil of light which is used; but in crystals with two axes, the axes change their position according to the color of the light employed, so that the inclination of the two axes varies with differently colored rays.

Until recently it was supposed that the number of optic axes never exceeds two;* but Dr. Brewster has lately discovered an example of a mineral (*analcime*) which has an indefinite number of axes of double refraction, in the direction of which, light suffers no separation, although when passing through the body in any other direction, it undergoes double refraction.

A cylinder of glass, first heated red hot, and then rolled on a plate of metal until it is cold, acquires a permanent doubly refracting structure. If, instead of heating the glass cylinder, we had placed it in a vessel, and surrounded it with boiling oil or boiling water, it would have acquired the same doubly refracting structure, when the heat had reached the axis; but this structure is only transient, as it disappears when the cylinder is uniformly heated. Analogous structures may be produced by pressure, and by the induration of soft solids, such as animal jellies, isinglass, &c.

* Herschel on Light, Sec. 781.

809. If the cylinder in the preceding article is not a regular one, but has its section perpendicular to the axis everywhere an *ellipse* instead of a *circle*, it will have two axes of double refraction. In like manner, if we use *rectangular plates* of glass, instead of cylinders, in the preceding experiment, we shall have plates with *two planes* of double refraction; a positive structure being on one side of each plane and a negative on the other. If we use perfect *spheres*, there will be axes of double refraction along every diameter, and consequently an infinite number of them. The crystalline lenses of the eyes of almost all animals, whether their figures be those of lenses, spheres, or spheroids, have one or more axes of double refraction.

POLARIZATION OF LIGHT.

810. Polarization of light *is a change which light undergoes after certain refractions or reflexions, by which a ray acquires* polarity, *or different properties on different sides.*

This quality of light, which is one of the most remarkable of all its properties, was discovered by Huygens, during his investigations into the cause of double refraction as exhibited in the Iceland crystal; but the attention of opticians was more particularly directed toward it by the discoveries of Malus, in 1810.* The knowledge of this singular property of light has afforded an explanation of many of the most intricate phenomena in optics.

811. With respect to the light of the sun, whether it be direct or reflected, whether it be white light or one of the prismatic colors, no such difference of properties exists in the different sides of a ray; and the same is true of the light of a candle or any self-luminous body. But if instead of employing a ray emitted directly from the sun or from any self-luminous source, we subject to examination a ray that has undergone double refraction, or a certain kind of reflexion to be more particularly described hereafter, or that has been in any one of a great variety of ways subjected to the action of material bodies, it seems to have acquired *sides;* a right and a left, a front and a back; and the *intensity*, or brightness, though not the *direction* of the reflected or transmitted portion, depends materially on the position with respect to these sides, in which the plane of incidence lies, though every thing else remains precisely the same.†

812. We may understand something of the nature of the

---

* See an interesting history of the progress of these discoveries, in the Edinburgh Philosophical Journal, Vols. I, II, &c.

† Herschel

changes produced in light by polarization from the following experiments.

First, we place on a horizontal table a piece of black paper, and draw on it two fine lines at right angles to each other, having a white point or dot at their intersection. For the convenience of reference, we will suppose that the direction of one of these lines is north and south, and that of the other east and west. Over these lines we place a crystal of Iceland spar, (Fig. 285,) having its *principal section* (which it must be recollected is the plane passing through both the rays into which the incident light is divided) in the meridian. Agreeably to what has been said in Art. 805, we shall see two images of the white spot formed by double refraction. The light thus divided will, when transmitted through another medium producing double refraction, become polarized. For, if we place a second crystal of the same kind directly over the first, having their principal sections parallel, or coincident, the two images will appear as before, except farther asunder. But, on turning the upper crystal round, on its vertical axis, two new images begin to make their appearance, being very faint at first, but growing brighter, while the two original images grow weaker, until the crystal is turned through the first quadrant, when the four images become equally bright. Continuing to turn the crystal through the second quadrant, the reverse takes place, that is, the new images grow fainter and the original ones brighter, until at the end of this quadrant, when the crystal has changed sides, only the two original images appear, as at first. Similar changes occur in turning the crystal through the third and fourth quadrants. Now the only change of circumstances involved in the foregoing process, is that the plane of incidence is successively presented to the *different sides* of the rays of light, the effect being greatest when it is applied to the sides directly opposite to each other; and as a particle of iron filings, however small, acquires, when magnetized, different and peculiar properties in two opposite points called the poles, so a ray of light, by the foregoing process, acquires properties on the opposite sides somewhat analogous to polarity, and hence is said to be *polarized*.*

813. Secondly, for an example of polarization by *reflexion*, we may take two tubes, a larger and a smaller, the latter turning within the former, like two tubes of a hand telescope. The compound tube thus formed being open at both ends, we may attach to each a glass reflecting plate, in such a manner, that by turning round the smaller tube the two reflecting plates may be placed in various positions with respect to each other. Thus, (Fig. 288,) let MNP be the tube, A and C plates of glass, so situ-

---

* Arago, Encyc. Brit., Sup Vol. VI.

ated that a ray of light RA, incident at A, at an angle of 56°, may be reflected along the axis of the tube AC, and striking on C at the same angle of incidence of 56°, be reflected to the eye

Fig. 288.

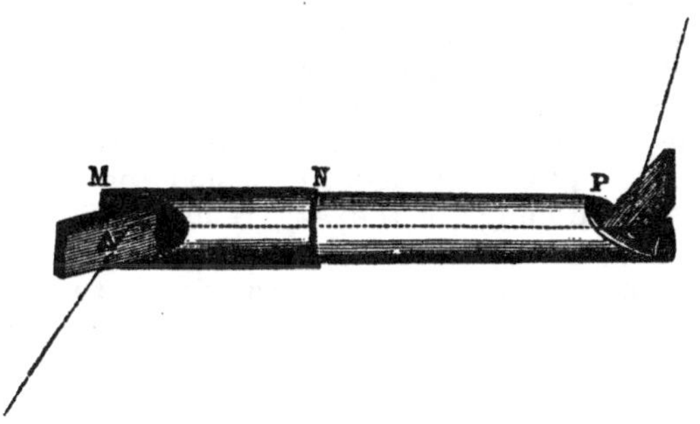

at E. Then in the position shown in the figure, where the first reflexion is made in a horizontal plane RAC, and the second in a vertical plane ACE, the image of R will be scarcely visible; but on turning round the tube NP, the image will grow brighter and brighter, until the plate C has been turned round ninety degrees, when it becomes the brightest possible. On being turned through the second quadrant, the image grows fainter and fainter, until, at 180° from the original position of C, the image almost disappears. Similar changes occur in the third and fourth quadrants. Now, at the commencement of the experiment, the mirror C is applied to the *under* side of the ray, and at the end of 180° it is applied to the *upper* side, at both which points there is scarcely any reflexion; but when the same plate is applied *laterally*, the reflexion is the same as for common light. Hence this light had acquired peculiar properties on its opposite sides, in consequence of its previous reflexion from the first plate of glass, which, as before, from its analogy to magnetic polarity, is denominated polarization.*

814. The angle of 56° is found by experiment to be that at which this effect was produced by glass, and hence this is called the polarizing angle for glass. Other substances have different polarizing angles. Thus, while the two reflecting plates are in the position shown in the figure, in which case there is scarcely any reflexion from the second plate, and consequently no image formed at the eye, yet on moistening the plate C, even with the breath, the image instantly appears, since the polarizing angle for water is different from that of glass.

Although the images formed by polarized light are variable in

* Library of Useful Knowledge.

brightness, yet the *direction* of such rays is the same as in common light, and consequently no change is produced by polarization in the *place* of the image.*

815. The relation of *color* to polarized light is highly interesting, the most gorgeous and varied hues being developed in experiments on this subject. The following is a simple example.

Let two plates of glass be placed as in Fig. 289, so that the light, which may be that of the sky, will be polarized by reflexion from the plate A, in consequence of which the eye situated at E, will not see the image of the sky reflected from C, as it would do in the case of common light, but in the place of it will see a dark, undefined spot. But on interposing between the two plates a thin film of selenite,† or mica, the eye being still at E

Fig. 289.

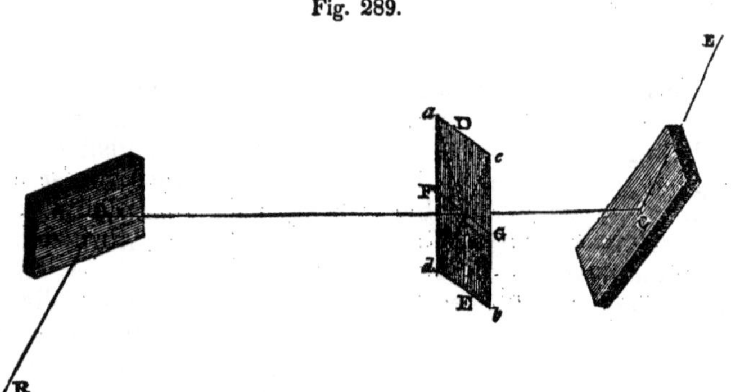

we shall see the surface of the interposed plate covered with the most brilliant colors. If its thickness is perfectly uniform throughout, the tint will be uniform; but if it has different thicknesses, every different thickness will display a different color,—some red, some green, some blue, some yellow, and all of the most brilliant description. If we turn the plate EDFG round, keeping it perpendicular to the polarized beam, the colors will become less or more bright, without changing their nature, and two lines ED, FG, at right angles to each other, will be found such, that when either of them is in the plane of reflexion RAC, no colors whatever are perceived, but the undefined dark spot before mentioned returns again. On continuing the rotation of the plate, the colors reappear, and reach their greatest brightness when either of the lines *ab* or *cd*, which are inclined 45° to the lines ED, FG, is in the plane of reflexion ACE.

Let the plate EDFG be now fixed in the position where it

* See a good explanation of these phenomena in Herschel's Treatise on Light, Sects. 816—819.

† A variety of sulphate of lime or plaster of Paris

gives the brightest color, which suppose to be *red*, and let the plate C be made to revolve. The brightness of the red color will gradually decline until the plate has turned round 45°, when it will wholly disappear, and the dark spot be seen again. Beyond 45° a *green* color will gradually make its appearance, and will become brighter and brighter until it reaches its maximum at 90°. Beyond 90° the green fades, and disappears at 135°, when the dark spot returns; but beyond this the red reappears, and reaches its maximum at 180°. Hence, when the plate EDFG alone revolves, only *one* color is seen, and when C only revolves, *two* colors are seen during each half of its revolution. If we repeat these experiments with plates of different thicknesses in different parts, giving different colors, we shall find that the two colors are always *complementary* of each other, or together make white light.* The experiment may be varied so as to evolve the red and green colors at the same time, making one overlap the other: the parts thus united form a perfect white.†

Instead of the plate of selenite or mica, by exposing doubly refracting crystals of different substances to the action of polarized light, an endless variety of beautiful colors, arranged in the most fanciful forms, may be seen, specimens of which are usually represented by figures, in works that treat at large of this subject.‡

CHAPTER VIII.

OF VISION.

816. As a preparation for studying the optical structure of the eye and the laws of vision, it will be useful first to learn in what way images of external objects are formed in a dark room, by light admitted through a hole in the window shutter.

817. *A beam of light from the sun, entering into a dark room through a small orifice, and striking upon an opposite wall or screen, forms a circular image on the wall, whatever be the shape of the orifice.*

We will suppose the orifice to be comparatively large, as an inch in diameter, and of a triangular or of an irregular shape; the image formed on the wall will still be circular. For, suppose the orifice to be reduced to a very small circular hole, as a pin-

* Brewster's Optics, 159. † Library of Useful Knowledge.

‡ See, especially, Pouillet, de Phys., t. II.

hole, (which may easily be done by placing over the orifice a metallic plate, as a sheet of lead, pierced by a pin;) then the rays of the sun, passing through this small opening, would of course be circular. But the large irregular orifice may be considered as made up of such smaller apertures, or the metallic plate may be conceived to be pierced with an indefinite number of pin holes, and the entire image formed upon the wall may be conceived to be made up of an assemblage of all these images of the sun blended with each other, and therefore, as bounded by innumerable curve lines, composed of the individual circles.

If the screen be brought near to the orifice, however, the image will be of the same figure as the orifice; for the rays, after they have passed the orifice, must have diverged considerably before the sections that form the image shall afford circles so large, that their blended circumferences shall compose a circular figure. (See Fig. 290.)

Fig. 290.

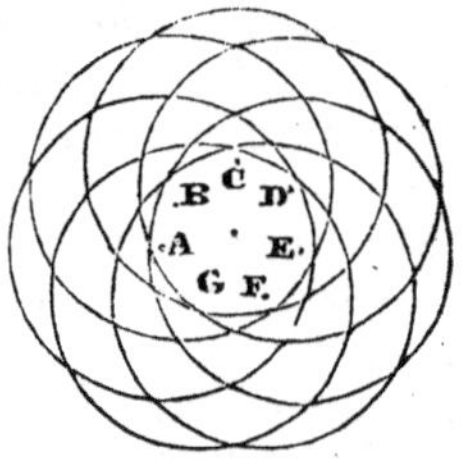

If the plane which receives the image be not parallel to the orifice, then the image will be elliptical, being the section of a cone oblique to its axis.

Circular images of the sun are sometimes projected on the ground, through the small openings among the leaves of the trees. During an eclipse of the sun, these images copy the figure of the eclipse.

If there are various orifices near to each other, *three*, for example, through which a beam of the sun shines into a dark room, we shall observe at first, at a certain distance, three distinct luminous circles. At a greater distance, these three circles begin to be blended, and finally, on enlarging sufficiently, they unite to form a single circle.*

818. *If, instead of a beam of solar light, we admit into a dark room, through an opening in the shutter, the light reflected from various objects without, an inverted picture of these objects will be formed on the opposite wall.*

A room fitted for exhibiting such a picture is called a *Camera Obscura.*

From what has been before explained, it will be readily understood, that from every point in the object, innumerable rays of light proceed and fall upon the window shutter. Of these, however, none can enter the aperture, except such as are very near to each other, all others diverging too far to enter a small opening. It is essential to the *distinctness* of the picture, that rays which proceed from every point in the object, should be collected into corresponding points in the image, and should exist

* Barlow, in Encyc. Metropol. Art. *Optics*

there free from any mixture of rays from any other point; and it is essential to the *brightness* of the picture, that as many rays as possible should be conveyed from each point in the object to its corresponding point in the image. To render the picture distinct, therefore, the opening in the window shutter must be small, else the pencils of rays from different points will *overlap* each other, and confuse the picture; but as the orifice is diminished, the brightness of the picture is impaired, since, in this case, a smaller number of rays are conveyed from the object to the image.

These modifications of the picture according to the size of the aperture, may be easily exhibited by beginning with a circular aperture, two or three inches in diameter, and reducing its size gradually, by covering it with a piece of board, or a metallic plate, perforated with holes of different sizes.*

819. *If, instead of passing through the naked orifice, the rays be received on a convex lens, two or three inches in diameter, fixed in the window shutter, a very bright and distinct picture of the external landscape will be formed on a screen, placed at the focal distance of the lens.*

The image is *brighter* and more *distinct* than when formed without the aid of the lens; first, because the diameter of the lens may be so great as to receive and transmit a much larger portion of the rays which proceed from each point of the object, than would be compatible with distinctness, if so large a naked aperture were employed; secondly, because the rays of each pencil are brought more accurately to a separate focus; and thirdly, because the picture being formed nearer to the window shutter, it is smaller, and of course the light, being spread over less space, is more intense.

A convex lens fixed in a ball is used for this purpose, which is so attached to the opening in the shutter, as to be capable of being turned toward different parts of the landscape, like the eye-ball in its socket. Such a lens, with its accompanying parts, is called a *Scioptic ball.*

In a bright sunny day, when the sun is on the side of the house opposite to the shutter, and of course illuminating the sides of objects which face the window, we may form, either with or without the aid of the scioptic ball, a very striking and beautiful

* A small room, ten feet square for example, having a window opening toward an unobstructed landscape, may easily be converted into a camera obscura. The perforation in the shutter must be made equidistant from the sides of the room; and from the aperture as a center, with a radius equal to the distance of the opposite wall, describe an arc of a circle, upon which, as a base, a new concave wall is to be constructed, finished with stucco. The other walls and ceiling are to be colored a dead black, while the concave wall, for receiving the image, is made as white as possible. On admitting the light through an aperture half an inch in diameter, a beautiful and distinct picture will be formed on the opposite wall

picture of external objects, exhibiting each in its relative situation, of a size and brightness corresponding to its distance, with all the colors and the most delicate motions of the landscape. The name *camera obscura*, which appropriately belongs to such a chamber, is also extended to certain boxes, in which similar pictures are formed, with peculiar devices for rendering the image erect instead of inverted. The *Daguerreotype* is an instrument by means of which the image formed on the principle of the camera obscura, is fixed so as to be permanent.

The eye is a camera obscura, and the analogy existing between its principal parts and the contrivances employed to form a picture of external objects, as in the preceding experiments, will appear very striking on comparison.

820. The EYE* is an assemblage of lenses which concentrate the rays emanating from each point of external objects on a delicate tissue of nerves, called the *retina*, there forming an image or exact representation of every object, which is the thing immediately perceived or *felt* by the retina. Figure 291, is a section of the human eye through its axis, in a horizontal plane. Its figure is, generally speaking, spherical, but in front considerably more prominent than the corresponding portion of a sphere. The eye consists of three principal chambers, filled with media of perfect transparency, whose refractive powers differ considerably among themselves, but none of them is greatly different from pure water. The first of these media, A, occupying the anterior chamber, is called the *aqueous humor*, and consists, in fact, chiefly of pure water, holding in solution a little common salt and gelatine, with a trace of albumen. Its refractive index, (Art. 747,) is almost precisely that of water, viz. 1.337, that of water being 1.336. The cell in which the aqueous humor is contained, is bounded, on its anterior side, by a strong, horny, and delicately transparent coat, *aa*, and is called the *cornea*, the figure of which is an ellipsoid, produced by the revolution of an ellipse about its major axis.

Fig. 291.

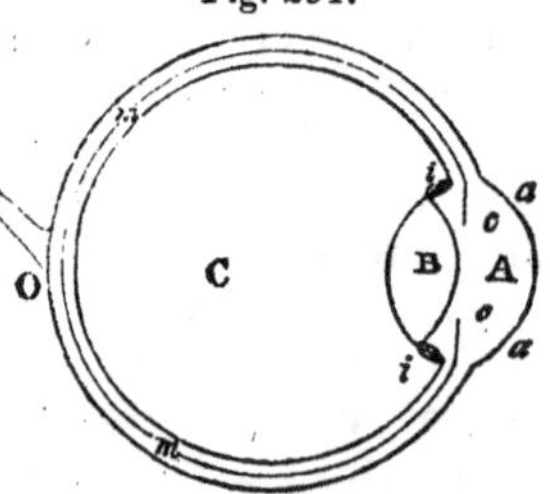

821. We have seen, (Art. 761,) that convex lenses of a spherical curvature do not bring rays of light accurately to a focus, but spread the light over a space of greater or less extent, which is called the spherical aberration of the lens. It has also appeared, (Art. 763,) that if the lens be made of the figure pro-

* The subjoined description of the eye is taken chiefly from Herschel's Treatise on Light

duced by the revolution of an ellipse on its major axis,—an ellipse whose major axis is to the distance between the foci, as the sine of incidence is to the sine of refraction; then the rays will be brought accurately to a focus, and no spherical aberration will take place. We have before us, in the aqueous humor, an example of this construction. Its figure is such an ellipsoid, the ratio of whose major axis to the distance between the foci, is almost precisely the same with that which exists between the sines of incidence and refraction; the former ratio being expressed by 1.3, and the latter by 1.337; hence parallel rays incident on the cornea in the direction of its axis, are made to converge to a focus situated behind the cornea, with almost mathematical precision, the aberration, which would have occurred had the external surface been a spherical figure, being almost completely destroyed.

822. At the posterior surface of the chamber A of the aqueous humor, is the *iris cc*, which is a kind of circular opake screen, or diaphragm, consisting of muscular fibres, by whose contraction or expansion, an aperture in its center, called the *pupil*, is diminished or dilated, according to the intensity of the light. In very strong lights, the opening of the pupil is greatly contracted, so as not to exceed twelve hundredths of an inch in the human eye, while in feebler illuminations it dilates to an opening not exceeding twenty-five hundredths, or double its former diameter. The use of this is evidently to moderate and equalize the illumination of the image on the retina, which might otherwise injure its sensibility. In animals, as the cat, which see well in the dark, the pupil is almost totally closed in the daytime, and reduced to a very narrow line; but in the human eye, the form of the aperture is always circular. The contraction of the pupil is involuntary, and takes place by the effect of the stimulus of the light itself; a beautiful piece of self-adjusting mechanism, the play of which may be easily seen by bringing a candle near to the eye, while directed to its own image in a looking glass. Immediately behind the opening of the iris, lies the *crystalline lens* B, enclosed in its capsule, which forms the posterior boundary of the chamber A. The figure of the crystalline lens is a solid of revolution, having its anterior surface much less curved than the posterior. Both figures are ellipsoids of revolution about their lesser axes; but the axes of the two figures are neither exactly coincident in direction with each other, nor with that of the cornea. This deviation would be fatal to distinct vision, were the crystalline lens very much denser than the others, or were the whole refraction performed by it. This, however, is not the case: for the mean refractive index of the lens is only 1.384, while that of the aqueous humor, as we have seen, is 1.337; and that of the *vitreous* humor C, which occupies the third chamber,

is 1.339; so that the whole amount of bending which the rays undergo at the surface of the crystalline, is small in comparison with the inclination of the surface at the point where the bending takes place; and since near the vertex, a material deviation in the direction of the axis can produce but a very minute change in the inclination of the ray to the surface, the cause of error is so weakened in its effect, as probably to produce no appreciable aberration. The consistence of the crystalline is that of a hard jelly, and it is purer and more transparent than the finest rock crystal.

823. In the crystalline, a very curious and remarkable contrivance is adopted for overcoming or preventing the spherical aberration, which (Art. 761) belongs to lenses of this form, that refract the rays more toward their marginal than near their central parts, and hence do not bring all the rays belonging to one pencil to the same focus. Here the difficulty is obviated by giving to the central portions of the crystalline a proportionately *greater density*, thus increasing its refractive power so as exactly to correspond to that of the other portions of the lens.

The posterior chamber C of the eye is filled with the *vitreous humor*, a fluid differing neither in specific gravity nor in chemical composition, in any sensible respect, from the aqueous; and as we have already seen, having a refractive index but little superior to it. Its name is derived from its supposed resemblance to melted glass; it is a clear, gelatinous fluid, very much resembling the white of an egg. Rays of light diverging from various objects without, on passing through the aqueous humor, (which is a concavo-convex lens,) have their divergency much diminished, or even, in most cases, are rendered converging, and in this state are transmitted through the crystalline, which has precisely such a degree of refractive power as enables it to bring them to a focus at the distance of the retina, which, as a screen, is spread out to receive the image. The retina, as its name imports, is a kind of white net-work, like gauze, formed of inconceivably delicate nerves, all branching from one great nerve O, called the *optic nerve*, which enters the eye obliquely at the inner side of the orbit, next the nose. The retina lines the whole of the cavity C up to *ii*, where the capsule of the crystalline commences. Its nerves are in contact with, or immersed in the *pigmentum nigrum*, a dark velvety matter, which covers the *choroid membrane*, *mm*, and whose office is to absorb and stifle all the light which enters the eye as soon as it has done its office of exciting the retina; thus preventing internal reflexions, and consequent confusion of vision. The whole of these humors and membranes are contained in a thick tough coat, called the *sclerotica*, to which is joined the cornea, which forms what is called the *white of the eye*.

824. Such, in general, is the structure by which *parallel* rays, and those coming from very distant objects, are brought to a focus on the retina. But there are *special contrivances* suited to particular purposes, which are no less evincive of design and skill than the general organization of the eye. Some of the most remarkable of these we proceed to mention. The cornea, by protruding, collects the rays of light that come to the eye laterally, and guides them into the eye, thus enlarging the range of vision. It answers to an appendage of the microscope, which will hereafter be described under the name of *field glass.* The motion of the eye-ball, by means of which the pupil may be turned in different directions, conduces to the same purpose. Hence, notwithstanding the minuteness of the aperture which admits the light, (and it must be small, otherwise the image will not be distinct,) the eye may take in at once, without moving the head, a horizontal range of 110° and a vertical range of 120°; namely, 50° above, and 70° below a horizontal line.*

825. As the radiant approaches the lens, the image recedes from it on the other side; and in our experiments on the formation of images, we are obliged either to change the place of the screen every time the distance of the radiant is altered, or to substitute a new lens, which will either throw back the image as much as the increased distance of the radiant brings it forward, or which brings the image as much nearer as the altered place of the radiant tends to carry it off. How then is the distinctness of the image maintained in the eye, notwithstanding the immense variety in the distances of objects? We can conceive of but two ways in which this can be accomplished: either by lengthening or shortening the diameter of the eye in the direction of its axis, so as to alter the distance of the retina from the cornea and crystalline, or by altering the curvature of the refracting lenses themselves, increasing their convexity for near objects, and lessening it for objects that are more remote. Perhaps both causes may operate; but the effect is believed to be produced chiefly by the latter cause, namely, change of figure in the refracting lenses. On this subject, Sir J. Herschel remarks, that it is the boast of science to have been able to trace so far the refined contrivances of this most admirable organ; not its shame to find something still concealed from its scrutiny; for however anatomists may differ on points of structure, or physiologists dispute on modes of action, there is that in what we *do* understand of the formation of the eye so similar, and yet so infinitely superior, to a product of human ingenuity,—such thought, such care, such refinement, such advantage taken of the properties of natural agents used as mere instruments for accomplishing a

* Brewster.

given end, as force upon us a conviction of deliberate choice and premeditated design, more strongly, perhaps, than any single contrivance to be found, whether in art or nature, and render its study an object of the deepest interest.

826. Writers on comparative anatomy, express the highest admiration of the adaptation of the eyes of different animals to the media in which they respectively live, and to the peculiar wants or habits of each. Thus the crystalline lens of the fish is formed with peculiar reference to the refracting properties of water. In the human eye, this lens has a refractive power only a little greater than that of water; but since the light passes out of a much rarer medium, (air,) such a density is sufficient to bring the rays to a focus; but were the density of the crystalline lens in the eye of the fish no greater than in the human eye, receiving the light from a medium (water) almost as dense as itself, it would be unable to give that change of direction to the rays which would be essential to distinct vision. But provision is made for this exigency by giving to the crystalline lens a much greater density, and of course a higher refracting power, which enables it completely to fulfil its purpose.

Animals which have occasion to see in the dark, as the owl and the cat, have the power of opening or closing the pupil to a much greater extent than man. By this means, they are enabled in the dark, to collect a far greater number of rays of light. But as such an expansion of the pupil would, in broad daylight, endanger the safety of eyes of such peculiar delicacy, the iris closes over the aperture and diminishes it with every increase in the intensity of the light, a change which is involuntary on the part of the animal. In animals, as birds which pounce upon their prey, the pupil of the eye is elongated perpendicularly, while in those that ruminate, as the ox, it is elongated horizontally; being, in each case, exactly adapted to the circumstances of the animal.

827. The images of external objects are of course formed *inverted* on the retina, and may be seen there by dissecting off the posterior coats of the eye of a newly killed animal, as an ox, and exposing the retina, like the image on a transparent screen, seen from behind. The appearance is particularly striking and beautiful when the eye is fixed like the scioptic ball, in the window shutter of a dark room. It is this image, and this only, which is *felt* by the nerves of the retina, on which the rays of light act as a stimulus; and the impressions therein produced are thence conveyed along the optic nerve to the sensorium, in a manner which we must rank at present among the profound mysteries of physiology, but which appear to differ in no respect from that in which the impressions of the other senses are transmitted

Thus, a paralysis of the optic nerve produces, while it lasts, total blindness, though the eye remains open, and the lenses retain their transparency; and some very curious cases of half blindness have been successfully referred to an affection of one of the nerves without the other.* On the other hand, while the nerves retain their sensibility, the degree of perfection of vision is exactly commensurate with that of the image formed on the retina. In cases of *cataract*, when the crystalline lens loses its transparency, the light is prevented from reaching the retina, or from reaching it in a proper state of regular concentration; being stopped, confused, and scattered, by the opake or semi-opake portions it encounters in its passage. The image, in consequence, is either altogether obliterated, or rendered dim and indistinct. If the opake lens be extracted, the full perception of light returns; but one principal instrument for producing the convergence of the rays being removed, the place of the image, instead of being *on* the retina, is considerably *behind* it, and the rays being received on it in a state of convergence, before they are brought to a focus, produce no regular picture, and therefore no distinct vision. But if we give to the rays, before they enter the eye, only a moderate degree of divergence, by the application of a convex lens, so as to render the lenses of the eye capable of finally effecting the exact convergence of the rays upon the retina, distinct vision is the immediate result. This is the reason why persons who have undergone the *operation for the cataract*, (which consists either in totally removing, or in putting out of the way an opake crystalline,) wear spectacles unusually convex. Such glasses perform the office of an artificial crystalline. An imperfection of vision similar to that produced by the removal of the crystalline, is the ordinary effect of old age, and its remedy is the same. In aged persons, the cornea loses something of its convexity, or becomes flatter. The refracting power of the eye is by this means diminished, and a perfect image can no longer be formed on the retina, the point to which the converging rays tend being beyond the retina. The deficient power is supplied by a convex lens, in a pair of spectacles, which are so selected and adapted to the eye, as exactly to compensate for the want of refracting power in the eye itself, and thus the rays are brought to a focus at the retina, where alone a distinct image can be formed.

828. Short-sighted persons have their eyes too convex, forming the image too soon, or before the rays reach the retina. Concave glasses counteract this effect. Rare cases have occurred where the cornea was so very prominent as to render it impossible to apply, conveniently, a lens sufficiently concave to coun-

* Wollaston, Phil. Trans 1824.

teract its action. Such cases would be accompanied with immediate blindness, but for that happy boldness, justifiable only by the certainty of our knowledge of the true nature and laws of vision, which in such a case has suggested the opening of the eye and removal of the crystalline lens, though in a perfectly sound state.* Other defects of eyesight, whose cause has been ascertained to depend on malconformation of the cornea, or some other part of the eye, have sometimes been remedied by adapting to them glasses of a peculiar construction, possessing optical properties suited to the particular defects they were required to remedy.

829. *The impression made by light remains on the eye for a short time after the light itself is withdrawn.*

Fig. 292.

The case of a stick ignited at the end and whirled in the air has already been noticed. (Art. 779.) Upon the same principle, the spokes of a wheel, and other parts of machinery in rapid motion, exhibit continuous surfaces, although made up of parts which are separated from each other by large intervals. Lightning, also, and fiery meteors, appear to describe long lines of light merely because their passage through the atmosphere is so rapid, that the eye does not lose the impression of the first portions until the last are added. The amusing toy called the *Thaumatrope*,† depends on the same principle. An example of it is exhibited in the preceding cut, (Fig. 292,) which represents a circular card, on one side of which is inscribed a chariot, and on the other the charioteer. To opposite sides of the circumference of the card are attached strings, by means of which, taken between the thumb and finger of each hand, a rapid revolution is given to the card, bringing the figures on the opposite sides in quick succession before the eye. When the motion is so swift that the eye retains the impression of both, the two appear united, or the charioteer appears in his proper place, driving the chariot. The *Phantasmascope* consists of disks bearing on their margin a variety of figures, which are so related to each other, that each succeeding figure shall afford a continuation of the preceding, and the whole taken together, when put in rapid revolution, shall exhibit a single figure performing some singular or amusing feat. Thus the figure might commence with a player holding a violin, and a bow which he is just beginning to draw; the second view

* Herschel on Light, Sec. 350—358.

† Θαυμα, *a wonder*, and τρεπω, *to turn.*

might represent the bow as drawn a little; the third still more; and the whole views would then exhibit the usual motions of the bow. In a similar manner are performed dances, feats of horsemanship, and the like.

830. As we have two eyes, and a separate image of every external object is formed in each, it may be asked, *why we do not see double?*

When we look at an object, we direct towards it the optic axis* of each eye, and see most distinctly the point where this axis produced meets the body. In looking at the same point with both eyes, we incline them so as to make the two axes meet in that point: we therefore see this point *in the same place* with both eyes, and it appears as one, the image being brighter than when seen with one eye. If, by any means, the optic axes are prevented from meeting in the same point, double vision is the consequence. Thus we make surrounding objects appear double by pressing the ball of one eye sideways with the finger. Those who have one eye distorted by a blow, see double, though they sometimes learn by habit to correct the defect, even while the distortion remains. The sense of touch is subject to similar distortion: if we lay the middle finger across the fore finger, and apply the ends of both fingers to any object, as a small ball, or the end of the nose, the object appears double. A similar separation of the optic axes, with a similar result, takes place when we hold a small object, as a pin, in front of the eyes, and then direct them to some distant object: the pin appears double. The same effect is produced, when we look intently at an object near the eye, and attempt at the same time to catch a view of a remote object: the latter appears double.

831. The reason why objects appear erect, while their images on the retina are inverted, has given rise to much discussion. It seems, however, a point not difficult to comprehend, that objects and the parts of objects, should appear in the direction in which the rays of light emitted from them come to the eye; and, accordingly, that those which come from the top and bottom of the object should be referred to those points respectively, just as one sound would be known to proceed from the top and another from the bottom of a high tower, merely by the different sensations which they excited in the ear, although the chain of vibrations from the top should strike the bottom, and those from the bottom the top of the ear. Indeed, this very circumstance might be that which determined the relative positions of the two points: and

* The optic axis is the axis of the crystalline lens, or a line passing through the center of the crystalline perpendicular to both its surfaces.

if these sounds presented to the mind a picture of the tower, they would represent it in its natural erect position.

Very minute objects, which cannot be seen by direct vision, may sometimes be rendered visible by looking a little way from them, so that their light strikes the eye obliquely. Thus, astronomers, in viewing the smallest stars or satellites with the telescope, have sometimes been able in this manner to catch a glimpse of them, when they could not otherwise be seen.

832. *The estimation of the* DISTANCES *and* MAGNITUDES *of objects is not dependent on optical principles alone, but the information afforded by the eye, is taken in connection with various circumstances that influence the mind in judging of these particulars.*

In the first place, we judge of the distance of an object by the *inclination of the optic axes,* which is greater for nearer objects and less for objects more remote. But beyond a certain distance, this method is very indeterminate, since great intervals among remote objects would scarcely affect the inclination of these axes. In the second place, we judge of distance by the *apparent magnitude of known objects;* as when a ship of large size, or a high mountain, appears comparatively small, we refer it to a great distance. We are also frequently deceived in our estimate of distance when we are approaching large objects, as a great city or a lofty mountain: we fancy they are nearer than they actually are. In the third place, we estimate the distance of objects by the degree of *distinctness of the parts* or *brightness of the colors.* Thus, a smoky mountain is referred to a great distance;* a mountain whose sides are precipitous and bare, (especially where the rocks have a new and fresh appearance in consequence of having been quarried for use,) appears nearer than the reality; vessels, or steamboats, seen through a mist in the night, have sometimes run foul of each other, being supposed by the pilots to be much further off, in consequence of the indistinctness of their appearance. In the fourth place, our estimate of distance is affected by the *number of intervening objects.* Hence, distances upon uneven ground do not appear so great as upon a plain; for the valleys, rivers, and other objects that lie low, are many of them lost to the sight. On this principle, the breadth of a river appears less when viewed from one side than from the center; a ship appears nearer than the truth to one unaccustomed to judge of distances on the water; and the horizontal distance of the sky appears much greater than the vertical distance, whence the aërial vault does not present the appear-

* This appearance exhibits the true color of the atmosphere, becoming visible in consequence of the extent of the medium, and the dark ground which the mountain affords upon which to view it.

ance of a hollow hemisphere. but of such a hemisphere much flattened in the zenith, and spread out at the horizon.

833. A similar variety of circumstances affects our estimate of the *magnitudes* of bodies seen at different distances. First, the *visual angle*, that is, *the angle subtended by the object at the eye*, determines the size of objects that are near; but it is scarcely any guide to the dimensions of remote objects, since all such objects subtend angles at the eye comparatively very small. Thus, on this principle, a fly, within a few inches of the eye, would appear larger than a ship of war seen at some distance on the water. A giant, nine feet in height, thirty feet off, would appear no larger than a child three feet high, seen at the distance of ten feet. But as this result is not conformable to experience, it is evident that we must have means of judging of the magnitude of objects, beside that derived from the visual angle. If the giant were to remove from the distance of ten feet from the eye to that of thirty feet, his image on the retina would be only one-third as long as before; but, on the other hand, the distance is trebled, and the sort of combination that takes place in us of the two impressions, the one of magnitude, the other of distance, is like the constant product of two quantities, of which one increases in the same ratio as the other diminishes; whence the giant would appear constantly of the same height, at whatever distance from us he was seen.*

834. This corrected result, however, we can make only in cases where we are familiar with the actual size of the body. When not thus familiar, we rely too much on the visual angle, and are thus often greatly deceived. A speck on the window being at the instant supposed to be an object on a distant eminence, is magnified, in our estimation, into a body of extraordinary size, (as a line half an inch long into a May-pole;) or distant objects supposed to be very near, appear of an exceedingly diminutive size. Secondly, the effect of *contrast* is visible in our estimation of the magnitudes of bodies, a given object appearing much below its ordinary size, when seen by the side of those of very great magnitude. Men quarrying stone at the base of a high mountain, sometimes appear, at a little distance, like pigmies, partly from the effect of contrast, but more, perhaps, from the impression which the mountain gives us of their being nearer than they actually are. Thirdly, objects seen at an angle considerably above or below us, as a man on the top of a spire, or a river in a deep valley seen from the top of a mountain, appear greatly diminished. In these cases, since there are no intervening objects to aid us in estimating the distance, we estimate it too low,

* Haüy.

and hence (Art. 832) the object appears less than the reality. Moreover, being seen *obliquely*, its apparent dimensions are diminished on this account, the apparent diameter being determined by the line into which the object is projected perpendicular to the axis of vision. Hence, children judge much less accurately both of distances and magnitudes than adults; and blind persons, suddenly restored to sight, have usually displayed an utter inability to judge of these particulars.

CHAPTER IX

OF MICROSCOPES

835. The *Microscope is an optical instrument, designed to aid the eye in the inspection of* MINUTE *objects.**

Telescopes, on the other hand, assist the eye in the examination of *distant* bodies. These two instruments have probably more than any other extended the boundaries of human thought, and no small part of the labor which has been bestowed upon the science of optics, has had for its ultimate object their improvement and perfection.

With the hope of making the learner well acquainted with the principles of the microscope, we shall begin with those varieties of the instrument which are the most simple in their construction, and successively advance to others of a more complicated structure.

836. The simplest microscope is a double convex lens. This, it is well known, when applied to small objects, as the letters of a book, renders them larger and more distinct. Let us see in what manner these effects are produced. When an object is brought nearer and nearer to the eye, we finally reach a point within which vision begins to grow imperfect. That point is called the *limit of distinct vision.* Its distance from the eye varies a little in different persons, but averages (for minute objects) about *five inches.* If the object is brought nearer than this distance, the rays come to the eye too diverging for the lenses of the eye to bring them to a focus soon enough, that is, so as to make the image fall exactly on the retina. Moreover, the rays which proceed from the extreme parts of the object meet the eye too obliquely to be brought to the same focus with those rays which meet it more directly, and hence contribute only to confuse the picture. We may verify these remarks by bringing gradually toward the eye a printed page with small letters.

* Μικρὸς, *small,* σκοπεω, *to see.*

When the letters are within two or three inches of the eye, they are blended together, and nothing is seen distinctly. If we now make a pin-hole through a piece of paper, (black paper is preferable,) and look at the same letters through this, we find them rendered far more distinct than before at nearer distances, and larger than ordinary. Their greater *distinctness* is owing to the exclusion of those oblique rays which, not being brought by the eye to an accurate focus with the central rays, only tend to confuse the picture formed by the latter.

As only the central rays of each pencil can enter so small an orifice, the picture is made up, as it were, of the *axes* of all the pencils. The *increased magnitude* of the letters is owing to their being seen nearer than ordinary, and thus under a greater angle, an increase of the visual angle having much influence in our estimate of the magnitude of near objects, though it has but little influence in regard to remote objects. (Art. 833.)

837. A convex lens acts on much the same principle, only it is still more effectual. It does not *exclude* the oblique rays, but it diminishes their obliquity so much, as to enable the eye to bring them to a focus upon the retina, and thus to make them contribute to the brightness of the picture. The object is magnified as before, because it is seen nearer, and consequently under a larger angle, which enables minute portions to be distinctly recognized by the eye, which were before invisible, because they did not occupy a sufficient space on the retina. The power of a lens to accomplish these purposes, will obviously depend on its refractive power ; and this (supposing the material of which the lens is made to remain the same) will depend on its increased sphericity, and diminished focal distance. Lenses of the smallest focal distance, therefore, other things being equal, have the greatest magnifying power, and spherules, or perfect spheres, have the highest magnifying power of all. When the radiant is situated in the focus of a lens, the rays go out parallel. When thus received by the eye, they are capable of being brought to a focus by it, and of forming a distinct image. Hence, by means of a lens, an object may be seen distinctly when it is exceedingly near to the eye, provided it be situated in the focus of the lens. The magnifying power of a lens, therefore, depends on the ratio between its focal distance and the limit of distinct vision. The latter being five inches, a lens whose focal distance is one inch, by bringing the object five times nearer, magnifies its linear dimensions in the same ratio, and its superficial dimensions in the ratio of the square. Thus in the case supposed, an object would appear five times as long and broad, and have twenty-five times as great a surface, when seen through the magnifier, as when seen by the naked eye. Lenses have been made capable of affording a distinct image of very minute ob-

jects, when their focal distances were only $\frac{1}{60}$ of an inch. In this case, the magnifying power would be $\frac{1}{60}$: 5, which is as 1 to 300, or as 1 to 90,000 in surface.

838. When, however, an object is so near to the eye, a very minute space covers the whole field of vision, and it is only the minutest objects, or the smallest parts of a body, that are visible in such microscopes. The extent of parts seen by a microscope is called the *field of view*. A microscope of small focal distance has a proportionally small field of view. Moreover, since, when the object is so near to the lens, the rays of light strike the lens extremely diverging, only the central rays of each pencil can be brought accurately to a focus. The more oblique rays, therefore, must be excluded by covering up all but the central portions of the lens, by which means the brightness of the image is diminished. The part of a lens through which the light is admitted, is called its *aperture*. The aperture of a lens of small focal distance and high magnifying powers, must of necessity be small, and one of the principal difficulties in the use of such microscopes, is the want of sufficient light. Hence microscopes of different focal distances are required for different purposes. Where we wish to view a large object at once, we must use a lens which has a large field of view, and of course but a comparatively small magnifying power. Such are the glasses used by watchmakers and other artists. Microscopes which magnify but little, yet afford a large field of view, are called *magnifiers*, or *magnifying glasses*. Such are the large lenses employed for viewing pictures. But for inspecting the minute parts of a small insect, we require a much higher power; and, the object being very small, a large field of view is not necessary. The only difficulty to be obviated is the want of light; and this evil is remedied, either by placing the object in the sun, or by condensing upon it a still stronger light, by means of apparatus specially adapted to that purpose, which will be described hereafter.*

839. Among the most distinguished achievements of philosophical artists, in our own times, has been the formation of microscopes out of the hardest precious gems, especially the *diamond* and the *sapphire*. The diamond seems to unite in itself almost every desirable quality for this purpose. It will be recollected that this substance is distinguished for its high refractive powers, its index of refraction being 2.439, while that of crown glass is only 1.530, (Art. 748;) hence a given refracting, and of course magnifying, power may be attained with a lens of less curvature, and consequently (Art. 761) subject to less *spherical*

* A convenient pocket microscope is sometimes sold in the shops, consisting of a slide of ivory or horn, two or three inches in length, in which are set three or four lenses of different powers, adapted to various purposes.

aberration than glass lenses of the same power. Indeed, it is estimated that the indistinctness arising from spherical aberration in a diamond lens, is only $\frac{1}{20}$th as great as in a glass lens of equivalent power. The sapphire has analogous properties, as also the garnet; and pure rock crystal (quartz) is much esteemed for refracting lenses; but some of the pellucid gems are unsuitable for this purpose on account of their possessing the property of double refraction. The comparative curvatures and thicknesses of three lenses of the same refracting power, made respectively of glass, sapphire, and diamond, are exhibited in the following diagrams.

Fig. 293.

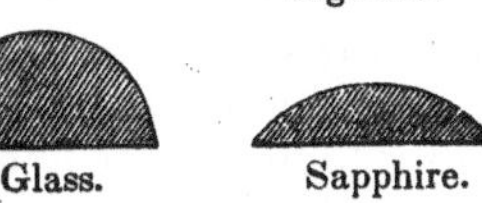

Since, moreover, a diamond lens admits of being made much thinner than a glass lens of the same power, the loss of light by absorption is far less, and the *brightness* of the image is proportionally augmented.

840. Another distinguishing and valuable property of the diamond is, that it combines with a high refractive, a *low dispersive power*. By dispersive power is meant *the power of separating the different colored rays*, that is, *of decomposing common light into its prismatic elements.* Diamond lenses are naturally nearly *achromatic*, or afford images which are destitute of color. But while these favorable qualities were known to appertain to the diamond, which, taken in connection with its great transparency and purity of structure, were observed to fit it admirably for microscopes of great magnifying powers, yet the extreme hardness of the substance, seemed to render the difficulty of grinding it into the requisite shape almost insuperable. This difficulty has, however, within a few years, been completely overcome by Mr. Pritchard, an eminent English artist, who has constructed a number of diamond and sapphire microscopes, whose performances have equalled the most sanguine expectations.

The following table exhibits the different magnifying powers of Pritchard's sapphire lenses, corresponding to different focal distances, the limit of distinct vision being taken at $\frac{1}{10}$ of an inch.

Parts of an inch.	Magnifying power. Linear.	Magnifying power. Superficial.
$\frac{1}{10}$	100	10,000
$\frac{1}{15}$	150	22,500
$\frac{1}{20}$	200	40,000
$\frac{1}{30}$	300	90,000
$\frac{1}{40}$	400	160,000
$\frac{1}{50}$	500	250,000
$\frac{1}{60}$	600	360,000
$\frac{1}{70}$	700	490,000
$\frac{1}{100}$	1000	1,000,000

841. A drop of transparent liquor may be easily converted into a magnifier, constituting a *Fluid Microscope*. The simplest kind of fluid microscope is formed by drilling a small hole in a plate of brass or lead, and applying to it a drop of water from the point of a pin. If the plate be hollowed out on both sides around the aperture, the water will spontaneously assume the shape of a convex lens. Water, however, possessing only a comparatively low refracting power, (Art. 748,) is less adapted to this purpose than several other fluids, particularly certain transparent balsams and aromatic oils. Sulphuric acid and castor oil answer well, but turpentine varnish and Canada balsam are preferred, especially because as they dry they become indurated, and form permanent microscopes. Instead of the aperture in a metallic plate above described, a small plate of glass may be employed, in which case it is only necessary to drop the varnish or balsam on the surface of the plate; and it will assume the figure of a plano-convex lens. The power of the microscope may be varied by employing a larger or a smaller drop, or by suffering it to spread itself on the upper or on the under surface, since the curvature of the drop, and of course its focal distance, is modified by each of these circumstances.

842. The Perspective Glass, which is used for viewing pictures, affords another example of the application of the simple microscope. It consists of a large double convex lens fixed in a frame in a vertical position, from the top of which, on the back side, proceeds a plane mirror, which is fixed at an angle of 45° with the horizon, and of course it makes the same angle with the lens. Pictures to be viewed are placed in an inverted position, (that is, with the top toward the spectator,) on a table at the foot of the instrument. The mirror being set at an angle of 45° with the horizon, renders horizontal objects erect. (Art. 771.) Its office, therefore, is merely to give a proper *direction* to the rays of light from the picture as they enter the lens, causing them, in fact, to come to the lens in the same manner as they would do were the mirror removed and the picture set up in a vertical position, parallel to the lens, at a distance from the lens equal to the length of any ray, measured from the picture to the mirror and from the mirror to the lens. (Art. 729.) Again, in order that the image may be erect, it is necessary that the picture should be placed with its top toward the observer; for since the image of every point in the picture is just as far behind the mirror as the point is before it, those parts of the picture which are designed to occupy the highest part of the image must be farthest below the mirror. This will be understood from the following diagram.

AA, a convex lens fixed vertically in a frame.

BB, a plane mirror making with the horizon an angle of 45°

C, an object placed horizontally upon the table, the upper part being toward the observer.

The object will be reflected by the mirror into a perpendicular position, and its rays will, therefore, fall on the lens in the same manner as they would were no mirror employed, and the object were situated perpendicularly behind the lens, at a distance from it equal to the sum of the distances from the object to the mirror, and from the mirror to the lens. Consequently, if the distance of C from the lens be equal to the focal distance of the lens, the rays will come to the eye parallel, and a distinct and magnified image will be formed. If the distance be greater than the focus, (as it may be rendered by depressing C to a lower level,) then the rays will come to the eye converging, and the image will be more magnified but less distinct. If the distance of C be less than the focus, the image will be less magnified, but it will be distinct within certain limits. The reasons of these several modifications, will be evident by reflecting on principles already expounded. (Art. 754.)

Fig. 294.

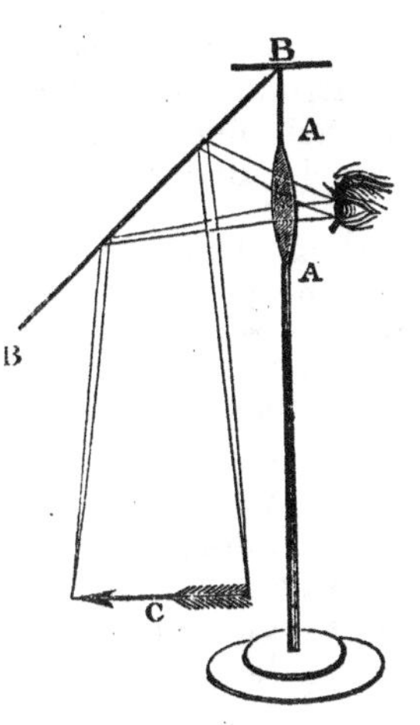

When the glass is of good quality, and the picture executed agreeably to the rules of perspective, the various parts are exhibited in their natural positions, and at their relative distances, so as greatly to improve the view. The greater distinctness of the parts and more natural distribution of light and shade than what attends the naked view, is owing not only to the increased magnitude and to the greater quantity of the light emitted from the picture which is collected by the lens and conveyed to the eye, but also to the separation of this portion of light from that which proceeds from various other objects. The lens both conveys more of the light of the picture to the eye than would otherwise reach it, and conveys it unmingled with extraneous light. The importance of the latter circumstance is manifested even by looking at the picture through an open tube, or through the hand so curved as to form a tube.

843. The microscopes hitherto examined are such as are designed to be interposed between the eye and the object to be viewed, the latter being placed in the focus of parallel rays of the lens, or a little nearer to the lens than that focus, so that the rays of the same pencil may come to the eye either parallel or with so small a degree of divergency, that the lenses of the eye shall be competent to make them converge and form an image on the retina. In this case, as the rays come to the eye in the same manner as rays from larger objects, at a greater distance, seen without the aid of a lens, the position of the object is not changed, that is, it is seen erect. Single microscopes, however, are also employed to form a magnified image on a wall or screen, which is seen by the eye instead of the object itself. Two celebrated instruments, the Magic Lantern and the Solar Microscope, magnify their objects in this manner, in the construction of which, the principles under review are happily exemplified.

844. From what has been already learned respecting lenses, the following points will be readily comprehended, being, for the most part, a recapitulation of principles already explained and demonstrated.

If, in a dark room, we place before a convex lens any luminous object, as a candle, we shall observe the following phenomena. (See Art. 754.)

1. If the radiant be placed nearer to the lens than its focus, since the rays will go out diverging, no image will be formed on the other side of the lens.

2. Even when the radiant is in the focus, so that the rays go out parallel, they never meet in a focus, and of course never form an image.*

3. But when the radiant is further from the lens than its focus, the rays converge on the other side, those of each pencil which proceed from the same point in the object, being accurately united in one point of the image, and occupying that point alone, without the interference of rays from any other point.

4. The axes of the rays from the extreme parts of the object cross each other in the center of the lens. Hence, they form an image *inverted* with respect to the object; and although the rays which make up any individual pencil are made to *converge* by the lens, yet the axes (which determine the magnitude of the picture) diverge from each other after crossing at the center of the lens, and hence the image is greater in proportion as it is formed at a greater distance from the lens. When the object is only a little further off from the lens than its focus, the image is thrown to a great distance, and is proportionally magnified. As the object is separated further from the lens, (wh   may be effected either by withdrawing the object from the lens or the lens from the object,) the image is formed at a less distance, and is of a diameter proportionally less. (See Art. 760.) Suppose now that we employ a magnifier of so small focal distance, that when the object is placed within one tenth of an inch of the lens, the image is formed on the other side upon a screen or wall, at the

* It will be remarked, that when the single microscope is used as an eye-glass, the eye itself brings the parallel rays to a focus and forms the image.

distance of twenty feet; the object will be magnified in the ratio of $\frac{1}{10}$ to $(20\times12=)$ 240; that is, the image will be 2,400 times greater than the object in diameter, and 5,760,000 times greater in surface. It would seem, therefore, as if nothing more were necessary, in order to form magnified images of objects, than a dark room, a convex lens, and a screen or wall for the reception of the picture. It must be remarked, however, that when the light which proceeds from the object is diffused over so great a space, its intensity must be greatly diminished, so as to be either incapable of affording a picture which shall be visible at all, or at least sufficiently bright for the purposes of distinct vision. This difficulty is remedied by *illuminating the object;* and it is for this purpose that most of the contrivances employed in the magic lantern and solar microscope are designed.

845. The Magic Lantern consists of a large tin canister, either cylindrical or cubical in its figure, having an opening near the bottom into which air may enter freely to supply the lamp, and a chimney proceeding from the top, and bent over so as to prevent the light of the lamp from shining into the room. The lantern has a door in the side, which shuts close, the object being throughout to prevent any light from escaping into the room, except what attends the picture. The room itself is made as dark as possible: or, what is better, the experiments are performed by night. In front of the lantern is fixed a large tube,

Fig. 295.

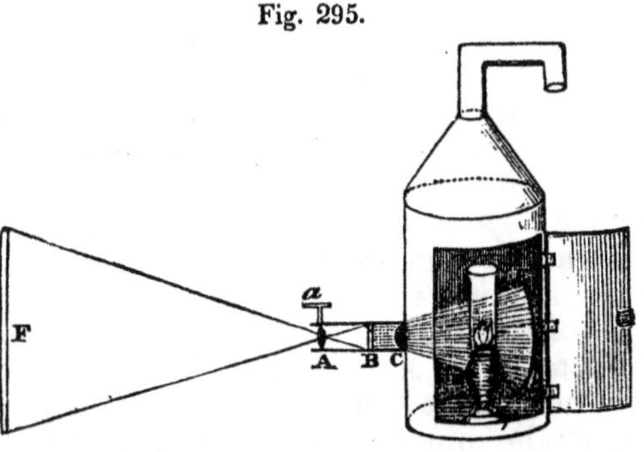

at the open end of which is placed the magnifying lens. In the same tube, at a distance from the lens somewhat greater than the focal distance, the object is introduced, which is usually some figure painted on glass in transparent colors, the other parts of the glass being blackened so that no light can pass through except that which falls on the object and illuminates it, by which means we shall have a luminous image projected on a black ground. For illuminating the object, an argand lamp is placed near the center of the lantern, the light of which is concentrated upon the object in two ways first, by means of a thick lens, usually plano-convex, so situated between the lamp and the object that the rays which diverge from the lamp shall be collected and condensed upon the object; and, secondly, by means of a concave reflector, situated behind the lamp, which serves a similar purpose.

A, the magnifying lens.

B, the object, introduced through an opening in the tube.

C, the condensing lens.

D, the lamp.

E, the concave mirror.

F, the image thrown on a screen, or a white wall, in a dark room.

a, a thumb-piece, by which the magnifier may be made to approach the object or to recede from it, and thus the image be thrown to a greater or less distance, according to the magnitude required. As the image is inverted with respect to the object, it is only necessary to introduce the object itself in an inverted position, and the image will be erect.

The objects employed in the Magic Lantern are very various, consisting of figures of men and animals; of caricatures; of representations of the passions; of landscapes; and of astronomical diagrams. When the last are employed, this apparatus becomes subservient to a useful purpose in teaching astronomy, and is frequently so employed by popular lecturers on that subject.

846. The Solar Microscope does not differ in principle from the Magic Lantern, only the object is illuminated by the concentrated light of the *sun* instead of that of a lamp.* And since a powerful illumination may thus be effected upon minute objects placed before a magnifier of great power, the solar microscope is usually employed to form very enlarged images of the most minute substances, as the smallest insects, the most delicate parts of plants, and other attenuated objects of natural history. For magnifiers several of different focal distances are employed, varying from an inch to the $\frac{1}{10}$ or $\frac{1}{20}$ of an inch, it being understood that those of the shortest focus and greatest magnifying powers can be used only for the minutest objects, since, when bodies of a larger size are brought so near a small lens, their light strikes the lens too obliquely to be transmitted through it. The magnifying lens is fixed into the mouth of a tube and the object placed near its focus, much in the same manner as in the magic lantern; but instead of the body of the lantern, (which contains the illuminating apparatus,) a mirror, about three or four inches wide, and from twelve to eighteen inches long, is attached to the

* The *oxy-hydrogen* microscope has recently been substituted for the solar, the intense flame resulting from the combustion of the gaseous elements of water, being used instead of the sun's light.

other end of the tube. This mirror is thrust through an opening in the window shutter of a dark room, and the mouth of the tube to which it is fixed is secured firmly to the shutter, so that the mirror is on the outside, and the tube with its lenses is on the inside of the shutter. By means of adjusting screws, the mirror is turned in such a way as to direct the sun's rays into the tube, where they are received by one or more of the lenses called *condensers*, which collect them and concentrate them upon the object, which thus becomes highly illuminated, and capable of affording an image sufficiently bright and distinct, though magnified many thousands or even millions of times. It will be observed that the magnitude of the image depends here, as in other cases of the simple microscope, upon the ratio between the distances of the object and the image from the center of the magnifier. If, for example, the object be within the tenth of an inch of the lens, and the image be thirty feet, or three hundred and sixty inches from it, then the image will be $360 \times 10 = 3600$ times as large as the object in diameter, and $(3600)^2 = 12{,}960{,}000$ times in surface. With a given lens, the size of the image depends wholly on the distance to which it is thrown; that is, on the distance of the wall or screen where it is formed.

847. When the solar microscope is well constructed, it affords the most wonderful results, and greatly enlarges our conceptions of the delicacy, perfection, and subtilty of the works of nature. In inspecting *vegetables*, the eye is delighted with the regularity and beauty which characterizes the texture and intricate structure of plants and flowers. The most delicate fibres of a leaf, the pores through which the vegetable fluids circulate, the downy covering of plants and foliage, as of certain mosses, which is too minute to disclose its figure to the naked eye,—objects of this kind, when expanded under the solar microscope, astonish and delight us by the symmetry of their structure. Their appropriate *colors* are not so well exhibited by this instrument, as by some other forms of the microscope to be described hereafter. In the *animal* kingdom, the solar microscope extends the range of vision in a manner no less surprising and instructive. The minutest insects we are acquainted with, are exhibited to us as animals of the largest size, and often of monstrous shapes, from the multiplicity of their parts and apparent disproportion; and animalcules, or those members of the animal creation which are too minute to be seen at all by the naked eye, are suddenly brought into life in countless numbers. The forms, the motions, and the habits of these beings, are among the most curious revelations of the solar microscope. The *circulation of the blood* may be seen in the fins of fishes and other transparent parts of animals, presenting a very curious and interesting spectacle. The *crystallization of salts*, which may be exhibited while the crystals

are forming and arranging themselves, (as many of them do with great precision and symmetry,) is among the finest representations of this instrument.

Since the light is transmitted through the object, it will of course be understood, that only such objects as are *transparent* can be employed in the manner already described. In some varieties of the solar microscope, there are special contrivances for exhibiting *opake* objects by means of reflected light.

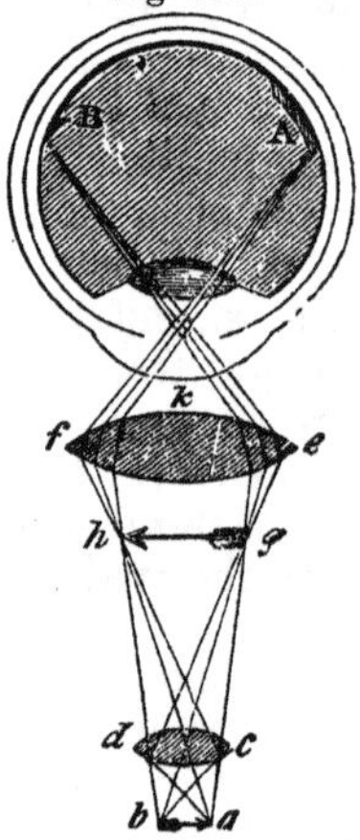

848. If we form an image of an object with the single microscope, (as is done in the magic lantern and solar microscope,) when that image is not too large, we may obviously apply to it a magnifier, as we would to an original object of the same size. This is the principle of the compound microscope.

Fig. 296.

The COMPOUND MICROSCOPE consists of at least two convex lenses, one of which, called the *object-glass*, is used to form an enlarged image of the object, and the other, called the *eye-glass*, is used to magnify the image still further.

Thus, let *ab*, (Fig. 296,) be the object, being placed a little further from the object-glass, *cd*, than the principal focus, then the rays of light emanating from it will be collected on the other side of the lens and form an image, *gh*, whose diameter is as much larger than that of the object as its distance from the lens is greater. (Art. 760.) Let *ef* be the eye-glass, which must be placed at such a distance from the image, that the latter shall be in the focus of parallel rays; then the rays proceeding from the image will go out parallel,* and come to the eye, situated behind the glass, in a state favorable for distinct vision.

849. The *magnifying* power of the Compound Microscope is estimated as follows. First, the diameter of the image will be to that of the object as their respective distances from the lens. Secondly, the image is magnified by the eye-glass according to the principles of the single microscope, (Art. 376,) namely, in the ratio of its focal distance to the limit of distinct vision. Thus, suppose the image is formed at ten times the distance of the object; it will of course be magnified ten times. Again, suppose

* It is to be remarked here, and in all similar cases, that it is only the rays of *each individual pencil* that are parallel; that is, those rays which come from the same point in the object. The rays of different pencils may cross each other variously, and the different pencils may converge or diverge among themselves; still, if the rays of each pencil are parallel to one another, the vision will be distinct.

the eye-glass has a focal distance of one inch, the limit of distinct vision being five inches; the image will be further magnified five times; by both glasses, therefore, the object will be magnified fifty times. If the first ratio be that of one to one hundred, then the instrument will magnify the linear dimensions five hundred times, and the surface two hundred and fifty thousand times. From this double magnifying process, it might be supposed that, by means of the compound microscope, it would be easy to attain a much higher magnifying power than by the single microscope; but this is not the fact, for, in the first place, we cannot form an image of a size beyond certain moderate limits, without making it too large for the eye-glass to cover; or, if an eye-glass of very large field of view be employed, its focal distance must be great, and consequently its magnifying power small. We are, therefore, unable to employ so high a magnifier for our object-glass as we may apply to the naked eye, and we can employ only a microscope of still inferior power for our eye-glass.

850. On account of the necessity of using a large eye-glass to view the magnified image, compound microscopes require to have the tube which contains the glasses larger toward the eye-glass than toward the object-glass. Sometimes the magnifiers are contained in a box of pyramidal shape, the reason of which is obvious. Of the latter figure is the *Lucernal Microscope*, a variety of the compound microscope, which admits of being used with the light of a *lamp* instead of day light, and is furnished with a reflector and a condensing lens, by one or the other of which the light of the lamp may be concentrated upon the object. The lucernal microscope is furnished with a piece of ground glass, upon which the image may be received as upon a screen. The object being illuminated by a lamp, and the image being seen in a dark room, this arrangement is very convenient for drawing insects, flowers, &c. Although the compound does not possess higher magnifying powers than the simple microscope, yet it commands a much greater field of view. We view the image with the eye-glass in the same manner as we view the object with a single microscope; but having already a magnified representation of the object, we have no occasion to apply to the eye so high a magnifier, and therefore we may employ one of greater focal distance, which consequently takes in a greater field of view. The field of view is still further improved in some compound microscopes by interposing a *field-glass*, which is a convex lens, introduced between the eye-glass and the place of the image, and near the latter, (as a little above *gh*, Fig. 296,) the effect of which is to diminish the divergency of the pencils of rays, and thus to bring into the range of the eye-glass those pencils, which would otherwise diverge too much to

fall within it. It has been before remarked, that the cornea performs a similar office for the crystalline lens of the eye. (Art. 823.)

851. Instead of employing a convex lens for the purpose of forming an image of the object, we may use a concave mirror for the same purpose. On this principle are constructed REFLECTING MICROSCOPES. The object being placed before the mirror, at a distance a little greater than the focal distance, a magnified image will be formed on the other side of the center, as in Fig. 256. To this image we may obviously apply an eye-glass, in the same manner as in the common compound microscope. Reflecting microscopes are supposed to have some advantage over the refracting, but they have not come into general use. By making the concave reflector of a parabolic figure, spherical aberration is prevented, and reflectors are not liable, like lenses, to form colored images in consequence of the decomposition of the light into its prismatic rays, called *chromatic aberration.* These difficulties, however, when they occur, admit of being obviated by peculiar contrivances, which will be more particularly described in connection with telescopes.

852. Dr. Brewster gives the following rules for making microscopic observations.

1. The eye should be protected from all extraneous light, and should not receive any of the light which proceeds from the illuminating body, excepting what is transmitted through, or is reflected from, the object.

2. Delicate observations should not be made when the fluid which lubricates the cornea is in a viscid state.

3. The best position for microscopical observations, is when the observer is lying horizontally on his back. This arises from the perfect stability of his head, and from the equality of the lubricating film of fluid which covers the cornea. The worst of all positions is that in which we look downward vertically.

4. If we stand straight up and look horizontally, parallel markings or lines will be seen most perfectly when their direction is vertical; viz. the direction in which the lubricating fluid descends over the cornea.

5. Every part of the object should be excluded except that which is under immediate observation.

6. The light which illuminates the object, should have a very small diameter. In the daytime it should be a single hole in the window shutter of a darkened room, and at night an aperture placed before an argand lamp.

7. In all cases particularly when high powers are used, the

natural diameter of the illuminating light should be diminished, and its intensity increased, by optical contrivances.*

The microscope is sometimes employed to form images for the purposes of drawing. In this manner landscapes are represented, objects of natural history are delineated, and artificial pictures are reduced and copied. The two instruments particularly employed for this purpose, are the Portable Camera Obscura and the Camera Lucida.

853. The PORTABLE CAMERA OBSCURA, which is used for delineating landscapes, and much more of late for taking likenesses by the Daguerreotype process, consists of a wooden box, (answering to the dark chamber, Art. 818,) with which is connected a convex lens, so exposed to the landscape as to receive the rays of light from the various objects in it, and form a picture of them on a screen placed within the box at the focal distance of the lens. Such is a general description of the instrument, of which there are several different forms. The following diagram represents a common convenient form.

Fig. 297.

ABCD, (Fig. 297,) a box usually made of thin pieces of mahogany.

ad, a plano-convex lens, this form being preferred because it has less aberration than a double convex. (Art. 762.)

ED, a plane mirror, turning on a hinge at D, and capable of being raised or lowered, so as to admit more or less of the landscape.

bc, a piece of pasteboard, covered with a sheet of fine white paper, and bent so as to form a concave screen, and placed at the focal distance of the lens. A casting of stucco, of the figure of a concave portion of a sphere, affords the most perfect picture.

The rays of light from external objects, falling upon the mirror ED, are conveyed to the lens in the same manner as though they came directly from objects at the same distance behind the mirror. Passing through the lens, they are brought to a focus, and form a picture of the landscape on the screen, which may be viewed by an opening in the side of the box at F, and may be copied by a hand introduced into the box by an opening below.

Although the image is inverted with respect to the objects, yet as the spectator, in looking into the box, stands with his back to the landscape, the picture appears erect.

* Brewster's Optics, 345

854. The Camera Lucida is an instrument of more recent origin, having been invented by the late Dr. Wollaston. It consists of a prism so contrived that its surfaces, by their reflecting properties,* give the proper direction to the rays of light, and finally project an image of the object in a convenient position for copying, as is represented in the following diagram.

Fig. 298.

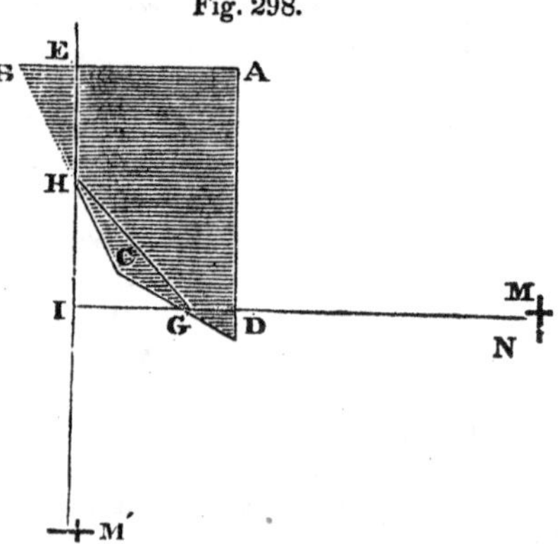

ABCD, (Fig. 298,) is a glass prism, having the angle at A 90°, the angle at D 67½°, the angle at C 135°. In taking an observation the prism is set with the side AD parallel to the object M. A ray of light ND, falling perpendicularly upon AD, suffers no refraction, but proceeds on to the second surface DC, where it makes with DC an angle of 22½°, (the complement of the angle at D.) Of course the angle CGH is 22½°, and these two angles, subtracted from 180°, leave NGH=135°. Again, since GCH=135°, and CGH=22½°, therefore CHG and BHE each equal 22½°, and therefore GHE=135°. Produce NG till it meets HM′ in I; then the angles IGH and IHG will be severally 45°, and consequently HIG (which is the angle made by the incident and emergent rays) will be 90°. Therefore, the perpendicular object, MN, will appear to the eye on a horizontal plane at M′, as far behind the reflecting surface as M is before it. (Art. 729.) Now if the prism is so formed, that the emergent rays shall be very near the angular point B, the eye situated at E may take in at once the image and the paper on which it is projected, seeing the former through the prism and the latter by direct vision; and thus the image may be very perfectly sketched. This beautiful instrument is usually mounted in a case, and has various appendages, which severally contribute to its utility, but we aim only to convey an idea of its *principle*.†

* It will be observed, in the following illustration, that the rays of light strike the surfaces of the prism at such an angle as to undergo *total reflexion*. (Art. 748.)

† For a more extended description of the Camera Lucida, see Nicholson's Phil Journal, and Tilloch's Phil. Magazine, for 1807

CHAPTER X.

OF TELESCOPES.

855. The *Telescope is an optical instrument, designed to aid the eye in viewing distant objects.**

The construction of this noblest of instruments, in its different forms, involves the application of all the leading principles of the science of Optics. The study of the Telescope is therefore the study of the science, and a distinct enunciation of the principles involved in it, will serve as a recapitulation of the most useful principles of Optics. The advantage which the student will derive from reviewing these points, as exemplified in their application, will justify us in bringing up distinctly to view various principles already unfolded.

856. The leading principle of the Telescope may be thus enunciated:

By means of either a convex lens, or a concave mirror, an image of the object is formed, which is viewed and magnified with a microscope.

Thus, let ABCD represent the tube of a telescope. At the front end, or the end which is directed toward the object, (which we will suppose to be the moon,) is inserted a convex lens L, which receives the rays of light from the moon, and collects them into the focus *a*, forming an image of the moon. This image is viewed by a magnifier attached to the end BC.

Fig. 299.

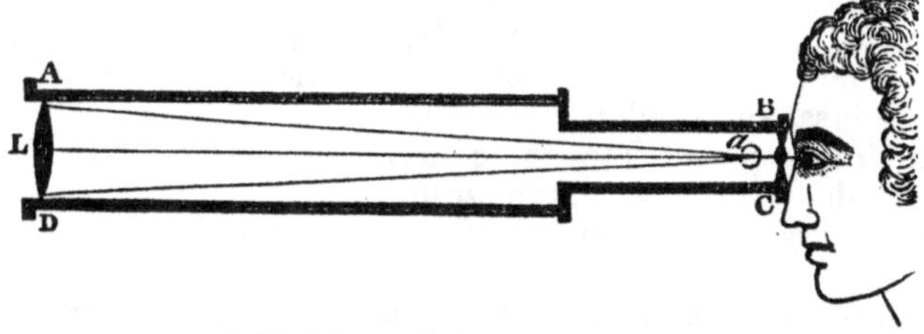

The most general division of the instrument is into Refracting and Reflecting Telescopes: of which the former produce their image by means of a convex lens, and the latter by means of a concave mirror. The instrument, according to the uses to which it is applied, receives the farther denominations of the Astronomical and the Terrestrial Telescope; and also telescopes are

* Τῆλε, *at a distance*, σκοπέω, *to see*

named after their several inventors, Galileo's, Newton's, Gregory's, Herschel's, &c.

THE ASTRONOMICAL TELESCOPE.

857. We begin with this variety because it is one of the most simple, and because, in connection with it, we may conveniently study the theory of the instrument at large.

The Astronomical Telescope has essentially but two glasses: these are usually fixed in a tube of brass, one at one end, and the other at the other end. The glass at the end of the tube which is directed to the object, is called the *object-glass;* and that at the end to which the eye is applied, is called the *eye-glass.* The object-glass is a convex lens which forms an image of a distant object, as a star, in its focus of parallel rays, and the eye-glass is a microscope with which we view the image, at a distance equal to its focus of parallel rays. Of course, the distance of the two glasses from each other is equal to the sum of their focal distances. See the annexed figure.

Fig. 300.

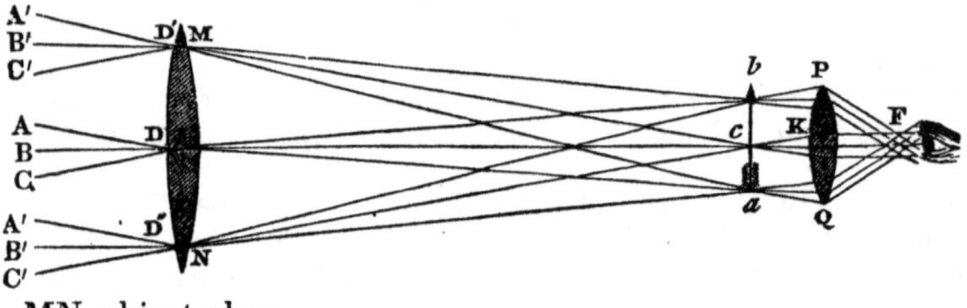

MN, object-glass.

PQ, eye-glass.

A′D′, AD, A″D″, parallel rays from the top of the object.

B′D′, BD, B″D″, " " " center do.

C′D′, CD, C″D″, " " " bottom do.

ba, inverted image formed in the focus of parallel rays.

*b*PF, a pencil of rays, proceeding from the top of the image to the eye-glass, and rendered parallel.

*c*KF, a similar pencil from the center.

*a*QF, do. from the bottom.

F, point where the different pencils cross the axis.

858. In this instrument we observe a striking resemblance to the Compound Microscope. (Fig. 296.) In the microscope, however, since the object is nearer than the image, the image is greater than the object; but in the telescope, since the object is removed to a great distance, the image is formed much nearer to the lens than the object, and is proportionally smaller. Hence, compound microscopes have their tubes enlarged in diameter toward the eye-glass, while telescopes have their tubes diminished

in that direction. Since the vertical angles at D, subtended on the one side by the object, and on the other by the image, are equal, were the eye situated at the center of the object-glass, it would see the object and the image under the same visual angle, and, consequently, both would appear of the same magnitude. Moreover, were the eye placed at the same distance from the image on the other side of it, it would be apparently of the same size as before, and therefore of the same apparent diameter as the object. But by means of a microscope, such as the eye-glass in fact is, we may view it at a much nearer distance, and of course magnify it to any extent, as was fully shown in explaining the principles of the simple microscope. (Art. 837.) Hence the magnifying power of the telescope depends on the ratio between the focal distances of the object-glass and the eye-glass. If, as in the figure, the common focus is ten times nearer the eye-glass than the object-glass, the instrument will magnify ten times; if one hundred times nearer, one hundred times; and so in all other cases. Hence we may increase the magnifying power of the instrument, either by employing an object-glass of very small curvature, which throws its image to a great distance, or an eye-glass of high curvature and small focal distance. Suppose, for example, the object-glass has a focal distance of forty feet, or four hundred and eighty inches, and the eye-glass has a focal distance of one tenth of an inch, then the magnifying power of this instrument would be four thousand and eight hundred in diameter, and the square of this number in surface.

859. As the sphericity of the eye-glass may be increased indefinitely, and its focal distance diminished to the same extent, it would seem possible to apply very high magnifying powers in very short telescopes. For example, suppose the focal distance of the object-glass is twenty-four inches; by using a microscope of $\frac{1}{10}$th of an inch focus, we have a power of two hundred and forty. But it must be kept in mind, that such microscopes command only an exceedingly small field of view, and would, therefore, not enable us to see any thing more than a minute portion of an object of any considerable size; and not sufficient light would be transmitted through such an aperture to answer the purposes of vision. Since the image is inverted with respect to the object, and is viewed in this situation by the glass, objects seen through Astronomical Telescopes appear inverted. By the addition of several more lenses, they may be made to appear erect, as will be shown in the description of the Day Glass, or Terrestrial Telescope; but at every new refraction a certain portion of light is extinguished, a loss which it is important to avoid in instruments designed to be used at night; while, in regard to celestial objects, it is not essential whether they are seen erect or inverted.

The place for the eye to view the image with the best advantage is at F, where the pencils of parallel rays meet.

860. The *difficulties* to be overcome in the construction of a perfect Refracting Telescope, (some of which are very formidable,) are chiefly the following: 1. Spherical aberration; 2. Chromatic aberration; 3. Want of sufficient light; 4. Want of a field of view sufficiently ample; 5. Imperfections of glass. Each of these particulars we will briefly consider.

861. *Spherical Aberration*, it will be recollected, occasions indistinctness in images formed by lenses, in consequence of the different rays of the same pencil not being all brought to a focus at the same point, those which fall upon the extreme parts of the lens being more refracted and coming to a focus sooner than those which are nearer to the axis. (See Art. 761.) The amount of this error is found to depend on two circumstances, namely, the diameter of the lens, or what is technically called its *aperture*,* and its focal distance, increasing rapidly as the aperture is increased, and diminishing as the focal distance is increased.† *Small apertures and flat or thin lenses are, therefore, most free from spherical aberration.* But if we use small apertures we cannot have a strong light, which is a circumstance of the greatest importance in astronomical observations, since it is of little consequence to enlarge the dimensions of an object if we have not light enough to render it visible. Indeed, many astronomical objects, as small stars, are rendered visible by the telescope, not in consequence of any apparent increase of size, but because this instrument collects and conveys to the eye a much larger beam of light from them than would otherwise enter it. While the diameter of the beam which falls upon the naked eye is only the fraction of an inch, that collected by the telescope may be several inches, or even several feet, according to the size of the instrument. Hence the advantage of large apertures is obvious. Again, we cannot wholly remedy the error in question, though we may diminish it by using very flat lenses which have great focal distances; but the tendency of this expedient is to render the instrument inconveniently long. Other expedients, therefore, become necessary for correcting spherical aberration in refracting telescopes.

* The aperture, strictly speaking, is the diameter of that part of the lens through which, in a given case, light is admitted, whether it be the whole surface or only a part of it.

† It is found by opticians, that the *longitudinal aberration* of lenses increases as the square of the aperture, with a given curvature, and is inversely as the focal distance, with a given aperture, and that the *lateral aberration* increases as the cube of the aperture, with a given radius, or inversely as the square of the radius with a given aperture.

862. In the eye-glasses, which are liable to the same difficulty, where the lens has a great curvature, as is the case with such as have high magnifying powers, the aperture is necessarily reduced very much, by excluding all the light except what passes through the central parts of the lens. At least this is the case where glass lenses are used. But the microscopes made of diamond, sapphire, and other gems, have not only high refractive powers, but are less subject to spherical aberration than similar lenses of glass. (Art. 839.) Thus, if three lenses were ground in the same tool, one of plate glass, one of sapphire, and the other of diamond, their respective magnifying powers and aberrations would be as follows :*

	Magnifying power.	Longitudinal aberration.
Glass, ...	150	1.167
Sapphire, ...	250	1.005
Diamond, ...	400	0.950

This difference in aberration will be much greater if the lenses be so formed as to give the same magnifying powers; for then the diamond and sapphire lenses may be made so much thinner as greatly to reduce the aberration.

But although eye-pieces, on account of their small size, may sometimes be made of the precious gems, yet this can rarely be the case on account of the great expense attending them. It is obvious, also, that they cannot be employed for the object lenses. The most successful method of diminishing spherical aberration in eye-pieces of glass, is by a combination of plano-convex lenses, by means of which a given refracting power may be attained with far greater distinctness than by a single lens of the same power. Thus, when two plano-convex lenses are placed as in Fig. 301, it is found that the image has four times the distinctness of a double convex lens of equivalent power.† Here F is a lens which would bring the parallel rays to a focus and form the image at the distance of G; but E is another similar lens, which, receiving them in a converging state, makes them converge more, and come to a focus at H. The double convex lens D, would do the same, but with much greater spherical aberration. It appears, indeed, that the spherical aberration may be wholly removed by combining a meniscus with a double convex lens of certain curvatures.‡

Fig. 301.

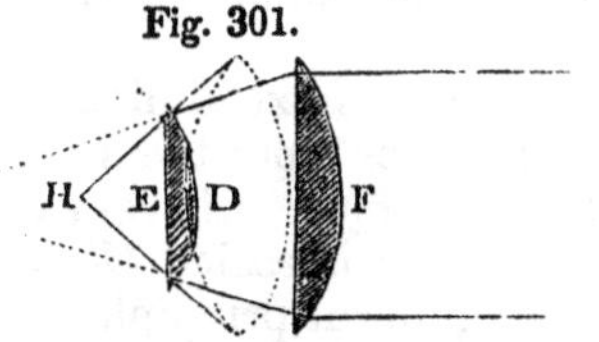

* The figure of the lens is supposed to be plano-convex, the convex side being turned toward parallel rays.

† The Scioptic Ball, used in the camera obscura, (Art. 819,) is formed of two such lenses.

‡ See Brewster's Optics, p. 58, or Herschel on Light, Sec. 316.

863. In object-glasses, which, on account of their smaller curvatures, are not so subject to error from spherical aberrations as eye-glasses are, the most advantageous form is that of a double convex lens of unequal curvatures, the radii of the opposite surfaces being as one to six, (Art. 764,) and the less convex side being turned toward the parallel rays.

In short, it appears, that in order to avoid the errors arising from spherical aberration in large lenses, they must be made as thin as convenience will permit; that where it is practicable, they may be most advantageously formed of the precious gems, particularly the diamond; that a plano-convex lens with its convex side toward the parallel rays has less aberration than a double convex lens of equivalent power; that two plano-convex lenses may be so combined as to have only $\frac{1}{4}$ as much aberration as the double lens, and a meniscus may be so united to a double convex lens as wholly to prevent aberration; and finally, that the aberration may be reduced to a very small error simply by employing a double convex lens whose curvatures on the opposite sides are as one to six.

Since lenses having the curvature of one of the conic sections are free from spherical aberration, Sir Isaac Newton ground an object-glass into the figure of a paraboloid. This was free from the error in question, but involved another still more formidable, since it decomposed the light and gave an image tinged with the colors of the rainbow. On observing this, Sir Isaac pronounced the further improvement of the *refracting* telescope to be hopeless, and betook himself to exclusive efforts for improving the *reflecting* telescope. But the combined ingenuity of philosophers and artists, has nearly overcome this error also.

864. The next difficulty therefore to be considered is that which arises from the separation of the prismatic colors, in consequence of the different refrangibility of the different rays, an error which is called *Chromatic Aberration.*

The general principles of Chromatic Aberration, will be readily comprehended by calling to mind, that distinct images are formed only when the rays of the same pencil which flow from any point in the object are collected into one and the same point in the image, unmixed with rays from any other point; that the prismatic rays which compose white light have severally different degrees of refrangibility, some being more turned out of their course than others, in passing through the same medium; that, consequently, the different colored rays of the same pencil would meet in different points, each set of colored rays forming its own image, but all these images becoming blended with one another, and thus composing a confused, colored picture.

To illustrate these principles let LL be a lens of crown glass, and RL, RL, rays of white light incident upon it, parallel to its axis R*r*. Let the extreme *violet* rays, whose index of refraction is 1.54666, be refracted so as to meet the axis in *v*; then the extreme *red*, whose index of refraction is only 1.5258, will meet the axis at some point more distant from the lens, as at *r*. C*v* and C*r* are the focal distances of the lens for the violet and the red rays respectively. The distance *vr* is the chromatic aberration, and the circle whose diameter is *ab*, which passes through the focus of the mean refrangible rays at *o*, is called the *circle of least aberration.*

Fig. 302.

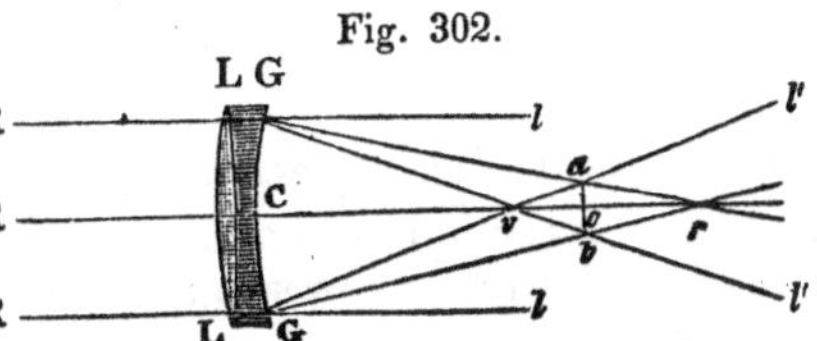

865. These effects may be shown experimentally by exposing the lens LL, (Fig. 302,) to the parallel rays of the sun. If we receive the image of the sun on a piece of paper placed between *o* and C, the luminous circle on the paper will have a *red* border, because it is a section of the cone L*r*L, the exterior rays of which L*a*, L*b*, are red; but if the paper is placed at any greater distance than *o*, the luminous circle on the paper will have a *violet* border, because it is a section of the cone *lv'l'*, the exterior rays of which *al'*, *bl'* are violet, being a continuation of the violet rays L*v*, L*v*. As the spherical aberration of the lens is here combined with its chromatic aberration, the undisguised effect of the latter will be better seen by taking a large convex lens LL, and covering up all the central part, leaving only a small rim round its circumference at LL, through which the rays of light may pass. The refraction of the differently colored rays will then be finely displayed by viewing the image of the sun on the different sides of *ab*.

It is clear from these observations, that the lens will form a violet image of the sun at *v*, a red image at *r*, and images of the other colors of the spectrum at intermediate points between *r* and *v*; so that if we place the eye behind these images, we shall see a confused image, possessing none of that sharpness and distinctness which it would have had if formed only by one kind of rays.*

The separation of white light into its prismatic colors, is called *dispersion*; and the comparative power of effecting this separation, possessed by different media, is called the *dispersive power*. The dispersive power is measured by the ratio which, in any case, the separation of the red and violet rays bears to the mean refraction of the compound ray. Thus if a ray of solar light on passing through a lens, is turned out of its original direction 27°, and the red and violet rays are separated from each other 1°,

* Brewster's Optics, p 79

then the dispersive power is said to be $\frac{1}{27}$, which is usually expressed in the form of a decimal fraction, $.037=\frac{1}{27}$.

866. *Different bodies possess different dispersive powers.*

The dispersive powers of a few of the most important substances in relation to the subject before us, are exhibited in the following table.

	Dispersive power.		Dispersive power.
Oil of Cassia, . . .	0.139	Plate Glass, . .	0.032
Sulphuret of Carbon, .	0.130	Sulphuric Acid, . .	0.031
Oil of Bitter Almonds,	0.079	Alcohol, ...	0.029
Flint Glass, ...	0.052	Rock Crystal, . .	0.026
Muriatic Acid, . . .	0.043	Blue Sapphire, .	0.026
Diamond, ...	0.038	Fluor Spar, ...	0.022
Crown Glass, (green,) .	0.036		

From this table it appears, that the transparent substances which have the highest dispersive power, are the oil of cassia and the sulphuret of carbon, both of which fluids have been made to perform an important service in the construction of achromatic telescopes; that flint glass, as that used for decanters, has a much higher dispersive power than crown glass, or that which is analogous to window glass; that the diamond has a low dispersive power, but is exceeded in this respect by rock crystal, the sapphire, and fluor spar, which last bodies have the least dispersive power of any known substances.

867. With these facts in view, we may now inquire *by what means the object-glass of the telescope is rendered achromatic.*

If we place behind LL (Fig. 302) a concave lens GG of the same glass, and having its surfaces ground to the same curvature, such a lens having properties directly opposite to those of the convex lens, will neutralize its effects. Consequently, the rays which were separated into their prismatic colors by the convex lens, will be reunited by the concave lens, and reproduce white light. But though such a combination of the two lenses will correct the color, yet it also destroys the power of the convex lens to form an image, on which its use solely depends. Could we find a concave lens which would correct all the color and yet not destroy this refracting power, the two lenses would evidently form the achromatic combination sought for. Now this is what is actually done: by making the concave lens of a substance which has a *higher dispersive power* than that of which the convex lens is made, the curvature of the concave lens will not need to be so great as that of the convex lens, and of course the two together, constituting the compound lens, will be equivalent in refracting power to a single lens, whose convexity is equal to the difference of their curvatures. The most common combination is that of flint glass with crown glass, the concave lens being made of flint glass, and the convex of crown. By the table in Art. 866, it will be seen that the dispersive power of flint glass is 52, while that of crown glass is 36, which numbers are nearly as 3 to 2, and these numbers, therefore, may be employed for the sake of illustration. Since the power of the concave lens to reunite the prismatic rays, is so much greater than that of the convex lens to separate them, we shall not require a refractive power to effect this equivalent to that of the convex lens; that is, a concave lens of less curvature and proportionally greater focal distance, will serve our purpose. Therefore,

An achromatic lens is formed by the union of a convex and a concave lens, whose dispersive powers are respectively proportional to their focal distances.

868. A telescope furnished with an object-glass thus formed, is called an *Achromatic Telescope.* The spherical aberration being corrected by the methods pointed out in Art. 762, and the chromatic aberration being destroyed in the manner above described, the Refracting Telescope becomes an instrument of great perfection, and is reckoned among the greatest works of art. Until recently, it was rare to meet with refracting telescopes of an aperture of more than from three to five inches. For we have already seen that the errors of spherical and chromatic aberration increase rapidly as the size of the aperture is augmented.

If it be asked, what is the *use* of a large aperture, since the magnifying power does not depend upon the diameter of the object-glass, but upon the ratio between the focal distance of the object-glass and the focal distance of the eye-glass, (Art. 858,) we answer, that the use of a large aperture is to admit, condense, and finally convey to the eye, a larger beam of light, and thus to render many objects, as the smaller stars, or Jupiter's belts, visible, which otherwise would not be so, on account of the feebleness of the light which they transmit to us. *Want of light* is in fact one of the greatest difficulties that the telescope has to contend with; for, in the first place, the object-glasses of most telescopes are comparatively small, and are necessarily so on account of the difficulty of procuring suitable glass for those of a larger size; and in the second place, of the light admitted through the object-glass, a great proportion is intercepted and wasted in various ways, many instruments being able to save only the central rays without rendering the image indistinct and colored. Thus, when very high magnifiers are applied, (which of course have very small focal distances,) the rays proceed from the focus and fall upon the microscope so obliquely, that only those which pass through the central parts of the lens can be ved since such as fall upon the marginal parts of the lens are too much affected by spherical and chromatic aberration, to form with the others a distinct and colorless image.

869. *Want of field of view* is another difficulty to be surmounted. When we use an object-glass of short focus with a high magnifier, the microscope must have a focus proportionally short, and of course the field of view will be very limited and the light but feeble. This difficulty may be obviated by using an object glass of very great focal distance. If, for example, the focal distance of the object-glass were only 12 inches, in order to attain a magnifying power of 120, we must employ a microscope whose focal distance is only $\frac{1}{10}$ of an inch. But if the focal distance of the object-glass were 10 feet, or 120 inches, then our microscope might have a focal distance of 1 inch, which would give a larger field and a stronger light. With a view of obviating several of the foregoing difficulties, the earlier astronomers who used the telescope, employed for their object-glasses lenses whose focal lengths were very great. Cassini, an Italian astronomer, constructed telescopes eighty, one hundred, and one hundred and thirty-six feet long; and Huygens employed such as were nearly the same length. The latter astronomer dispensed with the tube, fixing his object-glass, contained in a short tube, to the top of a high pole, and forming the image in the air, near the level of the eye, which image he viewed with an eye-glass, as usual. With telescopes of this description several of the satellites of Saturn were discovered.

870. But one of the most formidable difficulties hitherto encountered in the construction of large refracting telescopes, has arisen from the *imperfections of glass.*

The difficulty of obtaining glass of a perfectly homogeneous composition and structure, is thus set forth by Mr. Faraday—

"Although every part of the glass may in itself be as good as possible, yet without this condition [a perfectly homogeneous structure] the parts do not act in uniformity with each other; the rays of light are deflected from the course which they ought to pursue, and the piece of glass becomes useless. The streaks, striæ, veins or tails, which are seen within glass otherwise perfectly good, result from a want of this equality; they are visible only because they bend the rays of light which pass through them from their rectilinear course, and are constituted of a glass having either a greater or a smaller refractive power than the neighboring parts. When these irregularities are so powerful as to render their effects observable by the naked eye, it may easily be supposed to what an injurious extent their influence must extend, in the construction of telescopes and other instruments of a similar nature, where these faults are not only magnified many times, but where the effect is to give an equally magnified erroneous representation of the object looked at, when the very point to be attained is to examine that object with the utmost accuracy; and it is accordingly found that these striæ are the most fatal faults of glass intended for optical purposes. Besides this, not only do the striæ themselves occasion harm, but there is every reason to believe that they rarely occur in glass otherwise homogeneous. Sometimes, it is true, a grain of sand, in passing through, and at the same time dissolving in glass, will give a streak of different composition from the rest of the substance; and again, a bubble ascending may lift a line of heavy or more refractive matter into a lighter and less refractive portion above. Many a disk, which upon the most careful examination has appeared perfectly free from striæ, and quite uniform, has, when worked into an object-glass, been found incapable of giving a good image, on account of the existence of irregularities in the mass, which, though not sudden or strong enough to occasion striæ, still produce a confused effect; and if this happens with glass approaching so near to perfection, it happens still more frequently, and to a much stronger degree, with such as contains visible irregularities."*

871. These irregularities are much more frequent in flint glass than in crown; and by far the greatest obstacle to be overcome in constructing a large refracting telescope, is to procure a suitable piece of flint glass for the concave part of the achromatic object-glass. (Art. 867.) This want of uniformity arises chiefly from the *different specific gravities* of the materials that compose the glass. Oxide of lead, a very heavy substance, enters into the composition of flint glass to the amount of about one-third of its weight. The oxide of lead is so heavy a material, and at the same time so fusible, that it melts and sinks to the bottom, leaving the lighter materials to accumulate at the top; and so imperfect are the means of mixture, under ordinary circumstances, that glass of very different specific gravity, is produced from the bottom and the top of the same crucible.

These circumstances we have thought worthy of being recited, in order to impress on the mind of the learner the formidable nature, as well as the great number, of the difficulties to be overcome in the construction of a large achromatic telescope. Yet they have, in several instances, been completely surmounted. Fraunhofer executed two telescopes with achromatic object-glasses, the one nine inches and nine-tenths, and the other twelve inches in diameter; and at the period of his death he was purposing to undertake one eighteen inches in diameter. That of 9.9 inches aperture was made for the Russian government, for the use of the observatory at Dorpat, where, under the direction

* *Faraday*, Phil. Trans., 1830.

of M. Struve, a distinguished astronomer, it has already achieved numerous valuable discoveries in astronomy. The object-glass has a focal length of 14 feet.

THE TERRESTRIAL OR DAY TELESCOPE.

872. As the Astronomical Telescope represents objects inverted, it requires to be so modified for terrestrial views, that objects may appear erect. This is effected by the addition of two more lenses of similar figure to that of the eye-glass, and of the same focal length. The first of these additional glasses, forms a second image of the object, inverted with respect to the first image, and therefore erect with respect to the object. This image is viewed by the second glass as by any simple microscope. Thus, AB, the object-glass, forms an inverted image *nm* of the object MN. Instead of viewing this image by the eye placed at L, as in the common astronomical telescope, we suffer the pencils of parallel rays to cross each other at L and fall upon a second lens

Fig. 303.

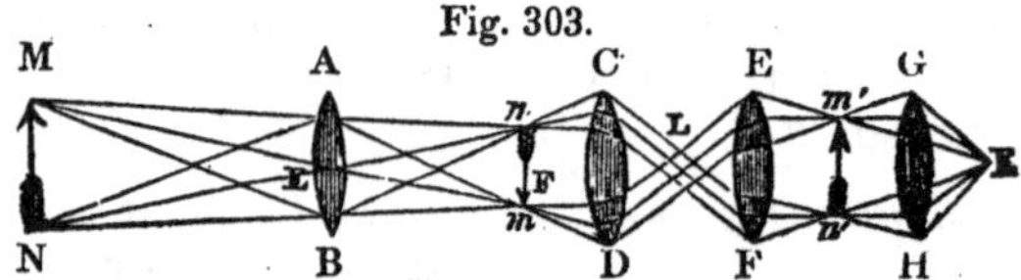

EF (similar in all respects to CD) which collects them into an image *m'n'* in its focus of parallel rays, which image is viewed by the eye-glass GH in the same manner as the object itself would be.

As some portion of the light is reflected, and some absorbed and dissipated by passing through these additional lenses, they of course diminish the brightness of the view; but in the daytime there will usually be light enough for distinct vision after this loss is sustained, while it is more agreeable and convenient to have the objects presented to us in their natural positions than inverted. It will be remarked that the additional lenses do not magnify, the focal length of each being the same as that of the first eye-glass. Were they rendered smaller for the purpose of magnifying, the field of view and the light would both be impaired.

873. We usually find in telescopes, particularly those designed for terrestrial objects, some contrivance, as a draw-tube, by which the eye-glass can be brought nearer to, or withdrawn from the object-glass. This is to accommodate the instrument to objects, at different distances. When it is directed to very near objects, the image is thrown further back, and therefore in order that it may be in the focus of the eye-glass, (which is essential to distinct vision,) the latter must be drawn backward; but where the

object is remote, the image is formed nearer to the object-glass, and then the eye-glass must be moved forward, till its focus of parallel rays comes to the place of the image. For a similar reason, near-sighted persons require the eye-glass to be brought nearer than usual to the object-glass; for then the image will be nearer to the eye-glass than its focus of parallel rays, and the rays will meet the eye diverging, a condition favorable to eyes naturally too convex. For a contrary reason, long-sighted persons, who usually wear convex spectacles, may adjust the telescope to suit their eyes without spectacles, by removing the eye-glass further back than usual.

Most terrestrial telescopes contain a greater number of glasses than are represented in figure 303. Such a number are used for the purpose of correcting spherical and chromatic aberration, these errors being less in several flat and thin lenses than in a smaller number of equivalent lenses of greater curvature.

Astronomical telescopes are easily adapted to terrestrial observations, by removing the eye-glass and substituting a tube containing the additional glasses for rendering the view erect.

GALILEO'S TELESCOPE.

874. This instrument was the first astronomical telescope, having been invented by Galileo, as the name implies. It differs from the common astronomical telescope, in having for the eye-glass a *concave* instead of a convex lens, which receives the pencils of light, as they are converging to form an image, at such a distance from the focus to which they tend, as to render them parallel. Thus, the object-glass AB collects the rays of light as they proceed from the object MN, and makes them converge toward the focus at E. But the concave lens CD is interposed at such a point as to render these converging rays parallel, and in this way they come to the eye situated behind the lens.

Fig. 304.

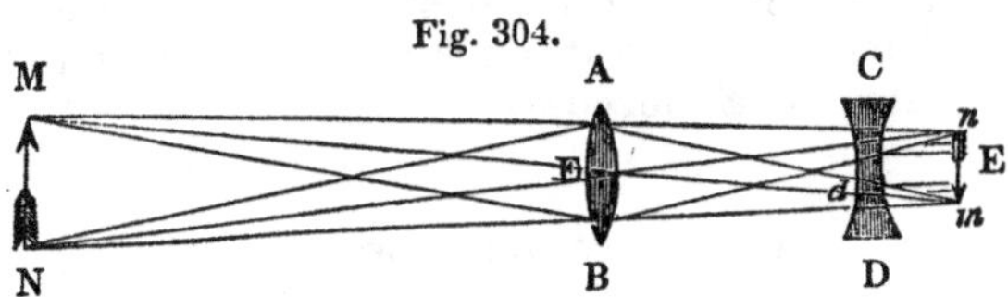

Since the concave lens restores the rays to that state of parallelism which they had before they passed through the object-glass, the learner may not readily see how this instrument aids the eye. That it does so, however, will be apparent from the following considerations.

First, a much broader beam of light falls upon the object-glass than upon the naked pupil of the eye, the greater part of which is collected and conveyed to the eye. By this means the *brightness* of objects is greatly increased.

Secondly, as in the astronomical telescope, (Art. 857,) were the eye situated at the center of the object-glass, the object and the image formed by the object-glass would have the same apparent dimensions; and inasmuch as the eye-glass enables us to view this image much nearer, it increases its apparent dimensions in the same ratio. But when we use a concave lens situated as in the Galilean telescope, the effect is the same as that of a convex lens situated in the same manner on the other side of the focus, so that the rays would come to the eye parallel. Hence, in the Galilean as in the common astronomical telescope, the magnifying power is as the ratio of the focal distance of the object-glass to that of the eye-glass.

This form of the telescope has several advantages and several disadvantages, when compared with the ordinary form. In the first place, requiring but two glasses to present objects erect, it occasions less loss of light than the ordinary form, and presents objects with proportionally greater brightness. In the second place, the eye-glass being *between* the object-glass and the image, instead of *beyond* it, the instrument admits of being made short and compact, a circumstance which fits it for the purposes of an *opera-glass*, to which use it is frequently applied. In the third place, the concave lens corrects the chromatic aberration of the convex lens, and where a proper proportion is observed between the curvatures of the two lenses, the instrument is easily rendered achromatic. The chief disadvantage attending the instrument, is its limited *field of view*. For the pencils of parallel rays, after passing through the concave eye-glass, diverge from one another, those toward the marginal parts of the lens being turned from those that are contiguous to the axis, and therefore not entering the pupil of the eye. And since only those near the axis at E, (Fig. 304,) can enter the pupil, the field of view must depend on the dimensions of the pupil, and cannot be increased by increasing the length of the instrument, as in the refracting telescope. This defect has caused this kind of telescope to fall into disuse for astronomical purposes.

REFLECTING TELESCOPE.

875. Reflecting Telescopes differ in principle from those already described only in forming their image by a *concave reflector* instead of a convex object-glass. The most common form of the Reflecting Telescope is the *Gregorian*, so called from the inventor, Dr. James Gregory, of Scotland. The general principles of this instrument may be explained as follows:

In the Gregorian Telescope, the light (supposed to come in

parallel rays) is first received by a large concave speculum, by which it is brought to a focus, and made to form an inverted image. On the opposite side of this image, and facing the large speculum, is placed a small concave speculum, of greater curvature, at such a distance from the image that the rays proceeding from it and falling on the speculum are made to converge to a focus situated a small distance behind the large speculum, passing through a circular aperture in the center of it. This second image is magnified by a microscope, as in the Refracting Telescope. This description may now be applied to the annexed figure.

ABCD, a large open tube of brass, iron, or mahogany, to contain the reflectors.

abcd, a smaller tube, to receive the second image and the eye-glass.

EE, a large concave speculum, usually composed of a metallic compound called *speculum metal.*

Fig. 305.

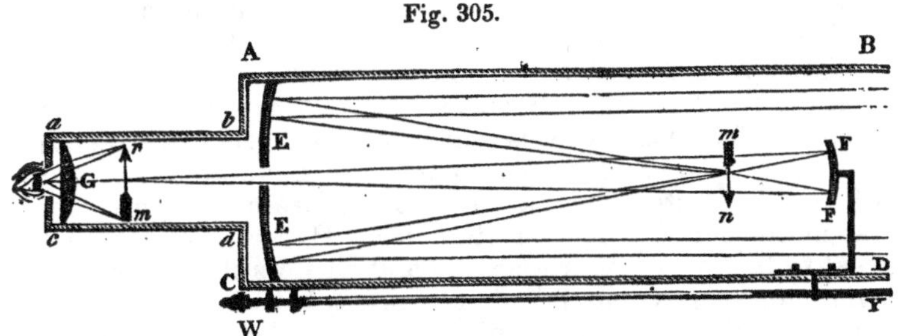

FF, small concave speculum.

mn, image formed by the large reflector.

nm, image formed by the small reflector.

G, eye-glass.

WY, a metallic rod having a screw connected with the small reflector, by means of which this reflector is made to approach the first image, or to recede from it.

Some of the pencils of rays necessary to form the respective images are omitted in the figure, to prevent confusion.

876. From the foregoing construction it is evident, first, that the image viewed by the eye being in the same position with the object, the latter will appear *erect;* secondly, that since the mirrors may be formed of a parabolic figure,* all *spherical aberration* may be easily prevented, (Art. 761;) thirdly, that since light is not decomposed by reflexion, reflecting telescopes are not subject to *chromatic aberration;* and, hence, that it is not neces-

* An elliptical figure has the same property.

sary to lengthen the tube as the aperture is increased, as is the case in refracting telescopes; but since the light will depend, chiefly, on the size of the large reflector, a strong light may be obtained with a comparatively short tube. The achromatic telescope, however, with all the latest improvements, is deemed a more perfect and more convenient instrument than the reflecting telescope; and it is supposed that there will be no occasion hereafter to construct reflectors of such enormous dimensions as those of Dr. Herschel. Some account of his forty feet reflector may form a suitable close to this sketch of optical instruments.

877. Under the munificent patronage of George III, Sir William Herschel began, in 1785, to construct a telescope forty feet long, and in 1789, on the day when it was completed, he discovered with it the sixth satellite of Saturn. The great speculum was more than *four feet* in diameter, and weighed two thousand one hundred and eighteen pounds. Its focal length was forty feet. The tube which contained it was made of sheet iron.

The *light* afforded by this instrument was astonishingly great. The largest fixed stars, as Sirius, shone in it with the splendor of the sun. The reason of this will be obvious, when we reflect that it collected and conveyed to the eye, in the place of the small beam that enters the naked organ, a beam of light from the star more than four feet in diameter. Hence, it was suited to reveal to the eye numberless stars and clusters of stars, which preceding telescopes had failed to exhibit, because they could not collect a sufficient quantity of their light. To economize the light to the best advantage, the small mirror employed in the Gregorian telescope, (see Fig. 305,) was dispensed with, since every successive reflexion dissipates a considerable portion of the light, and the image was thrown near to the open mouth of the tube, where it was viewed by the eye-glass directly, the observer being seated so as to look into the mouth in front. In order to prevent the head from obstructing too much of the light, the image was formed near one side of the tube.* Its greatest magnifying power was six thousand four hundred and fifty; but this was used only for the smallest stars.

The greatest telescope hitherto constructed, is the one recently built in Ireland by Lord Oxmanton, (Earl of Ross,) which has a focal distance of 50 feet, and an aperture of 6 feet. This gigantic instrument, it is expected, will reveal to us many secrets of the skies hitherto hidden from human view.

* Holcomb's telescopes are constructed in the same manner.

FINIS.

PROFESSOR COFFIN'S WORKS.

COFFIN'S ECLIPSES.

Solar and Lunar Eclipses, Familiarly Illustrated and Explained, with the Method of Calculating them according to the Theory of Astronomy, as taught in the New England Colleges.

COFFIN'S CONIC SECTIONS.

Elements of Conic Sections and Analytical Geometry.

These works, by Prof. James H. Coffin, of Lafayette College, Pa., have received the highest recommendations from qualified authority. Of the second named, (Conic Sections, &c.) Prof. Loomis, (College of New Jersey,) Prof. Sadler, Prof. Mason, and other distinguished gentlemen, unite in approval.

BY LYMAN PRESTON.

PRESTON'S DISTRICT SCHOOL BOOK-KEEPING,

Affording an interesting and profitable exercise for youth, being especially designed for Classes in our Common Schools.

PRESTON'S BOOK-KEEPING BY SINGLE ENTRY.

Adapted to the use of Retailers, Farmers, Mechanics, and Common Schools.

PRESTON'S TREATISE ON BOOK-KEEPING;

A common-sense guide to a common-sense mind. In two parts—the first the Single Entry method, the second being arranged more particularly for the instruction of young men who contemplate the pursuit of mercantile business, showing the method of keeping accounts by Double Entry, and embracing a variety of useful forms. This is the most popular work upon the subject, and is in very extensive use throughout the country.

ADAMS'S ARITHMETICAL SERIES.

ADAMS'S PRIMARY ARITHMETIC,

Or Mental Operationsin Numbers, an Introduction to Adams's New Arithmetic, revised edition. An excellent work for young learners.

ADAMS'S NEW ARITHMETIC. Revised edition.

By Daniel Adams, M.D. A new edition of this superior work, with various improvements and additions, as the wants of the times demand. A Key is published separately.

ADAMS'S MENSURATION, MECHANICAL POWERS, AND MACHINERY.

This admirable work is designed as a sequel to the Arithmetic.

ADAMS'S BOOK-KEEPING.

A Treatise on Book-keeping by Single Entry, accompanied by blanks for the use of learners.

ABBOTT'S SERIES.

ABBOTT'S READERS.

The Mount Vernon Reader for Junior Classes. 18mo.
The Mount Vernon Reader for Middle Classes. 18mo.
The Mount Vernon Reader for Senior Classes. 12mo.

ABBOTT'S FIRST ARITHMETIC.

The Mount Vernon Arithmetic. Part First: Elementary.
The Mount Vernon Arithmetic. Part Second: Fractions.

ABBOTT'S ABERCROMBIE'S PHILOSOPHY.

Inquiries concerning the Intellectual Powers and the Investigation of Truth. By John Abercrombie, M.D. F.R.S. Edited, with additions, by Jacob Abbott.

The Philosophy of the Moral Feelings. By John Abercrombie. Edited by Jacob Abbott.

ABBOTT'S DRAWING CARDS.

In three Sets, 40 Cards to each.

WHELPLEY'S COMPEND OF HISTORY.

With Questions by Joseph Emerson. 12mo.

BADLAM'S WRITING BOOKS.

The Common School Writing Books. A New Series of Nine Numbers. By Otis G. Badlam.

ADDICK'S ELEMENTS OF THE FRENCH LANGUAGE.

An excellent elementary work.

GIRARD'S ELEMENTS OF THE SPANISH LANGUAGE.

A practical work to Teach, to Speak, and Write Spanish. By J. F. Girard.

KIRKHAM'S GRAMMAR.

English Grammar in Familiar Lectures, for the Use of Schools By Samuel Kirkham. The work is very extensively used.

DAY'S MATHEMATICS.

A Treatise for Students, by President Day, of Yale College.

THE GOVERNMENTAL INSTRUCTOR.

A Brief and Comprehensive View of the Government of the United States, and of the State Governments, in easy lessons, for the use of Schools. By J. B. Shurtleff.

THE OXFORD DRAWING-BOOK.

Containing Progressive Lessons on Sketching, Drawing and Coloring Landscape Scenery, Animals and the Human Figure. With 100 Lithographic Drawings. 4to.

JOHN RUBENS SMITH'S DRAWING-BOOK.

The Rudiments of the Art explained in Easy Progressive Lessons, embracing Drawing and Shading. With numerous Copper-plate Engravings.

STUDIES IN FLOWER PAINTING.

By Jas. Andrews. Colored plates.

DYMOND'S ESSAYS.

On the Principles of Morality, and the Private and Political Rights and Obligations of Mankind. By J. Dymond.

KEMPIS.

The Imitation of Christ. By Thomas à Kempis. Translated by John Payne, with Introductory Essay by Thomas Chalmers.

ÆSOP'S FABLES.

A New Version, by Rev. Thomas James. Illustrated by James Tenniel. The best edition published in the country.

OUR COUSINS IN OHIO.

By Mary Howitt. From the Diary of an American Mother. 18mo.

GABRIEL.

A Story of Wichnor Wood. By Mary Howitt,

THE AMERICAN SCHOOL PRIMER

Illustrated. 46 pp., 12mo.

THE CHILD'S PRIMER.

Illustrated. 36 pp., 18mo.

JOHN WOOLMAN'S JOURNAL.

DAVID SANDS' JOURNAL.

THE NEW TESTAMENT.

8vo. Large print.

RECOMMENDATIONS OF "COFFIN'S CONIC SECTIONS."

FROM PROF. MASON, BETHANY COLLEGE, VA.

I have examined "Coffin's Conic Sections and Analytical Geometry," and hesitate not to say, that its size, plan and general arrangement, ought to procure for it a place in the hands of every teacher who wishes to give a liberal knowledge of Mathematics, without abridging the other branches of a good education. I am happy to say that your book is better adapted to my present wants than any I have seen. I have therefore concluded to adopt it for my next class.

FROM J. S. LEBAR, PRINCIPAL OF THE HIGH SCHOOL, HACKETTSTOWN, N. J.

By bringing out this work you have advanced the cause of science, merited the thanks of the lovers of learning, and done a lasting good to the public..... As a general principle, I am opposed to the shortening of books in order to shorten the course of study. The spirit of the day is to get along most *rapidly*, not most thoroughly; by bestowing the least possible labor, both in physical and mathematical science. But from the examination I have been able to make of your work, I find it contains some important matter, not found in other works of a like character, whilst nothing that is valuable is omitted. The compass is contracted, but the whole circle is there. It is just the work we need, and I shall adopt it in our Academy in preference to other treatises upon these subjects..... Certainly by your combining the geometrical and analytical methods, you present the subject more clearly..... Your demonstrations are admirable; they are so concise, yet so simple, that no student can pass over them without fully comprehending them.

FROM PROF. SUDLER, DICKINSON COLLEGE, PENN.

I have received and read with great pleasure your "Elements of Conic Sections and Analytical Geometry," and cannot hesitate to commend it as an excellent system. It embraces in a parallel view, the Geometrical Theory and the Analytical investigation of the properties of the Conic Sections—the method which I have always pursued in the instruction in my department, but have felt the want of a proper text-book on the subject. This want has been met by your Treatise, and I have adopted it as a text-book with pleasure.

FROM PROF. FOUCHÉ, ST. JOHN'S COLLEGE, N. Y.

Your work was examined, I do not say by me, but by judges more competent than I am. We are unanimous in saying that your demonstrations, with respect to perspicuity, are to be highly praised. You are clear, concise, and, whilst you avoid obscurity, you find the means of being both exact and brief. This is no small merit, and you are entitled to the gratitude of students, to whom you spare useless difficulties and exertions.

FROM PROF. LOOMIS, COLLEGE OF NEW JERSEY, PRINCETON, N. J.

Your treatise on "Analytical Geometry" appears to me to possess advantages over any other treatise which I have examined. One of these advantages I consider to be, the division into Propositions distinctly enunciated, so as to keep constantly before the mind of the student the object of his investigation. Another advantage consists in the introduction of numerical examples, which afford a useful exercise to the student, and present the best test of the clearness of his conceptions.

FROM J. S. ALVERSON, PRINCIPAL OF THE GENESEE WESLEYAN SEMINARY, N. Y.

I have received and examined a copy of your recent work on "Conic Sections."... The manner in which you have introduced both the geometrical and analytical demonstrations, ought to secure success to your work. You present what I think is needed by the American student. In the selection of problems and in the elucidation of parts that often prove unreasonably obscure to beginners, you have been very happy..... I recommend the work with confidence to the attention of Instructors and Students.

RECOMMENDATIONS.

FROM PROF. STEVENS, OAKLAND COLLEGE, MISS.

I sent to New Orleans for a copy of your treatise on the "Conic Sections and Analytical Geometry," which I have received and have examined with some care and much satisfaction. I think after the perusal I have given it, I may safely pronounce your treatise eminently suited for use in colleges; it is sufficiently elaborate, the arrangement is good, and there is a plainness and simplicity about it which I admire very much.

MESSRS. COLLINS & BROTHER.—Gent.: Illness has prevented me from examining Prof. Coffin's "Treatise on Conic Sections" till very recently. I am greatly pleased with the work, and think that the author has succeeded in developing the most interesting and important properties of these curves with unusual simplicity and brevity in the demonstrations. To this, the common property of the curves, assumed as their fundamental characteristic, is, by the skill and tact of the author, made to contribute very much. This property, while it unites the three curves in a common bond, and gives them a common source, is that on which the more useful properties seem most immediately to depend, and from which they are most readily and naturally deduced. In the Geometrical portion there is a very just medium preserved, in the extent to which the discussion of the curves is carried, the investigations being limited to the most essential and practically useful properties, and requiring no useless expenditure of time on points merely speculative or curious. The second part, devoted to "Analytical Geometry," is a very important addition to the work, and one, I think, without which the student will be very illy prepared to make his knowledge of these curves, readily and extensively available in their application to astronomical investigations. The work I regard as a very decided improvement on the old systems and treatises, and think in its preparation and publication a very valuable contribution has been made to educational instrumentalities.

Very respectfully yours,

WESLEYAN UNIVERSITY — AUG. W. SMITH.

FROM THE METHODIST QUARTERLY REVIEW.

"*Elements of the Conic Sections and Analytical Geometry, by* JAMES H. COFFIN, A. M. *Professor of Mathematics, &c., in Lafayette College.*"

In this treatise the doctrine of the Conic Sections is taught first geometrically; and in a second part, the student is taught how to represent lines, curves, and surfaces analytically, and to solve problems relating to them. We have examined the work with care, and testify to the skill, tact, and neatness of its expositions. Most books of Analytical Geometry are blind to scholars; most of them never learn, unless they have a teacher unusually skilful and diligent, how to interpret algebraical expressions, or how to make practical use of equations. It is precisely for its clearness, its practical character, and its adaptation to the work of the recitation room, that we heartily commend this volume.

FROM J. F. JENKINS, PRINCIPAL OF THE NORTH SALEM ACADEMY.

I have recently had an opportunity of examining your treatise on "Conic Sections and Analytical Geometry," and am happy to state that, in my opinion, it is a work of much merit. The selection and arrangement of the Propositions appear to me judicious; and the definition which you adopt, at the commencement, has certainly the advantage of supplying more simple and obvious demonstrations, in place of those whose prolixity and abstruseness have hitherto rendered this very important portion of Mathematics so formidable to learners. The first part appears to contain all the properties of the Sections which are required in the branches of Mathematical Philosophy usually embraced in our College courses; and the second part furnishes a valuable introduction to Analytical Geometry, rendered more valuable by its immediate connection with what precedes, and by the practical nature of the problems proposed for investigation by analysis. I intend introducing it as a text-book in this Institution.

FROM JAMES T. DORAN, TEACHER, MANALAPAN, N J

think it decidedly the best work in print upon the subject.

www.ingramcontent.com/pod-product-compliance
Lightning Source LLC
LaVergne TN
LVHW010740120826
845150LV00009B/175